Springer Studium Mathematik – Master

The series „Springer Studium Mathematik" is aimed at students of all areas of mathematics, as well as those studying other subjects involving mathematics, and anyone working in the field of applied mathematics or in teaching. The series is designed for Bachelor's and Master's courses in mathematics, and depending on the courses offered by universities, the books can also be made available in English.

More information about this series at http://www.springer.com/series/13893

Torsten Wedhorn

Manifolds, Sheaves, and Cohomology

Torsten Wedhorn
Technische Universität Darmstadt
Darmstadt, Germany

Springer Studium Mathematik – Master
ISBN 978-3-658-10632-4   ISBN 978-3-658-10633-1 (eBook)
DOI 10.1007/978-3-658-10633-1
Library of Congress Control Number: 2016947452

Mathematics Subject Classification (2010): 14-0, 18-01

Springer Spektrum

Editor: Ulrike Schmickler-Hirzebruch

Printed on acid-free paper

This Springer Spektrum imprint is published by Springer Nature
The registered company is Springer Fachmedien Wiesbaden

Preface

The language of geometry has changed drastically in the last decades. New and fundamental ideas such as the language of categories, sheaves, and cohomology are now indispensable in many incarnations of geometry, such as the theory of complex analytic spaces, algebraic geometry, or non-archimedean geometry. This book is intended as an introduction to these ideas illustrating them by example with the most ubiquitous branch of geometry, the theory of manifolds.

In its contemporary form, a "geometric object" is usually defined as an "object" that "locally" "looks like" a "standard geometric object". Depending on the geometry that one is interested in, there will be very different "standard geometric objects" as the basic building blocks. For the theory of (finite-dimensional) manifolds one chooses open subsets of finite-dimensional $\mathbb{R}$- or $\mathbb{C}$-vector spaces together with their "differentiable structure".

To make the notion of a geometric object precise, one proceeds in general as follows. First one introduces the language of categories yielding the notions of objects and the precise meaning of "looks like" as being isomorphic in that category. Next one has to find a (maybe very large) category C that contains the "standard geometric objects" as a subcategory and in which it makes sense to use the word "locally". Then finally one can give the precise definition of a *geometric object* as an object of C that is locally isomorphic to an object in the subcategory of standard geometric objects.

In this textbook we choose for such a category C the category of ringed spaces, although other choices such as ringed topoi might even be more natural. But for the sake of an introduction, ringed spaces seem to be the most accessible choice and they are also adequate for many geometric theories such as differential geometry, complex geometry, or the theory of schemes in algebraic geometry. Moreover a good grasp of ringed spaces will help a great deal in understanding more abstract concepts such as ringed topoi.

For the theory of manifolds the basic building blocks are open subsets X of finite-dimensional $\mathbb{R}$- or $\mathbb{C}$-vector spaces together with the collection of their "functions", where a "function" will be an α-fold continuously differentiable function for $\alpha \in \mathbb{N}_0 \cup \{\infty\}$ or an analytic function (also called C^ω-function) on some open subspace of X. Then a continuous map between standard geometric objects is a C^α-map ($\alpha \in \mathbb{N}_0 \cup \{\infty, \omega\}$) if and only if the composition with f sends a C^α-function to a C^α-function. This allows one to view such standard geometric objects and their structure preserving maps as a subcategory of (locally) ringed spaces over $\mathbb{R}$ or $\mathbb{C}$. Hence one obtains the notion of a geometric object by the general procedure explained above. These are called premanifolds.

A manifold will then be a premanifold whose underlying topological space has certain good properties (being Hausdorff and second countable). Let me briefly digress on this choice of terminology. First of all I follow the classical terminology. But an even more compelling reason to restrict the class of manifolds by asking for these topological properties is a multitude of techniques and results where these topological properties are indispensable hypotheses (such as embedding results or the theory of integration on manifold). In this textbook this is not that visible as many of such results are not covered here. Hence we will more often encounter premanifolds than manifolds and it would have been tempting to change terminology, if only to get rid of the annoying "pre-" everywhere. But I decided against this to remind the reader that the contents of this book are only the very beginning of a journey into the wondrous world of differential geometry – a world in which very often manifolds and not premanifolds are the central objects.

A fundamental idea in modern mathematics is the notion of a sheaf. Sheaves are needed to define the notion of a ringed space but their usefulness goes far beyond this. Sheaves embody the principle for passing from local to global situations – a central topic in mathematics. In the theory of smooth manifolds one can often avoid the use of sheaves as there is another powerful tool for local-global constructions, namely partitions of unity. Their existence corresponds to the fact that the sheaf of smooth functions is soft (see Chap. 9). But this is particular to the case of real C^{α}-manifolds with $\alpha \leq \infty$. In all other geometries mentioned before the theory of sheaves is indispensable. Hence sheaves will be a central topic of this book.

Together with sheaves and manifolds (as ringed spaces), the third main topic is the cohomology of sheaves. It is the main tool to make use of sheaves for local-global problems. As a rule, it allows one to consider the obstruction for the passage from local to global objects as an element in an algebraic cohomology object of a sheaf, usually a group. Moreover, the formalism of cohomology also yields a powerful tool to calculate such obstructions. In addition, many interesting objects (such as fiber bundles) are classified by the cohomology of certain sheaves.

It is not the goal of this book – and also would have been clearly beyond my abilities – to give a new quick and streamlined introduction to differential geometry using clever arguments to obtain deep results with minimal technical effort. Quite the contrary, the focus of this book is on the technical methods necessary to work with modern theories of geometry. As a principle I tried to explain these techniques in their "correct generality" (which is certainly a very subjective notion) to provide a reliable point of departure towards geometry.

There are some instances where I deviate from this principle, either because of lack of space or because I think that the natural generality and abstractness would seriously conceal the underlying simple idea. This includes the following subjects:

1. One might argue that the natural framework for sheaves are sheaves on an arbitrary site (i.e., a category endowed with a Grothendieck topology) or even general topos theory. But in my opinion this would have seriously hampered the accessibility of the theory.
2. Instead of working with manifolds modeled on open subsets of $\mathbb{R}^n$ or $\mathbb{C}^n$ one might argue that it is more natural (and not much more difficult) to model them on open subsets of arbitrary Banach spaces. I decided against this because the idea of the book is to demonstrate general techniques used in geometry in the most accessible example: finite-dimensional real and complex manifolds.
3. In the chapter on cohomology I do not use derived categories, although I tried to formulate the theory in such a way that a reader familiar with the notion of a derived category can easily transfer the results into this language[1].

Moreover, there are several serious omissions due to lack of space, among them Lie algebras, manifolds with corners (or, more generally, singular spaces), and integration theory – just to name a few. The educated reader will find many more such omissions.

Prerequisits
The reader should have knowledge of basic algebraic notions such as groups, rings and vector spaces, basic analytic and topological notions such as differentialbility in several variables and metric spaces.

Further prerequisites are assembled in five appendix chapters. It is assumed that the reader knows some but not all of the results here. Therefore many proofs and many examples are given in the appendices. These appendices give (Chap. 12) a complete if rather brisk treatment of basic concepts of point set topology, (Chap. 13) a quick introduction to the language of categories focused on examples, (Chap. 14) some basic definitions and results of abstract algebra, (Chap. 15) those notions of homological algebra necessary to cope with the beginning of cohomology theory, and (Chap. 16) a reminder on the notion of differential and analytic functions on open subsets of finite-dimensional $\mathbb{R}$- and $\mathbb{C}$-vector spaces.

Outline of contents
The main body of the text starts with two preliminary chapters. The first chapter (Chap. 1) introduces more advanced concepts from point set topology. The main notions of the first three sections are paracompact and normal spaces, covering important techniques like Urysohn's theorem, the Tietze extension theorem and the Shrinking Lemma for paracompact Hausdorff spaces. The last two sections of this chapter focus on separated and proper maps. In Chap. 2 basic notions of algebraic topology used in the sequel are introduced. Here we restrict the contents to those absolutely necessary (but with complete proofs) and ignore all progress made in the last decades.

---

[1] Of course, for most readers that are familiar with derived categories the cohomology chapter will not contain that many new results anyway.

In the third chapter (Chap. 3) we introduce the first main topic of the book: sheaves. We introduce two (equivalent) definitions of sheaves. The first one is that of a rule attaching to every open set of a topological space a set of so-called sections such that these sections can be glued from local to global objects. This is also the definition that generalizes from topological spaces to more abstract geometric objects such as sites. There is another point of view of sheaves that works just fine for sheaves on topological spaces, namely étalé spaces. It is proved that both concepts are equivalent and explained that some constructions for sheaves are more accessible via the first definition (such as direct images) and some are more accessible via étalé spaces (such as inverse images).

In Chap. 4 we introduce the class of geometric objects that will be studied in this book: manifolds. As explained above, we start by defining the very general category of ringed spaces over a fixed ring. Then we explain how to consider open subsets of real or complex finite-dimensional vector spaces as ringed spaces over $\mathbb{R}$ or $\mathbb{C}$. This yields our class of standard geometric objects. Then a premanifold is by definition a ringed space that is locally isomorphic to such a standard geometric object.

The central topic of Chap. 5 is that of linearization. We start by linearizing manifolds by introducing tangent spaces. Then the derivative of a morphism at a point is simply the induced map on tangent spaces and we can think of it as a pointwise linearization of the morphism. Next we study morphisms of (pre)manifolds that can even be locally linearized. These are precisely the morphisms whose derivatives have a locally constant rank. Examples are immersions, submersions, and locally constant maps. These linearization techniques are then used in the remaining sections of the chapter to study submanifolds, their intersections (or, more generally, fiber products of manifolds), and quotients of manifolds by an equivalence relation.

Chapter 6 introduces the symmetry groups in the theory of manifolds, the Lie groups. It focuses on actions of Lie groups on manifolds. The construction of quotients of manifolds in Chap. 5 then yields the existence of quotients for proper free actions of Lie groups.

In Chap. 7 we start to study local-global problems by introducing the first cohomology of a sheaf of (not necessarily abelian) groups via the language of torsors or, equivalently, by the language of Čech cocycles.

This is used in Chap. 8 to classify fiber bundles that are important examples of locally but not globally trivial objects. We start the chapter by introducing the general notion of a morphism that looks locally like a given morphisms p. Very often it is useful to restrict the classes of local isomorphisms. This yields the notion of a twist of p with a structure sheaf that is a subsheaf of the sheaf of all automorphisms of p. Specializing to the case that p is a projection and the structure sheaf is given by the faithful action of a Lie group G on the fiber then yields the notion of a fiber bundle with structure group G. Specializing further we obtain the important notions of G-principal bundles and vector bundles. We explain that vector bundles can also be described as certain modules over the sheaf of functions of the manifold. In the last two sections we study the most important examples of vector bundles on a manifold, the tangent bundle and the bundles of differential forms. In particular we will obtain the de Rham complex of a manifold.

As mentioned above, for real C^α-manifolds with $\alpha \leq \infty$ it is often possible to use other techniques than sheaves for solving local-global problems. The sheaf-theoretic reason for this lies in the softness of the sheaf of C^α-functions. This notion is introduced in Chap. 9. We deduce from the softness of the structure sheaf the existence of arbitrary fine partitions of unity. In the last section we show that the first cohomology of a soft sheaf is trivial and deduce immediately some local-global principles. For real C^α-manifolds with $\alpha \leq \infty$ all these results can also be obtained via arguments with a partition of unity. But these examples illustrate how to use the triviality of certain cohomology classes also in cases where partitions of unity are not available, for instance for certain complex manifolds.

After giving the rather ad hoc definition of the first cohomology in Chap. 7 (that had the advantage also to work for sheaves of not necessarily abelian groups) we now introduce cohomology in arbitrary degree in Chap. 10. After a quick motivation on how to do this, it becomes clear that it is more natural not to work with a single sheaf but with a whole complex of sheaves. This is carried out in the first three sections. Applying the whole formalism of cohomology to the de Rham complex we obtain de Rham's theorem relating de Rham cohomology and cohomology of constant sheaves in Sect. 10.4. We conclude the chapter by proving an other important result: the theorem of proper base change, either in the case of arbitrary topological spaces and proper separated maps (giving the theorem its name) or for metrizable spaces and closed maps.

In the last chapter (Chap. 11) we focus on the cohomology of constant sheaves. We show that for locally contractible spaces this is the same as singular cohomology, which is quickly introduced in Sect. 11.1. In particular we obtain the corollary that we can describe the de Rham cohomology on a manifold also via singular cohomology. Then we use the proper base change theorem to prove the main result of this chapter, the homotopy invariance of the cohomology of locally constant sheaves for arbitrary topological spaces. We conclude with some quick applications.

As with almost every mathematical text, this book contains a myriad of tiny exercises in the form of statements where the reader has to make some straightforward checks to convince herself (or himself) that the statement is correct. Beyond this, all chapters and appendices end with a group of problems. Some of the problems are sketching further important results that were omitted from the main text due to lack of space, and the reader should feel encouraged to use these problems as a motivation to study further literature on the topic.

Acknowledgments

This book grew out of a lecture I gave for third year bachelor students in Paderborn and I am grateful for their motivation to get a grip on difficult and abstract notions. Moreover, I thank all people who helped to improve the text by making comments on a preliminary version of the text, in particular Benjamin Schwarz and Joachim Hilgert. Special thanks go to Christoph Schabarum for TeXing (and improving) some of the passages of the book, to Jean-Stefan Koskivirta for designing lots of exercises for my class of which almost all of them can now be found in the problem sections, and to Joachim Hilgert for giving me access to his collection of figures.

Paderborn
January, 2016   *Torsten Wedhorn*

Standard Notation

For a set I and elements $i, j \in I$ we use the "Kronecker delta"

$$\delta_{ij} := \begin{cases} 1, & \text{if } i = j; \\ 0, & \text{if } i \neq j. \end{cases}$$

We denote by $\mathbb{N}$ the set $\{1, 2, \ldots\}$ of natural numbers (without 0), $\mathbb{N}_0 := \mathbb{N} \cup \{0\}$. $\mathbb{Z}$ denotes the ring of integers, $\mathbb{Q}$ the field of rational numbers, $\mathbb{R}$ (respectively $\mathbb{C}$) the field of real (respectively complex) numbers endowed with the metric given by the standard absolute value. The symbol $\mathbb{K}$ always either denotes $\mathbb{R}$ or $\mathbb{C}$.

To unify the notation for n-fold continuously differentiable, smooth, and analytic maps we define

$$\widehat{\mathbb{N}}_0 := \mathbb{N}_0 \cup \{\infty, \omega\}, \qquad \widehat{\mathbb{N}} := \mathbb{N} \cup \{\infty, \omega\}.$$

We extend the usual total order on $\mathbb{N}_0$ to $\widehat{\mathbb{N}}_0$ by requesting $n < \infty < \omega$ for all $n \in \mathbb{N}_0$. We also define $\infty \pm \alpha := \infty$ and $\omega \pm \alpha := \omega$ for all $\alpha \in \mathbb{N}_0$.

A set I is said to be *countable* if there exists an injective map $I \hookrightarrow \mathbb{N}$. In particular, all finite sets are countable.

By a *monoid* we mean a set M endowed with an associative binary operation $M \times M \to M$, $(m, m') \mapsto mm'$, having an identity element e_M. A *monoid homomorphism* is a map $\varphi \colon M \to N$ such that $\varphi(mm') = \varphi(m)\varphi(m')$ for all $m, m' \in M$ and $\varphi(e_M) = e_N$. A monoid M is called *commutative* if $mm' = m'm$ for all $m, m' \in M$.

All *rings* are assumed to have a unit, usually denoted by 1, and ring homomorphisms are assumed to preserve the unit. All *fields* are assumed to be commutative.

If R is a ring, by an R-module we mean a left R-module if not stated otherwise (for a reminder on some notions about modules see Appendix Section 14.1). If R is commutative, we will not distinguish between left R-modules and right R-modules.

In a metric space (X, d) we denote for $x_0 \in X$ and $r \in \mathbb{R}^{>0}$ the open and closed ball by

$$B_r(x_0) := \{x \in X \; ; \; d(x, x_0) < r\},$$
$$B_{\leq r}(x_0) := \{x \in X \; ; \; d(x, x_0) \leq r\}.$$

Let V be a finite-dimensional $\mathbb{K}$-vector space. If not otherwise stated, we endow a subset X of V always with the topology induced by some norm on V. As all norms on V are equivalent, this topology does not depend on the choice of the norm. See Appendix 12 for a reminder on topological spaces.

Contents

Topological Preliminaries

In this chapter we prove some results on topological spaces that will be needed later and go beyond the basic topological results and notions assembled in the Appendix Chap. 12. The chapter consists of two independent parts.

In the first part (Sects. 1.1–1.3) we introduce, after a quick review of countability properties, paracompact spaces. This is one of the central topological notions in this book. We show that the following classes of topological spaces are paracompact: Metrizable spaces (Proposition 1.13) and locally compact, second countable Hausdorff spaces (Proposition 1.10), see also Remark 1.14. Then we show that paracompact Hausdorff spaces are normal (Proposition 1.18). Hence Urysohn's separation theorem, the Tietze extension theorem (Theorem 1.15), and the shrinking lemma (Proposition 1.20, Corollary 1.21) are available for paracompact spaces.

The second part (Sects. 1.4 and 1.5) introduces relative versions of Hausdorff spaces and compact spaces: separated maps and proper maps.

1.1 Countability Properties for Topological Spaces

Recall that we call a set M *countable* if there exists an injective map $M \to \mathbb{N}$ (equivalently, $M = \emptyset$ or there exists a surjective map $\mathbb{N} \to M$). Hence any finite set is countable. Every subset of a countable set is again countable. Countable unions and finite products of countable sets are again countable.

Definition 1.1 (Countability properties). Let X be a topological space.

1. X is called *first countable* if every point of x has a countable neighborhood basis.
2. X is called *second countable* if the topology has a countable basis.
3. X is called a *Lindelöf space* if every open covering of X has a countable subcovering.

© Springer Fachmedien Wiesbaden 2016
T. Wedhorn, *Manifolds, Sheaves, and Cohomology*, Springer Studium Mathematik – Master,
DOI 10.1007/978-3-658-10633-1_1

4. X is called *separable* if it contains a countable dense subset.
5. X is called *σ-compact* if it is the union of countably many compact subspaces.

Example 1.2. Let X be a topological space.

1. Every metrizable space X is first countable. Let d be a metric inducing the given topology on X. For $x \in X$ the sets

$$\left\{ B_{1/n}(x) := \{\, y \in X \; ; \; d(x, y) < 1/n \,\} \; ; \; n \in \mathbb{N} \right\}$$
$$\left\{ B_{\leq 1/n}(x) := \{\, y \in X \; ; \; d(x, y) \leq 1/n \,\} \; ; \; n \in \mathbb{N} \right\}$$

both form countable neighborhood bases of x.
2. The space $\mathbb{R}^n$ is second countable and σ-compact for all $n \geq 0$: Choose a norm on $\mathbb{R}^n$ yielding a metric d. Then a countable basis is given by

$$\left\{ B_{1/m}(x) \; ; \; x \in \mathbb{Q}^n, m \in \mathbb{N} \right\},$$

and $\mathbb{R}^n$ is the union of the countably many closed balls $B_{\leq N}(0)$, $N \in \mathbb{N}$.
3. Let X be an uncountable set endowed with the discrete topology. Then X is first countable but not second countable.

Remark 1.3.

1. Every subspace Z of a second countable space X is second countable: If $\mathcal{B}_X$ a countable basis of the topology of X, then $\{\, B \cap Z \; ; \; B \in \mathcal{B}_X \,\}$ is a countable basis for Z. In particular, every subspace of $\mathbb{R}^n$ is second countable.
2. Let $f : X \to Y$ be a surjective open continuous map. If $\mathcal{B}$ is a basis of the topology of X, then $\{\, f(U) \; ; \; U \in \mathcal{B} \,\}$ is a basis of the topology of Y. In particular we see that if X is second countable, then Y is second countable.
3. Let $(X_n)_n$ be a countable family of second countable spaces. Then $\prod_n X_n$ is second countable: If $\mathcal{B}_n$ is a countable basis for X_n, then

$$\left\{ \prod_n U_n \; ; \; U_n \in \mathcal{B}_n, U_n = X_n \text{ for all but finitely many } n \right\}$$

is a countable basis for $\prod_n X_n$.
4. Suppose that X has a countable open covering $(U_n)_n$ such that the subspace U_n is second countable for all n. Then X is second countable: If $\mathcal{B}_n$ is a countable basis for U_n, then $\bigcup_n \mathcal{B}_n$ is a countable basis for X.

Proposition 1.4. *Let X be a topological space.*

1. *If X is second countable, then X is first countable, separable, and a Lindelöf space.*
2. *Suppose that X is a metrizable Lindelöf space. Then X is second countable.*

Problem 1.4 shows that a separable metrizable space is also second countable.

Proof. 1. Let $\mathcal{B} = \{ B_n \; ; \; n \in \mathbb{N} \}$ be a countable basis for X. Then for every point $x \in X$ the set $\{ B \in \mathcal{B} \; ; \; x \in B \}$ is a countable neighborhood basis of x. For every non-empty $B \in \mathcal{B}$ choose $x_B \in B$ and set $Q := \{ x_B \; ; \; B \in \mathcal{B} \}$. Then Q is dense in X because it meets every open subset of X. Hence X is first countable and separable.

It remains to show that X is Lindelöf. Let $(U_i)_{i \in I}$ be an open covering of X. Let M be the set of $n \in \mathbb{N}$ such that there exists $i_n \in I$ with $B_n \subseteq U_{i_n}$. As $(U_i)_i$ is an open covering and $\mathcal{B}$ is a basis, $(B_n)_{n \in M}$ is an open covering. Hence $(U_{i_n})_{n \in M}$ is a countable subcovering of $(U_i)_i$.

2. Fix $n \in \mathbb{N}$. As X is Lindelöf, we can choose a countable subcovering $\mathcal{U}_n$ of $\{ B_{1/n}(x) \; ; \; x \in X \}$. Then the union of all $\mathcal{U}_n$ is a countable basis for X. $\qquad\square$

Proposition 1.5. *Let G be a topological group.*

1. *Suppose that G is connected and that the neutral element of G has a compact neighborhood C. Then G is σ-compact.*
2. *Suppose that G is σ-compact. Then there exists for every neighborhood U of the neutral element e of G a sequence $(g_n)_n$ in G such that $G = \bigcup_n g_n U$.*
3. *If G is σ-compact and first countable, then G is second countable.*

Proof. 1. As G is connected, G is generated by C (Appendix Corollary 12.59). Replacing C by the compact subset $C \cup C^{-1}$ (with $C^{-1} := \{ g^{-1} \; ; \; g \in C \}$), we see $G = \bigcup_{n \in \mathbb{N}} C^n$ with $C^n = \{ g_1 g_2 \cdots g_n \; ; \; g_i \in C \}$. But C^n is compact, as the image of the compact n-fold product $C \times \cdots \times C$ under the n-fold multiplication map $G \times \cdots \times G \to G$.

2. Replacing U by its interior we may assume that U is open. Let $G = \bigcup_{n \in \mathbb{N}} K_n$ with $K_n \subseteq G$ compact subspace. For each n the open subsets kU for $k \in K_n$ cover the compact set K_n. Therefore there exist $k_{n,1}, \ldots, k_{n,r_n} \in K_n$ with $K_n \subseteq \bigcup_i k_{n,i} U$ and hence $G \subseteq \bigcup_{n \in \mathbb{N}} \bigcup_{1 \le i \le r_n} k_{n,i} U$. Renumbering the countably many $k_{n,i}$ yields the desired result.

3. Let $(U_n)_{n \in \mathbb{N}}$ be a fundamental system of neighborhoods of e. We first claim that for all $n \in \mathbb{N}$ there exists $m \in \mathbb{N}$ such that $U_m^{-1} U_m \subseteq U_n$. Indeed, consider the continuous map $\alpha \colon G \times G \to G, (g, h) \mapsto g^{-1} h$. Then $\alpha^{-1}(U_n)$ is a neighborhood of (e, e) in $G \times G$.

Hence there exist $m, m' \in \mathbb{N}$ such that $U_m \times U_{m'} \subseteq \alpha^{-1}(U_n)$. Replacing m and m' both by some $r \in \mathbb{N}$ with $U_r \subseteq U_m \cap U_{m'}$ we may assume that $m = m'$. This proves the claim.

By 2. there exist for all $n \in \mathbb{N}$ sequences $(g_{n,k})_k$ in G such that $G = \bigcup_{k \in \mathbb{N}} g_{n,k} U_n$. We show that $\{\, g_{n,k} U_n \;;\; k, n \in \mathbb{N} \,\}$ is a basis of the topology of G. In fact, let V be an open subset of G and $g \in V$. Then there exists $n \in \mathbb{N}$ such that $g U_n \subseteq V$. By our claim above we may choose $m \in \mathbb{N}$ such that $U_m^{-1} U_m \subseteq U_n$. If we choose $k \in \mathbb{N}$ such that $g \in g_{m,k} U_m$, then $g_{m,k} U_m \subseteq g U_m^{-1} U_m \subseteq g U_n \subseteq V$. $\square$

Combining 1. and 3. of Proposition 1.5 we obtain:

Corollary 1.6. *A connected locally compact first countable topological group is second countable.*

1.2 Paracompact Spaces

Definition 1.7 (Refinement). Let X be a topological space and let $\mathcal{U} = (U_i)_{i \in I}$ be a covering. A covering $(V_j)_{j \in J}$ is called a *refinement of* $\mathcal{U}$ if for all $j \in J$ there exists $i \in I$ with $V_j \subseteq U_i$. In other words, there exists a map $\alpha \colon J \to I$ such that $V_j \subseteq U_{\alpha(j)}$ for all $j \in J$.

A refinement $(V_j)_j$ of $\mathcal{U}$ is called *open*, if V_j is open in X for all $j \in J$.

Any subcovering is a refinement.

Definition 1.8 (Paracompact spaces). A topological space X is called *paracompact* if every open covering has an open refinement that is locally finite.

We do *not* suppose that paracompact spaces are Hausdorff (as it is often done, for instance in [BouGT1]).

Example 1.9.
1. Every compact space is paracompact.
2. Every discrete topological space X is paracompact: The covering $(\{x\})_{x \in X}$ is a locally finite open refinement of every open covering.

Proposition 1.10. *Let X be a locally compact and second countable Hausdorff space. Then X has the following properties:*

1. *There exists a sequence $(K_n)_n$ of compact subspaces of X such that $X = \bigcup_n K_n$ and $K_n \subseteq K_{n+1}^{\circ}$ for all $n \in \mathbb{N}$. In particular X is σ-compact.*

2. *Every open covering of X has a countable locally finite open refinement $(V_l)_{l \in \mathbb{N}}$ such that $\overline{V_l}$ is compact for all l. In particular X is paracompact.*

Proof. We first construct a countable base $\mathcal{B}$ such that $\overline{B}$ is compact for all $B \in \mathcal{B}$. Let $\mathcal{B}'$ be a countable base of X, C_x a compact neighborhood of $x \in X$ (automatically closed in X because X is Hausdorff). Define

$$\mathcal{B} := \{ V \cap W \; ; \; V \in \mathcal{B}' \text{ and } W \in \mathcal{B}' \text{ with } W \subseteq C_x \text{ for some } x \in X \}.$$

Then $\mathcal{B}$ is a countable basis because $X = \bigcup_x C_x$. The closure of each set $B \in \mathcal{B}$ is a closed subset of C_x for some x and hence compact. Write $\mathcal{B} = \{B_1, B_2, \dots\}$.

   Proof of 1. We construct compact subspaces K_n of X and $i(n) \in \mathbb{N}_0$ with $i(n) > i(n-1)$ such that $B_1 \cup \cdots \cup B_{i(n)} \subseteq K_n \subseteq K_{n+1}^\circ$ for all $n \in \mathbb{N}_0$ (then $\bigcup_{n \in \mathbb{N}} K_n \supseteq \bigcup_{i \in \mathbb{N}} B_i = X$).

   Define K_n and $i(n)$ inductively. Set $K_0 := \emptyset$ and $i(0) := 0$. Now let $n \geq 1$. As K_{n-1} is compact, we find $i(n) \in \mathbb{N}$ with $i(n) > i(n-1)$ and $K_{n-1} \subseteq \bigcup_{1 \leq j \leq i(n)} B_j$. Set $K_n := \bigcup_{1 \leq j \leq i(n)} \overline{B_j}$. Then K_n is compact and

$$K_{n-1} \subseteq \bigcup_{1 \leq j \leq i(n)} B_j \subseteq K_n^\circ.$$

   Proof of 2. Let $\mathcal{U} = (U_i)_i$ be an open covering of X. Let $\mathcal{W} = (W_j)_{j \in J}$ be the refinement of $\mathcal{U}$ consisting of those $W \in \mathcal{B}$ with $W \subseteq U_i$ for some i. Fix $n \in \mathbb{N}$ and set $W_{j,n} := W_j \cap (K_{n+2}^\circ \setminus K_{n-1})$. Then

$$\bigcup_{j \in J} W_{j,n} \supseteq \underbrace{K_{n+1} \setminus K_n^\circ}_{\text{compact}}.$$

Let $J(n) \subseteq J$ be finite such that $\bigcup_{j \in J(n)} W_{j,n} \supseteq K_{n+1} \setminus K_n^\circ$. Let $\mathcal{V} = (V_l)_l$ be the open covering consisting of $W_{n,j}$ for $n \in \mathbb{N}$ and $j \in J(n)$ (it is a covering because $\bigcup_n K_n = X$). Then $\mathcal{V}$ is countable and a refinement of $\mathcal{U}$ by construction. We have $\overline{W_{n,j}} \subseteq K_{n+2}$ and hence $\overline{V_l}$ is compact for all l. Finally, $\mathcal{V}$ is locally finite: For $x \in X$ let $n \in \mathbb{N}$ with $x \in K_n$. Then K_{n+1}° is an open neighborhood of x, which intersects only the finitely many $W_{j,m}$ with $j \in J(m)$ and $m \leq n + 1$ (by the definition of $W_{j,m}$). $\qquad\square$

Remark 1.11. Let X be a paracompact space. Then every closed subspace Y of X is paracompact. Indeed, the proof is the same as for compact spaces. Let $\mathcal{U}$ be an open covering of Y. Hence $\mathcal{U} = (U_i \cap Y)_{i \in I}$ for open subsets U_i of X with $Y \subseteq \bigcup_i U_i$. Adding $X \setminus Y$ to $(U_i)_i$, we obtain an open covering of X. This has a locally finite refinement $(V_j)_j$ because X is paracompact. Then $(V_j \cap Y)_j$ is a locally finite refinement of $\mathcal{U}$.

   Open subspaces of paracompact spaces are not paracompact in general ([BouGT1] Chap. I, §9, Exercise 11). Hence we introduce the following terminology.

Remark and Definition 1.12 (Hereditarily paracompact). A topological space X is called *hereditarily paracompact* if the following equivalent conditions are satisfied:

(i) Every open subspace of X is paracompact.
(ii) Every subspace of X is paracompact.

Indeed, suppose that every open subspace is paracompact and let $Y \subseteq X$ be an arbitrary subspace. An open covering of Y is of the form $(Y \cap U_i)_i$, where $(U_i)_{i \in I}$ is a family of open subsets with $Y \subseteq U := \bigcup_i U_i$. As U is paracompact, there exists a locally finite open refinement $(V_j)_j$ of the covering $(U_i)_i$ of U. Then $(Y \cap V_j)_j$ is a locally finite open refinement of $(Y \cap U_i)_i$.

Proposition 1.13. *Every metrizable topological space is hereditarily paracompact.*

There exist hereditarily paracompact spaces that are not metrizable (Problem 1.13).

Proof. As every subspace of a metrizable space is again metrizable via the restricted metric, it suffices to show that every metric space (X, d) is paracompact. Let $(U_i)_{i \in I}$ be an open covering. Choose a well ordering on I (Appendix Proposition 13.27). For $n \in \mathbb{N}$ define for all $i \in I$ open subsets $V_{i,n}$ of X by induction on $n \in \mathbb{N}$. Let $X_{i,n}$ be the set of $x \in X$ with

(a) $x \in U_i \setminus \bigcup_{j < i} U_j$,
(b) $B_{3 \cdot 2^{-n}}(x) \subseteq U_i$,
(c) $x \notin V_{j,m}$ for $m < n$ and for all $j \in I$.

Now define $V_{i,n} := \bigcup_{x \in X_{i,n}} B_{2^{-n}}(x)$. We claim that $(V_{i,n})_{i \in I, n \in \mathbb{N}}$ is a locally finite open refinement of $(U_i)_i$.

Clearly we have $V_{i,n} \subseteq U_i$ by (b). To see that $\bigcup_{i,n} V_{i,n} = X$, let $x \in X$. Let $i \in I$ be the smallest element such that $x \in U_i$ and let $n \in \mathbb{N}$ with $B_{3 \cdot 2^{-n}}(x) \subseteq U_i$. If $x \in X_{i,n}$, then $x \in V_{i,n}$. Otherwise (c) does not hold, hence there exists $j \in I$ and $m < n$ such that $x \in V_{j,m}$.

It remains to show that $(V_{i,n})$ is locally finite. Let $x \in X$ and let $i_0 \in I$ be the smallest element such that $x \in V_{i_0,m}$ for some $m \in \mathbb{N}$. Choose $p \in \mathbb{N}$ such that $B_{2^{-p}}(x) \in V_{i_0,m}$. We claim that $B_{2^{-m-p}}(x)$ meets only finitely many of the $V_{i,n}$. More precisely, we claim:

(i) If $n \geq m + p$, then $B_{2^{-m-p}}(x) \cap V_{i,n} = \emptyset$ for all $i \in I$.
(ii) If $n < m + p$, then there is at most one $i \in I$ with $B_{2^{-m-p}}(x) \cap V_{i,n} \neq \emptyset$.

Let us show (i). Let $i \in I$ and $n \geq m + p$. Since $n > m$, (c) implies that for $y \in X_{i,n}$ one has $y \notin V_{i_0,m}$. As $B_{2^{-p}}(x) \in V_{i_0,m}$ we find $d(x, y) \geq 2^{-p}$ for all $y \in X_{i,n}$. Hence

$B_{2^{-m-p}} \cap B_{2^{-n}}(y) = \emptyset$ for all $y \in X_{i,n}$ because $n \geq p+1$ and $m+p \geq p+1$. This proves (i).

To show (ii) let $n < m+p$. Suppose we have $y \in V_{i,n}$ and $z \in V_{j,n}$ with $i < j$. It suffices to show that $d(y,z) > 2^{-m-p+1}$. By definition of the $V_{i,n}$ we find $\tilde{y} \in X_{i,n}$ and $\tilde{z} \in X_{j,n}$ with $y \in B_{2^{-n}}(\tilde{y})$ and $z \in B_{2^{-n}}(\tilde{z})$. We have $B_{3 \cdot 2^{-n}}(\tilde{y}) \subseteq U_i$ by (b) and $\tilde{z} \notin U_i$ by (a). Hence $d(\tilde{y},\tilde{z}) \geq 3 \cdot 2^{-n}$ and hence $d(y,z) > 2^{-n} \geq 2^{-m-p+1}$. $\square$

Remark 1.14. One can also show that locally compact, second countable Hausdorff spaces are metrizable (e.g., [Br1] Chap. 1, Theorem 12.12).

1.3 Normal Spaces

Theorem 1.15 (Urysohn's theorem/Tietze extension theorem). *Let X be a topological space. Then the following assertions are equivalent:*

(i) *For any two closed subsets $A, B \subseteq X$ with $A \cap B = \emptyset$ there exists a continuous function $f: X \to [0,1]$ such that $f(a) = 0$ for all $a \in A$ and $f(b) = 1$ for all $b \in B$.*

(ii) *For all disjoint closed sets A and B of X there exist open disjoint sets U and V such that $A \subseteq U$ and $B \subseteq V$ (see Fig. 1.1).*

(iii) *For every closed subset A and every neighborhood W of A there exists an open neighborhood U of A such that $\bar{U} \subseteq W$.*

(iv) *For every closed subspace A of X and continuous map $f: A \to \mathbb{R}$ there exists a continuous function $\tilde{f}: X \to \mathbb{R}$ such that $\tilde{f}_{|A} = f$.*

(v) *For every closed subspace A of X and continuous map $f: A \to [-1,1]$ there exists a continuous function $\tilde{f}: X \to [-1,1]$ such that $\tilde{f}_{|A} = f$.*

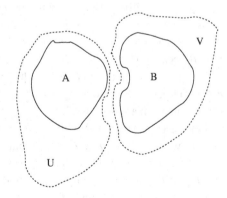

Figure 1.1 Separation of two closed subsets A und B

Of course, one can replace in (i) and in (v) the compact interval by any space homeomorphic to it, for example any other compact interval in $\mathbb{R}$ or the extended real line $\overline{\mathbb{R}} = \mathbb{R} \cup \{\pm\infty\}$ (Appendix Example 12.9).

Proof. We prove

$$(i) \Rightarrow (ii) \Rightarrow (iii) \Rightarrow (i) \Rightarrow (v) \Rightarrow (iv) \Rightarrow (i).$$

We start with the easy implications. To see "(i) $\Rightarrow$ (ii)" take $U = \{ x \in X ; f(x) < 1/2 \}$ and $V = \{ x \in X ; f(x) > 1/2 \}$. To see "(ii) $\Rightarrow$ (iii)", we may assume that W is an open neighborhood of A in X. Let $B := X \setminus W$. Then (ii) implies that there exists an open neighborhood U of A such that $\bar{U} \cap B = \emptyset$ and hence $\bar{U} \subseteq W$.

Next we show "(iv) $\Rightarrow$ (i)". Let A and B disjoint closed subsets of X. Then $f: A \cup B \to \mathbb{R}$ with $f(A) = \{0\}$ and $f(B) = \{1\}$ is continuous. Hence (iv) implies that f can be extended to a continuous map $\tilde{f}: X \to \mathbb{R}$. Then $g := \min(\max(\tilde{f}, 0), 1)$ is a continuous map on X with values in $[0, 1]$, equal to 0 on A and equal to 1 on B.

"(iii) $\Rightarrow$ (i)". Let A and B be disjoint closed subsets of X and denote by $D \subseteq \mathbb{Q}$ the set of dyadic rational numbers in $[0, 1]$. First, we will construct open subsets $U_q \subseteq X$ for every $q \in D$ with the following properties:

1. For each $q \in D$, $A \subseteq U_q$.
2. $U_1 = X$ and for each $q \in D$ with $q < 1$, $B \cap U_q = \emptyset$.
3. For $q, p \in D$ with $q < p$, $\bar{U}_q \subseteq U_p$.

Set $U_1 := X$. Then $X \setminus B$ is a neighborhood of A and by (iii) we find an open subset $U_0 \subseteq X$ such that $A \subseteq U_0$ and $\bar{U}_0 \subseteq X \setminus B$.

We will continue this construction by recursion. Let $m \in \mathbb{N}$. Assume that U_q has been constructed for $q = a/2^n \in [0, 1]$ for all $n < m$ and $a = 0, \ldots, 2^n$. For $p = b/2^m \in [0, 1]$ with $2 \nmid b$, set $q_1 := ((b + 1)/2)/2^{m-1} \in [0, 1]$ and $q_2 := ((b - 1)/2)/2^{m-1} \in [0, 1]$ so that $p = (q_1 + q_2)/2$. Therefore U_{q_1} and U_{q_2} have already been constructed and we have $\bar{U}_{q_1} \subseteq U_{q_2}$. By (iii) we find an open subset $U_p \subseteq X$ such that $\bar{U}_{q_1} \subseteq U_p$ and $\bar{U}_p \subseteq U_{q_2}$.

We now define $f: X \to [0, 1]$ by $f(x) := \inf\{ q \in D ; x \in U_q \}$, which is well-defined because $U_1 = X$. Then $f(a) = 0$ for $a \in A$ because $A \subseteq U_0$ and $f(b) = 1$ for $b \in B$ because $B \subseteq U_1$ and $B \cap U_q = \emptyset$ for $q \in D$ with $q < 1$. It is left to show that f is continuous. Notice that the sets $[0, c) \subseteq [0, 1]$ for $c \in (0, 1]$ and $(c, 1] \subseteq [0, 1]$ for $c \in [0, 1)$ generate the topology on $[0, 1]$. Hence it suffices to show that the preimage of sets of this form are open in X.

Let $c \in (0, 1]$ and $x \in X$ with $f(x) \in [0, c)$. Since $D \subseteq [0, 1]$ is dense, we can choose $q \in D$ with $f(x) < q < c$. By definition of f there exists $p \in D$ with $p < q$ and $x \in U_p$ so that we have $x \in U_q$. For $y \in U_q$ we obviously have $f(y) \leq q < c$ so that $x \in U_q \subseteq f^{-1}([0, c))$ is an open neighborhood of x.

Let $c \in [0, 1)$, $x \in X$ with $f(x) \in (c, 1]$ and choose $q \in D$ with $c < q < f(x)$. Then there exists $p \in D$ with $q < p$ and $x \notin U_p$ so that $x \notin \bar{U}_q$ because $\bar{U}_q \subseteq U_q$. Hence,

$X \setminus \bar{U}_q$ is an open neighborhood of x. Furthermore, for $y \in X \setminus \bar{U}_q$ we have $y \notin U_p$ for $p \leq q$ because $U_p \subseteq \bar{U}_q$ so that $c < q < f(y)$ and $X \setminus \bar{U}_q \subseteq f^{-1}((c, 1])$.

"$(v) \Rightarrow (iv)$". Suppose first that $f \geq 0$. By choosing a homeomorphism $[-1, 1] \cong [0, \infty] \subset \bar{\mathbb{R}}$ it follows from (v) that we find a continuous extension g_1 of f with values in $[0, \infty]$. Then $B := g_1^{-1}(\infty)$ is closed and $A \cap B = \emptyset$. Define a continuous map $h: A \cup B \to \mathbb{R}_{\geq 0}$ by $h_{|A} := f$ and $h_{|B} = 0$ and let g_2 be a continuous extension of h with values in $[0, \infty]$. Then $\tilde{f} := \min(g_1, g_2)$ is a continuous extension of f to X with values in $[0, \infty)$.

For a general f extend $\max(f, 0)$ and $\max(-f, 0)$ to continuous functions g_1 and g_2, respectively, on X with values in $[0, \infty)$. Then set $\tilde{f} := g_1 - g_2$.

"$(i) \Rightarrow (v)$". Let $A \subseteq X$ be a closed subset and $f: A \to [-1, 1]$ a continuous map. We find $f^{-1}([-1, -1/3]) \subseteq A$ and $f^{-1}([1/3, 1]) \subseteq A$ to be disjoint and closed in A and therefore also closed in X. Due to (i) with target interval $[-1/3, 1/3]$ instead of $[0, 1]$ there exists a continuous function $g: X \to [-1/3, 1/3]$ with $g(x) = -1/3$ for x with $-1 \leq f(x) \leq -1/3$ and $g(x) = 1/3$ for x with $1/3 \leq f(x) \leq 1$. This implies $|f(x) - g(x)| \leq 2/3$ for $x \in A$. We are now going to construct a sequence $(g_n)_{n \in \mathbb{N}_0}$ of continuous maps $g_n: X \to [-1, 1]$ with the following properties:

1. $|g_n(x)| \leq (1/3)(2/3)^n$ for $x \in X$.
2. $|\sum_{m=0}^n g_m(x) - f(x)| \leq (2/3)^{n+1}$ for $x \in A$.

For $n = 0$ set $g_0 := g$. Let $n \in \mathbb{N}$ and assume that g_m has already been constructed for $m < n$. Then $(3/2)^n (\sum_{m=0}^{n-1} g_m - f)$ maps into $[-1, 1]$ by assumption. Hence, the above construction yields a continuous map $g': X \to [-1/3, 1/3]$ with $|(3/2)^n (\sum_{m=0}^{n-1} g_m(x) - f(x)) - g'(x)| \leq 2/3$ for $x \in A$. Set $g_n := (2/3)^n g'$ so that $|g_n(x)| = (2/3)^n |g'(x)| \leq (1/3)(2/3)^n$ for $x \in X$. Using the inductive hypothesis, we have $|\sum_{m=0}^n g_m(x) - f(x)| = |\sum_{m=0}^{n-1} g_m(x) - f(x) - (2/3)^n g'(x)| \leq (2/3)^{n+1}$ for $x \in A$.

Now $|g_n| \leq (1/3)(2/3)^n$ for $n \in \mathbb{N}_0$ and $(1/3) \sum_{m=0}^{\infty} (2/3)^m$ is an absolute convergent series with limit 1. Hence, $\sum_{m=0}^{\infty} g_n$ converges absolutely and uniformly to a continuous map $g: X \to [-1, 1]$. For $x \in A$ we have $|\sum_{m=0}^n g_m(x) - f(x)| \leq (2/3)^{n+1}$ and therefore $\sum_{m=0}^{\infty} g_m(x) = f(x)$ so that $g_{|A} = f$. $\qquad \square$

Definition 1.16 (Normal spaces). A topological space X is called *normal* if it satisfies the equivalent conditions of Theorem 1.15.

Let X be a normal space. Every closed subspace of X is normal (use (ii) of Theorem 1.15).

Very often some further separation axioms are included in the definition of "normal". We will state these properties always explicitly. For instance we will use the following property.

Definition 1.17. A topological space X is called T_1-*space* if $\{x\}$ is closed in X for all $x \in X$.

Every Hausdorff space is a T_1-space. Conversely, if X is a normal T_1-space, then X is Hausdorff.

Let $(U_i)_i$ be an open covering of a topological space X. As being a closed subset is a local property (Appendix Proposition 12.32), the space X is a T_1-space if and only if each U_i is a T_1-space.

Proposition 1.18. *Every paracompact T_1-space is normal and Hausdorff.*

There exist normal spaces that are not paracompact ([BouGT2] IX, §4, Exercise 27).

Proof. As every normal T_1-space is Hausdorff, it suffices to show that X is normal. Let X be a paracompact space, $A, B \subseteq X$ be closed with $A \cap B = \emptyset$.

Claim: Let $Y, Z \subseteq X$ be closed with $Y \cap Z = \emptyset$. If for all $y \in Y$ there exists $y \in V_y \subseteq X$ open and $Z \subseteq W_y \subseteq X$ open with $V_y \cap W_y = \emptyset$, then there exists an open neighborhood T of Y and an open neighborhood U of Z with $U \cap T = \emptyset$.

Let us first show that the claim implies that X is normal. As X is a T_1-space, we may apply the claim to $Y = A$ and $Z = \{x\}$ for some $x \in B$. Hence there exists an open neighborhood T of A and U of x with $T \cap U = \emptyset$. Then we can apply our claim again to $Z = A$ and $Y = B$ to see that there exist open neighborhoods $\tilde{U}$ of A and $\tilde{V}$ of B with $\tilde{U} \cap \tilde{V} = \emptyset$. Hence X is normal.

Now let us show the claim. As X is paracompact, we find a locally finite open refinement $(T_i)_i$ of the open cover of X formed of $X \setminus Y$ and of the V_y for $y \in Y$. Let $I_Y := \{i \in I \; ; \; Y \cap T_i \neq \emptyset\}$. Being a refinement implies that for $i \in I_Y$ one has

$$\exists \, y_i \in Y \colon T_i \subseteq V_{y_i}. \tag{*}$$

Let $T := \bigcup_{i \in I_Y} T_i$. Then $T \subseteq X$ open and $Y \subseteq T$.

It remains to find U. For all $z \in Z$ there exists $z \in S_z \subseteq X$ open such that S_z intersects only finitely many T_i because $(T_i)_i$ is locally finite. In particular $J_z := \{i \in I_Y \; ; \; T_i \cap S_z \neq \emptyset\}$ is finite. Set $U_z := S_z \cap \bigcap_{i \in J_z} W_{y_i}$. Then U_z is an open neighborhood of z. For $i \in J_z$ we have $T_i \subseteq V_{y_i}$ by (*) and hence $W_{y_i} \cap T_i = \emptyset$. For $i \in I_Y \setminus J_z$ one has $S_z \cap T_i = \emptyset$. Hence $U_z \cap T = \emptyset$. Set $U := \bigcup_{z \in Z} U_z$. $\qquad\square$

Proposition 1.19. *Every metrizable topological space X is normal.*

By Propositions 1.18 and 1.13 there are the following implications:

"metrizable" $\Rightarrow$ "hereditarily paracompact and Hausdorff" $\Rightarrow$ "normal and Hausdorff".

As the proofs were somewhat delicate, we give also a simple direct proof.

Proof. Let d be a metric on X inducing its topology. For $Y \subseteq X$ closed, the map $X \in \mathbb{R}$, $x \mapsto d(x, Y) := \inf_{y \in Y} d(x, y)$, is continuous with $d(x, Y) = 0$ if and only if $x \in Y$.

Let $A, B \subseteq X$ be closed subsets with $A \cap B = \emptyset$. Then

$$f : X \to [0, 1], \qquad f(x) := \frac{d(x, A)}{d(x, A) + d(x, B)}$$

is a continuous function with $f(a) = 0$ for all $a \in A$ and $f(b) = 1$ for all $b \in B$. $\qquad \square$

We conclude this section with two versions of an important technical tool, the shrinking lemma.

Proposition 1.20 (Shrinking lemma). *Let X be a normal space and let $(U_i)_{i \in I}$ be an open covering such that for all $x \in X$ there exist only finitely many i with $x \in U_i$. Then there exists an open covering $(V_i)_{i \in I}$ such that $\overline{V_i} \subseteq U_i$ for all i.*

The hypothesis on $(U_i)_i$ is of course satisfied if $(U_i)_i$ is locally finite.

Proof. We choose a well ordering on I (Appendix Proposition 13.27) and show by transfinite recursion (Appendix Proposition 13.26) that there exists a family $(V_i)_{i \in I}$ of open subsets such that for all i one has $\overline{V_i} \subseteq U_i$ and $X = \bigcup_{j < i} V_j \cup \bigcup_{j \geq i} U_j$.

Let $i \in I$ and assume that we have already constructed V_j for $j < i$. Set

$$A_i := U_i \setminus \left(\left(\bigcup_{j < i} V_j \right) \cup \left(\bigcup_{j > i} U_j \right) \right).$$

Then $A_i \cap U_j$ is closed in U_j for all $j \geq i$ and $A_i \cap V_j$ is closed in V_j for all $j < i$. Hence A_i is closed in X because being closed is a local property (Appendix Proposition 12.32). As X is normal, there exists an open neighborhood V_i of A_i with $\overline{V_i} \subseteq U_i$ (Theorem 1.15 (iii)). Clearly we have $X = \bigcup_{j \leq i} V_j \cup \bigcup_{j > i} U_j$.

It remains to show that $X = \bigcup_{i \in I} V_i$. By the assumption on $(U_i)_i$ we find for all $x \in X$ an $i_x \in I$ such that $x \notin U_j$ for $j \geq i_x$. Now $x \in X = \bigcup_{j < i_x} V_j \cup \bigcup_{j \geq i_x} U_j$ and hence $x \in \bigcup_{j < i_x} V_j$. $\qquad \square$

If X is a paracompact Hausdorff space, we can omit the hypothesis on $(U_i)_i$.

Corollary 1.21 (Shrinking lemma). *Let X be a paracompact Hausdorff space and let $(U_i)_{i \in I}$ be an open covering of X.*

1. *There exists an open covering $(T_j)_{j \in J}$ such that $(\overline{T_j})_{j \in J}$ is a locally finite refinement of $(U_i)_i$.*
2. *There exists an open covering $(V_i)_{i \in I}$ such that $\overline{V_i} \subseteq U_i$ for all $i \in I$.*

Proof. As X is paracompact there exists a locally finite open covering $(W_j)_{j \in J}$ and a map $\alpha \colon J \to I$ such that $W_j \subseteq U_{\alpha(j)}$ for all $j \in J$. As X is normal (Proposition 1.18) we can apply Proposition 1.20 to $(W_j)_j$. Hence we find an open covering $(T_j)_{j \in J}$ with $\overline{T_j} \subseteq W_j$ for all j. In particular $(\overline{T_j})_{j \in J}$ is locally finite again.

For all $i \in I$ set $V_i := \bigcup_{j \in \alpha^{-1}(i)} T_j$. Then $(V_i)_{i \in I}$ is an open covering of X. As $(\overline{T_j})_j$ is locally finite, $\bigcup_{j \in \alpha^{-1}(i)} \overline{T_j}$ is closed and hence

$$\overline{V_i} \subseteq \bigcup_{j \in \alpha^{-1}(i)} \overline{T_j} \subseteq \bigcup_{j \in \alpha^{-1}(i)} W_j \subseteq U_i. \qquad \square$$

1.4   Separated Maps

We now come to a relative version of Hausdorff for continuous maps, so-called "separated maps". To characterize separated maps (Proposition 1.25) we first introduce the following notion for subspaces.

Definition 1.22 (Relatively Hausdorff subspace). Let X be a topological space. A subspace A of X is called *relatively Hausdorff* if any two distinct points in A have disjoint neighborhoods in X.

If A is a relatively Hausdorff subspace of a topological space X, then A is Hausdorff. The converse does not necessarily hold (Problem 1.16). If X is Hausdorff, then any subspace of X is relatively Hausdorff.

Proposition 1.23. *Let A be a relatively Hausdorff subspace of X and let C and C' be compact disjoint subspaces of A. Then there exist disjoint open neighborhoods U of C and U' of C' in X.*

Proof. Suppose first that $C' = \{x'\}$ for some $x' \in X$. As C is compact and A relatively Hausdorff in X we find a finite open covering $(U_1, \dots, U_n)$ of C in X and for each $i = 1, \dots, n$ an open neighborhood U_i' of x' in X such that $U_i' \cap U_i = \emptyset$ for all i. We set $U' = \bigcap_i U_i'$ and $U = \bigcup_i U_i$.

Now let C' be arbitrarily compact. As we have just seen we find for each point $x' \in C'$ an open disjoint neighborhoods $U_{x'}'$ of x' and $U_{x'}$ of C in X. As C' is compact, there exists a finite subset $F \subseteq C'$ such that $(U_{x'}')_{x' \in F}$ is a covering of C'. Now we set $U := \bigcap_{x' \in F} U_{x'}$ and $U' = \bigcup_{x' \in F} U_{x'}'$. $\qquad \square$

For the definition of separated maps recall that for a continuous map $f \colon X \to Y$ of topological spaces the diagonal of f (Appendix Example 12.29 5) is the map

$$\Delta_f \colon X \to X \times_Y X := \{ (x, x') \in X \times X \; ; \; f(x) = f(x') \}, \qquad x \mapsto (x, x).$$

Definition 1.24 (Separated maps). A continuous map $f: X \to Y$ is called *separated* if the diagonal map $\Delta_f: X \to X \times_Y X$ is a closed topological embedding (or, equivalently by Appendix Example 12.29 5, if $\Delta_f(X)$ is closed in $X \times_Y X$).

Every injective continuous map is separated by Appendix Example 12.29 5. This also follows from (ii) of the following characterization of separated maps.

Proposition 1.25. *For a continuous map $f: X \to Y$ of topological spaces the following assertions are equivalent:*

(i) *The map f is separated.*
(ii) *All fibers of f are relatively Hausdorff subspaces in X.*
(iii) *For every relatively Hausdorff subspace B of Y the inverse image $f^{-1}(B)$ is relatively Hausdorff in X.*

Proposition 1.25 (ii) shows in particular that the property "separated" is local on the target.

Proof. "(i) $\Rightarrow$ (ii)". Let $y \in Y$ and $x_1, x_2 \in f^{-1}(y)$ with $x_1 \neq x_2$. Then (x_1, x_2) is in $(X \times_Y X) \setminus \Delta_f(X)$. Because $\Delta_f(X) \subseteq X \times_Y X$ is closed there exists an open neighborhood U_i of x_i in X ($i = 1, 2$) such that $U_1 \times U_2 \cap X \times_Y X$ is contained in $(X \times_Y X) \setminus \Delta_f(X)$. Hence $U_1 \cap U_2 = \emptyset$.

"(ii) $\Rightarrow$ (i)". Let $(x_1, x_2) \in (X \times_Y X) \setminus \Delta_f(X)$ and $y := f(x_1) = f(x_2)$. As $f^{-1}(y)$ is relatively Hausdorff, there exist neighborhoods U_1 of x_1 and U_2 of x_2 such that $U_1 \cap U_2 = \emptyset$. Therefore $W := (U_1 \times U_2) \cap (X \times_Y X)$ is a neighborhood of (x_1, x_2) in $X \times_Y X$ and we have $W \cap \Delta_f(X) = \emptyset$. Thus $(X \times_Y X) \setminus \Delta_f(X)$ is open in $X \times_Y X$.

"(ii) $\Rightarrow$ (iii)". Let $x_1, x_2 \in f^{-1}(B)$ with $x_1 \neq x_2$ and $y_1 := f(x_1)$, $y_2 := f(x_2)$. Suppose first that $y_1 \neq y_2$. Since $B \subseteq Y$ is relatively Hausdorff, there exist neighborhoods V_1 of y_1 and V_2 of y_2 in Y such that $V_1 \cap V_2 = \emptyset$. Then $f^{-1}(V_1)$ and $x_2 \in f^{-1}(V_2)$ are disjoint neighborhoods of x_1 and x_2 in X. Now suppose that $y_1 = y_2$. Then x_1 and x_2 are points of the same fiber of f. In this case the assertion follows from (ii).

"(iii) $\Rightarrow$ (ii)". Clearly every one-point subspace is relatively Hausdorff. $\square$

Corollary 1.26. *Let $f: X \to Y$, $g: Y \to Z$ be continuous maps.*

1. *If f and g are separated, $g \circ f$ is separated.*
2. *If $g \circ f$ is separated, f is separated.*

Proof. Assertion 1 follows immediately from Proposition 1.25 (iii).

To show 2 let y be a point of Y. Then $f^{-1}(y)$ is a subspace of $(g \circ f)^{-1}(g(y))$, which is relatively Hausdorff in X. Hence $f^{-1}(y)$ is relatively Hausdorff in X. $\square$

Choosing for Z a point we obtain the following result.

Corollary 1.27. *Let* $f: X \to Y$ *be a continuous map.*

1. *If* Y *is Hausdorff and* f *is separated, then* X *is Hausdorff.*
2. *If* X *is Hausdorff, then* f *is separated.*

1.5   Proper Maps

There is also a relative version for continuous maps of the compactness notion. These are the proper maps. We first prove a characterization for compact spaces. We start with a lemma.

Lemma 1.28. *Let* X *and* Y *be topological spaces,* $A \subseteq X$ *and* $B \subseteq Y$ *compact subsets and* $W \subseteq X \times Y$ *open with* $A \times B \subseteq W$. *Then there exist open neighborhoods* U *of* A *in* X *and* V *of* B *in* Y *such that* $U \times V \subseteq W$.

Proof. For every $(a, b) \in A \times B$ there exist open subsets $U_{(a,b)} \subseteq X$ and $V_{(a,b)} \subseteq Y$ such that $(a, b) \in U_{(a,b)} \times V_{(a,b)} \subseteq W$. Fixing $b \in B$ yields an open covering $(U_{(a,b)})_{a \in A}$ of the compact set A so that there exist $a_1, \ldots, a_n$ with $A \subseteq U_{(a_1,b)} \cup \ldots \ldots \cup U_{(a_n,b)}$. We define open sets $U_b := U_{(a_1,b)} \cup \ldots \cup U_{(a_n,b)}$ and $V_b := V_{(a_1,b)} \cap \ldots \cap V_{(a_n,b)}$ and observe $A \subseteq U_b$, $b \in V_b$. Furthermore $U_b \times V_b$ is contained in W. Now $(V_b)_{b \in B}$ is an open covering of the compact set B so that there exist $b_1, \ldots, b_n$ with $B \subseteq V_{b_1} \cup \ldots \cup V_{b_n}$. For the open subsets $U := U_{b_1} \cap \ldots \cap U_{b_n}$ and $V := V_{b_1} \cup \ldots \cup V_{b_n}$ we find $A \subseteq U$, $B \subseteq V$ and $U \times V \subseteq W$ as before. $\square$

Proposition 1.29. *A topological space* X *is compact if and only if the projection* $Z \times X \to Z$ *is closed for every topological space* Z.

Proof. For every topological space Z denote by $p_1: Z \times X \to Z$ the projection. Suppose that X is not compact. Then there exists an open covering $(U_i)_{i \in I}$ of X that has no finite subcovering. In particular I is infinite. Set

$$Z := \{ J \subseteq I \ ; \ J \text{ finite and non-empty} \} \cup \{ I \}.$$

Endow Z with the topology generated by the sets $V_K := \{ J \in Z \ ; \ K \subseteq J \}$ for every finite subset $K \subseteq I$. For finite subsets $K, K' \subseteq I$ we have $V_K \cap V_{K'} = V_{K \cup K'}$ and $V_\emptyset = Z$ so that these sets form a basis of the topology of Z (Appendix Remark 12.5).

We will show that $p_1: Z \times X \to Z$ is not closed. Consider the subset

$$A := \left\{ (J, x) \in Z \times X \ ; \ x \in X \setminus \bigcup_{i \in J} U_i \right\}$$

of the product space $Z \times X$. If $(J, x) \notin A$ there exists $j \in J$ with $x \in U_j$ so that $(J, x) \in V_{\{j\}} \times U_j$. Also for every $(J', x') \in V_{\{j\}} \times U_j$ we have $x' \in U_j \subseteq \bigcup_{i \in J'} U_i$, hence the open subset $V_{\{j\}} \times U_j$ of $Z \times X$ is contained in $(Z \times X) \setminus A$. Therefore A is closed.

Now $p_1(A) = \{ J \in Z ; X \setminus \bigcup_{i \in J} U_i \neq \emptyset \} = Z \setminus \{I\}$ because $\bigcup_{i \in J} U_i \neq X$ for every finite subset $J \subseteq I$ and $\bigcup_{i \in I} U_i = X$ by assumption. If $Z \setminus \{I\}$ was closed, $\{I\}$ would be open in Z and therefore an element of the basis of topology given above. But this is absurd because I is infinite. Hence $p_1(A)$ is not closed proving that $p_1 : Z \times X \to Z$ is not closed.

Conversely, assume that X is compact. Let Z be a topological space, let $A \subseteq Z \times X$ be a closed subset, and let $z \in Z \setminus p_1(A)$. Then $p_1^{-1}(z) = \{z\} \times X \subseteq (Z \times X) \setminus A$ and X and $\{z\}$ are compact. Therefore we may apply Lemma 1.28 to find an open neighborhood U of z in Z with $U \times X \subseteq (Z \times X) \setminus A$. Then $U \subseteq Z \setminus p_1(A)$. Because z was arbitrary we see that $Z \setminus p_1(A)$ is open. $\square$

We can now formulate the main result of this section and define the notion of a proper map.

Definition and Theorem 1.30 (Proper maps). *A continuous map $f : X \to Y$ between topological spaces is called* proper *if it satisfies the following equivalent properties:*

(i) *For every topological space Z the map $\mathrm{id}_Z \times f : Z \times X \to Z \times Y$ is closed.*
(ii) *The map f is closed and for every compact subspace V of Y the preimage $f^{-1}(V)$ is a compact subspace of X.*
(iii) *The map f is closed and $f^{-1}(y)$ is compact for all $y \in Y$.*
(iv) *For every continuous map $g : Y' \to Y$, the projection $Y' \times_Y X \to Y'$ is closed.*

Sometimes a continuous map is defined to be proper if inverse images of compact subspaces are again compact. In general this is strictly weaker than our definition (Problem 1.21). For locally compact Hausdorff spaces both definitions are equivalent (Problem 1.20).

Proof. "(i) $\Rightarrow$ (ii)". Taking for Z the topological space consisting of a single point shows that (i) implies that f is closed.

Let $V \subseteq Y$ be compact and let Z be a topological space. As $Z \times X \to Z \times Y$ is closed, the restriction $Z \times f^{-1}(V) \to Z \times V$ is also closed. Due to Proposition 1.29 the projection $Z \times V \to Z$ is closed because V is compact and therefore the composition $Z \times f^{-1}(V) \to Z \times V \to Z$ is closed. This composition is the projection $Z \times f^{-1}(V) \to Z$. As Z was arbitrary, Proposition 1.29 shows that $f^{-1}(V)$ is compact.

"(ii) $\Rightarrow$ (iii)". A topological space with a single point is compact.

"(iii) $\Rightarrow$ (iv)". Let $g : Y' \to Y$ be a continuous map and $f' : Y' \times_Y X \to Y'$ the projection. First we note that $\mathrm{Im}(f') = \{ y' \in Y' ; g(y') \in \mathrm{Im}(f) \} = g^{-1}(f(X))$ is closed in Y' because f is closed by assumption. Let $A \subseteq Y' \times_Y X$ be a closed subset and $y' \in Y' \setminus f'(A)$. We have to find a neighborhood of y' in Y' that does not meet $f'(A)$. Now we have two different cases.

If $y' \notin \mathrm{Im}(f')$, then $Y' \setminus \mathrm{Im}(f')$ is an open neighborhood of y' contained in $Y' \setminus f'(A)$.

If $y' \in \mathrm{Im}(f')$, we have $(f')^{-1}(y') = \{y'\} \times f^{-1}(g(y')) \subseteq (Y' \times_Y X) \setminus A$. Since $Y' \times_Y X \subseteq Y' \times X$ is a subspace, there exists $A' \subseteq Y' \times X$ closed with $A = A' \cap (Y' \times_Y X)$. By hypothesis, $\{y'\} \times f^{-1}(g(y'))$ is compact and we may apply Lemma 1.28 to $\{y'\} \times f^{-1}(g(y')) \subseteq (Y' \times X) \setminus A'$. This yields $V' \subseteq Y'$ and $U \subseteq X$ open with $\{y'\} \times f^{-1}(g(y')) \subseteq V' \times U$ and $V' \times U \subseteq (Y' \times X) \setminus A'$. Then the set $V' \cap g^{-1}(Y \setminus f(X \setminus U))$ is open in Y' since f is closed, it contains y', and has an empty intersection with $f'(A)$. Thus $f'(A)$ is closed.

"(iv) $\Rightarrow$ (i)". Let Z be a topological space and $g: Z \times Y \to Y$ the projection. Then the projection $(Z \times Y) \times_Y X \to Z \times Y$ is closed by (iv) and it is equal to $\mathrm{id}_Z \times f: Z \times X \to Z \times Y$. $\qquad\square$

Example 1.31. Characterization (iii) shows that an injective continuous map is proper if and only if it is a closed embedding.

Proposition 1.32. *Let $f: X \to Y$ and $g: Y \to Z$ be continuous maps.*

1. *If f and g are proper, then $g \circ f$ is proper.*
2. *Suppose that g is separated and that $g \circ f$ is proper. Then f is proper.*
3. *Let $g \circ f$ be proper and f surjective. Then g is proper.*

Proof. Assertion 1 is clear by Theorem 1.30 (ii).

Let us show 2. By Appendix Example 12.29 4, $\Gamma_f: X \mapsto X \times_Z Y, x \mapsto (x, f(x))$ is an embedding. Its image is the inverse image of the diagonal $Y \times_Z Y$ under $f \times \mathrm{id}_Y$, and the diagonal is closed in $Y \times_Z Y$ because g is separated. Hence Γ_f is a closed embedding and hence proper (Example 1.31). Now f is the composition of Γ_f followed by the projection $X \times_Z Y \to Y$, which is proper by Theorem 1.30 (iv) because $g \circ f$ is proper. Hence f is proper by 2.

To show 3 let $z \in Z$. As f is surjective, it induces a surjective map $f^{-1}(g^{-1}(z)) \to g^{-1}(z)$. As $g \circ f$ is proper, $f^{-1}(g^{-1}(z))$ is compact. Hence $g^{-1}(z)$ is compact (Appendix Proposition 12.51). Now let $B \subseteq Y$ be closed. As f is surjective, we have $g(B) = g(f(f^{-1}(B)))$, which is closed because $g \circ f$ is closed. Therefore g is closed. $\qquad\square$

We obtain a new proof of Appendix Proposition 12.52.

Corollary 1.33. *Let $f: X \to Y$ be a continuous map. Suppose that X is compact and that Y is Hausdorff. Then f is proper and the subspace $f(X)$ of Y is closed in Y and compact.*

Proof. Applying Proposition 1.32 2 to the special case that Z consists of a single point shows that f is proper. Hence $f(X)$ is closed in Y. Being the image of a compact space, $f(X)$ is compact (Appendix Proposition 12.51 2). $\square$

Proposition 1.34. *Let* $(f_i \colon X_i \to Y_i)_{i \in I}$ *be a finite family of proper continuous maps. Then the product map* $(x_i)_i \mapsto (f_i(x_i))_i$ *is proper.*

One can show that the product of an arbitrary family of proper maps is again proper ([BouGT1], I, §10.2, Corollary 3 of Theorem 1).

Proof. By induction we may assume that $I = \{1, 2\}$. Let Z be a topological space. Then $f_1 \times f_2 \times \mathrm{id}_Z$ is the composition of the proper maps $f_1 \times \mathrm{id}_{X_1} \times \mathrm{id}_Z$ and $\mathrm{id}_{Y_2} \times f_1 \times \mathrm{id}_Z$. Hence it is proper by Proposition 1.32. $\square$

1.6 Problems

Problem 1.1. Let X be a Lindelöf space and let $f \colon X \to Y$ be a surjective continuous map. Show that Y is again a Lindelöf space.

Problem 1.2. Show that the product of a Lindelöf space and a compact space is again a Lindelöf space.

Problem 1.3. Let X be a topological space.

1. Suppose that X is a Lindelöf space. Show that every closed subspace of X is again a Lindelöf space.
2. Show that every subspace of X is Lindelöf if and only if every open subspace of X is Lindelöf. Such spaces are called *hereditarily Lindelöf spaces*.

Problem 1.4. Let X be a metrizable space. Show that the following assertions are equivalent:

 (i) X is second countable.
 (ii) X is a Lindelöf space.
 (iii) X is separable.

Problem 1.5. A topological space X is called *regular* if X is a T_1-space and if for all $x \in X$ and closed subsets $A \subseteq X$ with $x \notin A$ there exist disjoint open neighborhoods of x and of A.

1. Show that every Hausdorff normal space is regular.
2. Show that a T_1-space X is regular if and only if the closed neighborhoods of any $x \in X$ form a fundamental system of neighborhoods of x.
3. Show that every subspace of a regular space is again regular.
4. Show that in a regular Lindelöf space X every open covering has a countable locally finite open refinement. In particular X is paracompact.

Problem 1.6. Show that every locally compact Hausdorff space is regular (Problem 1.5).

Problem 1.7. Show that every compact Hausdorff space is normal.

Problem 1.8. Let X be a locally compact Hausdorff space. Show that X is paracompact if and only if X is the sum of a family of σ-compact spaces.

Problem 1.9. Let X be a topological space. A subset A of X is called *meager* if it is a union of countably many nowhere dense subsets (Appendix Problem 12.35). Show that the following assertions are equivalent:

 (i) Every intersection of countably many open dense subsets of X is dense in X.
(ii) Every countable union of closed subsets without interior points has no interior point in X.
(iii) The complement of every meager subset is dense in X.

If X satisfies these conditions, then X is called a *Baire space*.

Problem 1.10. Show that every locally compact Hausdorff space is a Baire space (Problem 1.9).
Hint: Let $(U_n)_{n \in \mathbb{N}}$ be a family of open dense subsets, $V \subseteq X$ open. Construct by induction open subsets V_n with $V_1 = V$, $\overline{V_{n+1}} \subseteq V_n \cap U_n$, and $\overline{V_n}$ compact for $n \geq 2$ (use Problem 1.6). Show that $V \cap \bigcap_n U_n \supseteq \bigcap_n \overline{V_n} \neq \emptyset$.

Problem 1.11. Let $X \neq \emptyset$ be a locally compact Hausdorff space and let $(A_n)_n$ be a countable family of closed sets such that $X = \bigcup_n A_n$. Show that there exists an $n \in \mathbb{N}$ such that $A_n^\circ \neq \emptyset$.
Hint: Problem 1.10.

Problem 1.12. Let X be a paracompact space and let Y be a compact space. Show that $X \times Y$ is paracompact.
  Remark: The product of two paracompact spaces is in general not paracompact (Problem 1.13).

Problem 1.13. Let X be the topological space whose underlying set is the set $\mathbb{R}$ of real numbers and whose topology is generated by $\mathcal{B} := \{\, (a,b] \,;\, a,b \in \mathbb{R}, a < b \,\}$.

1. Show that $\mathcal{B}$ is a basis of the topology, that the closed intervals are closed in this topology, and that the topology is finer then the left order topology (Appendix Problem 13.3) on $\mathbb{R}$.
2. Show that every open subset of X is a countable union of intervals of the form $(a,b]$ and (a,b) that are pairwise disjoint. Deduce that X is normal and Hausdorff.
3. Show that X is hereditarily Lindelöf (Problem 1.3). Deduce that X is hereditarily paracompact (use Problem 1.5).
4. Show that $X \times X$ is not normal.
5. Deduce that $X \times X$ is not paracompact and that X is not metrizable.
6. Show that every compact subspace of X is countable. Deduce that X is not σ-compact.

Problem 1.14. Show that every Hausdorff locally compact topological group is paracompact.

Problem 1.15. Let $(X, \leq)$ be a totally ordered set and endow X with the order topology (Appendix Problem 12.24). Show that X is normal.

Problem 1.16. Let X be a set endowed with the particular point topology with respect to some $p \in X$ (Appendix Problem 12.7). Show that any subspace A of $X \setminus \{p\}$ is discrete (in particular Hausdorff) but that A is not relatively Hausdorff in X if A consists of more than one point.

Problem 1.17. Let $f : X \to Y$, $g : Y \to Z$ be continuous maps. Suppose that $g \circ f$ is proper.

1. Let f be surjective. Show that g is proper.
2. Let g be injective. Show that f is proper.

Problem 1.18. Let $f : X \to Y$ be a continuous map. Let $(B_i)_i$ be a family of subsets of Y and suppose: (a) that the interiors cover Y *or* (b) that $(B_i)_i$ is a locally finite closed covering. Show that f is proper if and only if its restriction $f^{-1}(A_i) \to A_i$ is proper for all i.

Hint: Use Appendix Problem 12.13.

Problem 1.19. Let $f : X \to Y$ be a surjective map of topological spaces such that X carries the inverse image topology of Y. Show that f is proper and open.

Problem 1.20. Let X and Y be Hausdorff spaces and suppose that Y is locally compact. Show that a continuous map $f : X \to Y$ is proper if and only if for every compact subspace C of Y the inverse image $f^{-1}(C)$ is compact.

Hint: Problem 1.18.

Problem 1.21. Let $(X_i)_{i \in I}$ be a non-countable family of discrete topological spaces each having at least two points. Endow that product set $\prod_{i \in I} X_i$ with the topology generated by the subsets $\prod_{i \in I} M_i$ with $M_i \subseteq X_i$ for all i and $M_i = X_i$ for all but countably many $i \in I$.

1. Show that in X every countable intersection of open subsets is again open. Deduce that no point of X has a countable fundamental system of neighborhoods.
2. Show that X is regular and Hausdorff.
3. Show that any compact subspace of X is finite.
4. Let X' be the set X endowed with the discrete topology and let $f : X' \to X$ be the identity. Show that f is continuous and that $f^{-1}(C)$ is compact for every compact subspace C of X. Show that f is not proper.

Algebraic Topological Preliminaries

In this chapter we briefly introduce some elementary notions and results on homotopy, fundamental groups, and covering spaces that are used throughout the book.

2.1 Homotopy

We start with the notion of homotopy that is a central notion in topology.

Definition 2.1 (Homotopy). Two continuous maps $f, g: X \to Y$ of topological spaces are called *homotopic* if there exists a continuous map $H: X \times [0, 1] \to Y$ such that $H(x, 0) = f(x)$ and $H(x, 1) = g(x)$ for all $x \in X$. Then H is called a *homotopy between f and g*. We then write $f \simeq g$ or $H: f \simeq g$.

One should think of $H: X \times [0, 1] \to Y$ as a continuously varying family of maps $H_t: X \to Y$, $x \mapsto H(x, t)$ parametrized by $t \in [0, 1]$ interpolating between $f = H_0$ and $g = H_1$.

Remark and Definition 2.2. Let X and Y be topological spaces. Then the homotopy relation $\simeq$ is an equivalence relation on the set of all continuous maps $X \to Y$:

1. Clearly, one has $f \simeq f$ via the homotopy $H: X \times [0, 1] \to Y$, $(x, t) \mapsto f(x)$.
2. Given a homotopy $H: f \simeq g$, then $H^-: g \simeq f$ via the *inverse homotopy* $H^-: (x, t) \mapsto H(x, 1 - t)$.
3. Let $H: f \simeq g$ and $K: g \simeq h$ be given. Then $H * K: f \simeq h$ via the *product homotopy*

$$(H * K)(x, t) := \begin{cases} H(x, 2t), & 0 \le t \le 1/2; \\ K(x, 2t - 1), & 1/2 \le t \le 1. \end{cases}$$

© Springer Fachmedien Wiesbaden 2016

T. Wedhorn, *Manifolds, Sheaves, and Cohomology*, Springer Studium Mathematik – Master, DOI 10.1007/978-3-658-10633-1_2

The equivalence class of $f: X \to Y$ is denoted by $[f]$ and called the *homotopy class of* f. We denote by $[X, Y]$ the set of homotopy classes of continuous maps $X \to Y$.

A continuous map $X \to Y$ is called *null homotopic* if it is homotopic to a constant map.

Remark and Definition 2.3 (Homotopy category). We define the *homotopy category* (h-Top) as follows:

(a) Objects are topological spaces.
(b) For two objects X and Y we define $\operatorname{Hom}_{(\text{h-Top})}(X, Y) := [X, Y]$, the set of homotopy classes of continuous maps $X \to Y$.
(c) The identity of a topological space in (h-Top) is the homotopy class of id_X.
(d) For $[f] \in [X, Y]$ and $[g] \in [Y, Z]$ we define the composition by $[g] \circ [f] := [g \circ f]$. This is well defined. Indeed, let $f, f': X \to Y$ and $g, g': Y \to Z$ be continuous maps, and let $H: f \simeq f'$ and $K: g \simeq g'$ be homotopies. Then $g \circ f \simeq g' \circ f'$ via the homotopy $(x, t) \mapsto K(H(x, t), t)$.

A continuous map $f: X \to Y$ whose homotopy class is an isomorphism in (h-Top) is called a *homotopy equivalence*. This means that there exists a continuous map $g: Y \to X$ such that $g \circ f \simeq \operatorname{id}_X$ and $f \circ g \simeq \operatorname{id}_Y$. Topological spaces X and Y are called *homotopy equivalent* if they are isomorphic in (h-Top).

Example 2.4. Let $n \geq 1$ be an integer and let $\| \ \|$ be the Euclidean norm on $\mathbb{R}^n$. Then the $(n - 1)$-dimensional sphere

$$S^{n-1} := \{ x \in \mathbb{R}^n \ ; \ \|x\| = 1 \}$$

is homotopy equivalent to $\mathbb{R}^n \setminus \{0\}$: The inclusion $i: S^{n-1} \to \mathbb{R}^n \setminus \{0\}$ and the map $p: \mathbb{R}^n \setminus \{0\} \to S^{n-1}$, $x \mapsto \frac{x}{\|x\|}$ are mutually inverse homotopy equivalences. Indeed $p \circ i = \operatorname{id}_{S^{n-1}}$ and $i \circ p \simeq \operatorname{id}_{\mathbb{R}^n \setminus \{0\}}$ via the homotopy

$$(\mathbb{R}^n \setminus \{0\}) \times [0, 1] \to \mathbb{R}^n \setminus \{0\}, \qquad (x, t) \mapsto (1 - t)x/\|x\| + tx.$$

This is in particular an example of homotopy equivalent spaces that are not homeomorphic (S^{n-1} is compact, $\mathbb{R}^n \setminus \{0\}$ not).

Definition 2.5 ((Locally) contractible spaces). A topological space X is called *contractible*, if it is homotopy equivalent to a point (i.e., to the unique topological space consisting of a single point). It is called *locally contractible* if every point has a fundamental system of open contractible neighborhoods.

In other words, X is contractible if and only if there exists a point $x_0 \in X$ such that the constant map $X \to X$, $x \mapsto x_0$ and id_X are homotopy equivalent, i.e., if and only if id_X is null homotopic.

Example 2.6. Let V be a normed real or complex vector space.

1. Recall that a subset S of V is called *star-shaped* with *star center* $s_0 \in S$ if for all
   $s \in S$ the line segment $\{ ts + (1-t)s_0 ; 0 \leq t \leq 1 \}$ from s to s_0 is contained in S.
   Then the subspace S of V is contractible: $H : (s, t) \mapsto ts + (1-t)s_0$ is a homotopy
   between the constant map $s \mapsto s_0$ and id_S.
2. Every open ball in V is star-shaped. As the open balls form a basis of the topology on
   V, we see that every open subspace of V is locally contractible.

2.2 Paths

We quickly introduce some notation for paths. In this section, let $a, b \in \mathbb{R}$ with $a < b$.

Definition 2.7 (Paths). Let X be a topological space.

1. A continuous map $\gamma : [a, b] \to X$ is called *path in X*. The point $\gamma(a)$ is called the *start
   point*, $\gamma(b)$ is called the *end point* of γ. We say that γ is a *path from $\gamma(a)$ to $\gamma(b)$*. We
   also set:
   $$\{\gamma\} := \gamma([a, b]) \subseteq X.$$
2. A path $\gamma : [a, b] \to X$ is called *closed* or a *loop* if $\gamma(a) = \gamma(b)$.

Let $\gamma : [a, b] \to X$ be a map. Let $\varphi : [0, 1] \to [a, b]$, $\varphi(t) = a + (b-a)t$. Then γ is
continuous if and only if $\gamma \circ \varphi$ is continuous and $\{\gamma\} = \{\gamma \circ \varphi\}$. This allows us usually to
assume that $[a, b] = [0, 1]$.

Definition 2.8. Let X be a topological space.

1. Let $\gamma : [0, 1] \to X$ be a path. Define

   $$\gamma^- : [0, 1] \to X, \gamma^-(t) = \gamma(1-t)$$

   the *inverse path*.
2. Let $\gamma, \delta : [0, 1] \to X$ be two paths with $\gamma(1) = \delta(0)$. Define the *product* or the *con-
   catenation of paths* by

   $$\gamma \cdot \delta : [0, 1] \to X, \qquad t \mapsto \begin{cases} \gamma(2t), & 0 \leq t \leq 1/2; \\ \delta(2t-1), & 1/2 \leq t \leq 1, \end{cases}$$

   i.e., $\gamma \cdot \delta$ is the path where one first "walks along γ and then along δ each time with
   double velocity".

2.3 Path Connected Spaces

Let X be a topological space. We write $x \sim y$ if there exists a path with start point x and end point y. Then $\sim$ is an equivalence relation on X as shown by the construction in Definition 2.8.

Definition 2.9. An equivalence class with respect to $\sim$ is called a *path component of* X. The set of path components of X is denoted by $\pi_0(X)$. The space X is called *path connected* if there exists a single path component, i.e., $\pi_0(X)$ consists of a single point.

A topological space X is called *locally path connected* if every point of X has fundamental system of path connected neighborhoods.

The empty space is not path connected by definition.

Remark 2.10. Let $f: X \to Y$ be a surjective continuous map. If X is path connected, then Y is path connected.

Indeed, let $y, y' \in Y$. Choose $x, x' \in X$ with $f(x) = y$ and $f(x') = y'$ and a path γ in X connecting x and x'. Then $f \circ \gamma$ is a path connecting y and y'.

Example 2.11.
1. A discrete space X is locally path connected, but it is not path connected if it contains more than one point.
2. Conversely, there are path connected topological spaces that are not locally path connected (Problem 2.7).
3. Every open subspace of of a normed $\mathbb{R}$-vector space is locally path connected.

Proposition 2.12. *Every path connected topological space X is connected.*

Proof. Assume that X is not connected. Then X is the disjoint union of two non-empty open subsets U and V. Choose $x \in U$ and $y \in V$ and let $\gamma: [0, 1] \to X$ be a path with $\gamma(0) = x$ and $\gamma(1) = y$. Then $[0, 1]$ is the disjoint union of the open non-empty subsets $\gamma^{-1}(U)$ and $\gamma^{-1}(V)$. This is a contradiction, because $[0, 1]$ is connected (Appendix Proposition 12.43). $\qquad\square$

In general there exist connected spaces that are not path connected (Problems 2.6 and 2.7). But for locally path connected spaces this cannot happen. More precisely we have:

Proposition 2.13. *Let X be a topological space such that every point has a path connected neighborhood. Then a subset of X is a path component if and only if it is a connected component. All path components are open and closed in X, i.e., X is the sum of its path components.*

Proof. By Proposition 2.12 every path component A of X is connected (and hence contained in a connected component). Let $x \in X$ and let A be the path component of X containing x. By hypothesis there exists a path connected neighborhood V of x. Then $A \cup V$ is path connected and hence contained in A. This shows that A is open in X. As the complement of A is a union of path components, A is also closed in X.

Proposition 2.12 shows that every connected component B of X is a union of path components. Each of theses path components is open and closed in X and hence in B. As B is connected, B can only contain a single path component. $\square$

Remark 2.14. Suppose that X is a locally path connected space. Then Proposition 2.13 holds also for every open subspace of X. This shows in particular that every point of X has a fundamental system of *open* path connected neighborhoods.

Remark 2.15. Let $f: X \to Y$ be a continuous map. If $x, x' \in X$ are connected by some path γ, then $f(x)$ and $f(x')$ are connected by the path $f \circ \gamma$. Hence f induces a map $\pi_0(f): \pi_0(X) \to \pi_0(Y)$. We obtain a functor $\pi_0: (\text{Top}) \longrightarrow (\text{Sets})$. Moreover, if $f, g: X \to Y$ are homotopic continuous maps, then $\pi_0(f) = \pi_0(g)$ and we even obtain a functor

$$\pi_0: (\text{h-Top}) \longrightarrow (\text{Sets}).$$

Hence every homotopy equivalence $f: X \to Y$ induces a bijection $\pi_0(X) \overset{\sim}{\to} \pi_0(Y)$. In particular, we see that every (locally) contractible space is (locally) path connected.

2.4   Fundamental Group

Definition 2.16. Let X and Y be topological spaces, let $A \subseteq X$ be a subspace, and let $f, g: X \to Y$ be continuous maps. A homotopy $H: f \simeq g$ is said to be *relative to A* if $H(a, t) = f(a) = g(a)$ for all $a \in A, t \in [0, 1]$. We write $H: f \simeq g$ (rel A) in this case.

The existence of such a homotopy implies in particular that $f_{|A} = g_{|A}$. The same constructions as in Remark 2.2 show that "homotopy relative to A" is an equivalence relation on the set of all continuous maps $X \to Y$.

Definition 2.17. Let X be a topological space and let $\gamma, \delta: [0, 1] \to X$ be two paths with the same start and end points. A *homotopy of paths* between γ and δ is a homotopy $H: \gamma \simeq \delta$ (rel $\{0, 1\}$), i.e., H is a homotopy that leaves the start point and the end point fixed.

Remark 2.18. Let $\gamma, \gamma_1, \gamma_2, \gamma_3, \delta_1, \delta_2: [0, 1] \to X$ be paths in X.

1. Let $\varphi: [0, 1] \to [0, 1]$ be continuous with $\varphi(0) = 0$ and $\varphi(1) = 1$. Then $\gamma \simeq \gamma \circ \varphi$ (rel $\{0, 1\}$).

2. Set $x_0 := \gamma(0)$, $x_1 := \gamma(1)$, and let ε_i be the constant path $t \mapsto x_i$. Then

$$\varepsilon_0 \cdot \gamma \simeq \gamma \simeq \gamma \cdot \varepsilon_1 \ (\text{rel } \{0, 1\}).$$

3. Assume $\gamma_i(1) = \delta_i(0)$ for $i = 1, 2$. Then $\gamma_1 \simeq \gamma_2 \ (\text{rel } \{0, 1\})$ and $\delta_1 \simeq \delta_2 \ (\text{rel } \{0, 1\})$ imply that $\gamma_1 \cdot \delta_1 \simeq \gamma_2 \cdot \delta_2 \ (\text{rel } \{0, 1\})$.
4. If one has $\gamma_1 \simeq \gamma_2 \ (\text{rel } \{0, 1\})$, then $\gamma_1 \cdot \gamma_2^-$ is null-homotopic.
5. Assume $\gamma_1(1) = \gamma_2(0)$ and $\gamma_2(1) = \gamma_3(0)$. Then

$$\gamma_1 \cdot (\gamma_2 \cdot \gamma_3) \simeq (\gamma_1 \cdot \gamma_2) \cdot \gamma_3 \ (\text{rel } \{0, 1\}).$$

Proof. 1. A homotopy $\gamma \circ \varphi \simeq \gamma$ is given by $(s, t) \mapsto \gamma(ts + (1-t)\varphi(s))$.

2.,5. One has $\varepsilon_0 \cdot \gamma = \gamma \circ \varphi$ with $\varphi(t) = 0$ for $0 \le t \le \frac{1}{2}$ and $\varphi(t) = 2t - 1$ for $\frac{1}{2} \le t \le 1$. Hence $\varepsilon_0 \cdot \gamma \simeq \gamma \ (\text{rel } \{0, 1\})$ by 1. Similarly one shows $\gamma \simeq \gamma \cdot \varepsilon_1 \ (\text{rel } \{0, 1\})$ and 5.

3. If $H: \gamma_1 \simeq \gamma_2 \ (\text{rel } \{0, 1\})$ and $G: \delta_1 \simeq \delta_2 \ (\text{rel } \{0, 1\})$, then $\gamma_1 \cdot \delta_1 \simeq \gamma_2 \cdot \delta_2$ $(\text{rel } \{0, 1\})$. via the homotopy $(s, t) \mapsto H(2s, t)$ for $0 \le s \le \frac{1}{2}$ and $(s, t) \mapsto G(2s - 1, t)$ for $\frac{1}{2} \le s \le 1$.

4. One has $\gamma_1 \simeq \gamma_2 \ (\text{rel } \{0, 1\})$ if and only if $\gamma_1^- \simeq \gamma_2^- \ (\text{rel } \{0, 1\})$. Hence we may assume that $\gamma_1 = \gamma_2$ by 3. Then $(s, t) \mapsto \gamma_1(2s(1-t))$ for $0 \le s \le \frac{1}{2}$ and $(s, t) \mapsto \gamma_1(2(1-s)(1-t))$ for $\frac{1}{2} \le s \le 1$ is a homotopy between $\gamma \cdot \gamma^-$ and the constant path with value $\gamma(0)$. $\square$

Definition 2.19. Let X be a topological space, $x \in X$. Define the *fundamental group of* X *with base point x* by

$$\pi_1(X, x) := \{ \gamma: [0, 1] \to X \ ; \ \gamma \text{ path with } \gamma(0) = \gamma(1) = x \} / (\simeq \ (\text{rel } \{0, 1\}),$$

i.e., $\pi_1(X, x)$ is the set of homotopy classes $[\gamma]$ of closed paths γ starting (and ending) in x. Define a multiplication on $\pi_1(X, x)$ by

$$[\gamma][\delta] := [\gamma \cdot \delta].$$

By Remark 2.18 this is well defined and yields a group structure on $\pi_1(X, x)$. The neutral element is the constant path with value x and the inverse of $[\gamma] \in \pi_1(X, x)$ is $[\gamma^-]$.

In the sequel we will often write γ instead of $[\gamma]$ for elements in $\pi_1(X, x)$.

Remark 2.20. Let (Toppt) be the category whose objects are pointed topological spaces, i.e., pairs (X, x) consisting of a topological space X and a point $x \in X$. Morphisms $(X, x) \to (Y, y)$ are continuous maps $f: X \to Y$ with $f(x) = y$. Composition in (Toppt) is given by composition of maps.

If $f: (X, x) \to (Y, y)$ is a morphism in (Toppt), then $\gamma \mapsto f \circ \gamma$ defines a group homomorphism $\pi_1(f): \pi_1(X, x) \to \pi_1(Y, y)$. It is easy to check that we obtain a functor

$$\pi_1: (\text{Toppt}) \to (\text{Grp}).$$

Remark 2.21. Let X be a topological space, $x_0, x_1 \in X$ be points and let σ be a path in X with start point x_0 and end point x_1. Then

$$\sigma_*\colon \pi_1(X, x_0) \to \pi_1(X, x_1), \qquad \gamma \mapsto \sigma^- \cdot \gamma \cdot \sigma$$

is an isomorphism of groups: Remark 2.18 shows that σ_* is well defined and that for $\gamma, \gamma' \in \pi_1(X, x_0)$ one has

$$\sigma_*(\gamma)\sigma_*(\gamma') = \sigma^- \cdot \gamma \cdot \sigma \cdot \sigma^- \cdot \gamma' \cdot \sigma \simeq \sigma^- \cdot \gamma \cdot \gamma' \cdot \sigma \simeq \sigma^- \cdot \gamma \cdot \gamma' \cdot \sigma = \sigma_*(\gamma\gamma').$$

Hence σ_* is a group homomorphism. An inverse is given by $\sigma_*^-\colon \delta \mapsto \sigma \cdot \delta \cdot \sigma^-$.

In particular we see that for a path connected space X the fundamental group $\pi_1(X, x_0)$ is up to isomorphism independent of the chosen point $x_0 \in X$.

Definition 2.22. A topological space X is called *simply connected* if X is path connected and if $\pi_1(X, x) = 1$ for all (equivalently, for one) $x \in X$.

In other words, X is simply connected if and only if every closed path in X is homotopic relative $\{0, 1\}$ to a given constant path.

Remark 2.23. Let X and Y be topological spaces, $x_0 \in X$, $y_0 \in Y$, let p and q be the projections from $X \times Y$ to X and Y, respectively. Then

$$(\pi_1(p), \pi_1(q))\colon \pi_1(X \times Y, (x_0, y_0)) \longrightarrow \pi_1(X, x_0) \times \pi_1(Y, y_0)$$

is an isomorphism of groups. An inverse is given by $\pi_1(i)\pi_1(j)$, where $i\colon X \to X \times Y$, $x \mapsto (x, y_0)$ and $j\colon Y \to X \times Y$, $y \mapsto (x_0, y)$.

We conclude this section by showing that homotopic maps induce isomorphic maps on fundamental groups.

Proposition 2.24. *Let $f_0, f_1\colon X \to Y$ be continuous maps and let $H\colon f_0 \simeq f_1$ be a homotopy. Let $x \in X$ and define $y_i := f_i(x)$ for $i = 0, 1$. Then $\sigma(t) := H(x, t)$ is a path from y_0 to y_1 and the following diagram commutes:*

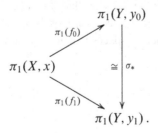

Proof. Let $[\gamma] \in \pi_1(X, x)$. We have to show that

$$(f_0 \circ \gamma) \cdot \sigma \simeq \sigma \cdot (f_1 \circ \gamma) \ (\text{rel } \{0, 1\}).$$

Let $h: [0, 1] \times [0, 1] \to Y$, $(s, t) \mapsto H(\gamma(s), t)$. Let $a, b, c, d: [0, 1] \to [0, 1] \times [0, 1]$, $a(t) = (t, 0)$, $b(t) = (1, t)$, $c(t) = (0, t)$, $d(t) = (t, 1)$ be the sides of the square. Then $(f_0 \circ \gamma) \cdot \sigma = h \circ (a \cdot b)$ and $\sigma \cdot (f_1 \circ \gamma) = h \circ (c \cdot d)$. Clearly there is a homotopy $a \cdot b \simeq c \cdot d$ in $[0, 1] \times [0, 1]$ relative $\{0, 1\}$ (e.g., a linear homotopy $(s, t) \mapsto (1-t)(a \cdot b)(s) + t(c \cdot d)(s)$) and this homotopy induces the desired homotopy by composition. □

Corollary 2.25. *Let* $f: X \to Y$ *be a homotopy equivalence.*

1. *Let* $x \in X$. *Then* $\pi_1(f): \pi_1(X, x) \to \pi_1(Y, f(x))$ *is a group isomorphism.*
2. *The topological space* X *is simply connected if and only if* Y *is simply connected.*

Proof. Let $g: Y \to X$ be a homotopy inverse map of f. As $g \circ f$ (respectively $f \circ g$) is homotopic to the identity, Proposition 2.24 implies that $\pi_1(f)$ is injective (respectively surjective). This shows (1).

Assertion (2) follows from (1) and because X is path connected if and only if Y is path connected (Remark 2.15). □

Corollary 2.26. *Every contractible space is simply connected.*

Altogether we obtain the following implications for various notions of connectivity (where the first two notions have been only defined for subspaces of normed vector spaces):

$$\text{"convex"} \Rightarrow \text{"star shaped"} \Rightarrow \text{"contractible"} \Rightarrow \text{"simply connected"}$$
$$\Rightarrow \text{"path connected"} \Rightarrow \text{"connected"}.$$

Each of these implications is a proper implication: The union of the coordinate axes $\{(x, y) \in \mathbb{R}^2 \ ; \ xy = 0\}$ is star shaped but not convex, $\mathbb{R}^2 \setminus \{(x, y) \ ; \ y = x^2, x \geq 0\}$ is not star shaped but contractible (Problem 2.8), $\mathbb{R}^n \setminus \{0\}$ is simply connected for $n \geq 3$ (Problem 2.15) but not contractible (see Corollary 11.21 below), $\mathbb{R}^2 \setminus \{0\}$ is path connected (Problem 2.5) but not simply connected (see Corollary 2.44 below), and Problems 2.6 and 2.7 give examples of connected spaces that are not path connected.

2.5   Covering Spaces

Definition 2.27 (Covering space). Let X be a topological space.

1. A continuous map $p\colon \tilde{X} \to X$ is called a *covering space of X* or a *covering map* if for all $x \in X$ there exists $x \in U \subseteq X$ open such that

$$p^{-1}(U) = \coprod_{i \in I} \tilde{U}_i \quad \text{sum of topological spaces, } I \neq \emptyset \text{ some set} \qquad (2.1)$$

   for open sets $\tilde{U}_i \subseteq X$ such that $p_{|\tilde{U}_i}\colon \tilde{U}_i \to U$ is a homeomorphism for all $i \in I$ (see Fig. 2.1).
2. A covering map $p\colon \tilde{X} \to X$ is called a *universal covering map* if $\tilde{X}$ is simply connected.
3. Define the category $(\mathrm{CovSp}(X))$ of covering spaces of X as follows. Objects are covering spaces $p\colon \tilde{X} \to X$. A morphism between two covering spaces $p_1\colon \tilde{X}_1 \to X$ and $p_2\colon \tilde{X}_2 \to X$ is a continuous map $\alpha\colon \tilde{X}_1 \to \tilde{X}_2$ such that $p_2 \circ \alpha = p_1$.

For a justification of the notion of a "universal covering" see Remark 2.38 below.

A covering map p is always a surjective map: For every U as in (2.1) and for every $y \in U$ we have a bijection $p^{-1}(\{y\}) \leftrightarrow I$ and $I \neq \emptyset$ by definition. The cardinality of the set $p^{-1}(x)$ is locally constant in x. If it is constant (e.g., if X is connected), we call the cardinality of a fiber the *degree of the covering p*.

Sometimes we consider *pointed covering spaces* $p\colon (\tilde{X}, \tilde{x}) \to (X, x)$ of pointed topological spaces. By this we mean a covering map $p\colon \tilde{X} \to X$ with $p(\tilde{x}) = x$. One has the obvious notion of a *morphism of pointed covering spaces*.

Remark 2.28. Let X be a topological space. Let $E \neq \emptyset$ be a set considered as a discrete topological space. Then the projection $X \times E \to X$ is a covering map. Covering maps that are isomorphic to a covering map of this form are called *trivial covering maps*.

A continuous map $p\colon \tilde{X} \to X$ is a covering map if and only if it is locally on X a trivial covering map.

Figure 2.1 Triviality of the covering over U

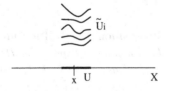

Example 2.29.

1. The function $p : \mathbb{R} \to S^1 = \{ z \in \mathbb{C} ; |z| = 1 \}$, $x \mapsto e^{2\pi i x}$ is a universal covering. Indeed, $\mathbb{R}$ is simply connected because it is convex. The map is a covering: For $j = -1, 1$ let $U_j = S^1 \setminus \{ j \}$. Then $U_{-1} \cup U_1 = S^1$. One has $p^{-1}(U_1) = \coprod_{n \in \mathbb{Z}} (n, n+1)$ and $p|_{(n,n+1)}$ is a homeomorphism $(n, n+1) \xrightarrow{\sim} U_1$; similarly for $p^{-1}(U_{-1}) = \coprod_{n \in \mathbb{Z}} (n - \frac{1}{2}, n + \frac{1}{2})$.
2. The function $\exp : \mathbb{C} \to \mathbb{C}^{\times}$ is a universal covering (use (1) and the polar decomposition of complex numbers).
3. For $n \geq 2$, the map $f : \mathbb{C} \to \mathbb{C}$, $z \mapsto z^n$ is surjective, but there exists no open neighborhood U of 0 such that $\# f^{-1}(\{ z_0 \}) = \# f^{-1}(\{ 0 \}) = 1$ for all $z_0 \in U$. Hence f is not a covering. However its restriction to $\mathbb{C} \setminus \{ 0 \}$ is a covering of degree n (Problem 2.17).

Remark 2.30. The fibers of a covering map $p : \tilde{X} \to X$ are relatively Hausdorff, hence every covering map is separated (Proposition 1.25). In particular, if X is Hausdorff, then $\tilde{X}$ is Hausdorff (Corollary 1.27).

Let $p : \tilde{X} \to X$ and let $f : Z \to X$ be a continuous map. A continuous map $\tilde{f} : Z \to \tilde{X}$ such that $p \circ \tilde{f} = f$ is called a *lifting of f along p*.

Proposition 2.31 (Uniqueness of liftings). *For $i = 1, 2$ let $\tilde{f}_i : Z \to \tilde{X}$ be liftings of f along a covering map $p : \tilde{X} \to X$. Suppose that Z is connected and that there exists $z_0 \in Z$ with $\tilde{f}_1(z_0) = \tilde{f}_2(z_0)$. Then $\tilde{f}_1 = \tilde{f}_2$.*

Proof. As $p \circ \tilde{f}_i = f$, $(\tilde{f}_1, \tilde{f}_2)$ yields a map $\tilde{f} : Z \to \tilde{X} \times_X \tilde{X}$. Let $\Delta_p \subseteq \tilde{X} \times_X \tilde{X}$ be the diagonal. As p is a separated by Remark 2.30 and a local homeomorphism, Δ_p is closed and open in $\tilde{X} \times_X \tilde{X}$ (Proposition 1.25 and Appendix Remark 12.37). Hence $\tilde{f}^{-1}(\Delta_p) = \left\{ z \in Z ; \tilde{f}_1(z) = \tilde{f}_2(z) \right\}$ is closed and open in Z. It contains z_0 and hence is equal to Z because Z is connected. $\qquad\square$

The proof shows that the same assertion remains true for separated local homeomorphisms p.

Proposition 2.32 (Lifting homotopies). *Let $p : \tilde{X} \to X$ be a covering map. Let Z be a topological space, let $H : Z \times [0, 1] \to X$ be a homotopy and let $\tilde{f} : Z \to \tilde{X}$ be a continuous map such that $p(\tilde{f}(z)) = H(z, 0)$. Then there exists a unique homotopy $\tilde{H} : Z \times [0, 1] \to \tilde{X}$ such that $p \circ \tilde{H} = H$ and such that $\tilde{f}(z) = \tilde{H}(z, 0)$ for all $z \in Z$.*

The proof will show that if H is a homotopy relative to a subset $A \subseteq Z$, then $\tilde{H}$ is also a homotopy relative to A.

Proof. If $\tilde{H}$ exists, it is uniquely determined (apply Proposition 2.31 to $H_z(t) := H(z,t)$ for all $z \in Z$). Hence it suffices to construct $\tilde{H}$ locally, i.e., to show that for all $z \in Z$ there exists an open neighborhood W of z in Z and $\tilde{H}: W \times [0,1] \to \tilde{X}$ such that $p \circ \tilde{H} = H_{|W \times [0,1]}$ and such that $\tilde{f}(z) = \tilde{H}(z,0)$ for all $z \in W$.

Let $(U_\alpha)_\alpha$ be an open covering of X such that for all α, $p^{-1}(U_\alpha)$ is the sum of open subspaces each of which is mapped homeomorphically to U_α. As H is continuous, every point (z,t) has a neighborhood of the form $W_t \times (a_t, b_t)$ in $Z \times [0,1]$ such that $H(W_t \times (a_t, b_t)) \subseteq U_\alpha$. As $[0,1]$ is compact, we find $0 = t_0 < t_1 \cdots < t_r = 1$ and a neighborhood W of z such that for all $i = 1, \ldots, r$, $H(W \times [t_{i-1}, t_i])$ is contained in some U_{α_i}.

By induction we may assume that $\tilde{H}_{|W \times [0, t_{i-1}]}$ has already been constructed. As p is a covering map, there exists an open subspace $\tilde{U}_{\alpha_i} \subseteq \tilde{X}$ containing $\tilde{H}(z, t_{i-1})$ such that $p_i := p_{|\tilde{U}_{\alpha_i}}: \tilde{U}_{\alpha_i} \to U_{\alpha_i}$ is a homeomorphism. After shrinking W we may assume that $\tilde{H}(W \times \{t_{i-1}\}) \subseteq \tilde{U}_{\alpha_i}$. Now we can extend $\tilde{H}$ to $W \times [0, t_i]$ by $\tilde{H}_{|W \times [t_{i-1}, t_i]} := p_i^{-1} \circ H_{|W \times [t_{i-1}, t_i]}$. $\qquad\square$

Taking Z to be a point yields the path lifting property for covering spaces.

Corollary 2.33 (Lifting of paths). *Let $p: \tilde{X} \to X$ be a covering map. Let $\gamma: [0,1] \to X$ be a path and let $\tilde{x}_0 \in p^{-1}(\gamma(0))$. Then there exists a unique path $\tilde{\gamma}: [0,1] \to \tilde{X}$ with $p \circ \tilde{\gamma} = \gamma$ and $\tilde{\gamma}(0) = \tilde{x}_0$.*

Corollary 2.34. *Let $p: \tilde{X} \to X$ be a covering map. For every topological space Z, composition with p yields an injective map*

$$\mathrm{Hom}_{(\text{h-Top})}(Z, \tilde{X}) \hookrightarrow \mathrm{Hom}_{(\text{h-Top})}(Z, X).$$

In other words, a covering map is an epimorphism in the category (h-Top). The image of $\mathrm{Hom}_{(\text{h-Top})}(Z, \tilde{X}) \to \mathrm{Hom}_{(\text{h-Top})}(Z, X)$ can be described in terms of the fundamental group.

Proposition 2.35. *Let $p: \tilde{X} \to X$ be a covering map, $\tilde{x}_0 \in \tilde{X}$, $x_0 := p(\tilde{x}_0)$. Let $f: (Z, z_0) \to (X, x_0)$ be a morphism of pointed topological spaces, where Z is path connected and locally path connected.*

Then there exists a morphism $\tilde{f}: (Z, z_0) \to (\tilde{X}, \tilde{x}_0)$ of pointed topological spaces with $p \circ \tilde{f} = f$ if and only if $\pi_1(f)(\pi_1(Z, z_0)) \subseteq \pi_1(p)(\pi_1(\tilde{X}, \tilde{x}_0))$.

The morphism $\tilde{f}$ is unique by Proposition 2.31 if it exists.

Proof. If there exists a morphism $\tilde{f}: (Z, z_0) \to (\tilde{X}, \tilde{x}_0)$ of pointed topological spaces with $p \circ \tilde{f} = f$, then $\pi_1(f) = (\pi_1(p) \circ \pi_1(\tilde{f}))$ and therefore clearly $\pi_1(f)(\pi_1(Z, z_0)) \subseteq \pi_1(p)(\pi_1(\tilde{X}, \tilde{x}_0))$.

Now assume $\pi_1(f)(\pi_1(Z, z_0)) \subseteq \pi_1(p)(\pi_1(\tilde{X}, \tilde{x}_0))$. Let $z \in Z$ and let γ be a path from z_0 to z in Z, which exists because Z is path connected. Then $f \circ \gamma$ becomes a path from x_0 to $f(z)$ in X so that Corollary 2.33 yields a unique path $\tilde{\gamma}$ in $\tilde{X}$ starting at $\tilde{x}_0$ such that $p \circ \tilde{\gamma} = f \circ \gamma$. Set $\tilde{f}(z) = \tilde{\gamma}(1)$ so that $(p \circ \tilde{f})(z) = (p \circ \tilde{\gamma})(1) = (f \circ \gamma)(1) = f(\gamma(1)) = f(z)$. We will show that this is well defined (independent of the choice of γ) and defines a continuous map.

Let δ be another path from z_0 to z in Z and let $\tilde{\delta}$ be the unique path in $\tilde{X}$ starting at $\tilde{x}_0$ such that $p \circ \tilde{\delta} = f \circ \delta$. Furthermore set $\sigma := \gamma \cdot \delta^-$, which is a closed path in Z starting at z_0, and let $\tilde{\sigma}$ be the unique path in $\tilde{X}$ starting at $\tilde{x}_0$ such that $p \circ \tilde{\sigma} = f \circ \sigma$. By assumption we have $\pi_1(f)([\sigma]) \in \pi_1(p)(\pi_1(\tilde{X}, \tilde{x}_0))$ so that there exists a closed path $\tilde{\sigma}'$ in $\tilde{X}$ starting at $\tilde{x}_0$ with $\pi_1(p)([\tilde{\sigma}']) = \pi_1(f)([\sigma])$. Therefore there exists a homotopy of paths $H: (p \circ \tilde{\sigma}') \simeq (f \circ \sigma)$ (rel $\{0,1\}$) and applying Proposition 2.32 yields a unique homotopy of paths $\tilde{H}: \tilde{\sigma}'' \simeq \tilde{\sigma}$ (rel $\{0,1\}$) for a path $\tilde{\sigma}''$ in $\tilde{X}$ such that $p \circ \tilde{H} = H$. This implies $p \circ \tilde{\sigma}'' = p \circ \tilde{\sigma}'$ and $\tilde{\sigma}''(0) = \tilde{\sigma}(0) = \tilde{x}_0 = \tilde{\sigma}'(0)$ so that $\tilde{\sigma}'' = \tilde{\sigma}'$ by the uniqueness of lifts. In particular $\tilde{\sigma}$ has to be a closed path and due to the uniqueness of lifts we have $\tilde{\sigma} = \tilde{\gamma} \cdot \tilde{\delta}^-$ and therefore $\tilde{\delta}(1) = \tilde{\gamma}(1)$. Hence, $\tilde{f}: Z \to \tilde{X}$ is a well-defined map.

Now let $z \in Z$, $\tilde{x} := \tilde{f}(z)$ and $x := p(\tilde{x}) = f(z)$. By definition there exist $x \in U \subseteq X$ open and $\tilde{x} \in \tilde{U} \subseteq \tilde{X}$ such that $p_{|\tilde{U}}: \tilde{U} \to U$ is a homeomorphism. Since Z is locally path connected, there exists $z \in V \subseteq f^{-1}(U)$ open and connected. It is sufficient to show that $\tilde{f}(V) \subseteq \tilde{U}$ so that $\tilde{f}_{|V} = (p_{|\tilde{U}})^{-1} \circ f_{|V}$ has to be continuous. Let $v \in V$, let γ be a path from z to v in V and δ a path from z_0 to z in Z so that $\delta \cdot \gamma$ becomes a path from z_0 to v. Furthermore let $\tilde{\delta}$ be a path from $\tilde{x}_0$ to $\tilde{x}$ with $p \circ \tilde{\delta} = f \circ \delta$ and set $\tilde{\gamma} := (p_{|\tilde{U}})^{-1} \circ f \circ \gamma$. Then $(\tilde{\delta} \cdot \tilde{\gamma})(0) = \tilde{x}_0$ and $p \circ (\tilde{\delta} \cdot \tilde{\gamma}) = (p \circ \tilde{\delta}) \cdot (p \circ \tilde{\gamma}) = (f \circ \delta) \cdot (f \circ \gamma) = f \circ (\delta \cdot \gamma)$ so that $\tilde{\delta} \cdot \tilde{\gamma}$ lifts $f \circ (\delta \cdot \gamma)$. Therefore $\tilde{f}(v) = (\tilde{\delta} \cdot \tilde{\gamma})(1) = ((p_{|\tilde{U}})^{-1} \circ f \circ \gamma)(1) = (p_{|\tilde{U}})^{-1}(f(v)) \in \tilde{U}$. $\qquad \square$

Corollary 2.36. *The map $\tilde{f}$ always exists if Z is in addition simply connected.*

Corollary 2.37. *Let X be path connected and locally path connected, $x \in X$, and for $i = 1, 2$ let $p_i: (\tilde{X}_i, \tilde{x}_i) \to (X, x)$ be covering maps such that $\tilde{X}_i$ is path connected. Set $C_i := \pi_1(p_i)(\pi_1(\tilde{X}_i, \tilde{x}_i)) \subseteq \pi_1(X, x)$. Then there exists a unique morphism of pointed covering spaces $\alpha: (\tilde{X}_1, \tilde{x}_1) \to (\tilde{X}_2, \tilde{x}_2)$ if and only if $C_1 \subseteq C_2$.*

If $C_1 = C_2$, then α is necessarily an isomorphism of covering spaces. By symmetry, there also exists a unique morphism of pointed covering spaces $\beta: (\tilde{X}_2, \tilde{x}_2) \to (\tilde{X}_1, \tilde{x}_1)$ and by uniqueness we have $\alpha \circ \beta = \mathrm{id}_{\tilde{X}_2}$ and $\beta \circ \alpha = \mathrm{id}_{\tilde{X}_1}$.

Remark 2.38. Let X be path connected and locally path connected, $x \in X$. If $p: \tilde{X} \to X$ is a universal covering map, $\tilde{x} \in p^{-1}(x)$, then there exists for each covering map $q: (\tilde{Y}, \tilde{y}) \to (X, x)$ a unique morphism of pointed covering spaces $(\tilde{X}, \tilde{x}) \to (\tilde{Y}, \tilde{y})$ (apply Corollary 2.37 to the path component of $\tilde{Y}$ containing $\tilde{y}$). In other words, a pointed universal covering $(\tilde{X}, \tilde{x})$ is an initial object in the category of pointed coverings of (X, x).

Remark 2.39. Let X be path connected and locally path connected, $x \in X$. Corollary 2.37 shows that the map

$$\begin{Bmatrix} \text{isomorphism classes of pointed} \\ \text{path connected covering spaces} \\ p \colon (\tilde{X}, \tilde{x}) \to (X, x) \end{Bmatrix} \longrightarrow \begin{Bmatrix} \text{subgroups} \\ \text{of } \pi_1(X, x) \end{Bmatrix}, \qquad (2.2)$$

$$(p \colon (\tilde{X}, \tilde{x}) \to (X, x)) \longmapsto \pi_1(p)(\pi_1(\tilde{X}, \tilde{x}))$$

is injective.

Let X be in addition semilocally simply connected (Problem 2.19, e.g., if X has an open covering by simply connected subspaces), then one can show that (2.2) is in fact bijective (Problem 2.20). In particular there exists a universal covering, corresponding to the trivial subgroup.

The map (2.2) can also be extended to a functor on the category of (not necessarily path connected) covering spaces as follows.

Remark 2.40. Let X be path connected and locally path connected and let $p \colon \tilde{X} \to X$ be a covering map. Fix $x \in X$, and let $F := p^{-1}(x)$ be the fiber. Then F is a discrete subspace of $\tilde{X}$. Define a right action

$$F \times \pi_1(X, x) \longrightarrow F$$

as follows. Let $\tilde{x} \in F$ and γ be a closed path representing an element in $\pi_1(X, x)$. By Corollary 2.36 there exists a unique lifting to a path $\tilde{\gamma}$ in $\tilde{X}$ with start point $\tilde{x}$. Let $\tilde{x} \cdot [\gamma] := \tilde{\gamma}(1)$ (this depends only on the homotopy class of γ by Proposition 2.32). Let $p' \colon \tilde{X}' \to X$ be a second covering map and let $g \colon \tilde{X} \to \tilde{X}'$ be a morphism of covering spaces. Then the induced map on fibers $g_x \colon p^{-1}(x) \to p'^{-1}(x)$ is equivariant with respect to the $\pi_1(X, x)$-action. We obtain a functor

$$\Phi_x \colon (\mathrm{CovSp}(X)) \longrightarrow (\mathrm{Sets}\text{-}\pi_1(X, x)), \qquad p \mapsto p^{-1}(x). \qquad (2.3)$$

Here (Sets-G) denotes the category of sets together with a right action by the group G.

The action of $\pi_1(X, x)$ on $p^{-1}(x)$ is transitive if and only if $\tilde{X}$ is path connected. Indeed, the condition is clearly necessary. Conversely suppose that $\tilde{X}$ is path connected. Let $\tilde{y} \in p^{-1}(x)$ and choose a path $\tilde{\gamma}$ in $\tilde{X}$ from $\tilde{x}$ to $\tilde{y}$. Then $\tilde{y} = \tilde{x} \cdot [p \circ \tilde{\gamma}]$.

If $p \colon (\tilde{X}, \tilde{x}) \to (X, x)$ is a pointed path connected covering space, then the stabilizer $\Gamma_{\tilde{x}}$ of $\tilde{x} \in p^{-1}(x)$ in the group $\pi_1(X, x)$ consists of those homotopy classes of paths whose lift to $\tilde{X}$ with start point $\tilde{x}$ has also end point $\tilde{x}$. Hence $\Gamma_{\tilde{x}} = \pi_1(p)(\pi_1(\tilde{X}, \tilde{x}))$ and we see that the functor (2.3) indeed generalizes (2.2) by attaching to a subgroup Γ of $\pi_1(X, x)$ the transitive $\pi_1(X, x)$-set of left cosets $\Gamma \backslash \pi_1(X, x)$.

If X has a universal covering space, then one can show that the functor Φ_x is an equivalence of categories (Problem 2.21). Here we prove only the following easy very special case.

Proposition 2.41. *Let X be a simply connected locally path connected space. Then every covering map of X is isomorphic to a trivial covering, i.e., to* $\mathrm{pr}_2\colon F \times X \to X$ *for a discrete topological space F.*

Proof. Fix $x_0 \in X$. Let $p\colon \tilde{X} \to X$ be a covering map and set $F := p^{-1}(x_0)$. By Corollary 2.36 there exists for every $z \in F$ a unique continuous map $i_z\colon X \to \tilde{X}$ such that $p \circ i_z = \mathrm{id}_X$ and $i_z(x_0) = z$. As p is separated (Remark 2.30), i_z is proper (Proposition 1.32 2) and hence a closed embedding. As p is a local homeomorphism, i_z is an open embedding (Appendix Remark 12.37 3). Therefore $i_z(X)$ is open and closed in $\tilde{X}$ and hence a connected component isomorphic to X (recall that connected components and path components are the same because $\tilde{X}$ is locally path connected [Proposition 2.13]). Let $\tilde{x} \in \tilde{X}$, choose a path γ from $p(\tilde{x})$ to x_0, and let $\tilde{\gamma}$ be the lift in $\tilde{X}$ with start point $\tilde{x}$. Then the end point of $\tilde{\gamma}$ is a point $z \in F$ and hence $\tilde{x}$ and z are in the same path component, i.e., $\tilde{x} \in i_z(X)$. This shows that $\tilde{X} = \coprod_{z \in F} i_z(X) \cong F \times X$. $\square$

2.6 Fundamental Groups and Coverings of Topological Groups

We now study the example of topological groups. We first show that their fundamental groups are always abelian. This relies on the following simple remark.

Remark 2.42. Let X be a topological space, $e \in X$, and let $m\colon X \times X \to X$ be a continuous map such that $m(x, e) = m(e, x) = x$ for all $x \in X$. By functoriality we obtain a homomorphism of groups

$$\pi_1(m)\colon \pi_1(X, e) \times \pi_1(X, e) = \pi_1(X \times X, (e, e)) \to \pi_1(X, e), (\gamma, \delta) \mapsto \gamma * \delta,$$

where the equality is by Remark 2.23.

We claim that $\pi_1(m)$ coincides with group multiplication in $\pi_1(X, e)$. Indeed, let ε be the homotopy class of the constant path with image e. Then one has

$$\gamma \cdot \delta \simeq (\gamma * \varepsilon) \cdot (\varepsilon * \delta) = (\gamma \cdot \varepsilon) * (\varepsilon \cdot \delta) \simeq \gamma * \delta.$$

As a group G is abelian if and only if the multiplication $G \times G \to G$ is a homomorphism of groups, one sees in particular that $\pi_1(X, e)$ is an abelian group, because $\pi_1(m)$ is a homomorphism of groups.

Proposition 2.43. *Let G be a topological group with neutral element $e \in G$ and let $p: \tilde{G} \to G$ be a covering map, where $\tilde{G}$ is a path connected and locally path connected topological space. Fix $\tilde{e} \in p^{-1}(e)$.*

1. *There exists a unique group structure on $\tilde{G}$ that makes $\tilde{G}$ into a topological group with neutral element $\tilde{e}$ such that p is a homomorphism of groups.*
2. *The topological group $\tilde{G}$ is abelian if and only if G is abelian.*
3. *One has an exact sequence of abelian groups*

$$1 \to \pi_1(\tilde{G}, \tilde{e}) \xrightarrow{\pi_1(p)} \pi_1(G, e) \xrightarrow{\partial_{\tilde{e}}} \mathrm{Ker}(p) \to 1. \qquad (2.4)$$

In particular we see that $\partial_{\tilde{e}}$ yields a group isomorphism $\pi_1(G, e) \xrightarrow{\sim} \mathrm{Ker}(p)$ if $\tilde{G}$ is simply connected.

Proof. (1) and (2). Denote by $m: G \times G \to G$ the group law on G. By Proposition 2.35 we can lift $m \circ (p \times p)$ along p to a (necessarily unique) continuous map $\tilde{m}: \tilde{G} \times \tilde{G} \to \tilde{G}$ with $\tilde{m}(\tilde{e}, \tilde{e}) = \tilde{e}$ if

$$(\pi_1(m) \circ \pi_1(p \times p))(\pi_1(\tilde{G} \times \tilde{G}, (\tilde{e}, \tilde{e}))) \subseteq \pi_1(p)(\pi_1(\tilde{G})).$$

But Remark 2.42 shows that $(\pi_1(m) \circ \pi_1(p \times p))([\gamma_1, \gamma_2]) = \pi_1(p)([\gamma_1 \cdot \gamma_2])$. Therefore $\tilde{m}$ exists. The uniqueness of liftings implies that $\tilde{m}$ is associative and that $\tilde{m}$ is commutative if m is commutative. The same argument shows that $G \to G$, $g \mapsto g^{-1}$ can be lifted to a map $\tilde{G} \to \tilde{G}$ mapping $\tilde{e}$ to $\tilde{e}$ and defining an inverse for the multiplication $\tilde{m}$.

As $\tilde{m}$ is a lifting of $m \circ (p \times p)$, p is a surjective homomorphism of groups. In particular G is abelian if $\tilde{G}$ is abelian.

(3). The homomorphism $\pi_1(p)$ is injective by Proposition 2.32. By Remark 2.40 we have a transitive $\pi_1(G, e)$-action on $\mathrm{Ker}(p)$, which yields a surjective map $\partial_{\tilde{e}}: \pi_1(G, e) \to \mathrm{Ker}(p)$, $[\gamma] \mapsto \tilde{e} \cdot [\gamma]$ with $\partial_{\tilde{e}}^{-1}(\tilde{e}) = \mathrm{Im}(\pi_1(p))$. The alternative description of the group law of $\pi_1(G, e)$ in Remark 2.42 shows that $\partial_{\tilde{e}}$ is a group homomorphism. $\qquad \square$

By Example 2.29 we deduce:

Corollary 2.44. *One has $\pi_1(S^1, 1) = \pi_1(\mathbb{C} \setminus \{0\}, 1) \cong \mathbb{Z}$.*

The first equality is given by the inclusion $S^1 \hookrightarrow \mathbb{C} \setminus \{0\}$, which is a homotopy equivalence by Example 2.4. Going through the construction of $\partial_{\tilde{e}}$ one sees that this isomorphism is given by attaching to $n \in \mathbb{Z}$ the closed path $t \mapsto e^{2\pi i n t}$. The spheres S^n for $n \geq 2$ are simply connected (Problem 2.15).

2.7 Problems

Problem 2.1. Let Y be a contractible space. Show that any two continuous maps $X \to Y$ are homotopic.

Problem 2.2. Let I be an index set.

1. Let $f_i, g_i \colon X_i \to Y_i$ be continuous maps of topological spaces. Show that if $f_i \simeq g_i$ for all $i \in I$, then the maps $\prod_i f_i$ and $\prod_i g_i$ from $\prod_i X_i$ to $\prod_i Y_i$ are homotopic.
2. Let X and Y be topological spaces, let Y be contractible. Show that the projection $X \times Y \to X$ is a homotopy equivalence.

Problem 2.3. Show that the inclusions $O(n) \hookrightarrow \mathrm{GL}_n(\mathbb{R})$ and $U(n) \hookrightarrow \mathrm{GL}_n(\mathbb{C})$ are homotopy equivalences.

  Hint: Use the polar decomposition and Problem 2.2.

Problem 2.4. Show that the union of the coordinate axes $(\mathbb{R} \times \{0\}) \cup (\{0\} \times \mathbb{R})$ in $\mathbb{R}^2$ is not homeomorphic to $\mathbb{R}$.

Problem 2.5. Let $n \geq 2$ be an integer, let $U \subseteq \mathbb{R}^n$ be an open subspace, and let $F \subseteq U$ be a finite subset. Show that if U is path connected, then $U \setminus F$ is path connected.

Problem 2.6. Let X be an infinite countable set endowed with the cofinite topology (Appendix Problem 12.16). Show that every non-empty open subspace is connected and that no open subspace is path connected.

Problem 2.7. Let X be the subspace

$$\{ (1/n, y) \; ; \; n \in \mathbb{N}, y \in [0,1] \} \cup \{ (0, y) \; ; \; y \in [0,1] \} \cup \{ (x, 0) \; ; \; x \in [0,1] \}$$

of $\mathbb{R}^2$, and let $X^0 := X \setminus \{0, 0\}$.

1. Show that X is path connected but not locally path connected.
2. Show that X^0 is connected but not path connected.

Problem 2.8. Show that $\mathbb{R}^2 \setminus \{ (x, y) \; ; \; y = x^2, x \geq 0 \}$ is not star shaped but contractible.

Problem 2.9. Let X be a topological space. The *fundamental groupoid* $\Pi(X)$ is the category whose objects are the points of X. For $x, y \in X$ the set of morphisms $\mathrm{Hom}_{\Pi(X)}(x, y)$ is the set of paths from x to y modulo homotopy relative $\{0, 1\}$. Composition is defined by $[\gamma] \circ [\delta] := \delta \cdot \gamma$.

1. Show that $\Pi(X)$ is a groupoid (Appendix Problem 13.7). For $x \in X$ show that the map $\mathrm{Aut}_{\Pi(X)}(x) \overset{\sim}{\to} \pi_1(X, x)$, $[\gamma] \mapsto [\gamma^-]$ is an isomorphism of groups.
2. Show that every continuous map $f \colon X \to Y$ induces a functor $\Pi(f) \colon \Pi(X) \to \Pi(Y)$.
3. Show that if f is a homotopy equivalence, then $\Pi(f)$ is an equivalence of categories.

Problem 2.10. Every group G can be considered as a category with a single object whose set of endomorphisms is the group G. Composition is defined by group multiplication (convince yourself that this is really a category and even a groupoid [Appendix Problem 13.7]).

Let X be a path connected, let $\Pi(X)$ be its fundamental groupoid (Problem 2.9). Fix a point $x \in X$ and consider the group $\pi_1(X, x)$ as a groupoid. Show that the natural inclusion functor $\pi_1(X, x) \to \Pi(X)$ is an equivalence of categories.

Problem 2.11. State and prove a similar result as Remark 2.23 for the fundamental groupoid of $X \times Y$ (Problem 2.9).

Problem 2.12. Let X be a topological space and fix $x \in X$. Consider the map

$$
\left\{ \begin{matrix} \sigma \colon S^1 \to X \text{ continuous} \\ \sigma(1) = x \end{matrix} \right\} \longrightarrow \left\{ \begin{matrix} \gamma \colon [0, 1] \to X \text{ closed path} \\ \gamma(0) = x \end{matrix} \right\}, \quad \sigma \mapsto \sigma \circ e,
$$

where $e \colon [0, 1] \to S^1$ is the map $t \mapsto e^{2\pi i t}$.

1. Show that this map induces an isomorphism $[(S^1, 1), (X, x)] \overset{\sim}{\to} \pi_1(X, x)$, where the left-hand side denotes the set of continuous maps $f \colon S^1 \to X$ with $f(1) = x$ up to homotopy relative $\{1\}$.
2. Let γ be a closed path in X with start point x and let σ be the corresponding continuous map $S^1 \to X$. Show that γ is null-homotopic if and only if σ can be extended to a continuous map $D := \{z \in \mathbb{C} \; ; \; |z| \leq 1\} \to X$.
3. Let G be a path connected topological group with neutral element e. Show that the map "forgetting the base point" yields a bijective map $[(S^1, 1), (G, e)] \to [S^1, G]$.

Problem 2.13. Let $n \geq 1$ be an integer. A *piece-wise affine* path in a subspace A of $\mathbb{R}^n$ is a path $\gamma \colon [0, 1] \to A$ such that there exist $0 = t_0 < t_1 < \cdots < t_r = 1$ and $v_1, \ldots, v_r \in \mathbb{R}^n$ such that $\gamma(t) = \gamma(t_{i-1}) + (t - t_{i-1})v_i$ for all t with $t_{i-1} \leq t \leq t_i$.

Let $U \subseteq \mathbb{R}^n$ be an open subspace. Show that every path in U is homotopic relative $\{0, 1\}$ in U to a piecewise affine path in U.

Hint: Use that every homotopy of paths is uniformly continuous.

Problem 2.14. Let $n \geq 1$ be an integer and let $0 \neq v \in \mathbb{R}^n$. Show that $\mathbb{R}^n \setminus \mathbb{R}^{\geq 0}v$ is contractible. Deduce that $S^{n-1} \setminus \{x_0\}$ is contractible for any $x_0 \in S^{n-1}$.

Problem 2.15. Let $n \geq 3$ be an integer. Show that $\mathbb{R}^n \setminus \{0\}$ and S^{n-1} are simply connected.

Hint: After replacing a closed path in $\mathbb{R}^n \setminus \{0\}$ by a piecewise affine path (Problem 2.13), show that there exists $0 \neq v \in \mathbb{R}^n$ such that $\mathbb{R}^{\geq 0}v \in \mathbb{R}^n \setminus \{\gamma\}$. Then use Problem 2.14.

Problem 2.16. Show that a covering map of finite degree is proper. Show that a covering map of degree 1 is a homeomorphism.

Problem 2.17. Let $n \geq 1$ be an integer. Show that $\mathbb{C} \setminus \{0\} \to \mathbb{C} \setminus \{0\}$, $z \mapsto z^n$ is a covering map of degree n.

Problem 2.18. Let X be a topological space.

1. Show that every morphism in $(\mathrm{CovSp}(X))$ is again a covering map.
2. Let $p: \tilde{X} \to X$ be a covering map with $\tilde{X}$ connected. Show that $A := \mathrm{Aut}_{(\mathrm{CovSp}(X))}(p)$ acts properly discontinuous on $\tilde{X}$, i.e., each $\tilde{x} \in \tilde{X}$ has an open neighborhood U such that $U \cap \alpha(U) = \emptyset$ for all $\alpha \neq \mathrm{id}_{\tilde{x}}$.

Problem 2.19. A topological space X is called *semilocally simply connected* if every point $x \in X$ has an open neighborhood U such that the map $\pi_1(U, x) \to \pi_1(X, x)$, induced by the inclusion, is trivial.

1. Show that if X has an open covering by simply connected subspaces, then X is semilocally simply connected.
2. Suppose that X has a universal covering. Show that X is semilocally simply connected.
3. For $n \in \mathbb{N}$ let $C_n := \partial B_{\frac{1}{n}}((\frac{1}{n}, 0)) \subset \mathbb{R}^2$ and set $X := \bigcup_n C_n$, considered as a subspace of $\mathbb{R}^2$. Show that X is not semilocally simply connected.

Problem 2.20. Let X be a path connected, locally path connected, semilocally simply connected (Problem 2.19) space. The goal of this problem is to show that the map (2.2) is surjective (and hence bijective). In particular X has a universal covering space.

Fix $x_0 \in X$. Let $\tilde{X}$ be the set of homotopy classes (relative $\{0, 1\}$) of paths in X starting at x_0 and let $p: \tilde{X} \to X$ be the map $[\gamma] \mapsto \gamma(1)$. Let $\tilde{x}_0 \in \tilde{X}$ be the homotopy class of the constant path with value x_0. Fix a subgroup C of $\pi_1(X, x_0)$.

1. Show that there exists a left action $\pi_1(X, x_0) \times \tilde{X} \to \tilde{X}$ given by $[\sigma] \cdot [\gamma] := [\sigma \cdot \gamma]$. Let $\tilde{X}_C := C \backslash \tilde{X}$ be the set of C-orbits in $\tilde{X}$ and $\tilde{x}_{C,0}$ be the image of $\tilde{x}_0$ in $\tilde{X}_C$. Show that p induces a map $p_C: \tilde{X}_C \to X$.

2. Let $\mathcal{U}$ be the set of open path connected subspaces $U \subseteq X$ such that $\pi_1(U, x) \to$ $\pi_1(X, x)$ is trivial for one (equivalently, for all) $x \in U$. Show that $\mathcal{U}$ is a basis of the topology of X.

3. For $U \in \mathcal{U}$ and for γ a path in X with start point x_0 and end point in U set $\tilde{U}_{[\gamma]} :=$ $\{ [\gamma \cdot \sigma] \, ; \, \sigma \text{ path in } U \text{ with } \sigma(0) = \gamma(1) \}$. Show that the $\tilde{U}_{[\gamma]}$ form a basis for a topology on $\tilde{X}$. Endow $\tilde{X}_C$ with the quotient topology.

4. Show that $p_C \colon (\tilde{X}_C, \tilde{x}_{C,0}) \to (X, x_0)$ is a pointed path connected covering.

5. Show that $\pi_1(p_C)(\pi_1(\tilde{X}_C, \tilde{x}_{C,0})) = C$. Deduce that $\tilde{X}$ is simply connected.

Problem 2.21. Let X be a topological space and a point x. We want to show that the functor Φ_x (2.3) is an equivalence of categories if X has a universal covering space (see Problem 2.20 for an existence criterion). Fix a pointed universal covering $\pi \colon (\tilde{X}, \tilde{x}) \to$ (X, x).

1. Let $[\gamma] \in \pi_1(X, x)$ and let $\tilde{\gamma}$ be a lift of γ in $\tilde{X}$ with start point $\tilde{x}$. Show that there exists a unique automorphism α of the covering space π such that $\alpha(\tilde{x}) = \tilde{\gamma}(1)$. Show that this construction yields a well-defined group isomorphism $\pi_1(X, x) \xrightarrow{\sim} \mathrm{Aut}(\pi)$. In particular, we obtain a left action of $\pi_1(X, x)$ on $\tilde{X}$.

2. For every set F (considered as a discrete topological space) with right $\pi_1(X, x)$-action let $\Gamma_\pi(F) := F \times^{\pi_1(X,x)} \tilde{X}$ be the quotient space of $F \times \tilde{X}$ by the $\pi_1(X, x)$ right action $(f, \tilde{x}) \cdot \gamma \mapsto (f\gamma, \gamma^{-1}\tilde{x})$. Show that $F \times \tilde{X} \xrightarrow{\mathrm{pr}_2} \tilde{X} \xrightarrow{\pi}$ induces a continuous map $p_F \colon \Gamma_\pi(F) \to X$ and that p_F is a covering map.

3. Show that for every morphism $\alpha \colon F_1 \to F_2$ of sets with right $\pi_1(X, x)$-action, the map $\alpha \times \mathrm{id}_{\tilde{X}}$ induces a morphism of covering spaces $\Gamma_\pi(\alpha) \colon \Gamma_\pi(F_1) \to \Gamma_\pi(F_2)$. Show that one obtains a functor

$$\Gamma_\pi \colon (\text{Sets-}\pi_1(X, x)) \longrightarrow (\text{CovSp}(X)).$$

4. Show that the functors Φ_x and Γ_π are mutually inverse equivalences of categories.

Problem 2.22. Let $\varphi \colon G \to H$ be a continuous homomorphism of topological groups. Show that φ is a covering map if and only if φ is open surjective and $\mathrm{Ker}(\varphi)$ is a discrete subgroup of G.

Problem 2.23. Let $z_0 \in \mathbb{C}$ and let $h \colon \mathbb{C} \setminus \{z_0\} \to S^1, z \mapsto \frac{z - z_0}{|z - z_0|}$.

1. Show that h is a homotopy equivalence and induces isomorphisms

$$W_{z_0} \colon [S^1, \mathbb{C} \setminus \{z_0\}] \xrightarrow{\sim} [S^1, S^1] \xrightarrow{\sim} [(S^1, 1), (S^1, 1)]$$

$$\xrightarrow{\sim} \mathrm{Hom}_{\mathbb{Z}}(\pi_1(S^1, 1), \pi_1(S^1, 1)) = \mathbb{Z}$$

(use Problem 2.12). For $\sigma \colon S^1 \to \mathbb{C} \setminus \{z_0\}$, $W_{z_0}(\sigma)$ is called the *winding number of σ around z_0*. We also write $W_{z_0}(\gamma)$ if γ is the closed path corresponding to σ.

2. Show that for a continuous map $\sigma \colon S^1 \to \mathbb{C}$, the map $\mathbb{C} \setminus \sigma(S^1) \to \mathbb{Z}$, $z \mapsto W_z(\sigma)$ is locally constant.

3. Let $\gamma \colon [0, 1] \to \mathbb{C} \setminus \{0\}$ be a closed path and let $\tilde{\gamma} \colon [0, 1] \to \mathbb{C}$ be a lifting along the covering $\mathbb{C} \to \mathbb{C} \setminus \{0\}$, $z \mapsto \exp(2\pi i z)$. Show that $W_0(\gamma) = \tilde{\gamma}(1) - \tilde{\gamma}(0)$.

4. Let $\gamma \colon [0, 1] \to \mathbb{C} \setminus \{z_0\}$ be a piecewise continuously differentiable closed path. Deduce from 3 that

$$W_{z_0}(\gamma) = \frac{1}{2\pi i} \int_{\gamma} \frac{dz}{z - z_0}.$$

Sheaves

3

In this chapter we introduce one of the central notions in this book: sheaves. They are the main tool to keep track systematically of locally defined data attached to an open subset of a topological space. Giving such data are compatible with restriction to smaller open subsets yields the notions of a presheaf. A sheaf is then a presheaf where such local data, which are given on an open covering $(U_i)_i$ and are compatible with restriction to pairwise intersections $U_i \cap U_j$, can be glued to a global datum. One basic example is to attach to an open set U the $\mathbb{R}$-algebra of continuous functions $U \to \mathbb{R}$: For an open covering $(U_i)_i$ of X giving a continuous function on X is the same as giving continuous functions f_i on all U_i such that $f_{i|U_i \cap U_j} = f_{j|U_i \cap U_j}$ for all i, j.

After defining presheaves and sheaves we make precise in the second section the notion of "defined in some unspecified neighborhood of a point" and introduce stalks of sheaves and germs of sections of sheaves. Section 3.3 explains how to make a presheaf into a sheaf. In Sect. 3.4 we will see that one may view sheaves on a topological space X also as local homeomorphisms $E \to X$, called étalé spaces. This second point of view for sheaves will come in handy in Sect. 3.5 where we define for continuous maps the direct image and inverse image of sheaves. The direct image can be easily defined in terms of sheaves. The inverse image is more easily understood in terms of étalé spaces. In the final section we construct limits and colimits of sheaves showing that the category of sheaves is complete and cocomplete.

Notation: In this chapter X always denotes a topological space.

3.1 Presheaves and Sheaves

Definition 3.1 (Presheaf). A *presheaf* $\mathcal{F}$ on a topological space X consists of the following data:

© Springer Fachmedien Wiesbaden 2016

T. Wedhorn, *Manifolds, Sheaves, and Cohomology*, Springer Studium Mathematik – Master,
DOI 10.1007/978-3-658-10633-1_3

(a) For every open set U of X a set $\mathcal{F}(U)$.
(b) For each pair of open sets $U \subseteq V$ of X a map $\mathrm{res}_U^V \colon \mathcal{F}(V) \to \mathcal{F}(U)$, called a *restriction map*,

such that the following conditions hold:

1. $\mathrm{res}_U^U = \mathrm{id}_{\mathcal{F}(U)}$ for every open set $U \subseteq X$.
2. For $U \subseteq V \subseteq W$ open sets of X, $\mathrm{res}_U^W = \mathrm{res}_U^V \circ \mathrm{res}_V^W$.

Let $\mathcal{F}_1$ and $\mathcal{F}_2$ be presheaves on X. A *morphism of presheaves* $\varphi \colon \mathcal{F}_1 \to \mathcal{F}_2$ is a family of maps $\varphi_U \colon \mathcal{F}_1(U) \to \mathcal{F}_2(U)$ (for all $U \subseteq X$ open) such that for all pairs of open sets $U \subseteq V$ in X the following diagram commutes:

$$
\begin{array}{ccc}
\mathcal{F}_1(V) & \xrightarrow{\ \varphi_V\ } & \mathcal{F}_2(V) \\
{\scriptstyle \mathrm{res}_U^V}\Big\downarrow & & \Big\downarrow{\scriptstyle \mathrm{res}_U^V} \\
\mathcal{F}_1(U) & \xrightarrow{\ \varphi_U\ } & \mathcal{F}_2(U).
\end{array}
$$

Composition of morphisms φ and ψ of presheaves is defined in the obvious way: $\varphi \circ \psi := (\varphi_U \circ \psi_U)_U$. We obtain the category $(\mathrm{PSh}(X))$ of presheaves on X.

If $U \subseteq V$ are open sets of X and $s \in \mathcal{F}(V)$ we will usually write $s_{|U}$ instead of $\mathrm{res}_U^V(s)$. The elements of $\mathcal{F}(U)$ are called *sections of $\mathcal{F}$ over U*. Very often we will also write $\Gamma(U, \mathcal{F})$ instead of $\mathcal{F}(U)$.

Remark 3.2. We can also describe presheaves as follows. Let (Open_X) be the category whose objects are the open sets of X and, for two open sets $U, V \subseteq X$, $\mathrm{Hom}(U, V)$ consists of the inclusion map $U \to V$ if $U \subseteq V$ and is empty otherwise. In other words, (Open_X) is the category attached to the set of open subsets of X partially ordered by inclusion (Appendix Remark 13.32). Composition of morphisms is the composition of the inclusion maps. Then a presheaf is the same as a contravariant functor $\mathcal{F}$ from the category (Open_X) to the category (Sets) of sets. A morphism of presheaves is the same as a morphism of functors.

Remark and Definition 3.3. By replacing (Sets) in Remark 3.2 by some other category C (e.g., the category of abelian groups, the category of rings, or the category of R-algebras, R a fixed ring) we obtain the notion of a *presheaf $\mathcal{F}$ with values in C* (e.g., a *presheaf of groups*, a *presheaf of rings*, a *presheaf of A-algebras for some commutative ring A*). This signifies that $\mathcal{F}(U)$ is an object in C for every open subset U of X and that the restriction maps are morphisms in C.

A morphism $\mathcal{F}_1 \to \mathcal{F}_2$ of presheaves with values in C is then simply a morphism of functors.

•

If $\mathcal{F}$ is a presheaf, one should think of $\mathcal{F}(U)$ as some kind of maps on open subsets U of X with a way to restrict to smaller open subsets. In fact most of the presheaves that we will encounter will be presheaves of functions in the following sense.

Example 3.4 (Presheaf of functions). Let E be a set. For each open subset $U \subseteq X$ let $\mathrm{Map}_E(U) := \mathrm{Map}_{X;E}(U)$ be the set of all maps $U \to E$. For an open subset $V \subseteq U$ we define $\mathrm{res}_V^U: \mathrm{Map}_E(U) \to \mathrm{Map}_E(V)$ as the usual restriction of maps. Then $\mathrm{Map}_{X;E}$ is a presheaf on X.

More generally, a family $\mathcal{F}$ of subsets $\mathcal{F}(U) \subseteq \mathrm{Map}_E(U)$, where U runs through the open subsets of X, is called a *presheaf of E-valued functions on X*, if it is stable under restriction, i.e., for all open sets $V \subseteq U$ and all $f \in \mathcal{F}(U)$ one has $f_{|V} \in \mathcal{F}(V)$. Then $\mathcal{F}$ together with the restriction maps is a presheaf of sets. If E is a group (respectively an R-module for some ring R, respectively an A-algebra for some commutative ring A), then $\mathcal{F}$ is a presheaf of groups (respective of R-modules, respective of A-algebras).

Example 3.5. Let X be a topological space.

1. Let Y be a topological space. For $U \subseteq X$ open set

$$\mathcal{C}_{X;Y}(U) := \{\, f: U \to Y \;;\; f \text{ continuous}\,\}.$$

   Then $\mathcal{C}_{X;Y}$ is a presheaf of Y-valued functions on X.
   If $Y = \mathbb{K}$ ($\mathbb{K} = \mathbb{R}$ or $\mathbb{K} = \mathbb{C}$), then $\mathcal{C}_{X;\mathbb{K}}(U)$ is a commutative $\mathbb{K}$-algebra and $\mathcal{C}_{X;\mathbb{K}}$ is a presheaf of $\mathbb{K}$-algebras.

2. Let V and W be finite-dimensional $\mathbb{R}$-vector spaces and let X be an open subspace of V. Let $\alpha \in \widehat{\mathbb{N}}_0$. For $U \subseteq X$ open set

$$\mathcal{C}^\alpha_{X;W}(U) := \{\, f: U \to W \;;\; f \text{ is a } \mathbb{R}\text{-}C^\alpha\text{-map}\,\}.$$

   Then $\mathcal{C}^\alpha_{X;W}$ is a presheaf of functions on X. It is a presheaf of $\mathbb{R}$-vector spaces. If $W = \mathbb{R}$ we simply write $\mathcal{C}^\alpha_X$. This is a presheaf of $\mathbb{R}$-algebras.

3. Let V and W be finite-dimensional $\mathbb{C}$-vector spaces. Let X be an open subspace of V. For $U \subseteq X$ open set

$$\mathcal{O}_{X;W}(U) := \mathcal{O}^{\mathrm{hol}}_{X;W}(U) := \{\, f: U \to W \;;\; f \text{ holomorphic}\,\}.$$

   Then $\mathcal{O}_{X;W}$ (with the usual restriction maps) is a presheaf of $\mathbb{C}$-vector spaces. For $W = \mathbb{C}$ we simply write $\mathcal{O}_X$ or $\mathcal{O}^{\mathrm{hol}}_X$. This is a sheaf of $\mathbb{C}$-algebras.

4. Let E be a set and set $\mathcal{F}(U) := E$ for all $U \subseteq X$ open. Define all restriction maps to be the identity id_E. Then $\mathcal{F}$ is a presheaf the *constant presheaf with value E*. We can also view this as the sheaf of constant E-valued functions on X.

Example 3.6. Let $\alpha \in \widehat{\mathbb{N}}$. For $X = \mathbb{R}$ and $U \subseteq \mathbb{R}$ open let $d_U : \mathcal{C}_{\mathbb{R}}^{\alpha}(U) \to \mathcal{C}_{\mathbb{R}}^{\alpha-1}(U)$ be the derivative $f \mapsto f'$. Then $(d_U)_U$ is a morphism of presheaves of $\mathbb{R}$-vector spaces (but not of $\mathbb{R}$-algebras because usually $(fg)' \neq f'g'$).

Definition 3.7 (Sheaf). Let X be a topological space. The presheaf $\mathcal{F}$ on X is called a *sheaf*, if for all open sets U in X and every open covering $U = \bigcup_{i \in I} U_i$ the following condition holds:

(Sh) Given $s_i \in \mathcal{F}(U_i)$ for all $i \in I$ such that $s_{i|U_i \cap U_j} = s_{j|U_i \cap U_j}$ for all $i, j \in I$. Then there exists a unique $s \in \mathcal{F}(U)$ such that $s_{|U_i} = s_i$ for all $i \in I$.

A *morphism of sheaves* is a morphism of presheaves.

We obtain the category of sheaves on the topological space X, which we denote by $(\mathrm{Sh}(X))$.

In the same way we can define the notion of a *sheaf of groups*, a *sheaf of rings*, a *sheaf of R-modules* (R a fixed ring), or a *sheaf of A-algebras* (A a fixed commutative ring). Instead of a (pre)sheaf of abelian groups (equivalently, of $\mathbb{Z}$-modules) we sometimes speak just of an *abelian (pre)sheaf*.

The uniqueness assertion in Condition (Sh) can be also phrased as follows.

Remark 3.8. Let X be a topological space, $\mathcal{F}$ a sheaf on X, let $U \subseteq X$ be open and let $U = \bigcup_i U_i$ be an open covering. Let $s, s' \in \mathcal{F}(U)$ such that $s_{|U_i} = s'_{|U_i}$ for all $i \in I$. Then $s = s'$.

Remark 3.9. If $\mathcal{F}$ is a sheaf on X, $\mathcal{F}(\emptyset)$ is a set consisting of one element (use the covering of the empty set $U = \emptyset$ with empty index set $I = \emptyset$ in Condition (Sh)).

In particular, if the only open subsets of X are X and $\emptyset$ (for instance if X consists of one point), then a sheaf $\mathcal{F}$ on X is already uniquely determined by $\mathcal{F}(X)$ and sometimes we identify $\mathcal{F}$ with $\mathcal{F}(X)$.

Remark 3.10 (Sheaf of functions). Let E be a set and let $\mathcal{F}$ be a presheaf of E-valued functions on X (Example 3.4). Then $\mathcal{F}$ is a sheaf if and only if the following condition holds:

(ShF) For every open subset U of X, for every open covering $(U_i)_i$ of U and for every map $f : U \to E$ such that $f_{|U_i} \in \mathcal{F}(U_i)$ for all i one has $f \in \mathcal{F}(U)$.

Remark 3.11 (Restriction of a sheaf to an open subspace). Let $\mathcal{F}$ be a presheaf on a topological space X and let U be an open subspace of X. Then every open subset V of U is also open in X and we define a presheaf $\mathcal{F}_{|U}$ on U by $\mathcal{F}_{|U}(V) := \mathcal{F}(V)$ with the same restriction maps as for $\mathcal{F}$. If $\mathcal{F}$ is a sheaf, then $\mathcal{F}_{|U}$ is a sheaf.

Example 3.12. Notations of Example 3.5. The presheaves $C_{X;Y}$, $C^{\alpha}_{X;W}$, and $\mathcal{O}^{\text{hol}}_X$ are sheaves.

The presheaf of constant functions with values in some set is in general not a sheaf: If U_1 and U_2 are disjoint non-empty open subsets and if $f_1\colon U_1 \to E$ and $f_2\colon U_2 \to E$ are constant maps that take different values, there exists no *constant* map f on $U = U_1 \cup U_2$ whose restriction to U_i is f_i for $i = 1, 2$. The problem is that "constant" cannot be checked locally, i.e., does not satisfy Condition (ShF). If one takes instead the sheaf of locally constant functions with values in some set E, then this is a sheaf.

To define a sheaf it suffices to give its values on a basis of the topology. For instance if an open covering $(U_i)_i$ of X is given, it suffices to give the value of a sheaf only on those open subsets that are contained in one of the U_is.

Remark 3.13. Let X be a topological space and let $\mathcal{F}$ be a sheaf on X. Let $\mathcal{B}$ a basis of the topology on X. If we know the value $\mathcal{F}(U)$ of a sheaf on every element U of $\mathcal{B}$, we can use the sheaf property to determine $\mathcal{F}(V)$ on an arbitrary open V of X. We simply cover V by elements of $\mathcal{B}$. Here is a more systematic way of saying this:

$$\mathcal{F}(V) \;=\; \left\{ (s_U) \in \prod_{\substack{U \in \mathcal{B} \\ U \subseteq V}} \mathcal{F}(U) \;;\; \text{for } U' \subseteq U \text{ both in } \mathcal{B}\colon s_{U|U'} = s_{U'} \right\} \tag{3.1}$$

$$\;=\; \varprojlim_{\substack{U \in \mathcal{B} \\ U \subseteq V}} \mathcal{F}(U). \tag{3.2}$$

Using this observation, we see that it suffices to define a sheaf on a basis $\mathcal{B}$ of open sets of the topology of a topological space X: Consider $\mathcal{B}$ as a full subcategory of (Open_X). Then a *presheaf on $\mathcal{B}$* is a functor $\mathcal{F}\colon \mathcal{B}^{\text{opp}} \to (\text{Sets})$. Every such presheaf $\mathcal{F}$ on $\mathcal{B}$ can be extended to a presheaf $\mathcal{F}'$ on X by using (3.2) as a definition, i.e., for V open in X,

$$\mathcal{F}'(V) = \varprojlim_{\substack{U \in \mathcal{B} \\ U \subseteq V}} \mathcal{F}(U). \tag{3.3}$$

The restriction maps are then given by the projections via the description (3.1). A morphism of presheaves on $\mathcal{B}$ is again defined as a morphism of functors.

Then the presheaf $\mathcal{F}'$ on X is a sheaf if and only if $\mathcal{F}$ satisfies the condition (Sh) of Definition 3.7 for every $U \in \mathcal{B}$ and for every open covering $(U_i)_i$ of U with $U_i \in \mathcal{B}$ for all i. In this case, we say that $\mathcal{F}$ is a *sheaf on $\mathcal{B}$*.

Attaching to a sheaf $\mathcal{F}$ on $\mathcal{B}$ the sheaf $\mathcal{F}'$ on X is clearly functorial in $\mathcal{F}$ and we obtain an equivalence between the category of sheaves on $\mathcal{B}$ and the category of sheaves on X.

A similar remark holds for sheaves of groups, of R-modules (R some ring), or of A-algebras (A some commutative ring).

3.2 Stalks

Let X be a topological space, $\mathcal{F}$ be a presheaf on X, considered as a functor $\mathcal{F}\colon (\mathrm{Open}_X)^{\mathrm{opp}} \to (\mathrm{Sets})$, and let $x \in X$ be a point. Let

$$\mathcal{U}(x) := \{\, U \subseteq X \text{ open} \,;\, x \in U \,\}$$

be the set of open neighborhoods of x, ordered by inclusion. We consider $\mathcal{U}(x)$ as a full subcategory of $\mathrm{Open}(X)$. By restricting $\mathcal{F}$ to $\mathcal{U}(x)$ we obtain a functor $\mathcal{F}\colon \mathcal{U}(x)^{\mathrm{opp}} \to (\mathrm{Sets})$. Note that the category $\mathcal{U}(x)^{\mathrm{opp}}$ is filtered (Appendix Definition 13.37): For any two neighborhoods U_1 and U_2 of x there exists a neighborhood V of x with $V \subseteq U_1 \cap U_2$ (e.g., $V = U_1 \cap U_2$).

Definition and Remark 3.14.

1. The colimit

$$\mathcal{F}_x := \operatorname*{colim}_{\mathcal{U}(x)} \mathcal{F} \tag{3.4}$$

    is called the *stalk of $\mathcal{F}$ in x*. More concretely (Appendix Example 13.33), one has

$$\mathcal{F}_x = \{\, (U, s) \,;\, U \text{ open neighborhood of } x, s \in \mathcal{F}(U) \} / \sim, \tag{3.5}$$

    where two pairs (U_1, s_1) and (U_2, s_2) are equivalent, if there exists an open neighborhood V of x with $V \subseteq U_1 \cap U_2$ such that $s_{1|V} = s_{2|V}$ (note that V might be smaller than $U_1 \cap U_2$).
2. For each open neighborhood U of x we have a canonical map

$$\mathcal{F}(U) \to \mathcal{F}_x, \qquad s \mapsto s_x, \tag{3.6}$$

    which sends $s \in \mathcal{F}(U)$ to the class of (U, s) in $\mathcal{F}_x$. We call s_x the *germ of s in x*.
3. If $\varphi\colon \mathcal{F} \to \mathcal{G}$ is a morphism of presheaves on X, we have an induced map

$$\varphi_x := \operatorname*{colim}_{U \ni x} \varphi_U \colon \mathcal{F}_x \to \mathcal{G}_x$$

    of the stalks in x. It sends the equivalence class of (U, s) in $\mathcal{F}_x$ to the class of $(U, \varphi_U(s))$ in $\mathcal{G}_x$. We obtain a functor $\mathcal{F} \mapsto \mathcal{F}_x$ from the category of presheaves on X to the category of sets. For every $x \in X$ and every open neighborhood one has a commutative diagram

$$
\begin{array}{ccc}
\mathcal{F}(U) & \xrightarrow{\;s \mapsto s_x\;} & \mathcal{F}_x \\
{\scriptstyle \varphi_U}\big\downarrow & & \big\downarrow{\scriptstyle \varphi_x} \\
\mathcal{G}(U) & \xrightarrow{\;t \mapsto t_x\;} & \mathcal{G}_x .
\end{array}
$$

If $\mathcal{F}$ is a presheaf of functions one should think of the stalk $\mathcal{F}_x$ as the set of functions defined in some unspecified open neighborhood of x.

Remark 3.15. If $\mathcal{F}$ is a presheaf on X with values in C, where C is any category in which filtered colimits exist (for instance the category of groups, of R-modules [R ring], or of A-algebras [A commutative ring]), then the stalk $\mathcal{F}_x$ is an object in C and we obtain a functor $\mathcal{F} \mapsto \mathcal{F}_x$ from the category of presheaves on X with values in C to the category C.

Let us make this more precise for a sheaf $\mathcal{G}$ of groups. The underlying set of $\mathcal{G}_x$ is given by (3.5). The group law on $\mathcal{G}_x$ is defined as follows. Let $g, h \in \mathcal{G}_x$ be represented by (U, s) and (V, t). Choose an open neighborhood W of x with $W \subseteq U \cap V$. Then $(U, s) \sim (W, s_{|W})$ and $(V, t) \sim (W, t_{|W})$ and the product gh is the equivalence class of $(W, (s_{|W})(t_{|W}))$.

In the same way addition and multiplication (respectively scalar multiplication) is defined on the stalk for a sheaf of rings (respective of R-modules).

Example 3.16. Let $X = \mathbb{K}$ and $\mathcal{O}_{\mathbb{K}}$ be the sheaf of analytic functions on X. Fix $z_0 \in \mathbb{K}$. Then two analytic functions f_1 and f_2 defined in open neighborhoods U_1 and U_2, respectively, of z_0 agree on some open neighborhood $V \subseteq U_1 \cap U_2$ if and only if they have same Taylor expansion around z_0. Therefore

$$\mathcal{O}_{\mathbb{K}, z_0} = \left\{ \sum_{n \geq 0} a_n (z - z_0)^n \text{ power series with radius of convergence} > 0 \right\}.$$

We have a similar description of the stalk of the sheaf of analytic functions on any open subset X of $\mathbb{K}^m$ for some $m \in \mathbb{N}$.

Example 3.17 (Stalk of the sheaf of continuous functions). Let X be a topological space, let $\mathcal{C}_X = \mathcal{C}_{X;\mathbb{R}}$ be the sheaf of continuous $\mathbb{R}$-valued functions on X, and let $x \in X$. Then

$$\mathcal{C}_{X,x} = \{ (U, f) \, ; \, x \in U \subseteq X \text{ open}, \, f : U \to \mathbb{R} \text{ continuous} \} / \sim,$$

where $(U, f) \sim (V, g)$ if there exists $x \in W \subseteq U \cap V$ open such that $f_{|W} = g_{|W}$. In particular we see that for $x \in V \subseteq U$ open and $f : U \to \mathbb{R}$ continuous one has $(U, f) \sim (V, f_{|V})$. As $\mathcal{C}_X$ is a sheaf of $\mathbb{R}$-algebras, $\mathcal{C}_{X,x}$ is an $\mathbb{R}$-algebra.

If the germ $s \in \mathcal{C}_{X,x}$ of a continuous function at x is represented by (U, f), then $s(x) := f(x) \in \mathbb{R}$ is independent of the choice of the representative (U, f). We obtain an $\mathbb{R}$-algebra homomorphism

$$\mathrm{ev}_x : \mathcal{C}_{X,x} \to \mathbb{R}, \qquad s \mapsto s(x),$$

which is surjective because $\mathcal{C}_{X,x}$ contains in particular the germs of all constant functions. Let $\mathfrak{m}_x := \ker(\mathrm{ev}_x) = \{ s \in \mathcal{C}_{X,x} \, ; \, s(x) = 0 \}$. Then $\mathfrak{m}_x$ is a maximal ideal because $\mathcal{C}_{X,x}/\mathfrak{m}_x \cong \mathbb{R}$ is a field. Let $s \in \mathcal{C}_{X,x} \setminus \mathfrak{m}_x$ be represented by (U, f). Then $f(x) \neq 0$. By shrinking U we may assume that $f(y) \neq 0$ for all $y \in Y$ because f is continuous. Hence $1/f$ exists and s is a unit in $\mathcal{C}_{X,x}$. Therefore the complement of $\mathfrak{m}_x$ consists of units of $\mathcal{C}_{X,x}$. This shows that $\mathcal{C}_{X,x}$ is a local ring with maximal ideal $\mathfrak{m}_x$.

The same argument shows that for every open subspace X of a finite-dimensional $\mathbb{R}$-vector space (respectively a finite-dimensional $\mathbb{C}$-vector space) the stalk $\mathcal{C}^{\alpha}_{X,x}$ (respectively $\mathcal{O}^{\mathrm{hol}}_{X,x}$) is a local ring with residue field $\mathbb{R}$ (respectively $\mathbb{C}$).

The following result will be used very often.

Proposition 3.18. *Let X be a topological space, $\mathcal{F}$ and $\mathcal{G}$ presheaves on X, and let $\varphi, \psi : \mathcal{F} \to \mathcal{G}$ be two morphisms of presheaves.*

1. *Assume that $\mathcal{F}$ is a sheaf. Then the induced maps on stalks $\varphi_x : \mathcal{F}_x \to \mathcal{G}_x$ are injective for all $x \in X$ if and only of $\varphi_U : \mathcal{F}(U) \to \mathcal{G}(U)$ is injective for all open subsets $U \subseteq X$.*
2. *If $\mathcal{F}$ and $\mathcal{G}$ are both sheaves, the maps φ_x are bijective for all $x \in X$ if and only if φ_U is bijective for all open subsets $U \subseteq X$.*
3. *If $\mathcal{F}$ and $\mathcal{G}$ are both sheaves, the morphisms φ and ψ are equal if and only if $\varphi_x = \psi_x$ for all $x \in X$.*

Proof. For $U \subseteq X$ open consider the map

$$\mathcal{F}(U) \to \prod_{x \in U} \mathcal{F}_x, \quad s \mapsto (s_x)_{x \in U}.$$

We claim that this map is injective if $\mathcal{F}$ is a sheaf. Indeed let $s, t \in \mathcal{F}(U)$ such that $s_x = t_x$ for all $x \in U$. Then for all $x \in U$ there exists an open neighborhood $V_x \subseteq U$ of x such that $s_{|V_x} = t_{|V_x}$. Clearly, $U = \bigcup_{x \in U} V_x$ and therefore $s = t$ by Remark 3.8.

Using the commutative diagram

$$\begin{array}{ccc}
\mathcal{F}(U) & \longrightarrow & \prod_{x \in U} \mathcal{F}_x \\
{\scriptstyle \varphi_U} \downarrow & & \downarrow {\scriptstyle \prod_x \varphi_x} \\
\mathcal{G}(U) & \longrightarrow & \prod_{x \in U} \mathcal{G}_x,
\end{array}$$

we see that (3) and the necessity of the condition in (1) are implied by the above claim.

Moreover, a filtered colimit of injective maps is always injective again. Indeed, this follows either from abstract nonsense (Appendix Example 13.50) or we can argue directly. Let $s_0, t_0 \in \mathcal{F}_x$ such that $\varphi_x(s_0) = \varphi_x(t_0)$. Let s_0 be represented by (s, U) and t_0 by (t, V). By shrinking U and V we may assume $U = V$. As

$$\varphi_U(s)_x = \varphi_x(s_0) = \varphi_x(t_0) = \varphi_U(t)_x,$$

there exists an open neighborhood $x \in W \subseteq U$ such that

$$\varphi_W(s_{|W}) = \varphi_U(s)_{|W} = \varphi_U(t)_{|W} = \varphi_W(t_{|W}).$$

As φ_W is injective, we find $s_{|W} = t_{|W}$ and hence $s_0 = s_x = t_x = t_0$. Therefore the condition in (1) is also sufficient.

Hence we are done if we show that the bijectivity of φ_x for all $x \in U$ implies the surjectivity of φ_U. Let $t \in \mathcal{G}(U)$. For all $x \in U$ we choose an open neighborhood U^x of x in U and $s^x \in \mathcal{F}(U^x)$ such that $(\varphi_{U^x}(s^x))_x = t_x$. Then there exists an open neighborhood $V^x \subseteq U^x$ of x with $\varphi_{V^x}(s^x_{|V^x}) = t_{|V^x}$. Then $(V^x)_{x \in U}$ is an open covering of U and for $x, y \in U$

$$\varphi_{V^x \cap V^y}(s^x_{|V^x \cap V^y}) = t_{|V^x \cap V^y} = \varphi_{V^x \cap V^y}(s^y_{|V^x \cap V^y}).$$

As we already know that $\varphi_{V^x \cap V^y}$ is injective, this shows $s^x_{|V^x \cap V^y} = s^y_{|V^x \cap V^y}$ and the sheaf condition (Sh) ensures that we find $s \in \mathcal{F}(U)$ such that $s_{|V^x} = s^x$ for all $x \in U$. Clearly, we have $\varphi_U(s)_x = t_x$ for all $x \in U$ and hence $\varphi_U(s) = t$. $\qquad\square$

Definition 3.19. We call a morphism $\varphi: \mathcal{F} \to \mathcal{G}$ of sheaves *injective* (respectively *surjective*, respectively *bijective*) if $\varphi_x: \mathcal{F}_x \to \mathcal{G}_x$ is injective (respectively surjective, respectively bijective) for all $x \in X$.

Problem 3.7 relates these notions to the categorical notions of monomorphism (respectively epimorphism, respectively isomorphism).

Remark 3.20. Let $\varphi: \mathcal{F} \to \mathcal{G}$ be a morphism of sheaves.

1. Then φ is injective (respectively bijective) if and only if $\varphi_U: \mathcal{F}(U) \to \mathcal{G}(U)$ is injective (respectively bijective) for all open $U \subseteq X$ (Proposition 3.18).
2. The morphism φ is surjective if and only if for all open subsets $U \subseteq X$ and every $t \in \mathcal{G}(U)$ there exist an open covering $U = \bigcup_i U_i$ (depending on t) and sections $s_i \in \mathcal{F}(U_i)$ such that $\varphi_{U_i}(s_i) = t_{|U_i}$, i.e., locally we can find a preimage of t. But the surjectivity of φ does *not imply that $\varphi_U: \mathcal{F}(U) \to \mathcal{G}(U)$ is surjective for all open sets U of X as Example 3.22 shows.*

Similarly, as we defined the notions of injectivity or surjectivity "stalkwise", we also define the notion of an exact sequence of sheaves of groups "stalkwise". Recall that a sequence $G \xrightarrow{\varphi} H \xrightarrow{\psi} K$ of homomorphisms of groups is called *exact* if $\mathrm{Im}(\varphi) = \mathrm{Ker}(\psi)$.

Definition 3.21. A sequence $\mathcal{F} \xrightarrow{\varphi} \mathcal{G} \xrightarrow{\psi} \mathcal{H}$ of homomorphisms of sheaves of groups on a topological space is called *exact* if for all $x \in X$ the induced sequence on stalks $\mathcal{F}_x \xrightarrow{\varphi_x} \mathcal{G}_x \xrightarrow{\psi_x} \mathcal{H}_x$ is an exact sequence of groups. A sequence $\cdots \to \mathcal{F}_{i-1} \to \mathcal{F}_i \to \mathcal{F}_{i+1} \to \ldots$ is called *exact* if $\mathcal{F}_{i-1} \to \mathcal{F}_i \to \mathcal{F}_{i+1}$ is exact for all i.

A homomorphism $\varphi: \mathcal{F} \to \mathcal{G}$ of sheaves of groups is injective (respectively surjective) if and only if $0 \to \mathcal{F} \xrightarrow{\varphi} \mathcal{G}$ (respectively $\mathcal{F} \xrightarrow{\varphi} \mathcal{G} \to 0$) is an exact sequence.

Example 3.22. Let $\mathcal{O}_X$ be the sheaf of holomorphic functions on an open subset X of $\mathbb{C}$. For every open subspace $U \subseteq X$ and $f \in \mathcal{O}_X(U)$ we let $D_U(f) := f'$ be the derivative. We obtain a morphism $D: \mathcal{O}_X \to \mathcal{O}_X$ of sheaves of $\mathbb{C}$-vector spaces. Then D is surjective, because locally every holomorphic function has a primitive.

But there exist open subsets U of X and functions f on U that have no primitive., for instance $U = B_r(z_0) \setminus \{z_0\}$ contained in X and $f: z \mapsto 1/(z - z_0)$. More precisely, by complex analysis we know that D_U is surjective if and only if every connected component of U is simply connected. The sufficiency of this condition will be also an immediate application of cohomological methods developed later (Example 7.6).

We obtain an exact sequence of sheaves of $\mathbb{C}$-vector spaces

$$0 \longrightarrow \mathbb{C}_X \overset{\iota}{\longrightarrow} \mathcal{O}_X \overset{D}{\longrightarrow} \mathcal{O}_X \longrightarrow 0,$$

where $\mathbb{C}_X$ denotes the sheaf of locally constant $\mathbb{C}$-valued functions on X and where ι_U is the inclusion for all $U \subseteq X$ open.

Remark 3.23. Let $\mathcal{F}$ and $\mathcal{G}$ be sheaves on a topological space X and let $(U_i)_i$ be an open covering of X. A morphism of sheaves $\varphi: \mathcal{F} \to \mathcal{G}$ is injective (respectively surjective, respectively bijective) if and only if its restriction $\varphi_{|U_i}: \mathcal{F}_{|U_i} \to \mathcal{G}_{|U_i}$ to morphisms of sheaves on U_i is injective (respectively surjective, respectively bijective) for all i because these notions are defined via the stalks.

But note that the existence of the morphism φ is crucial. There exist sheaves $\mathcal{F}$ and $\mathcal{G}$ such that $\mathcal{F}_{|U_i}$ is isomorphic to $\mathcal{G}_{|U_i}$ for all i and such that $\mathcal{F}$ and $\mathcal{G}$ are not isomorphic (see Example 3.42).

Definition 3.24. Let $\mathcal{F}$ be a sheaf of abelian groups (using additive notation) on a topological space X, $U \subseteq X$ open and $s \in \mathcal{F}(U)$ a section. The *support of s* is the set $\text{supp}(s) := \{x \in U \; ; \; s_x \neq 0\}$. The *support of $\mathcal{F}$* is the set $\text{supp}(\mathcal{F}) := \{x \in X \; ; \; \mathcal{F}_x \neq 0\}$.

Lemma 3.25. *Let $\mathcal{F}$ be a sheaf of abelian groups on a topological space, $U \subseteq X$ open, and $s \in \mathcal{F}(U)$ a section. Then $\text{supp}(s)$ is closed in U.*

Proof. For $x \in U \setminus \text{supp}(s)$ we have $s_x = 0$ so that there exists $V \subseteq U$ open with $s_{|V} = 0$. This implies that $s_{x'} = 0$ for every $x' \in V$ and therefore $V \subseteq U \setminus \text{supp}(s)$. Hence $U \setminus \text{supp}(s)$ is open. $\square$

Example 3.26. Let X be a topological space. Let $C_X = C_{X;\mathbb{R}}$ be the sheaf of continuous $\mathbb{R}$-valued functions on X. Let $U \subseteq X$ be open and $s \in C_X(U)$ a continuous function $U \to \mathbb{R}$. In the proof of Lemma 3.25 we have just seen that $U \setminus \text{supp}(s)$ is the interior of $\{x \in U \; ; \; s(x) = 0\}$. Therefore

$$\text{supp}(s) = \overline{\{x \in U \; ; \; s(x) \neq 0\}} \qquad \text{(closure in } U\text{)}.$$

3.3   Sheaves Attached to Presheaves

There is a functorial way to attach to a presheaf a sheaf.

Proposition and Definition 3.27 (Sheafification). *Let $\mathcal{F}$ be a presheaf on a topological space X. Then there exists a pair $(\tilde{\mathcal{F}}, \iota_{\mathcal{F}})$, where $\tilde{\mathcal{F}}$ is a sheaf on X and $\iota_{\mathcal{F}} : \mathcal{F} \to \tilde{\mathcal{F}}$ is a morphism of presheaves, such that the following holds: If $\mathcal{G}$ is a sheaf on X and $\varphi : \mathcal{F} \to \mathcal{G}$ is a morphism of presheaves, then there exists a unique morphism of sheaves $\tilde{\varphi} : \tilde{\mathcal{F}} \to \mathcal{G}$ with $\tilde{\varphi} \circ \iota_{\mathcal{F}} = \varphi$. The pair $(\tilde{\mathcal{F}}, \iota_{\mathcal{F}})$ is unique up to unique isomorphism.*
Moreover, the following properties hold:

1. *For all $x \in X$ the map on stalks $\iota_{\mathcal{F},x} : \mathcal{F}_x \to \tilde{\mathcal{F}}_x$ is bijective.*
2. *For every presheaf $\mathcal{G}$ on X and every morphism of presheaves $\varphi : \mathcal{F} \to \mathcal{G}$ there exists a unique morphism $\tilde{\varphi} : \tilde{\mathcal{F}} \to \tilde{\mathcal{G}}$ making the diagram*

$$
\begin{array}{ccc}
\mathcal{F} & \xrightarrow{\ \iota_{\mathcal{F}}\ } & \tilde{\mathcal{F}} \\
{\scriptstyle \varphi}\big\downarrow & & \big\downarrow{\scriptstyle \tilde{\varphi}} \\
\mathcal{G} & \xrightarrow{\ \iota_{\mathcal{G}}\ } & \tilde{\mathcal{G}}
\end{array}
\tag{3.7}
$$

commutative. In particular, $\mathcal{F} \mapsto \tilde{\mathcal{F}}$ is a functor from the category of presheaves on X to the category of sheaves on X.

The sheaf $\tilde{\mathcal{F}}$ is called the sheaf associated to $\mathcal{F}$ *or the* sheafification of $\mathcal{F}$.

We can reformulate the first part of Proposition 3.27 by saying that the sheafification functor is a left adjoint functor to the inclusion functor of the category of sheaves into the category of presheaves.

Proof. For $U \subseteq X$ open, elements of $\tilde{\mathcal{F}}(U)$ are by definition families of elements in the stalks of $\mathcal{F}$, which locally give rise to sections of $\mathcal{F}$. More precisely, we define

$$
\tilde{\mathcal{F}}(U) := \left\{ (s_x) \in \prod_{x \in U} \mathcal{F}_x ;\ \forall x \in U : \exists \text{ an open neighborhood } W \subseteq U \text{ of } x, \right.
$$
$$
\left. \text{and } t \in \mathcal{F}(W) : \forall w \in W : s_w = t_w \right\}.
$$

For $U \subseteq V$ the restriction map $\tilde{\mathcal{F}}(V) \to \tilde{\mathcal{F}}(U)$ is induced by the natural projection $\prod_{x \in V} \mathcal{F}_x \to \prod_{x \in U} \mathcal{F}_x$. Then it is easy to check that $\tilde{\mathcal{F}}$ is a sheaf. For $U \subseteq X$ open, we define $\iota_{\mathcal{F},U} : \mathcal{F}(U) \to \tilde{\mathcal{F}}(U)$ by $s \mapsto (s_x)_{x \in U}$. The definition of $\tilde{\mathcal{F}}$ shows that, for $x \in X$, $\tilde{\mathcal{F}}_x = \mathcal{F}_x$ and that $\iota_{\mathcal{F},x}$ is the identity.

Now let $\mathcal{G}$ be a presheaf on X and let $\varphi \colon \mathcal{F} \to \mathcal{G}$ be a morphism. Sending $(s_x)_x \in \tilde{\mathcal{F}}(U)$ to $(\varphi_x(s_x))_x \in \tilde{\mathcal{G}}(U)$ defines a morphism $\tilde{\mathcal{F}} \to \tilde{\mathcal{G}}$. By Proposition 3.18 (3) this is the unique morphism making the diagram (3.7) commutative.

If we assume in addition that $\mathcal{G}$ is a sheaf, then the morphism of sheaves $\iota_{\mathcal{G}} \colon \mathcal{G} \to \tilde{\mathcal{G}}$, which is bijective on stalks, is an isomorphism by Proposition 3.18 (2). Composing the morphism $\tilde{\mathcal{F}} \to \tilde{\mathcal{G}}$ with $\iota_{\mathcal{G}}^{-1}$, we obtain the morphism $\tilde{\varphi} \colon \tilde{\mathcal{F}} \to \mathcal{G}$. Finally, the uniqueness of $(\tilde{\mathcal{F}}, \iota_{\mathcal{F}})$ is a formal consequence. $\qquad\square$

Remark 3.28. If $\mathcal{F}$ is a presheaf of (abelian) groups, of rings, of R-modules, or of R-algebras, its associated sheaf is a sheaf of (abelian) groups, of rings, of R-modules, or of R-algebras.

Remark 3.29. From Proposition 3.18 (2), we get the following characterization of the sheafification. Let $\mathcal{F}$ be a presheaf and $\mathcal{G}$ be a sheaf. Then $\mathcal{G}$ is isomorphic to the sheafification of $\mathcal{F}$ if and only and if there exists a morphism $\iota \colon \mathcal{F} \to \mathcal{G}$ such that ι_x is bijective for all $x \in X$.

Example 3.30. Let E be a set and let $\mathcal{F}$ be a presheaf of functions with values in E. Then its sheafification is given by

$$\tilde{\mathcal{F}}(U) = \{f \colon U \to E \mid \exists \text{ open covering } (U_i)_i \text{ of } U$$
$$\text{such that } f_{|U_i} \in \mathcal{F}(U_i) \text{ for all } i\}.$$

Indeed, this is a sheaf by Remark 3.10 and the inclusions $\mathcal{F}(U) \hookrightarrow \tilde{\mathcal{F}}(U)$ for U open define a morphism of presheaves $\mathcal{F} \to \tilde{\mathcal{F}}$ that is bijective on stalks. Hence we can apply Remark 3.29.

Example 3.31. Let E be a set and denote by E_X the sheaf of locally constant functions, i.e., $E_X(U) = \{f \colon U \to E \text{ locally constant map}\}$. This is the sheafification of the presheaf of constant functions with values in E. The sheaf E_X is called the *constant sheaf with value E*.

If E is a group, then the multiplication in E makes E_X into a sheaf of groups. A similar remark applies if E is an R-module (R some ring) or an A-algebra (A some commutative ring).

3.4 Sheaves and Étalé Spaces

We now give a different description of sheaves via so-called étalé spaces.

Definition 3.32 (Étalé space). Let X be a topological space.

1. A pair (E, π), where E is a topological space and $\pi \colon E \to X$ is a local homeomorphism (Appendix Definition 12.36), is called an *étalé space* over X.
2. Let (E_1, π_1) and (E_2, π_2) be étalé spaces over X. A *morphism* $f \colon (E_1, \pi_1) \to (E_2, \pi_2)$ *of étalé spaces* is a continuous map $f \colon E_1 \to E_2$ such that $\pi_1 = \pi_2 \circ f$.

Denote by $(\text{Ét}/X)$ the category of étalé spaces over X.

One can visualize an étalé space E as a puff pastry lying over the base space X (see Fig. 3.1).

Definition and Remark 3.33. Let (E, π) be an étalé space over X. For $U \subseteq X$ open, a *section of E over U* is a continuous map $s \colon U \to E$ with $\pi \circ s = \mathrm{id}_U$. The *fiber* of E over $x \in X$ is the set $E_x := \pi^{-1}(x)$. A morphism of étalé spaces $f \colon (E_1, \pi_1) \to (E_2, \pi_2)$ gives rise to maps $f_x \colon (E_1)_x \to (E_2)_x$ for $x \in X$.

We will now show that the notion of an étalé spaces over X is equivalent to the notion of a sheaf (of sets) over X. More precisely, we will construct functors

$$\mathbf{F} \colon (\text{Ét}/X) \to (\mathrm{Sh}(X)), \qquad \mathbf{G} \colon (\mathrm{PSh}(X)) \to (\text{Ét}/X)$$

and we will show that $\mathbf{G} \circ \mathbf{F}$ is isomorphic to the identity functor and that $\mathbf{F} \circ \mathbf{G}$ is isomorphic to the sheafification functor. In particular, $\mathbf{F}$ and the restriction of $\mathbf{G}$ to $(\mathrm{Sh}(X))$ yield an equivalence of the categories $(\text{Ét}/X)$ and $(\mathrm{Sh}(X))$.

Figure 3.1 Étalé space

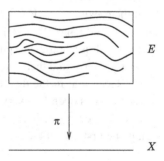

Construction 3.34. Let (E, π) be an étalé space over X. Define a presheaf $\mathcal{E}$ of E-valued functions by

$$\mathcal{E}(U) := \{s\colon U \to E \text{ section}\} = \{s\colon U \to E \text{ ; s continuous}, \pi \circ s = \mathrm{id}_U\}.$$

It is a sheaf of E-valued functions because being continuous and being a section of π is local on X (Remark 3.10).

Let $f\colon (E_1, \pi_1) \to (E_2, \pi_2)$ be a morphism of étalé spaces and $\mathcal{E}_1$, $\mathcal{E}_2$ the corresponding sheaves. For $U \subseteq X$ open and $s \in \mathcal{E}_1(U)$ we have $\pi_2 \circ (f \circ s) = \pi_1 \circ s = \mathrm{id}_U$ and because f and s were continuous we have $f \circ s \in \mathcal{E}_2(U)$. Now we are able to define the morphism $\hat{f}\colon \mathcal{E}_1 \to \mathcal{E}_2$ of sheaves by $\hat{f}_U(s) = f \circ s$ for $U \subseteq X$ open and $s \in \mathcal{E}_1(U)$. We obtain a functor denoted by

$$\mathbf{F}\colon (\text{Ét}/X) \to (\text{Sh}(X)).$$

The Lemma 3.35 gives a description of stalks and of the topology of an étalé space.

Lemma 3.35. *Let (E, π) be an étalé space over X and $\mathcal{E} = \mathbf{F}((E, \pi))$ the associated sheaf on X.*

1. *The maps $\tau_x\colon \mathcal{E}_x \to E_x, s \mapsto s(x)$ are bijective for every $x \in X$.*
2. *The topology of E is the finest topology such that every $s \in \mathcal{E}(U)$, for any $U \subseteq X$ open, is continuous (direct image topology).*

Proof. 1. For $x \in X$, the stalk $\mathcal{E}_x$ is the set of equivalence classes of pairs (U, s), where U is an open neighborhood of x and $s\colon U \to E$ is a section of π. Here (U, s) and (V, t) are equivalent if there exists $x \in W \subseteq U \cap V$ open such that $s_{|W} = t_{|W}$.

For $e \in E$, $x := \pi(e)$ (i.e., $e \in E_x$) there exists an open neighborhood $e \in V \subseteq E$ such that $\pi_{|V}$ is a homeomorphism onto its open image. Then $(\pi_{|V})^{-1}$ is obviously a section of π with $(\pi_{|V})^{-1}(x) = e$. This shows that τ_x is surjective for every $x \in X$.

Let $s\colon U \to E, s'\colon U' \to E$ be sections with $s(x) = s(x') =: e$ for a point $x \in U \cap U'$. Then there exists $e \in V \subseteq E$ open such that $\pi_{|V}\colon V \to \pi(V)$ is a homeomorphism onto its open image. We set $W := \pi(V) \cap U \cap U'$ and replace V by $\pi_{|V}^{-1}(W)$. Now $\pi_{|V}\colon V \to W$ is a homeomorphism again and $x \in W \subseteq U \cap U'$ is open. Then $s_{|W} \circ \pi_{|V} = \mathrm{id}_V = s'_{|W} \circ \pi_{|V}$ and therefore $s_{|W} = s'_{|W}$. Hence τ_x is injective for every $x \in X$.

2. Recall that in the direct image topology a subset W of E is open if and only if $s^{-1}(W) \subseteq X$ is open for every $U \subseteq X$ open and $s \in \mathcal{E}(U)$. This is true for the topology of E because every section is a homeomorphism onto its open image and we have already seen in the first part of the proof that E admits an open cover by images of sections. $\square$

Now we attach conversely to every presheaf an étalé space as follows.

Construction 3.36. Let $\mathcal{E}$ be a presheaf on X. Define $E := \coprod_{x \in X} \mathcal{E}_x$ as a set, the map $\pi: E \to X$, $\mathcal{E}_x \ni e \mapsto x$, and define the maps $f_{U,s}: U \to E$, $x \mapsto s_x$ for $U \subseteq X$ open, $s \in \mathcal{E}(U)$. Endow E with the finest topology such that all maps $f_{U,s}$ are continuous.

To show that (E, π) is an étalé space, let $e \in E$ and $x := \pi(e)$. Then $e \in \mathcal{E}_x$ so that there exist $x \in U \subseteq X$ open and $s \in \mathcal{E}(U)$ with $s_x = e$. Now $\pi \circ f_{U,s} = \mathrm{id}_U$ and with $V := f_{U,s}(U)$ this also implies $f_{U,s} \circ \pi_{|V} = \mathrm{id}_V$ so that $\pi_{|V}$ and $f_{U,s}$ are inverse to each other. Furthermore $(f_{U',s'})^{-1}(V) = \{y \in U' \cap V; s'_y = s_y\} \subseteq X$ is open for every $U' \subseteq X$ open, $s' \in \mathcal{F}(U)$ so that $e \in V \subseteq E$ is open. Therefore (E, π) is an étalé space over X.

Let $\hat{f}: \mathcal{E}_1 \to \mathcal{E}_2$ be a morphism of presheaves and let (E_1, π_1), (E_2, π_2) be the étalé spaces corresponding to $\mathcal{E}_1$ and $\mathcal{E}_2$, respectively. Define the map

$$f: E_1 \to E_2, \qquad (\mathcal{E}_1)_x \ni e \mapsto \hat{f}_x(e) \in (\mathcal{E}_2)_x$$

so that $\pi_1 = \pi_2 \circ f$. Then $(f \circ f_{U,s})(x) = f(s_x) = \hat{f}_x(s_x) = \hat{f}(s)_x = f_{U,\hat{f}(s)}(x)$ and therefore $f \circ f_{U,s} = f_{U,\hat{f}(s)}$ is continuous for $U \subseteq X$ open, $s \in \mathcal{E}(U)$. By definition of the topology of E_1, f is continuous and therefore a morphism of étalé spaces. We obtain a functor

$$\mathbf{G}: (\mathrm{PSh}(X)) \to (\text{Ét}/X).$$

Lemma 3.37. *There is a natural isomorphism from $\mathbf{F} \circ \mathbf{G}$ to the sheafification.*

Proof. Let $\mathcal{E}$ be a presheaf on X, $(E, \pi) = \mathbf{G}(\mathcal{E})$ and $\mathcal{E}' = \mathbf{F}((E, \pi))$. By construction $f_{U,s}: U \to E$, $x \mapsto s_x$ is a section of (E, π) and therefore an element of $\mathcal{E}'(U)$ for $U \subseteq X$ open, $s \in \mathcal{E}(U)$. We define a morphism of presheaves $\kappa: \mathcal{E} \to \mathcal{E}'$ by $\mathcal{E}(U) \to \mathcal{E}'(U)$, $s \mapsto f_{U,s}$.

Let $x \in X$ be an arbitrary point. By construction $E_x = \mathcal{E}_x$ and due to Lemma 3.35 there is a bijective map $\tau_x: \mathcal{E}'_x \to E_x = \mathcal{E}_x$, $\tilde{s} \mapsto \tilde{s}(x)$. For $U \subseteq X$ open, $s \in \mathcal{E}(U)$ we have $\tau_x(\kappa_x(s_x)) = \tau_x((f_{U,s})_x) = f_{U,s}(x) = s_x$ so that $\tau_x \circ \kappa_x = \mathrm{id}_{\mathcal{E}_x}$ and κ_x is bijective for every $x \in X$. It is straightforward to check that κ defines a natural transformation and due to Proposition 3.27 we attain an isomorphism from $\mathcal{E}'$ to the sheafification of $\mathcal{E}$ in a natural way making it a natural isomorphism. $\square$

Lemma 3.38. *There is a natural isomorphism $\mathbf{G} \circ \mathbf{F}$ to the identity functor.*

Proof. Let (E, π) be an étalé space over X, $\mathcal{E} = \mathbf{F}((E, \pi))$ and $(E', \pi') = \mathbf{G}(\mathcal{E})$. By Lemma 3.35 we have a bijection $\tau_x: \mathcal{E}_x \to E_x$ and by construction $E' = \coprod_{x \in X} \mathcal{E}_x$. This defines a bijective map $\tau: E' \to E$ with $\pi \circ \tau = \pi'$. For $U \subseteq X$ open and $s \in \mathcal{E}(U)$ we have $\tau(f_{U,s}(x)) = \tau(s_x) = s(x)$ for every $x \in X$ so that $\tau \circ f_{U,s} = s$. The topology of E' is the finest such that $f_{U,s}: U \to E'$ is continuous for every s, U and the topology of E is the finest such that $s: U \to E$ is continuous for every s, U. This implies that τ is a homeomorphism:

$$W \subseteq E \text{ is open} \Leftrightarrow s^{-1}(W) \subseteq X \text{ is open for every } s: U \to E$$
$$\Leftrightarrow f_{U,s}^{-1}(\tau^{-1}(W)) \subseteq X \text{ is open for every } s: U \to E$$
$$\Leftrightarrow \tau^{-1}(W) \subseteq E' \text{ is open.}$$

It is straightforward to check that this isomorphism is natural. $\square$

As the sheafification of sheaf is the sheaf itself we deduce from Lemma 3.37 and Lemma 3.38:

Proposition 3.39. *Let X be a topological space X. The functors* **F** *and* **G** *yield an equivalence between the category* $(\text{Ét}/X)$ *of étalé spaces over X and the category* $(\text{Sh}(X))$ *of sheaves on X.*

Example 3.40 (Étalé Spaces of constant sheaves). Let E be a set that we also consider as a discrete topological space. Let E_X be the constant sheaf with values in E on a topological space X (Example 3.31). Then the corresponding étalé space is $(X \times E, \text{pr}_1)$ because for $U \subseteq X$ open the sections of pr_1 over U are just the maps $x \mapsto (x, s(x))$, where $s: U \to E$ is locally constant.

Note that the map $\text{pr}_1: X \times E \to X$ is a trivial covering map. More generally, every covering map is an étalé space that is locally on X a trivial covering map (Remark 2.28). Hence we obtain the following result.

Definition and Proposition 3.41 (Locally constant sheaves). *A sheaf $\mathcal{F}$ on X is called* locally constant *if there exists an open covering $(U_i)_i$ of X such that $\mathcal{F}_{|U_i}$ is a constant sheaf.*

The equivalence of $(\text{Sh}(X))$ *and* $(\text{Ét}/X)$ *yields an equivalence between the full subcategory of locally constant sheaves on X and the category of covering spaces of X (Definition 2.27).*

Example 3.42. Let $\mathcal{L}$ be the sheaf of complex logarithms $\mathcal{L}$ on $\mathbb{C} \setminus \{0\}$,

$$\mathcal{L}(U) := \{ l: U \to \mathbb{C} \text{ holomorphic} ; \exp \circ l = \text{id}_U \}, \qquad U \subseteq \mathbb{C} \setminus \{0\} \text{ open.}$$

For every simply connected open subspace U of $\mathbb{C} \setminus \{0\}$ the choice of a logarithm l_0 on U yields an isomorphism of sheaves of abelian groups $(2\pi i \mathbb{Z})_U \cong \mathcal{L}_{|U}$: One attaches to a locally constant function t with values in $2\pi i \mathbb{Z}$ on an open subset V of U the logarithm $l_{0|V} + t$. Hence $\mathcal{L}$ is a locally constant sheaf of abelian groups. But it is not constant because $\mathcal{L}(\mathbb{C} \setminus \{0\}) = \emptyset$. The associated étalé space to $\mathcal{L}$ is the covering map $\exp: \mathbb{C} \to \mathbb{C} \setminus \{0\}$.

Suppose that X is path connected, locally path connected, and semilocally simply connected (Problem 2.19). This will for instance be the case if X is the underlying topological space of a connected premanifold (see Sect. 4.2). Then the functor Φ_x (2.3) is for all $x \in X$ an equivalence of categories (Problem 2.20 and Problem 2.21). Hence we see that the above equivalence also yields an equivalence between the category of locally constant sheaves and the category of sets endowed with a right $\pi_1(X, x)$-action.

There is an analogous equivalence for sheaves of R-modules (R some ring), see Problem 3.14.

3.5 Direct and Inverse Images of Sheaves

In this section $f: X \to Y$ denotes a continuous map of topological spaces. We will now see how to use f in order to attach to a sheaf on X a sheaf on Y (direct image) and to a sheaf on Y a sheaf an X (inverse image). We start with the direct image.

Definition 3.43 (Direct image of a sheaf). Let $f: X \to Y$ be a continuous map. Let $\mathcal{F}$ be a presheaf on X. We define a presheaf $f_*\mathcal{F}$ on Y by (for $V \subseteq Y$ open)

$$(f_*\mathcal{F})(V) = \mathcal{F}(f^{-1}(V))$$

the restriction maps given by the restriction maps for $\mathcal{F}$. Then $f_*\mathcal{F}$ is called the *direct image of $\mathcal{F}$ under f*. Whenever $\varphi: \mathcal{F}_1 \to \mathcal{F}_2$ is a morphism of presheaves, the family of maps $f_*(\varphi)_V := \varphi_{f^{-1}(V)}$ for $V \subseteq Y$ open is a morphism $f_*(\varphi): f_*\mathcal{F}_1 \to f_*\mathcal{F}_2$. We obtain a functor f_* from the category of presheaves on X to the category of presheaves on Y.

Example 3.44 (Direct image of constant sheaves). Let $p: X \to Y$ be a continuous map. Let E be a set and let E_X and E_Y be the sheaf of locally constant E-valued functions on X and Y respectively. For $V \subseteq Y$ open and for locally constant map $g: V \to E$ the composition $g \circ p: p^{-1}(V) \to E$ is locally constant. Hence we obtain a morphism of sheaves

$$\varphi: E_Y \longrightarrow p_*E_X.$$

Now suppose that p is surjective, that Y has the quotient topology of X, and that p has connected fibers (for instance if $p: Y \times I \to Y$ is the projection for a connected space I). For $V \subseteq Y$ open a locally constant map $h: p^{-1}(V) \to E$ is the same as a continuous map if we endow E with the discrete topology. The restriction of h to the fibers of p is constant and hence by the universal property of the quotient topology there exists a unique continuous map $g: V \to E$ such that $g \circ p = h$. Hence we see that φ is an isomorphism in this case.

The following properties are immediate.

Remark 3.45.

1. If $\mathcal{F}$ is a sheaf on X, $f_*\mathcal{F}$ is a sheaf on Y. Therefore f_* also defines a functor
   $f_* : (\mathrm{Sh}(X)) \to (\mathrm{Sh}(Y))$.
2. If $g : Y \to Z$ is a second continuous map, there exists an identity $g_*(f_*\mathcal{F}) = (g \circ f)_*\mathcal{F}$
   that is functorial in $\mathcal{F}$.

We now define the inverse image of a sheaf.

Definition 3.46 (Inverse image of a sheaf). Let $f : X \to Y$ be a continuous map and let $\mathcal{G}$ be a presheaf on Y. Define a presheaf $f^+\mathcal{G}$ on X by

$$U \mapsto \varinjlim_{\substack{V \supseteq f(U), \\ V \subseteq Y \text{ open}}} \mathcal{G}(V), \tag{3.8}$$

the restriction maps being induced by the restriction maps of $\mathcal{G}$. Let $f^{-1}\mathcal{G}$ be the sheafification of $f^+\mathcal{G}$. We call $f^{-1}\mathcal{G}$ the *inverse image of $\mathcal{G}$ under f*.

If f is the inclusion of a subspace X of Y, we also write $\mathcal{G}_{|X}$ instead of $f^{-1}\mathcal{G}$ and we write $\mathcal{G}(X) := (f^{-1}(\mathcal{G}))(X)$.

Note that even if $\mathcal{G}$ is a sheaf, $f^+\mathcal{G}$ is not a sheaf in general.

Remark 3.47. Let $f : X \to Y$ be a continuous map of topological spaces. Let $\mathcal{G}$ be a presheaf on Y. The construction of $f^+\mathcal{G}$ and hence of $f^{-1}\mathcal{G}$ is functorial in $\mathcal{G}$. Therefore we obtain a functor

$$f^{-1} : (\mathrm{PSh}(Y)) \to (\mathrm{Sh}(X)).$$

Remark 3.48. Let $f : X \to Y$ be an open continuous map and $\mathcal{G}$ a presheaf on Y. Then for $U \subseteq X$ open one has $f^+\mathcal{G}(U) = \mathcal{G}(f(U))$. In this case, if $\mathcal{G}$ is a sheaf, $f^+\mathcal{G}$ is a sheaf and hence $f^+\mathcal{G} = f^{-1}\mathcal{G}$.

In particular if f is the inclusion of an open subspace $U = X$ of Y. Then for every sheaf $\mathcal{G}$ on Y and $V \subseteq U$ open

$$(\mathcal{G}_{|U})(V) = \mathcal{G}(V).$$

Direct image and inverse image are functors that are adjoint to each other. More precisely:

Proposition 3.49. *Let* $f: X \to Y$ *be a continuous map of topological spaces, let* $\mathcal{F}$ *be a sheaf on* X *and let* $\mathcal{G}$ *be a presheaf on* Y. *Then there is a bijection*

$$\operatorname{Hom}_{(\operatorname{Sh}(X))}(f^{-1}\mathcal{G}, \mathcal{F}) \leftrightarrow \operatorname{Hom}_{(\operatorname{PreSh}(Y))}(\mathcal{G}, f_*\mathcal{F}),$$
$$\varphi \mapsto \varphi^\flat, \tag{3.9}$$
$$\psi^\sharp \leftarrow\!\shortmid \psi,$$

which is functorial in $\mathcal{F}$ *and* $\mathcal{G}$.

Proof. Let $\varphi: f^{-1}\mathcal{G} \to \mathcal{F}$ be a morphism of sheaves on X, and let $t \in \mathcal{G}(V)$, $V \subseteq Y$ open. Since $f(f^{-1}(V)) \subseteq V$, we have a map $\mathcal{G}(V) \to f^+\mathcal{G}(f^{-1}(V))$, and we define $\varphi^\flat_V(t)$ as the image of t under the map

$$\mathcal{G}(V) \to f^+\mathcal{G}(f^{-1}(V)) \longrightarrow f^{-1}\mathcal{G}(f^{-1}(V)) \xrightarrow{\varphi_{f^{-1}(V)}} \mathcal{F}(f^{-1}(V)) = f_*\mathcal{F}(V).$$

Conversely, let $\psi: \mathcal{G} \to f_*\mathcal{F}$ be a morphism of sheaves on Y. To define the morphism $\psi^\sharp$ it suffices to define a morphism of presheaves $f^+\mathcal{G} \to \mathcal{F}$, which we call again $\psi^\sharp$. Let U be open in X, and $s \in f^+\mathcal{G}(U)$. If V is some open neighborhood of $f(U)$, U is contained in $f^{-1}(V)$. Let V be such a neighborhood such that there exists $s_V \in \mathcal{G}(V)$ representing s. Then $\psi_V(s_V) \in f_*\mathcal{F}(V) = \mathcal{F}(f^{-1}(V))$. Let $\psi^\sharp_U(s) \in \mathcal{F}(U)$ be the restriction of the section $\psi_V(s_V)$ to U.

Clearly, these two maps are inverse to each other. Moreover, it is straightforward – albeit quite cumbersome – to check that the constructed maps are functorial in $\mathcal{F}$ and $\mathcal{G}$. $\square$

Example 3.50. Let X be a topological space, let $i: A \hookrightarrow X$ be the inclusion of a closed subspace and let $\mathcal{F}$ be a sheaf on A. As $X \setminus A$ is open in X, we have

$$i_*(\mathcal{F})_x = \begin{cases} \mathcal{F}_x, & \text{if } x \in A; \\ \{*\}, & \text{if } x \in X \setminus A. \end{cases} \tag{3.10}$$

Via (3.9) there corresponds to $\operatorname{id}_{i_*\mathcal{F}}$ a morphism of sheaves $i^{-1}(i_*\mathcal{F}) \to \mathcal{F}$. Looking at stalks, (3.10) and (3.12) shows that this is an isomorphism.

Remark 3.51 (Inverse image and composition). Let $f: X \to Y$, $g: Y \to Z$ be continuous maps.

1. Let $i: (\operatorname{Sh}(Z)) \to (\operatorname{PreSh}(Z))$ be the inclusion functor. By Proposition 3.49 the functor $f_*: (\operatorname{Sh}(X)) \to (\operatorname{Sh}(Y))$ is right adjoint to $f^{-1}: (\operatorname{Sh}(Y)) \to (\operatorname{Sh}(X))$, the functor $i \circ$

$g_*: (\mathrm{Sh}(Y)) \to (\mathrm{PreSh}(Z))$ is right adjoint to $g^{-1}: (\mathrm{PreSh}(Z)) \to (\mathrm{Sh}(Y))$, and the functor $i \circ (g \circ f)_*$ is right adjoint to $(g \circ f)^{-1}$. Hence the isomorphism of functors $i \circ (g \circ f)_* \cong (i \circ g_*) \circ f_*$ (Remark 3.45) yields an isomorphism of their left adjoints

$$(g \circ f)^{-1} \cong f^{-1} \circ g^{-1}. \tag{3.11}$$

2. If x is a point of X and $i: \{x\} \to X$ is the inclusion, the definition (3.8) shows that

$$i^{-1}\mathcal{F} = \mathcal{F}_x$$

for every presheaf $\mathcal{F}$ on X (more precisely: $i^{-1}(\mathcal{F})(\{x\}) = \mathcal{F}_x$).
3. In particular, (3.11) yields for each presheaf $\mathcal{G}$ on Y a functorial isomorphism

$$(f^{-1}\mathcal{G})_x \cong \mathcal{G}_{f(x)}. \tag{3.12}$$

The identification (3.11) and (3.12) will also follow immediately from the description of the inverse image in terms of étalé spaces (Proposition 3.55).

Remark 3.52. We will almost never use the concrete description of $f^{-1}\mathcal{G}$ in the sequel. Very often we are given f, $\mathcal{F}$, and $\mathcal{G}$ as in the Proposition 3.49, and a morphism of sheaves $\psi: \mathcal{G} \to f_*\mathcal{F}$. Then usually it will be sufficient to understand for each $x \in X$ the map

$$\psi_x^\sharp: \mathcal{G}_{f(x)} \overset{(3.12)}{=} (f^{-1}\mathcal{G})_x \longrightarrow \mathcal{F}_x$$

induced by $\psi^\sharp: f^{-1}\mathcal{G} \to \mathcal{F}$ on stalks. The proof of Proposition 3.49 shows that we can describe this map in terms of ψ as follows. For every open neighborhood $V \subseteq Y$ of $f(x)$, we have maps

$$\mathcal{G}(V) \overset{\psi_V}{\longrightarrow} \mathcal{F}(f^{-1}(V)) \longrightarrow \mathcal{F}_x,$$

and taking the colimit over all V we obtain the map $\psi_x^\sharp: \mathcal{G}_{f(x)} \to \mathcal{F}_x$.

Remark 3.53. Note that if $\mathcal{F}$ is a sheaf of rings (or of R-modules, or of A-algebras) on X, $f_*\mathcal{F}$ is a sheaf on Y with values in the same category. A similar statements holds for the inverse image. Finally, Proposition 3.49 holds (with the same proof) if we consider morphisms of sheaves of rings (or of R-modules, etc.).

We have already seen that there is a natural correspondence between sheaves and étalé space and that it is possible to describe the sheafification of a presheaf in terms of associated étalé spaces. We will now show that the formation of the inverse image of a presheaf has a simple description in terms of étalé spaces: The corresponding étalé space is given by the fiber product. Hence let us consider continuous maps $f: X' \to X$ and $\pi: E \to X$ of topological spaces. We form the fiber product (Appendix Definition 12.28) and obtain the following commutative diagram, where $g\colon E \times_X X' \to E$ and $\pi'\colon E \times_X X' \to X'$ are the projections:

$$
\begin{array}{ccc}
E \times_X X' & \xrightarrow{\ \pi'\ } & X' \\
\big\downarrow{\scriptstyle g} & & \big\downarrow{\scriptstyle f} \\
E & \xrightarrow{\ \pi\ } & X.
\end{array}
$$

For $x' \in X'$ the map g induces a homeomorphism $\pi'^{-1}(x') \xrightarrow{\sim} \pi^{-1}(f(x'))$. The inverse is given by $e \mapsto (e, x')$.

Lemma 3.54. *Suppose that π has one of the following properties:*

1. *homeomorphism,*
2. *open topological embedding, or*
3. *local homeomorphism.*

Then π' has the same property.

Proof. Assertion 1 is clear because $\pi \mapsto \pi'$ is functorial: If ϖ is a continuous inverse of π, then $\varpi'\colon x' \mapsto (\varpi(x), x')$ is a continuous inverse of π'. If π induces a homeomorphism $E \xrightarrow{\sim} U$ for $U \subseteq X$ open, then π' induces a homeomorphism $E' \xrightarrow{\sim} f^{-1}(U)$ by 1. This shows 2. Finally, if there exists an open covering $(W_i)_i$ of E such that $\pi_{|U_i}$ is an open embedding for all i, then $(g^{-1}(W_i))_i$ is an open covering of E' and $\pi'_{|g^{-1}(W_i)}$ is an open embedding by 2. This proves 3. $\qquad\square$

The fiber product construction above yields a functor $(\text{Ét}/X) \to (\text{Ét}/X')$ by sending a morphism $f\colon E_1 \to E_2$ of étalé spaces over X to the map $E_1 \times_X X' \to E_2 \times_X X'$ induced by $f \times \mathrm{id}_{X'}$.

Proposition 3.55 (Inverse image via étalé spaces). *Let $f\colon X \to Y$ be a continuous map of topological spaces, $\mathcal{E}$ a presheaf on Y and (E, π) the étalé space over Y associated to $\mathcal{E}$. The functor that sends $\mathcal{E}$ to the sheaf associated to the étalé space $(E \times_Y X, \pi')$ is naturally isomorphic to the inverse image functor f^{-1}.*

Proof. Denote by $\mathcal{E}' := \mathbf{F}(E, \pi)$ and $\mathcal{F} := \mathbf{F}(E \times_Y X, \pi')$ the associated sheaves on Y and X respectively.

For $U \subseteq X$ we have $\mathcal{F}(U) = \{\, s\colon U \to E \times_Y X \;;\; s \text{ section of } \pi' \,\}$ and

$$
f^+\mathcal{E}'(U) = \{\, (V, s) \;;\; V \subseteq Y \text{ open with } f(U) \subseteq V, s \in \mathcal{E}'(V) \,\} / \sim,
$$

where $(V, s) \sim (V', s')$ if there exists $W \subseteq V' \cap V$ open with $f(U) \subseteq W$ such that $s_{|W} = s'_{|W}$. Now let $[V, s] \in f^+\mathcal{E}'(U)$. Then $\pi \circ (s \circ f)_{|U} = f \circ i$ and we obtain a continuous map $\tilde{s}: U \to E \times_Y X, x \mapsto (s(f(x)), x)$. Then $\tilde{s}$ is a section of π' over U, i.e., $\tilde{s} \in \mathcal{F}(U)$. Notice that the construction of $\tilde{s}$ does only depend on the values of s on $f(U)$ and is hence independent of the choice of (V, s) within its equivalence class. We get a morphism of presheaves $\omega: f^+ \mathcal{E}' \to \mathcal{F}$ by $f^+ \mathcal{E}'(U) \to \mathcal{F}(U), s \mapsto \tilde{s}$.

As $\mathcal{E}'$ is the sheafification of $\mathcal{E}$ (Lemma 3.37) there is a morphism of presheaves $\kappa: \mathcal{E} \to \mathcal{E}'$ that is bijective on stalks. We claim that the morphism of presheaves $\omega \circ f^+(\kappa): f^+ \mathcal{E} \to \mathcal{F}$ is also bijective on stalks. Let $x \in X$, $y := f(x), x \in U \subseteq X, V \subseteq Y$ open with $f(U) \subseteq V$ and $s \in \mathcal{E}(V)$. By Lemma 3.35, the map $\tau_x: \mathcal{F}_x \to (E \times_Y X)_x$, $s \mapsto s(x)$ is bijective. Furthermore the map $f_x': (E \times_Y X)_x \to E_y, (e, x) \mapsto e$ is bijective as well and by construction we have $E_y = \mathcal{E}_y$, so that $f_x' \circ \tau_x: \mathcal{F}_x \to \mathcal{E}_y$ is bijective. Now

$$(\omega \circ f^+(\kappa))_x(s_y) = \omega_y((f_{V,s})_y) = (\tilde{f}_{V,s})_x$$

and

$$(f_x' \circ \tau_x)((\tilde{f}_{V,s})_x) = f_x'(\tilde{f}_{V,s}(x)) = f_x'((f_{V,s}(y), x)) = f_{V,s}(y) = s_y.$$

Therefore $(\omega \circ f^+(\kappa))_x$ and $(f_x' \circ \tau_x)$ are inverse to each other.

Since $f^{-1}\mathcal{E}$ is the sheafification of $f^+\mathcal{E}$ and $\omega \circ f^+(\kappa)$ is bijective on stalks, it induces an isomorphism $f^{-1}\mathcal{E} \xrightarrow{\sim} \mathcal{F}$. By construction it is functorial. $\qquad\square$

Example 3.56 (Inverse image of constant sheaves). Let $f: X \to Y$ be a continuous map. Let E be a set, let E_Y be the sheaf of locally constant E-valued functions on Y. The corresponding étalé space is the projection $E \times Y \to Y$, where we consider E as a discrete topological space. Then the projection $(E \times Y) \times_Y X \to E \times X$ is a homeomorphism compatible with the projections to X (an inverse is given by $(e, x) \mapsto ((e, f(x)), x)$). Hence $f^{-1}E_Y = E_X$.

Remark and Definition 3.57 (Pullback of sections). Let $f: X \to Y$ be a continuous map, let $\mathcal{G}$ be a sheaf on Y, and let $t \in \mathcal{G}(V)$, $V \subseteq Y$ open. Let $\pi: G \to Y$ be the étalé space corresponding to $\mathcal{G}$ and consider t as a continuous section $t: V \to G$ of π. Then

$$f^{-1}(t): f^{-1}(V) \to G \times_Y X, \qquad x \mapsto (t(f(x)), x)$$

is a continuous section of $G \times_Y X \to X$. Hence we obtain a *pullback map*

$$f^{-1}: \mathcal{G}(V) \mapsto (f^{-1}\mathcal{G})(f^{-1}(V)), \tag{3.13}$$

which is functorial in $\mathcal{G}$ and compatible with restrictions to smaller open subsets of Y.

Example 3.58. Let X be a topological space, let Z be a subspace of X, and denote by $i: Z \to X$ the inclusion. If $\mathcal{E}$ is a sheaf on X with corresponding étalé space $\pi: E \to X$, then the étalé space corresponding to $\mathcal{F}_{|Z}$ is the usual restriction $\pi_{|\pi^{-1}(Z)}: \pi^{-1}(Z) \to$

Z. The pullback $i^{-1}(s)$ of a continuous section $s\colon X \to E$ of π over X is simply the restriction $s_{|Z}$ and we usually write $s_{|Z}$ instead of $i^{-1}(s)$.

An application is a simple proof of the following fact (Proposition 3.59), which generalizes the sheaf property from open coverings to more general coverings.

Proposition 3.59. *Let $(A_i)_{i\in I}$ be a covering of a topological space X satisfying* one *of the following hypotheses:*

1. *The interiors A_i° of A_i cover X.*
2. *The covering $(A_i)_i$ is locally finite and A_i is closed in X for all i.*

Let $\mathcal{E}$ be a sheaf on X and let $s_i \in \mathcal{E}(A_i)$ be sections such that $s_{i|A_i\cap A_j} = s_{j|A_i\cap A_j}$ for all $i, j \in I$. Then there exists a unique section $s \in \mathcal{E}(X)$ such that $s_{|A_i} = s_i$ for all $i \in I$.

Proof. Let $\pi\colon E \to X$ be the étalé space corresponding to $\mathcal{E}$. Then s_i is a continuous section of π over A_i. By hypothesis, there exists a unique map $s\colon X \to E$ with $s_{|A_i} = s_i$. Hence s is a section of π. Moreover, s is continuous by Appendix Corollary 12.34 and hence defines an element of $\mathcal{E}(X)$. $\qquad\square$

3.6 Limits and Colimits of Sheaves

We fix a topological space X. We will describe limits and colimits of sheaves and show in particular that the category $(\mathrm{Sh}(X))$ is complete and cocomplete. Let $\mathcal{I}$ be a small category and let $\mathcal{I} \ni i \mapsto \mathcal{F}_i$ be an $\mathcal{I}$-diagram of sheaves on X.

Remark 3.60 (Limits of sheaves). Using the explicit construction of limits in the category of sets (Appendix Example 13.33), one sees that $U \mapsto \lim_i \mathcal{F}_i(U)$, $U \subseteq X$ open, defines a sheaf of sets. It is a limit in the category of presheaves on X and in particular in the category of all sheaves on X.

As special cases we obtain for a family of sheaves $(\mathcal{F}_i)_i$ the product of sheaves $\prod_i \mathcal{F}_i$ given by $U \mapsto \prod_i \mathcal{F}_i(U)$.

Remark 3.61 (Limits and stalks). Let $x \in X$ be a point. The maps $\mathcal{F}_i(U) \to (\mathcal{F}_i)_x$ yield maps $\lim_i \mathcal{F}_i(U) \to \lim_i (\mathcal{F}_i)_x$. Taking the (filtered) colimit over the open neighborhoods of x we obtain a map

$$(\lim_i \mathcal{F}_i)_x \longrightarrow \lim_i (\mathcal{F}_i)_x. \tag{3.14}$$

As filtered colimits commute with finite limits (Appendix Proposition 13.39), we deduce that (3.14) is an isomorphism if $\mathcal{I}$ is finite. In general, (3.14) is not bijective (Problem 3.15).

Remark 3.62 (Colimits of sheaves). The map $U \mapsto \mathrm{colim}_i \, \mathcal{F}_i(U)$, $U \subseteq X$ open, is a colimit of the $\mathcal{I}$-diagram $i \mapsto \mathcal{F}_i$ in the category of presheaves on X. Hence the universal property of the sheafification (Proposition 3.27) shows that the sheafification of $U \mapsto \mathrm{colim}_i \, \mathcal{F}_i(U)$ is a colimit in the category of sheaves on X.

Remark 3.63 (Colimits and stalks). Let $x \in X$ be a point. The maps $\mathcal{F}_i(U) \to (\mathcal{F}_i)_x$ yield maps $\mathrm{colim}_i \, \mathcal{F}_i(U) \to \mathrm{colim}_i(\mathcal{F}_i)_x$. As colimits commute with each other (Appendix Remark 13.36) and sheafification does not change the stalks, we obtain an isomorphism of sets

$$(\mathrm{colim}_i \, \mathcal{F}_i)_x \xrightarrow{\sim} \mathrm{colim}_i(\mathcal{F}_i)_x. \tag{3.15}$$

Remark 3.64. Let $f\colon X \to Y$ be a continuous map of topological spaces. Let $\mathcal{I}$ be a (small) category.

1. Let $i \mapsto \mathcal{F}_i$ be an $\mathcal{I}$-diagram of sheaves on X. As the direct image functor $f_*\colon (\mathrm{Sh}(X)) \to (\mathrm{Sh}(Y))$ has a left adjoint (namely the inverse image functor f^{-1}), it commutes with arbitrary limits (Appendix Proposition 13.47) and hence we obtain an isomorphism $f_*(\lim_i \mathcal{F}_i) \xrightarrow{\sim} \lim_i f_* \mathcal{F}_i$ of sheaves on Y. In particular, f_* is left exact (Appendix Definition 13.48).
2. Dually, let $i \mapsto \mathcal{G}_i$ be an $\mathcal{I}$-diagram of sheaves on Y. Then as f^{-1} has a right adjoint, we have an isomorphism $\mathrm{colim}_i f^{-1} \mathcal{G}_i \xrightarrow{\sim} f^{-1}(\mathrm{colim}_i \mathcal{G}_i)$ of sheaves on X. In particular, f^{-1} is right exact (Appendix Definition 13.48).

Remark 3.65 (Limits and colimits of sheaves in other categories). Instead of sheaves with values in the category of sets we can also consider sheaves with values in the category of groups or the category of R-modules (R a fixed ring). These categories are complete and cocomplete and the forgetful functor to the category of sets preserves limits. Hence the same constructions as above yield limits or colimits of sheaves with values in these categories.

We can also consider the category of commutative rings or the category of R-algebras (R a fixed commutative ring). These categories are complete and in them exist arbitrary filtered colimits. Again the forgetful functor to the category of sets preserves limits. Hence the constructions above yield limits and filtered colimits of sheaves in these categories.

3.7 Problems

Problem 3.1. Let $\mathcal{F}$ be the presheaf of bounded continuous functions on $\mathbb{R}^n$ ($n \in \mathbb{N}$) with values in $\mathbb{R}$. Show that $\mathcal{F}$ is not a sheaf. Show that its sheafification is the sheaf $C_{\mathbb{R}^n;\mathbb{R}}$ of continuous $\mathbb{R}$-valued functions.

Problem 3.2. Let X be a topological space and let $\mathcal{B}_X$ be the Borel-σ-algebra of X (i.e., the σ-algebra generated by the open subsets of X). For $U \subseteq X$ open let $\mathcal{M}_X(U)$ be the set of measurable functions with value in $\overline{\mathbb{R}}$. Show that $\mathcal{M}_X$ is a presheaf of functions, which is a sheaf if X is a Lindelöf space.

Problem 3.3. For $U \subseteq \mathbb{R}^n$ open, $n \in \mathbb{N}$, let $L^1(U)$ be the L^1-space of $\mathbb{R}$-valued Lebesgue integrable functions $U \to \mathbb{R}$ modulo the space of functions f such that $|f| = 0$ almost everywhere.

1. Show that $U \mapsto L^1(U)$ with the usual restriction map is a presheaf of $\mathbb{R}$-vector spaces on $\mathbb{R}^n$, which is not a sheaf.
2. Show that its sheafification is the sheaf $U \mapsto L^1_{\text{loc}}(U)$, where $L^1_{\text{loc}}(U)$ is the quotient of the space of Lebesgue measurable functions $f \colon U \to \mathbb{R}$ with $\int_K |f| \, dx < \infty$ for each compact subspace $K \subseteq U$ by the subspace of functions f such that $|f| = 0$ almost everywhere.

Problem 3.4. Let X be a topological space and let $\mathcal{F}$ be a presheaf. A *subpresheaf of $\mathcal{F}$* is a presheaf $\mathcal{G}$ such that $\mathcal{G}(U) \subseteq \mathcal{F}(U)$ for all $U \subseteq X$ open such that for all open subsets $V \subseteq U \subseteq X$ the restriction maps $\mathcal{G}(U) \to \mathcal{G}(V)$ are induced by the restriction maps for $\mathcal{F}$.

Suppose that $\mathcal{F}$ is a sheaf. Show that a subpresheaf $\mathcal{G}$ of $\mathcal{F}$ is a sheaf if and only if for every open subset U of X, for every open covering $(U_i)_i$ of U and for every $s \in \mathcal{F}(U)$ with $s_{|U_i} \in \mathcal{G}(U_i)$ one has $s \in \mathcal{G}(U)$.

Problem 3.5. Let X be a topological space and define a presheaf Ω on X by $\Omega(U) := \{V \subseteq U \; ; \; V \subseteq X \text{ open}\}$. For $U' \subseteq U$ the restriction $\Omega(U) \to \Omega(U')$ is defined by $V \mapsto V \cap U'$.

1. Show that Ω is a sheaf.
2. Show that Ω classifies subsheaves in the following sense. Let $\mathcal{F}$ be a sheaf on X and let $\Phi \colon \mathcal{F} \to \Omega$ be a morphism of sheaves. Then $U \mapsto \{s \in \mathcal{F}(U) \; ; \; \Phi_U(s) = U\}$ is a subsheaf (Problem 3.4) $\mathcal{G}_\Phi$ of $\mathcal{F}$ and one obtains a bijection, functorial in $\mathcal{F}$,

$$\text{Hom}_{(\text{Sh}(X))}(\mathcal{F}, \Omega) \leftrightarrow \{\text{subsheaves of } \mathcal{F}\}, \qquad \Phi \mapsto \mathcal{G}_\Phi.$$

Problem 3.6. Let X be a topological space, $x \in X$ and denote by $i_x \colon \{x\} \to X$ the inclusion. For every set E (considered as a sheaf on $\{x\}$) we call $(i_x)_*(E)$ the *skyscraper sheaf in x with value E*.

1. Show that $E \mapsto (i_x)_*(E)$ defines a functor (Sets) $\to$ (Sh(X)), which is right adjoint to the functor (Sh(X)) $\to$ (Sets) that sends a sheaf $\mathcal{F}$ to its stalk $\mathcal{F}_x$ in x.
2. Show that for the stalks of the skyscraper sheaf one has $(i_x)_*(E)_y = E$ for $y \in \overline{\{x\}}$ and that $(i_x)_*(E)_y$ is a singleton for $y \notin \overline{\{x\}}$.

Problem 3.7. Let $\varphi \colon \mathcal{F} \to \mathcal{G}$ be a morphism of sheaves on a topological space X. Show that the following assertions are equivalent:

(i) φ is a monomorphism (respectively an epimorphism, respectively an isomorphism) in the category of sheaves on X.

(ii) φ is injective (respectively surjective, respectively bijective).

Hint: To show "φ epimorphism $\Rightarrow$ φ surjective" use Problem 3.6.

Problem 3.8. Let X be a topological space. Show that the functor from the category of sets to the category of sheaves on X that sends a set E to the constant sheaf E_X is left adjoint to the functor $\mathcal{F} \mapsto \mathcal{F}(X)$.

Problem 3.9. Let $\mathcal{G}$ be a sheaf, let $f \colon X \to Y$ be a continuous map of topological spaces, and let $U \subseteq X$ be an open subset. Show that $(f^{-1}\mathcal{G})(U)$ can be described as the set of $s = (s_x)_{x \in U} \in \prod_{x \in U} \mathcal{G}_{f(x)}$ such that for all $x \in U$ the following condition holds: There exist $x \in W \subseteq U$ open, $V \subseteq Y$ open with $f(W) \subseteq V$, and $t \in \mathcal{G}(V)$ such that $t_{f(w)} = s_{f(w)}$ for all $w \in W$.

Problem 3.10. Let $f \colon X \to Y$ be a continuous map. Show that the functor $f^{+} \colon (\mathrm{PSh}(Y)) \to (\mathrm{PSh}(X))$ is left adjoint to the functor $f_{*} \colon (\mathrm{PSh}(X)) \to (\mathrm{PSh}(Y))$.

Problem 3.11. Let $f \colon X \to Y$ be a continuous map.

1. Let $\mathcal{G}$ be a sheaf on Y and let $s \in \mathcal{G}(Y)$. Show that $\mathrm{supp}(f^{-1}(s)) = f^{-1}(\mathrm{supp}(s))$.
2. Let $\mathcal{F}$ be a sheaf on X. Then a section $\tilde{s} \in (f_{*}\mathcal{F})(Y)$ is the same as a section $s \in \mathcal{F}(X)$. Show that $\mathrm{supp}(\tilde{s}) = \overline{f(\mathrm{supp}\, s)}$.

Problem 3.12. Let X be a topological space. One says that an abelian sheaf $\mathcal{F}$ on X *satisfies the principle of unique continuation* if every section of $\mathcal{F}$ over any open subset of X has open support.

1. Let $U \subseteq X$ be open and connected and $s, t \in \mathcal{F}(U)$. Suppose that $s_x = t_x$ for one point $x \in U$. Show that $s = t$.
2. Let $g \colon Z \to X$ be a continuous map. Show that $g^{-1}\mathcal{F}$ also satisfies the principle of unique continuation.
3. Show that every locally constant sheaf satisfies the principle of unique continuation.
4. Let X be an open subset of a finite-dimensional $\mathbb{K}$-vector space. Show that the sheaf of analytic functions on X satisfies the principle of unique continuation.
   Hint: Appendix Problem 16.3.

Problem 3.13. Let X be a topological space, let $\mathcal{F}$ be an abelian sheaf on X, and let (E, π) be the corresponding étalé space over X. Show that $\mathcal{F}$ satisfies the principle of unique continuation (Problem 3.12) if and only if π is separated.

Problem 3.14. Let X be a path connected, locally path connected, semilocally simply connected (Problem 2.19) space so that there exists a universal covering of X (Problem 2.20). Let R be a ring. A *local system of R-modules* is a locally constant sheaf of R-modules. Choose $x_0 \in X$. Show that the following categories are equivalent:

(i) The category of local systems of R-modules on X.
(ii) The category of functors $\Pi(X) \to (R\text{-Mod})$ (here $\Pi(X)$ is the fundamental groupoid of X, see Problem 2.9).
(iii) The category of representations of $\pi_1(X, x_0)$ on R-modules.

Problem 3.15. Let $(\mathcal{F}_i)_{i \in I}$ be a family of sheaves on X and let $x \in X$.

1. Show that the map (3.14) $(\prod_{i \in I} \mathcal{F}_i)_x \to \prod_{i \in I} \mathcal{F}_{i,x}$ is injective.
2. Let $X = \mathbb{R}$, $x = 0$ and $\mathcal{C}_{\mathbb{R}}$ be the sheaf of continuous $\mathbb{R}$-valued functions on $\mathbb{R}$. For $n \in \mathbb{N}$ let $f_n \in \mathcal{C}_{\mathbb{R},0}$ be the germ of the function $x \mapsto 1/(x - (1/n))$. Show that $(f_n)_{n \in \mathbb{N}}$ is not in the image of $(\prod_n \mathcal{C}_{\mathbb{R}})_0 \to \prod_n \mathcal{C}_{\mathbb{R},0}$.

Problem 3.16. Let $f : X \to Y$ be a continuous map and let $\mathcal{F}$ be a sheaf on X. Define the *proper direct image* $f_!\mathcal{F}$ as the sheaf on Y with ($V \subseteq Y$ open)

$$(f_!\mathcal{F})(V) := \left\{ s \in \mathcal{F}(f^{-1}(V)) \; ; \; f : \mathrm{supp}(s) \to V \text{ is proper} \right\}.$$

1. Show that $f_!\mathcal{F}$ is a subsheaf of $f_*\mathcal{F}$ and that $\mathcal{F} \mapsto f_!\mathcal{F}$ yields a functor $f_! : (\mathrm{Sh}(X)) \to (\mathrm{Sh}(Y))$. Show that $f_* = f_!$ if f is proper.
   Hint: Problem 1.18.
2. Let $g : Y \to Z$ be a second continuous map. Show that the identification $(g \circ f)_* = g_* \circ f_*$ of functors $(\mathrm{Sh}(X)) \to (\mathrm{Sh}(Z))$ induces $(g \circ f)_! = g_! \circ f_!$.
   Hint: Use Problem 3.11.

Problem 3.17. Let $Z \subseteq X$ be a locally closed subspace of a topological space X, let $i : Z \to X$ be the inclusion, and let $i_! : (\mathrm{Sh}(Z)) \to (\mathrm{Sh}(X))$ be the proper direct image functor (Problem 3.16).

1. Let $\mathcal{F}$ be a sheaf on Z. Show that for $x \in X$ one has

$$i_!(\mathcal{F})_x = \begin{cases} \mathcal{F}_x, & \text{if } x \in Z; \\ \{*\}, & \text{if } x \in X \setminus Z. \end{cases}$$

Show that $i_!$ yields an equivalence between $(\mathrm{Sh}(Z))$ and the full subcategory of $(\mathrm{Sh}(X))$ of sheaves $\mathcal{F}$ such that $\mathcal{F}_x = \{*\}$ for all $x \in X \setminus Z$. The quasi-inverse functor is induced by i^{-1}. Deduce that $i_! : (\mathrm{Sh}(Z)) \to (\mathrm{Sh}(X))$ is left adjoint to i^{-1}.

2. For a sheaf $\mathcal{F}$ on X define for $U \subseteq X$ open

$$\mathcal{F}^Z(U) = \{ s \in \mathcal{F}(U) \ ; \ \mathrm{supp}(s) \subseteq Z \}.$$

Show that $\mathcal{F}^Z$ is a subsheaf of $\mathcal{F}$ and show that $\mathcal{F} \mapsto i^!\mathcal{F} := i^{-1}\mathcal{F}^Z$ yields a functor $(\mathrm{Sh}(X)) \to (\mathrm{Sh}(Z))$. Show that $i^! = i^{-1}$ if Z is open in X.

3. Show that $i^!$ is right adjoint to $i_!$.

Problem 3.18. Let X be a topological space, $U \subseteq X$ open, $Z := X \setminus U$. Let $\mathcal{F}$ be an abelian sheaf on X. Let $j : U \to X$ and $i : Z \to X$ be the inclusions. Show that there is an exact sequence of abelian sheaves

$$0 \longrightarrow j_! j^{-1} \mathcal{F} \longrightarrow \mathcal{F} \longrightarrow i_* i^{-1} \mathcal{F} \longrightarrow 0.$$

Hint: Problem 3.17.

Manifolds

4

In this chapter we define a real (respectively complex) premanifold as a space together with a sheaf of functions that locally resembles $\mathbb{R}^n$ (respectively $\mathbb{C}^n$) for some n together with its sheaf of C^α-functions ($\alpha \in \mathbb{N}_0 \cup \{\infty, \omega\}$ fixed) (respective of holomorphic functions). Hence we first define the abstract notion of a space together with a sheaf of R-algebras for some ring R. These are called R-ringed spaces. Then a real (respectively complex) premanifold will be a special case of an $\mathbb{R}$-ringed (respectively a $\mathbb{C}$-ringed) space. A (real or complex) manifold is defined to be a premanifold whose underlying topological space is Hausdorff and second countable. This implies several other nice topological properties. In the last section we explain simple constructions for how to get new (pre)manifolds from old ones. The most important constructions will be the product and the gluing of (pre)manifolds.

4.1 Ringed Spaces

Ringed spaces formalize the idea of giving a geometric object by specifying its underlying topological space and the "functions" on all open subsets of this space. This is motivated by the observation that to say that a continuous map $F : U \to \mathbb{R}^n$ ($U \subseteq \mathbb{R}^m$ open) is a C^α-map if and only if for every C^α-function $f : \mathbb{R}^n \to \mathbb{R}$ the composition $f \circ F : U \to \mathbb{R}$ is a C^α-map (in fact it suffices to take for f the projections to the coordinates). These functions usually form some commutative algebra over some ring[1]. For instance the $\mathbb{R}$-valued C^α-functions on an open subset of $\mathbb{R}^m$ form a commutative $\mathbb{R}$-algebra. An immediate observation is that because of continuity the germs of such functions f at a point x either satisfy $f(x) = 0$ or $1/f$ exists in some open neighborhood of x. In other words, the stalk

---

[1] Notable exceptions are for instance algebras of differential operators, which are almost never commutative.

© Springer Fachmedien Wiesbaden 2016

T. Wedhorn, *Manifolds, Sheaves, and Cohomology*, Springer Studium Mathematik – Master, DOI 10.1007/978-3-658-10633-1_4

of the sheaf of functions in x is a local ring whose maximal ideal consists of germs f with $f(x) = 0$. This leads to the notion of a locally ringed space.

The notion of a (locally) ringed space also gives us the flexibility to consider very general notions of "functions": Any sheaf of rings will do. It is not necessarily a sheaf of functions. This is very useful in other geometric theories such as in the theory of complex analytic spaces or in algebraic geometry. But in this book, for almost all ringed spaces the underlying sheaf of rings will be a sheaf of functions.

We fix a commutative ring R. In the sequel, R will usually be the field $\mathbb{R}$ of real numbers or the field $\mathbb{C}$ of complex numbers. But occasionally we will consider other cases, for example $R = \mathbb{Z}$.

Definition 4.1 ((Locally) ringed spaces).

1. An *R-ringed space* is a pair $(X, \mathcal{O}_X)$, where X is a topological space and where $\mathcal{O}_X$ is a sheaf of commutative R-algebras on X. The sheaf of rings $\mathcal{O}_X$ is called the *structure sheaf* of $(X, \mathcal{O}_X)$.
2. A *locally R-ringed space* is an R-ringed space $(X, \mathcal{O}_X)$ such that the stalk $\mathcal{O}_{X,x}$ is a local ring for all $x \in X$. We then denote by $\mathfrak{m}_x$ the maximal ideal of $\mathcal{O}_{X,x}$ and by $\kappa(x) := \mathcal{O}_{X,x}/\mathfrak{m}_x$ its residue field.

As every ring has a unique structure as $\mathbb{Z}$-algebra, we simply say *(locally) ringed space* instead of (locally) $\mathbb{Z}$-ringed space. Usually we will denote a (locally) R-ringed space $(X, \mathcal{O}_X)$ simply by X.

Our principle example will be sheaves of real-valued C^α-functions (with $\alpha \in \widehat{\mathbb{N}}_0 := \mathbb{N}_0 \cup \{\infty, \omega\}$) or holomorphic functions. Basic notions and properties of such functions are recalled in Appendix 16.

Example 4.2. Recall that $\mathbb{K}$ denotes the field of real or of complex numbers. Let X be an open subset of a finite-dimensional $\mathbb{K}$-vector space. Let $\alpha \in \widehat{\mathbb{N}}_0$. We denote by $\mathcal{C}^\alpha_X$ the sheaf of C^α-functions, i.e.,

$$\mathcal{C}^\alpha_X(U) = \{ f : U \to \mathbb{K} \; ; \; f \text{ is } C^\alpha\text{-function}\}, \qquad U \subseteq X \text{ open}.$$

Then $\mathcal{C}^\alpha_X$ is a sheaf of $\mathbb{K}$-algebras.

For $\alpha = 0$ this is the sheaf of continuous $\mathbb{K}$-valued functions on X. If $\mathbb{K} = \mathbb{C}$ and $\alpha \geq 1$, then $\mathcal{C}^\alpha_X$ is the sheaf of holomorphic functions on X and we will denote this sheaf usually by $\mathcal{O}_X$ or $\mathcal{O}^{\text{hol}}_X$. If $\mathbb{K} = \mathbb{R}$ and $1 \leq \alpha \leq \infty$, then $\mathcal{C}^\alpha_X$ is the sheaf of $\mathbb{R}$-valued α-fold continuously differentiable functions. If $\mathbb{K} = \mathbb{R}$ and $\alpha = \omega$, then $\mathcal{C}^\omega_X$ is the sheaf of $\mathbb{R}$-valued real analytic functions on X.

The same argument as for sheaves of continuous functions (Remark 3.17) yields the following observation: For all $x \in X$ the stalk $\mathcal{C}^\alpha_{X,x}$ is a local ring. In particular $(X, \mathcal{C}^\alpha_X)$ is a locally $\mathbb{K}$-ringed space. The maximal ideal $\mathfrak{m}_x$ of $\mathcal{C}^\alpha_{X,x}$ consists of germs of C^α-functions with $f(x) = 0$. The evaluation map $f \mapsto f(x)$ induces an isomorphism of $\mathbb{K}$-algebras $\kappa(x) := \mathcal{C}^\alpha_{X,x}/\mathfrak{m}_x \xrightarrow{\sim} \mathbb{K}$, where we consider $\mathbb{K}$ as $\mathbb{K}$-algebra via $\mathrm{id}_\mathbb{K}$.

Definition and Remark 4.3 (Morphisms of (locally) ringed spaces). Let $X = (X, \mathcal{O}_X)$ and $Y = (Y, \mathcal{O}_Y)$ be R-ringed spaces. A *morphism of R-ringed spaces $X \to Y$* is a pair $(f, f^\flat)$, where $f \colon X \to Y$ is a continuous map of the underlying topological spaces and where $f^\flat \colon \mathcal{O}_Y \to f_* \mathcal{O}_X$ is a homomorphism of sheaves of R-algebras on Y.

The datum of $f^\flat$ is equivalent to the datum of a homomorphism of sheaves of R-algebras $f^\sharp \colon f^{-1} \mathcal{O}_Y \to \mathcal{O}_X$ on X by Proposition 3.49. Usually we simply write f instead of $(f, f^\sharp)$ or $(f, f^\flat)$.

Morphisms of *locally* ringed spaces have to satisfy an additional property. To state this property, observe that a morphism $f \colon X \to Y$ of R-ringed spaces induces morphisms on the stalks as follows. Let $x \in X$. Using the identification $(f^{-1}\mathcal{O}_Y)_x = \mathcal{O}_{Y,f(x)}$ established in (3.12), we get a homomorphism of R-algebrasF17@$\mathcal{F}_x$, stalk of a sheaf

$$f_x := (f^\sharp)_x \colon \mathcal{O}_{Y,f(x)} \to \mathcal{O}_{X,x}.$$

By Remark 3.52 there is the following more explicit description of this homomorphism: For U an open neighborhood of $f(x)$ one has a map

$$\mathcal{O}_Y(U) \xrightarrow{f^\flat_U} \mathcal{O}_X(f^{-1}(U)) \to \mathcal{O}_{X,x}.$$

These maps induce the map an stalks $f_x \colon \mathcal{O}_{Y,f(x)} = \operatorname{colim} \mathcal{O}_Y(U) \to \mathcal{O}_{X,x}$.

Now let X and Y be locally R-ringed spaces. We define a *morphism of locally R-ringed spaces $X \to Y$* to be a morphism $(f, f^\flat)$ of ringed spaces such that the homomorphism of local rings $f^\sharp_x \colon \mathcal{O}_{Y,f(x)} \to \mathcal{O}_{X,x}$ is *local* (i.e., $f^\sharp_x(\mathfrak{m}_{f(x)}) \subseteq \mathfrak{m}_x$).

In general there exist locally ringed spaces and morphisms of ringed spaces between them that are not morphisms of *locally* ringed spaces (Problem 4.1). For spaces with functions of C^α-functions such as the premanifolds defined below in Definition 4.13 we will see that every morphism of ringed spaces is automatically a morphism of locally ringed spaces (see Example 4.5).

Remark 4.4. The composition of morphisms of (locally) R-ringed spaces is defined in the obvious way using the compatibility of direct images with composition (i.e., $(g \circ f)_* = g_* \circ f_*$, see Remark 3.45 (2)). We obtain the category of (locally) R-ringed spaces.

In general, $f^\flat$ (or $f^\sharp$) is an additional datum for a morphism. For instance it might happen that f is the identity but $f^\flat$ is not an isomorphism of sheaves (Problem 4.5). We will usually encounter the simpler case that the structure sheaf is a sheaf of functions on open subsets of X and that $f^\flat$ is given by composition with f. The following special case and its globalization (Proposition 4.18) is the main example.

Example 4.5 (Principal example). Let $X \subseteq V$ and $Y \subseteq W$ be open subsets of finite-dimensional $\mathbb{K}$-vector spaces V and W. Let $\alpha \in \widehat{\mathbb{N}}_0$.

Every C^α-map $f \colon X \to Y$ defines by composition a morphism of locally $\mathbb{K}$-ringed spaces $(f, f^\flat) \colon (X, \mathcal{C}_X^\alpha) \to (Y, \mathcal{C}_Y^\alpha)$ by

$$f_U^\flat \colon \mathcal{C}_Y^\alpha(U) \to f_*(\mathcal{C}_X^\alpha)(U) = \mathcal{C}_X^\alpha(f^{-1}(U)), \qquad t \mapsto t \circ f$$

for $U \subseteq Y$ open.

The induced map on stalks $f_x \colon \mathcal{C}_{Y, f(x)}^\alpha \to \mathcal{C}_{X,x}^\alpha$ is then also given by composing a $\mathbb{K}$-valued C^α-function t, defined in some neighborhood of $f(x)$, with f, which yields a $\mathbb{K}$-valued C^α-function $t \circ f$ defined in some neighborhood of x.

Conversely, let $(f, f^\flat) \colon (X, \mathcal{C}_X^\alpha) \to (Y, \mathcal{C}_Y^\alpha)$ be any morphism of $\mathbb{K}$-ringed spaces. We claim:

1. $(f, f^\flat)$ is automatically a morphism of *locally* $\mathbb{K}$-ringed spaces.
2. For all $U \subseteq Y$ open the $\mathbb{R}$-algebra homomorphism $f_U^\flat \colon \mathcal{C}_Y^\alpha(U) \to \mathcal{C}_X^\alpha(f^{-1}(U))$ is automatically given by the map $t \mapsto t \circ f$. Note that then f is a C^α-map (choose a basis of W; considering for t projections to the coordinates shows that each component of f is a C^α-map).

To show 1 let $x \in X$. Set $\varphi := f_x^\sharp$, $B := \mathcal{C}_{X,x}^\alpha$, and $A := \mathcal{C}_{Y, f(x)}^\alpha$. Then $\varphi \colon A \to B$ is a homomorphism of local $\mathbb{K}$-algebras such that $A/\mathfrak{m}_A = \mathbb{K}$ and $B/\mathfrak{m}_B = \mathbb{K}$. We claim that φ is automatically local, equivalently that $\varphi^{-1}(\mathfrak{m}_B)$ is a maximal ideal of A. Indeed, φ induces an injective homomorphism of $\mathbb{K}$-algebras

$$A/\varphi^{-1}(\mathfrak{m}_B) \hookrightarrow B/\mathfrak{m}_B = \mathbb{K}.$$

As a homomorphism of $\mathbb{K}$-algebras, it is automatically surjective, hence $A/\varphi^{-1}(\mathfrak{m}_B) \cong \mathbb{K}$ is a field and hence $\varphi^{-1}(\mathfrak{m}_B)$ is a maximal ideal.

Let us show 2. Let $U \subseteq Y$ be open and $x \in f^{-1}(U)$. Consider the commutative diagram of $\mathbb{K}$-algebra homomorphisms

$$
\begin{array}{ccc}
\mathcal{C}_Y^\alpha(U) & \xrightarrow{\;f_U^\flat\;} & \mathcal{C}_X^\alpha(f^{-1}(U)) \\
{\scriptstyle t \mapsto t_{f(x)}} \downarrow & & \downarrow {\scriptstyle s \mapsto s_x} \\
\mathcal{C}_{Y, f(x)}^\alpha & \xrightarrow{\;f_x^\sharp\;} & \mathcal{C}_{X,x}^\alpha \\
{\scriptstyle \mathrm{ev}_{f(x)} \colon t \mapsto t(f(x))} \downarrow & & \downarrow {\scriptstyle \mathrm{ev}_x \colon s \mapsto s(x)} \\
\mathbb{K} & & \mathbb{K}.
\end{array}
$$

The evaluation maps are surjective. Hence there exists a homomorphism of $\mathbb{K}$-algebras $\iota \colon \mathbb{K} \to \mathbb{K}$ making the lower rectangle commutative if and only if one has $f_x^{\sharp}(\ker(\mathrm{ev}_{f(x)}))$ $\subseteq \ker(\mathrm{ev}_x)$. But this latter condition is satisfied because $f_x^{\sharp}$ is local by 1. Moreover, as a homomorphism of $\mathbb{K}$-algebras, one must have $\iota = \mathrm{id}_{\mathbb{K}}$. Therefore we find $f_U^{\flat}(t)(x) = t(f(x))$, which shows (b).

Remark 4.6. A morphism $f \colon X \to Y$ of R-ringed spaces is an isomorphism in the category of R-ringed spaces if and only if f is a homeomorphism and $f_x \colon \mathcal{O}_{Y,f(x)} \to \mathcal{O}_{X,x}$ is an isomorphism of R-algebras for all $x \in X$.

Indeed, $(f, f^{\flat})$ is an isomorphism if and only if f is a homeomorphism and $f^{\flat}$ is an isomorphism of sheaves of rings. And the description of f_x in Remark 4.3 shows that if f is a homeomorphism, then $f^{\flat}$ is an isomorphism if and only if f_x is an isomorphism for all $x \in X$.

Next we define the notion of an open embedding of ringed spaces and of a local isomorphism of ringed spaces.

Remark and Definition 4.7. Let X be a (locally) R-ringed space and let $U \subseteq X$ be open. Then $(U, \mathcal{O}_{X|U})$ is a (locally) R-ringed space, which we usually denote simply by U. Such a (locally) R-ringed space is called an *open subspace of X*. There is an *inclusion morphism* $i \colon U \to X$ of (locally) R-ringed spaces, where the continuous map $i \colon U \to X$ is the inclusion of the underlying topological spaces and where $i^{\flat}$ is given by the restriction $\mathcal{O}_X(V) \to i_*(\mathcal{O}_{X|U})(V) = \mathcal{O}_X(U \cap V)$ for all $V \subseteq X$ open. Then $i^{\sharp} \colon i^{-1}\mathcal{O}_X \to \mathcal{O}_{X|U}$ is the identity. In particular i_x is the identity for all $x \in U$.

For a morphism $f \colon X \to Y$ of (locally) R-ringed spaces we denote by $f_{|U} \colon U \to Y$ the composition $f \circ i$ of morphisms of (locally) ringed spaces.

Finally, a morphism $j \colon (Z, \mathcal{O}_Z) \to (X, \mathcal{O}_X)$ of (locally) R-ringed spaces is called *open embedding* if $U := j(Z)$ is open in X and j induces an isomorphism $(Z, \mathcal{O}_Z) \xrightarrow{\sim} (U, \mathcal{O}_{X|U})$.

Definition 4.8. A morphism $f \colon X \to Y$ of (locally) R-ringed spaces is called a *local isomorphism* if there exists an open covering $(U_i)_i$ of X such that $f_{|U_i} \colon U_i \to Y$ is an open embedding for all i.

In other words, a morphism f is a local isomorphism if there exists an open covering $(U_i)_i$ of X and for all $i \in I$ an open subspace V_i of Y such that f induces for all i an isomorphism $U_i \xrightarrow{\sim} V_i$ of ringed spaces.

Remark 4.9. Remark 4.6 shows that a morphism $f \colon X \to Y$ of R-ringed spaces is a local isomorphism if and only if f is a local homeomorphism and $f_x \colon \mathcal{O}_{Y,f(x)} \to \mathcal{O}_{X,x}$ is an isomorphism.

Example 4.10. Let $X \subseteq V$ and $Y \subseteq W$ be open subsets of finite-dimensional $\mathbb{K}$-vector spaces V and W, let $\alpha \in \widehat{\mathbb{N}}$, and let $f: X \to Y$ be a C^α-map, which we consider as a morphism of locally $\mathbb{K}$-ringed spaces $(X, \mathcal{C}_X^\alpha) \to (Y, \mathcal{C}_Y^\alpha)$ (Example 4.5). Then the inverse function theorem (Appendix 16.16) means that f is a local isomorphism if and and only if for all $x \in X$ the derivative $Df(x): V \to W$ is a $\mathbb{K}$-linear isomorphism.

For $\alpha \geq \infty$ it will follow from Proposition 5.11 that $Df(x)$ is an isomorphism if f_x is an isomorphism.

We conclude our short introduction to ringed spaces by generalizing the gluing procedure from topological spaces to (locally) ringed spaces. Suppose we have given a family $(U_i)_{i \in I}$ of locally R-ringed spaces, for all $i, j \in I$ an open subset $U_{ij} \subseteq U_i$ (considered as open R-ringed subspace of U_i), and for all $i, j \in I$ an isomorphism $\varphi_{ji}: U_{ij} \to U_{ji}$ of locally R-ringed spaces such that

(a) $U_{ii} = U_i$ for all $i \in I$,
(b) the *cocycle condition* holds: $\varphi_{kj} \circ \varphi_{ji} = \varphi_{ki}$ on $U_{ij} \cap U_{ik}$, $i, j, k \in I$.

In the cocycle condition we implicitly assume that $\varphi_{ji}(U_{ij} \cap U_{ik}) \subseteq U_{jk}$, such that the composition is meaningful. The cocycle condition implies for $i = j = k$ that $\varphi_{ii} = \mathrm{id}_{U_i}$, for $i = k$ that $\varphi_{ij}^{-1} = \varphi_{ji}$, and that φ_{ji} induces an isomorphism $U_{ij} \cap U_{ik} \xrightarrow{\sim} U_{ji} \cap U_{jk}$.

We call such a datum a *gluing datum of locally ringed spaces*.

Proposition 4.11. *There exists a locally R-ringed space X together with morphisms $\psi_i: U_i \to X$, such that:*

(a) *For all i the map ψ_i is an open embedding.*
(b) *$\psi_j \circ \varphi_{ji} = \psi_i$ on U_{ij} for all i, j.*
(c) *$X = \bigcup_i \psi_i(U_i)$.*
(d) *$\psi_i(U_i) \cap \psi_j(U_j) = \psi_i(U_{ij}) = \psi_j(U_{ji})$ for all $i, j \in I$.*

Furthermore, $(X, (\psi_i)_i)$ has the following universal property (which determines it uniquely up to unique isomorphism): If Z is a locally R-ringed space, and for all $i \in I$, $\xi_i: U_i \to Z$ is a morphism of locally R-ringed spaces, such that $\xi_j \circ \varphi_{ji} = \xi_i$ on U_{ij} for all $i, j \in I$, then there exists a unique morphism $\xi: X \to Z$ with $\xi \circ \psi_i = \xi_i$ for all $i \in I$.

In other words, the R-ringed space X is a colimit in the category of R-ringed spaces.

Proof. By forgetting the structure sheaves in the gluing datum, we obtain a gluing datum of topological spaces, which we glue to a topological space X (Appendix Proposition 12.27). It is endowed with open topological embeddings $\psi_i: U_i \to X$, we have

$X = \bigcup_i \psi_i(U_i)$, and a subset $U \subseteq X$ is open if and only if for all i the preimage $\psi_i^{-1}(U)$ is open in U_i.

To obtain an R-locally ringed space, we have to "glue" the structure sheaves on the U_i so as to define a sheaf $\mathcal{O}_X$ of R-algebras on X. The sheaf $\mathcal{O}_X$ is uniquely determined by its sections (and the corresponding restriction maps) on a basis of the topology (Remark 3.13). It is thus sufficient to define it on those open subsets $U \subseteq X$ that are contained in one of the $\psi_i(U_i)$, and to check that this is well defined and satisfies the sheaf axioms. For each such U, we fix an i with $U \subseteq \psi_i(U_i)$, and we set $\mathcal{O}_X(U) = \mathcal{O}_{U_i}(\psi_i^{-1}(U))$. If $U \subseteq U_i \cap U_j$, then we identify the rings $\mathcal{O}_{U_i}(\psi_i^{-1}(U))$ and $\mathcal{O}_{U_j}(\psi_j^{-1}(U))$ with $\mathcal{O}_{U_{ij}}(U)$ via φ_{ji}. This allows us to define restriction maps. We obtain a sheaf $\mathcal{O}_X$ of R-algebras on X that is independent of our choices. Since all the U_i are locally R-ringed spaces, the same is true for X.

Furthermore, with this definition the ψ_i are morphisms of locally R-ringed spaces; they identify U_i with $(\psi_i(U_i), \mathcal{O}_{X|\psi_i(U_i)})$.

The universal property follows from the universal property of the construction of the underlying topological space (Appendix Proposition 12.27), which yields a continuous map $\xi: X \to Z$, and that it suffices to give a morphism of sheaves $\xi^\sharp: \xi^{-1}\mathcal{O}_Z \to \mathcal{O}_X$ for those open sets that are contained in one of the $\psi_i(U_i)$ (Remark 3.13). □

Example 4.12 (Disjoint union). As a (trivial) special case of this construction we have the disjoint union of locally R-ringed spaces. We simply let $U_{ij} = \emptyset$ for all i, j, so the underlying topological space is indeed the sum of the topological spaces of the U_i. The structure sheaf is the unique sheaf of rings whose restriction to U_i is $\mathcal{O}_{U_i}$. We denote the disjoint union by $\coprod_{i \in I} U_i$ and call it the *sum* or *coproduct*. The universal property of the glued space means that this is indeed a coproduct in the category of locally R-ringed spaces.

Proposition 4.11 and Example 4.12 are also valid in the category of arbitrary R-ringed spaces (with the same proof).

4.2 Premanifolds and Manifolds

In this section we fix $\alpha \in \widehat{\mathbb{N}}_0 = \mathbb{N}_0 \cup \{\infty, \omega\}$. We now define premanifolds as locally $\mathbb{K}$-ringed spaces that are locally isomorphic to an open subspace of $\mathbb{K}^m$ endowed with either (for $\mathbb{K} = \mathbb{R}$) the sheaf of $\mathbb{R}$-valued C^α-functions or (for $\mathbb{K} = \mathbb{C}$) the sheaf of holomorphic functions.

Definition 4.13.

1. A locally $\mathbb{R}$-ringed space $(M, \mathcal{O}_M)$ is called *(real) C^α-premanifold* if there exists an open covering $M = \bigcup_{i \in I} U_i$ such that for all $i \in I$ there exist $m \in \mathbb{N}_0$, an open

subspace Y of $\mathbb{R}^m$, and an isomorphism of locally $\mathbb{R}$-ringed spaces $(U_i, \mathcal{O}_{M|U_i}) \xrightarrow{\sim}$ $(Y, \mathcal{C}_Y^\alpha)$ (called *chart*). Here m and Y may depend on i. In this case the structure sheaf is denoted by $\mathcal{C}_M^\alpha$ and is called the *sheaf of C^α-functions on M*.

We call a real C^∞-premanifold a *smooth premanifold* and a real C^ω-premanifold a *real analytic premanifold*.

2. A *morphism of C^α-premanifolds* $(M, \mathcal{O}_M) \to (N, \mathcal{O}_N)$ is defined as a morphism of locally $\mathbb{R}$-ringed spaces. Such a morphism is also called a *C^α-map*. Again, C^∞-maps are also called *smooth*, and C^ω-maps are called *real analytic maps*.

We obtain the category of C^α-premanifolds. A (local) isomorphism in the category of C^α-premanifolds is called a *(local) C^α-diffeomorphism*.

Definition 4.14.

1. A $\mathbb{C}$-ringed space $(M, \mathcal{O}_M)$ is called *complex premanifold* if there exists an open covering $M = \bigcup_{i \in I} U_i$ such that for all $i \in I$ there exist $m \in \mathbb{N}_0$, an open subspace Y of $\mathbb{C}^m$ (both dependent on i) and an isomorphism of locally $\mathbb{C}$-ringed spaces $(U_i, \mathcal{O}_{M|U_i}) \xrightarrow{\sim} (Y, \mathcal{O}_Y^{\text{hol}})$ (again called *chart*).

2. A *morphism of complex premanifolds* $(M, \mathcal{O}_M) \to (N, \mathcal{O}_N)$ is defined as a morphism of locally $\mathbb{C}$-ringed spaces. Such a morphism is also called a *holomorphic map*.

Again we obtain the category of complex premanifolds. A (local) isomorphism in the category of complex premanifolds is called a *(locally) biholomorphic map*.

Definition 4.15. A *(real) C^α-manifold* (respectively a *complex manifold*) is a C^α-premanifold (respectively a complex premanifold) whose underlying topological space is Hausdorff and second countable. A *morphism of manifolds* is a morphism of premanifolds.

Terminology: To ease the handling of the different types of (pre)manifolds we will use the following terminology. A *(pre)manifold* is

- either a real C^α-(pre)manifold where α will be always an element in $\hat{\mathbb{N}}_0$; in this case we set $\mathbb{K} := \mathbb{R}$. This will be called the *real case* or that the (pre)manifold is of *real type*;
- or a complex (pre)manifold. In this case we set $\mathbb{K} := \mathbb{C}$. This will be called the *complex case* or that the (pre)manifold is of *complex type*.

If not otherwise stated, the type of the premanifold is always the same within some statement. For instance, if we speak of a morphism f of (pre)manifolds, then both (pre)manifolds are real C^α and f is a C^α-map (with the same α everywhere) or both are complex (pre)manifolds and f is holomorphic.

All categories of real C^α-(pre)manifolds and complex (pre)manifolds will sometimes be generically denoted by (PMfd) respectively (Mfd).

If M is a premanifold of arbitrary type, we denote its structure sheaf by $\mathcal{O}_M$. A chart is usually denoted by (U, x) with $U \subseteq M$ open, $x: U \to \mathbb{K}^m$ an open embedding. Its components are denoted by $x^i: U \to \mathbb{K}$, $i = 1, \ldots, m$. We denote by $x^{-1}: x(U) \to U$ the inverse isomorphism of manifolds. For real C^α-premanifolds M the structure sheaf will also be denoted by $\mathcal{C}_M^\alpha$.

Remark and Definition 4.16 (Open submanifolds). Let M be a premanifold and let $U \subseteq M$ be an open subspace (Definition 4.7). Then $(U, \mathcal{O}_{M|U})$ is a premanifold, called an *open subpremanifold of M*. If M is a manifold, then $(U, \mathcal{O}_{M|U})$ is a manifold, called *open submanifold of M*, because every subspace of a Hausdorff (respectively a second countable) space is again Hausdorff (respectively second countable).

Description of the Structure Sheaf and of Morphisms of Premanifolds
We will now see that the structure sheaf of a premanifold is the sheaf of $\mathbb{K}$-valued morphisms of premanifolds and any morphism between premanifolds is given by the underlying map of topological spaces.

Let M and N be premanifolds. We denote by $\mathcal{O}_{M;N}$ the sheaf on M of N-valued morphisms, i.e., for an open subset U of M, $\mathcal{O}_{M;N}(U)$ is the set of all morphisms $U \to N$ of premanifolds. The restriction maps are given by the restriction of morphisms of locally $\mathbb{K}$-ringed spaces to open subspaces (Remark 4.7).

Let M be a premanifold. We claim that we have for its structure sheaf $\mathcal{O}_M = \mathcal{O}_{M;\mathbb{K}}$. More precisely, for every open subspace U of M, a morphism $t: U \to \mathbb{K}$ of premanifolds yields an element $t_{\mathbb{K}}^\flat(\mathrm{id}_{\mathbb{K}}) \in \mathcal{O}_M(U)$, where $\mathrm{id}_{\mathbb{K}} \in \mathcal{O}_{\mathbb{K}}(\mathbb{K})$ is the identity.

Proposition 4.17. *The map $t \mapsto t_{\mathbb{K}}^\flat(\mathrm{id}_{\mathbb{K}})$ defines an isomorphism of sheaves $\mathbb{K}$-algebras*

$$\iota: \mathcal{O}_{M;\mathbb{K}} \xrightarrow{\sim} \mathcal{O}_M.$$

Proof. Indeed, ι is clearly a morphism of sheaves of $\mathbb{K}$-algebras. To see that ι is an isomorphism, we may work locally (Remark 3.23) and hence we may assume that M has a global chart $x: M \xrightarrow{\sim} U \subseteq \mathbb{K}^m$. By composing with the isomorphism x^{-1}, we may assume that $M \subseteq \mathbb{K}^m$ is open. In this case, ι is an isomorphism by definition. $\square$

Next we show that morphisms between premanifolds M and N are given be the underlying continuous maps.

Proposition 4.18. *A morphism of premanifolds $f: M \to N$ is the same as a continuous map $f: M \to N$ satisfying the following condition:*

$$t \in \mathcal{O}_N(V) \Rightarrow t \circ f \in \mathcal{O}_M(f^{-1}(V)) \qquad \text{for all } V \subseteq N \text{ open.} \qquad (*)$$

Proof. Let $f: M \to N$ be a morphism of premanifolds. We have to show that $f^\flat$ is given by (*). As the equality of two morphisms of sheaves can be checked on stalks (Proposition 3.18), we have to see that $f_x: \mathcal{O}_{N,f(x)} \to \mathcal{O}_{M,x}$ is given for all $x \in X$ by composing a germ of a $\mathbb{K}$-valued function at $f(x)$ with f. This can again be done locally on M and N and hence we may assume that $M \subseteq \mathbb{K}^m$ and $N \subseteq \mathbb{K}^n$ are open. In this case we have seen in Example 4.5 that f_x has the desired form.

Conversely, let $f: M \to N$ be a continuous map satisfying (*), then $t \mapsto t \circ f$ defines by Proposition 4.17 a morphism $f^\flat: \mathcal{O}_N \to f_* \mathcal{O}_M$ of sheaves of $\mathbb{K}$-algebras and hence a morphism of premanifolds. $\qquad\square$

Dimension of Premanifolds

Let M be a premanifold and let $p \in M$. An integer $m \geq 0$ is called *dimension of M in p* (denoted $\dim_p(M)$) if there exists an open neighborhood U of p and a chart $x: U \xrightarrow{\sim} Y$, where $Y \subseteq \mathbb{K}^m$ is open.

The dimension is uniquely determined. Let $y: V \xrightarrow{\sim} Z \subseteq \mathbb{K}^n$ be a second chart and consider the change of charts

$$\varphi: x_{|U \cap V} \circ y^{-1}_{|y(U \cap V)}: y(U \cap V) \xrightarrow{\sim} x(U \cap V),$$

which is an isomorphism of manifolds of an open subset of $\mathbb{K}^n$ onto an open subset of $\mathbb{K}^m$. We claim that the existence of such an isomorphism implies $m = n$. Indeed in the complex case and in the real case for $\alpha \geq 1$ this is clear because we may form the derivative of φ in a point and obtain a $\mathbb{K}$-linear isomorphism $\mathbb{K}^n \xrightarrow{\sim} \mathbb{K}^m$. In the real case with $\alpha = 0$ this is more difficult but a standard result from algebraic topology (for instance [tDi] (10.3.8)).

The function

$$M \longrightarrow \mathbb{N}_0, \qquad p \mapsto \dim_p(M)$$

is by definition locally constant. If it is constant $= m$ (for instance if M is connected), then we say that M *has dimension m* and write $m = \dim M$.

Example 4.19 (Premanifolds of dimension 0). Let M be a set and consider it as a discrete topological space. Let $\mathcal{O}_M$ be the sheaf of all $\mathbb{K}$-valued maps on M, i.e., $\mathcal{O}_M(U)$ is the $\mathbb{K}$-algebra of all maps $U \to \mathbb{K}$ for every subset $U \subseteq M$ (automatically open). Then $(M, \mathcal{O}_M)$ is a C^α-premanifold for all $\alpha \in \hat{\mathbb{N}}_0$ for $\mathbb{K} = \mathbb{R}$ (respectively a complex manifold for $\mathbb{K} = \mathbb{C}$) of dimension 0.

Conversely, every 0-dimensional premanifold is of this form. Such a premanifold is automatically Hausdorff. It is second countable if and only if M is countable as a set.

Let M be any premanifold and let M_d be the underlying set of M endowed with the discrete topology and consider M_d as a 0-dimensional premanifold (of the same type as M) as explained above. Then $\mathrm{id}_M: M_d \to M$ is a morphism of premanifolds.

Premanifold Structure Defined by an Atlas

An important alternative (and more classical) way to construct premanifolds is by way of an atlas as follows.

Figure 4.1 Charts of a premanifold

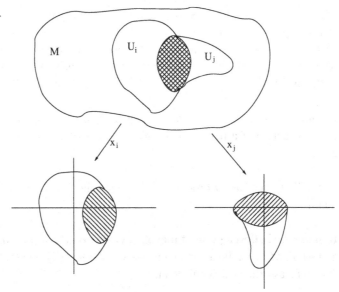

Let M be a topological space. Suppose that there exists an open covering $M = \bigcup_{i \in I} U_i$ and for all $i \in I$ homeomorphisms $x_i : U_i \xrightarrow{\sim} Y_i$ with $Y_i \subseteq \mathbb{K}^{n_i}$ open satisfy-ing the following condition. For all $i, j \in I$ the homeomorphism (*change of charts*)

$$x_i(U_i \cap U_j) \xrightarrow{x_i^{-1}} U_i \cap U_j \xrightarrow{x_j} x_j(U_i \cap U_j)$$

is a C^α-diffeomorphism (in the real case) respectively a biholomorphic map (in the com-plex case), see Fig. 4.1. Such a family $(U_i, x_i)_i$ is called an *atlas of M*. It defines the structure of a premanifold on M as follows. For $V \subseteq M$ open define $\mathcal{O}_M(V)$ as the set of maps $f : V \to \mathbb{K}$ such that $f_{|U_i \cap V} \circ x_i^{-1} : x_i(U_i \cap V) \to \mathbb{K}$ is C^α respectively holomor-phic for all i. This defines a sheaf of $\mathbb{K}$-algebras on M and we obtain a $\mathbb{K}$-ringed space $(M, \mathcal{O}_M)$.

To see that this is a premanifold, fix $i_0 \in I$. For $V \subseteq U_{i_0}$ open let $f : V \to \mathbb{K}$ be a map such that $f \circ x_{i_0}^{-1} : x_{i_0}(U_{i_0} \cap V) \to \mathbb{K}$ is a function (C^α respectively holo-morphic). Then $f \in \mathcal{O}_M(V)$ because the change of charts between i and i_0 are C^α-diffeomorphisms (respectively biholomorphic). Therefore x_{i_0} yields an isomorphism $(U_{i_0}, \mathcal{O}_{M|U_{i_0}}) \xrightarrow{\sim} (Y_{i_0}, \mathcal{O}_{Y_{i_0}})$, where $\mathcal{O}_{Y_{i_0}}$ is the sheaf of C^α respectively holomorphic func-tions on Y_{i_0}. Hence $(M, \mathcal{O}_M)$ is a ringed space that is locally isomorphic to a manifold. Hence it is a premanifold.

Conversely, if $(M, \mathcal{O}'_M)$ is a premanifold, we may choose a family of charts $(x_i : (U_i, \mathcal{O}'_{M|U_i}) \xrightarrow{\sim} (Y_i, \mathcal{O}_{Y_i}))_{i \in I}$ with $Y_i \subseteq \mathbb{K}^{n_i}$ open and with $\bigcup_{i \in I} U_i = M$. We obtain an atlas $(U_i, x_i)_i$ of M defining a premanifold structure $\mathcal{O}_M$ as above. For $V \subseteq M$ open, $\mathcal{O}_M(V)$ is the $\mathbb{K}$-algebra of maps $f : V \to \mathbb{K}$ such that $f_{|V \cap U_i}$ is a morphism of premanifolds $V \cap U_i \to \mathbb{K}$. Hence f is a morphism of premanifolds. This shows $\mathcal{O}_M = \mathcal{O}'_M$ by Proposition 4.17.

Altogether we see that for a topological space M the structure of a premanifold on M can be either given by defining the structure sheaf or by specifying an atlas for M. Different atlases giving rise to the same structure sheaf are called *equivalent*.

Example 4.20. If $U \subseteq \mathbb{K}^m$ is an open subspace, then (U, id_U) is an atlas defining on U the structure of a real C^α-manifold for all α (if $\mathbb{K} = \mathbb{R}$) or a complex manifold (if $\mathbb{K} = \mathbb{C}$).

Similarly one has by Proposition 4.18 the following description of morphisms of premanifolds.

Remark 4.21 (Description of morphisms via an atlas). Let M and N be premanifolds and let $f: M \to N$ be a continuous map of the underlying topological spaces. Then the following assertions are equivalent:

(i) The continuous map f is a morphism of premanifolds (Proposition 4.18).
(ii) For every chart (V, y) of N and for every chart (U, x) of M the composition

$$f_{(U,x),(V,y)}: x(f^{-1}(V) \cap U) \xrightarrow{x^{-1}} f^{-1}(V) \cap U \xrightarrow{f} V \xrightarrow{y} y(V)$$

is a C^α-map (respectively a holomorphic map).
(iii) There exists an atlas $(U_i, x_i)_i$ of M and an atlas $(V_j, y_j)_j$ of N such that the composition $f_{(U_i,x_i),(V_j,y_j)}$ is a C^α-map (respectively a holomorphic map) for all i and j.

Definition and Remark 4.22 (Charts of morphisms). Let M and N be premanifolds, let $F: M \to N$ be a continuous map and let $p \in M$. A *chart of F at p* consists of open neighborhoods $F(p) \in V \subseteq N$ and $p \in U \subseteq M$ and a commutative diagram of continuous map

$$
\begin{array}{ccc}
U & \xrightarrow{F_{|U}} & V \\
{\scriptstyle x}\downarrow{\scriptstyle \cong} & & {\scriptstyle \cong}\downarrow{\scriptstyle y} \\
\tilde{U} & \xrightarrow{\tilde{F}} & \tilde{V},
\end{array}
$$

where $x: U \xrightarrow{\sim} \tilde{U} \subseteq \mathbb{K}^m$ and $y: V \to \tilde{V} \subseteq \mathbb{K}^n$ are charts, $m = \dim_p(M)$, $n = \dim_{F(q)}(N)$.

Such a chart always exists: Choose a chart (V, y) at $F(p)$ and a chart (U, x) at p with $U \subseteq F^{-1}(V)$ and define $\tilde{F} := y \circ F_{|U} \circ x^{-1}$.

In addition one may assume:

1. $x(p) = 0$ and $y(F(p)) = 0$ (by composing x and y with a translation).
2. $\tilde{U} = B_1(0)$ and $\tilde{V} = B_1(0)$ for any chosen norms on $\mathbb{K}^m$ and $\mathbb{K}^n$ (first shrink $\tilde{V}$, then $\tilde{U}$ so that we may assume $\tilde{U} = B_\delta(0)$ and $\tilde{V} = B_\varepsilon(0)$, then compose x and y with multiplication by $1/\delta$ and $1/\varepsilon$, respectively).

In Theorem 5.22 we will give an important criterion for a morphism F to have a chart $(U, x, V, y, \tilde{F})$ where $\tilde{F}$ is the restriction of a linear map.

Changing of Structure

Remark 4.23 (Weakening of structure). Let $\alpha, \beta \in \widehat{\mathbb{N}}_0$ with $\alpha \leq \beta$ and let M be a real C^β-premanifold. Then every atlas of M is also an atlas of a C^α-structure on M because every C^β-change of charts is also a C^α-diffeomorphism. We obtain on M the structure of a C^α-premanifold M_α independent of the choice of the atlas. This structure is called the *underlying C^α-structure of M*. If M is a manifold, so is M_α because the underlying topological spaces are the same.

If M is an open submanifold of a finite-dimensional real vector space V, then $\mathcal{O}_{M_\alpha}$ is the sheaf of $\mathbb{R}$-valued C^α-functions on M (take the atlas consisting of a single chart $M \hookrightarrow V \cong \mathbb{K}^m$).

Every morphism of C^β-premanifolds $F: M \to N$ is also a morphism $F_\alpha: M_\alpha \to N_\alpha$ of C^α-premanifolds and we obtain a faithful functor $M \mapsto M_\alpha$ from the category of C^β-premanifolds to the category of C^α-manifolds. The inclusion $\iota: \mathcal{C}_M^\beta \to \mathcal{C}_{M_\alpha}^\alpha$ defines a morphism $(\mathrm{id}_M, \iota): M_\alpha \to M$ of locally $\mathbb{R}$-ringed spaces, called *canonical*, which is functorial in M.

Let $\beta' \in \widehat{\mathbb{N}}_0$ and let M' be a real $C^{\beta'}$-premanifold. Suppose $\alpha \leq \min\{\beta, \beta'\}$. Then a C^α-*map* $M \to M'$ is defined to be a continuous map $M \to M'$ that is a morphism of C^α-premanifolds $M_\alpha \to M'_\alpha$.

Conversely, one can show that for $\alpha \in \widehat{\mathbb{N}}$ and for every real C^α-premanifold M there exists for all $\beta \geq \alpha$ the structure of a C^β-premanifold $\tilde{M}$ on M such that $\tilde{M}_\alpha = M$ ([KoPu]). If M is a manifold, then any two such C^β-manifolds are isomorphic ([Whi2] and [Gra]). Both the existence and the uniqueness result are wrong for $\alpha = 0$ ([Mil], [Ker]). The uniqueness result is also wrong if M is not second countable or not Hausdorff (for all $\alpha, \beta \in \widehat{\mathbb{N}}$ with $\alpha < \beta$; see [KoPu] and Problem 4.11).

Remark 4.24 (Complex manifolds as real manifolds). Let M be a complex premanifold. By identifying $\mathbb{C}^m$ with $\mathbb{R}^{2m}$ we may consider every chart $x: U \to \mathbb{C}^m$ of M as a chart $x_\mathbb{R}: U \to \mathbb{R}^{2m}$. As every complex analytic map between open subsets of $\mathbb{C}^m$ is in particular real analytic, we see that for an atlas $(U_i, x_i)_i$ of M we obtain the structure of a real analytic premanifold $M_{\mathbb{R}}$ defined by the atlas $(U_i, (x_i)_{\mathbb{R}})_i$. This structure is independent of the choice of atlas and we call $M_{\mathbb{R}}$ the *underlying real analytic premanifold*. If M is a manifold, so is $M_{\mathbb{R}}$ because the underlying topological spaces are the same.

Every holomorphic map $F: M \to N$ between complex premanifolds is a real analytic map $F_{\mathbb{R}}: M_{\mathbb{R}} \to N_{\mathbb{R}}$ and we obtain a faithful functor $M \mapsto M_{\mathbb{R}}$ from the category of complex premanifolds to the category of real analytic premanifolds. The inclusions $\mathcal{O}_M \to \mathcal{O}_{M_{\mathbb{R}}:\mathbb{C}}$ (here we consider $\mathbb{C}$ as a real analytic manifold) and $\mathcal{O}_{M_{\mathbb{R}}} \to \mathcal{O}_{M_{\mathbb{R}}:\mathbb{C}}$ define functorial morphisms

$$M_{\mathbb{R}} \longleftarrow (M, \mathcal{O}_{M_{\mathbb{R}}:\mathbb{C}}) \longrightarrow M$$

of locally $\mathbb{R}$-ringed spaces (for the left arrow) and locally $\mathbb{C}$-ringed spaces (for the right arrow), which we call *canonical*.

A map $f: M \to N$ between complex premanifolds is called a *real C^α-map* ($\alpha \in \widehat{\mathbb{N}}_0$) if it is a C^α-map $M_{\mathbb{R}} \to N_{\mathbb{R}}$ (Remark 4.23).

4.3 Examples of Manifolds

Example 4.25 (1-dimensional real torus). Let $\mathbb{T}^1 := \mathbb{R}/\mathbb{Z}$ be the *1-dimensional real torus*. Endow $\mathbb{T}^1$ with the quotient topology with respect to the projection $\pi: \mathbb{R} \to \mathbb{T}^1$. Then $\bar{e}: \mathbb{T}^1 \to S^1 = \{z \in \mathbb{C} \; ; \; |z| = 1\}$, $\bar{e}(t) := \exp(2\pi i t)$ is a homeomorphism, in particular $\mathbb{T}^1$ is compact, Hausdorff, and second countable.

Define the structure of a real analytic manifold on $\mathbb{T}^1$ as follows:

1. We may use an atlas. Set $U_1 := \{x + \mathbb{Z} \in \mathbb{R}/\mathbb{Z} \; ; \; 0 < x < 1\}$ and $U_2 := \{x + \mathbb{Z} \in \mathbb{R}/\mathbb{Z} \; ; \; -1/2 < x < 1/2\}$. Then U_1 and U_2 are open in $\mathbb{T}^1$ and $U_1 \cup U_2 = \mathbb{T}^1$. Define

$$\Phi_1: U_1 \xrightarrow{\sim} (0, 1) \subseteq \mathbb{R}, \qquad x + \mathbb{Z} \mapsto x,$$
$$\Phi_2: U_2 \xrightarrow{\sim} (-1/2, 1/2) \subseteq \mathbb{R}, \qquad x + \mathbb{Z} \mapsto x.$$

Then Φ_1 and Φ_2 are homeomorphisms and the change of charts

$$(0, 1/2) = \Phi_1(U_1 \cap U_2) \xrightarrow{\Phi_2 \circ \Phi_1^{-1}} \Phi_2(U_1 \cap U_2) = (0, 1/2), \qquad x \mapsto x$$

is a real analytic diffeomorphism.

2. Alternatively, we can directly define the structure sheaf on $\mathbb{T}^1$. For $V \subseteq \mathbb{T}^1$ open define a sheaf of $\mathbb{R}$-algebras

$$\mathcal{C}^\omega_{\mathbb{T}^1}(V) := \left\{ f: V \to \mathbb{R} \; ; \; f \circ \pi: \pi^{-1}(V) \to \mathbb{R} \text{ is } C^\omega \right\}.$$

We obtain a $\mathbb{R}$-ringed space $\mathbb{T}^1$ that is a real analytic manifold. This can be checked locally on $\mathbb{T}^1$. Cover $\mathbb{T}^1$ by open subsets V such that there exists $U \subseteq \mathbb{R}$ with $\pi_U :=$

$\pi_{|U} \colon U \to V$ a homeomorphism (any open $V \subsetneq \mathbb{T}^1$ has this property). We claim that $\pi_{|U}$ is an isomorphism of $\mathbb{R}$-ringed spaces $(U, \mathcal{C}_U^\omega) \xrightarrow{\sim} (V, \mathcal{C}_{\mathbb{T}^1|V}^\omega)$. We have to show that for all $W \subseteq V$ open and $f \colon W \to \mathbb{R}$ map:

$$f \in \mathcal{C}_{\mathbb{T}^1}^\omega(W) :\Leftrightarrow f \circ \pi_{|\pi^{-1}(W)} \in \mathcal{C}_{\mathbb{R}}^\omega(\pi^{-1}(W))$$
$$\Leftrightarrow f \circ (\pi_{U|\pi_U^{-1}(W)}) \in \mathcal{C}_U^\omega(\underbrace{\pi_U^{-1}(W)}_{=U \cap W}).$$

But $\pi^{-1}(W) = \coprod_{n \in \mathbb{Z}}(U \cap W) + n$. Hence this equivalence is clear.

The projection $\pi \colon \mathbb{R} \to \mathbb{T}^1$ is a morphism of manifolds: Either one can use Proposition 4.18 and (2) or the description of morphisms via an atlas (Remark 4.21) using the atlas $(U_i, \Phi_i)_{i=1,2}$ defined above.

Example 4.26 (Projective space). Recall that for a field k the *projective n-space* is defined as follows:

$$\mathbb{P}^n(k) = (k^{n+1} \setminus \{0\})/\sim = \{L \subseteq k^{n+1} ; L \text{ 1-dim. sub-v.s.}\}$$
$$v \mapsto \langle v \rangle$$

where $v \sim v'$ if and only if there exists $\lambda \in k^\times$ such that $v' = \lambda v$. For $(x_0, \dots, x_n) \in k^{n+1} \setminus \{0\}$ we denote by $(x_0 : \dots : x_n)$ its equivalence class in $\mathbb{P}^n(k)$.

There is the following alternative description. Let

$$P := \{g = (a_{ij})_{0 \le i, j \le n} \in \mathrm{GL}_{n+1}(k) ; a_{10} = a_{20} = \dots = a_{n0} = 0\}$$

and let $L_0 = \langle e_0 \rangle = \langle (1, 0, \dots, 0) \rangle \in \mathbb{P}^n(k)$. Then

$$\mathrm{GL}_{n+1}(k)/P \xrightarrow{\sim} \mathbb{P}^n(k), \qquad g \mapsto g(L_0)$$

is bijective.

Now let $k = \mathbb{K}$. Endow $\mathbb{P}^n(\mathbb{K})$ with quotient topologies in three ways by surjective maps:

(i) $\mathbb{K}^{n+1} \setminus \{0\} \to \mathbb{P}^n(\mathbb{K}), (x_0, \dots, x_n) \mapsto (x_0 : \dots : x_n)$.

(ii) $S^n := \{x \in \mathbb{K}^{n+1} ; \|x\|_2 := \sum_{i=0}^n |x_i|^2 = 1\} \to \mathbb{P}^n(\mathbb{K})$,
$\quad (x_0, \dots, x_n) \mapsto (x_0 : \dots : x_n)$.

(iii) $\mathrm{GL}_{n+1}(\mathbb{K}) \to \mathbb{P}^n(\mathbb{K}), g \mapsto g(L_0)$.

Then we have a commutative diagram of continuous maps

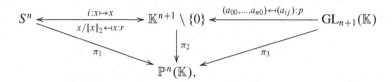

where the right horizontal map p is open (being the restriction of a projection map). Hence all three topologies on $\mathbb{P}^n(\mathbb{K})$ defined above are equal. Indeed, let $U \subseteq \mathbb{P}^n(K)$ be a subset. Then $\pi_1^{-1}(U) = i^{-1}(\pi_2^{-1}(U))$ is open if and only if $\pi_2^{-1}(U) = r^{-1}(\pi_1^{-1}(U))$ is open. Hence the quotient topologies of S^n and of $\mathbb{K}^{n+1} \setminus \{0\}$ are equal. Furthermore $\pi_2^{-1}(U) = p(\pi_3^{-1}(U))$ is open if and only if $\pi_3^{-1}(U) = p^{-1}(\pi_2^{-1}(U))$ is open. Therefore the quotient topologies of $\mathbb{K}^{n+1} \setminus \{0\}$ and of $\mathrm{GL}_{n+1}(\mathbb{K})$ are equal.

As P is closed in $\mathrm{GL}_n(\mathbb{K})$, Description (iii) shows that $\mathbb{P}^n(\mathbb{K})$ is Hausdorff (Appendix Proposition 12.58). By Description (ii), $\mathbb{P}^n(\mathbb{K})$ is compact because S^n is compact (Appendix Proposition 12.52).

We define the structure of a real analytic $\mathbb{R}$-manifold on $\mathbb{P}^n(\mathbb{R})$ (respectively a complex manifold on $\mathbb{P}^n(\mathbb{C})$) by an atlas $(U_i, \Phi_i)_{0 \leq i \leq n}$ with

$$\Phi_i \colon U_i := \{ (x_0 : \ldots : x_n) \in \mathbb{P}^n(\mathbb{K}) \; ; \; x_i \neq 0 \} \xrightarrow{\sim} \mathbb{K}^n,$$

$$(x_0 : \ldots : x_n) \mapsto \left(\frac{x_0}{x_i}, \ldots, \frac{x_{i-1}}{x_i}, \frac{x_{i+1}}{x_i}, \ldots, \frac{x_n}{x_i} \right).$$

Then $\Phi_i(U_i \cap U_j) = \big\{ (y_0, \ldots, y_{i-1}, y_{i+1}, \ldots, y_n) \; ; \; y_j \neq 0 \big\}$ and for $0 \leq i \neq j \leq n$ the change of charts is given by

$$\Phi_i(U_i \cap U_j) \to U_i \cap U_j$$
$$\to \Phi_j(U_i \cap U_j),$$
$$(y_0, \ldots, y_{i-1}, y_{i+1}, \ldots, y_n) \mapsto (y_0 : \ldots : y_{i-1} : 1 : y_{i+1} : \ldots : y_n)$$
$$\mapsto \left(\frac{y_0}{y_j}, \ldots, \underbrace{\frac{1}{y_j}}_{i}, \ldots, \frac{y_{j-1}}{y_j}, \frac{y_{j+1}}{y_j}, \ldots : \frac{y_n}{y_j} \right),$$

which is a real bianalytic (respectively a biholomorphic) map.

4.4   Topological Properties of (Pre)Manifolds

Remark 4.27. Let M be a premanifold.

1. Locally, M is homeomorphic to an open convex subspace of $\mathbb{R}^n$. Hence M is locally contractible (Definition 2.5), first countable, locally compact, and a T_1-space.

2. As M is locally contractible, it is in particular locally path connected. Hence in M connected components and path components are equal and all connected components are open and closed in M (Proposition 2.13).
3. If M has a countable atlas, then M is second countable (Remark 1.3 4). Therefore every Hausdorff premanifold with a countable atlas is a manifold.

As a premanifold also has an open covering consisting of simply connected subspaces, it is semilocally simply connected (Problem 2.19) and hence each connected premanifold has a universal covering space (Problem 2.20). We will not use this fact in the sequel. Proposition 4.31 then shows that the universal cover has a canonical structure of a premanifold.

Theorem 4.28. *Every manifold M has the following properties:*

1. *M is paracompact and Lindelöf. More precisely, for every open covering $(U_i)_i$ of M there exists a countable locally finite refinement $(V_n)_n$ such that for all n the closure $\overline{V_n}$ is compact and contained in some U_i.*
2. *There exists a sequence $(C_n)_{n \in \mathbb{N}}$ of compact subspaces of M such that $M = \bigcup_n C_n$ and $C_n \subseteq C_{n+1}^\circ$ for all n. In particular M is σ-compact.*
3. *M is normal.*

Moreover, Remark 1.14 (for which we did not give a proof) implies that every manifold is metrizable. We will not use this fact in the sequel.

Proof. All assertions of 1 hold by Proposition 1.10 except for the property that $\overline{V_n}$ is contained in some U_i for all n. This last property follows from the shrinking lemma 1.21. Assertion 2 also follows from Proposition 1.10. Finally, as M is paracompact and Hausdorff, M is normal by Proposition 1.18.   □

4.5   Basic Constructions and Further Examples of Manifolds

Products

Let M and N be premanifolds with atlases $(U_i, \Phi_i)_i$ of M and $(V_j, \Psi_j)_j$ of N. Then $(U_i \times V_j, \Phi_i \times \Psi_j)$ is an atlas on the product space $M \times N$ making $M \times N$ into a premanifold. In particular for $x \in M$ and $y \in N$ one has $\dim_{(x,y)}(M \times N) = \dim_x(M) + \dim_y(N)$.

If M and N are manifolds, then $M \times N$ is a manifold because the product of two Hausdorff (respectively second countable) spaces is again Hausdorff (respectively second countable).

Figure 4.2 Möbius band

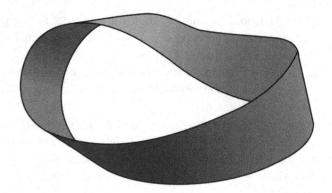

Example 4.29 (Real tori). In Example 4.25 we endowed $\mathbb{T}^1 := \mathbb{R}/\mathbb{Z}$ with the structure of a 1-dimensional real analytic manifold. Hence we obtain for all $n \geq 1$ the structure of an n-dimensional real analytic manifold on the *n-dimensional real torus* $\mathbb{T}^n := \mathbb{R}^n/\mathbb{Z}^n = (\mathbb{T}^1)^n$.

Gluing

Suppose that $((U_i)_{i \in I}, (U_{ij})_{i,j}, (\varphi_{ij})_{i,j})$ is a gluing datum of locally $\mathbb{K}$-ringed spaces such that the U_i are all premanifolds. Then the glued locally $\mathbb{K}$-ringed space (Proposition 4.11) is locally isomorphic to a premanifold. Hence it is a premanifold.

Example 4.30 (Möbius band). If $I = \{1, 2\}$ consists of two elements, then the non-redundant data of such a gluing datum consist of two premanifolds U_1 and U_2, open subpremanifolds $U_{12} \subseteq U_1$, $U_{21} \subseteq U_2$ and an isomorphism $\varphi \colon U_{12} \xrightarrow{\sim} U_{21}$.

For instance, identify $S^1 = \{ z \in \mathbb{C} \; ; \; |z| = 1 \}$. Then the (open) *Möbius band* is glued from the two rectangular strips $U_1 := (S^1 \setminus \{1\}) \times (-1, 1)$ and $U_2 := (S^1 \setminus \{-1\}) \times (-1, 1)$ along the open submanifolds $U_{12} := U_{21} := U_1 \cap U_2 = W \times (-1, 1)$ with $W = \{ z \in S^1 \; ; \; \text{Im}(z) \neq 0 \}$ using the real analytic isomorphism

$$U_{12} \ni (z, t) \mapsto \begin{cases} (z, t), & \text{if } \text{Im}(z) > 0; \\ (z, -t), & \text{if } \text{Im}(z) < 0, \end{cases} \in U_{21}$$

(see Fig. 4.2). We obtain a real analytic manifold.

Coverings of Manifolds

Proposition 4.31. *Let M be a premanifold, let $\tilde{M}$ be a topological space and let $f \colon \tilde{M} \to M$ be a local homeomorphism. Then there exists a unique structure of a premanifold on $\tilde{M}$ such that f is a local isomorphism of premanifolds.*

This proposition can in particular be applied if f is a covering map, for instance if $\tilde{M}$ is a universal covering of a connected premanifold M.

Proof. Existence: Define $\mathcal{O}_{\tilde{M}} := f^{-1}\mathcal{O}_M$ and let $f^{\sharp} := \mathrm{id}_{f^{-1}\mathcal{O}_M} : f^{-1}\mathcal{O}_M \to \mathcal{O}_{\tilde{M}}$. Then $(f, f^{\sharp}): \tilde{M} \to M$ is a morphism of $\mathbb{K}$-ringed spaces such that f is a local homeomorphism and such that f_x is an isomorphism. Hence it is a local isomorphism of $\mathbb{K}$-ringed spaces (Remark 4.9). Clearly, any $\mathbb{K}$-ringed space that is locally isomorphic to a premanifold is itself a premanifold (of the same type). Hence $\tilde{M}$ is a premanifold.

Uniqueness: If $\mathcal{O}'$ is another sheaf of $\mathbb{K}$-algebras defining a manifold structure on $\tilde{M}$ such that f is a local isomorphism, then $\mathrm{id}_{\tilde{M}}$ is a local isomorphism and a homeomorphism and hence an isomorphism. $\qquad\square$

Corollary 4.32. *We keep the notation of Proposition 4.31. If M is a manifold and f is a covering map with countable fibers, then $\tilde{M}$ is a manifold.*

One can show that if $\tilde{M}$ is a path connected topological space, then the fibers of any covering map $\pi: \tilde{M} \to M$ of a manifold M are automatically countable: By [LeeJo] Prop. 1.9 the fundamental group $\pi_1(M, p)$ is countable for every $p \in M$ and by Remark 2.40 there is a surjection $\pi_1(M, p) \to \pi^{-1}(p)$ if $\tilde{M}$ is path connected.

Proof. Every covering map is separated (Remark 2.30), hence $\tilde{M}$ is Hausdorff by Corollary 1.27. As M is a Lindelöf space (Theorem 4.28), we find a countable open covering $(U_n)_n$ by charts such that $\Phi^{-1}(U_n) = \coprod_{l \in L_n} \tilde{U}_{n,l}$ and $\Phi: \tilde{U}_{n,l} \to U_n$ is a homeomorphism for all n and l. By hypothesis, L_n is countable. The countably many $\tilde{U}_{n,l}$s form an open covering of $\tilde{M}$. As every U_n is homeomorphic to some open subset of $\mathbb{R}^m$, U_n and hence $\tilde{U}_{n,l}$ is second countable for all n and all $l \in L_n$. Therefore $\tilde{M}$ is second countable (Remark 1.3 4). $\qquad\square$

4.6 Problems

Problem 4.1. Let R be a local integral domain that is not a field and let K be its field of fractions. Show that the inclusion $R \to K$ is not a local homomorphism. Use this observation to construct locally ringed spaces whose underlying topological spaces consist of a single point and a morphism of ringed spaces between them that is not a morphism of locally ringed spaces.

Problem 4.2. Show that $\mathbb{R} \to \mathbb{R}$, $x \mapsto -x$, induces an automorphism of the real analytic manifold $\mathbb{T}^1$.

Problem 4.3. Let $\alpha \in \widehat{\mathbb{N}}$, $m, n \in \mathbb{N}$, and let $F: \mathbb{R}^{n+1} \setminus \{0\} \to \mathbb{R}^{m+1} \setminus \{0\}$ be a morphism of C^{α}-manifolds that is homogeneous of degree $d \in \mathbb{Z}$ (i.e., $F(\lambda x) = \lambda^d F(x)$ for all $x \in \mathbb{R}^{n+1} \setminus \{0\}$ and $\lambda \in \mathbb{R}^{\times}$). Show that F induces a morphism of C^{α}-manifolds $\mathbb{P}^n(\mathbb{R}) \to \mathbb{P}^m(\mathbb{R})$.

Remark: See also Problem 6.4 for a generalization.

Problem 4.4. Let P be the topological space consisting of a single point $*$ and define a sheaf of $\mathbb{K}$-algebras on P by $\mathcal{O}_P(P) = \mathbb{K}$.

1. Show that P is a final object in the category of $\mathbb{K}$-ringed spaces and in the category of locally $\mathbb{K}$-ringed spaces.
2. Let X be a locally $\mathbb{K}$-ringed space. Show that $f \mapsto f(*)$ yields a bijection between the set of morphisms of locally $\mathbb{K}$-ringed spaces $P \to X$ and the set of $x \in X$ such that $\kappa(x) = \mathbb{K}$.

Note that $P := (P, \mathcal{O}_P)$ is the 0-dimensional manifold consisting of a single point.

Problem 4.5. Let $\mathbb{K}[\varepsilon]$ be the $\mathbb{K}$-algebra $\mathbb{K}[X]/(X^2)$, where ε denotes the image of X in $\mathbb{K}[\varepsilon]$. Let $P[\varepsilon]$ be the $\mathbb{K}$-ringed space whose underlying topological space consists of a single point $*$ with $\mathcal{O}_{P[\varepsilon]}(P[\varepsilon]) = \mathbb{K}[\varepsilon]$. Let X be a locally $\mathbb{K}$-ringed space.

1. Show that $\mathbb{K}[\varepsilon]$ is a 2-dimensional $\mathbb{K}$-vector space with basis 1 and ε and that the multiplication is given by $(a_1 + b_1\varepsilon)(a_2 + b_2\varepsilon) = a_1a_2 + (a_1b_2 + a_2b_1)\varepsilon$. Deduce that $\mathbb{K}[\varepsilon]$ is a local ring with maximal ideal $\mathbb{K}\varepsilon := \{ a\varepsilon \; ; \; a \in \mathbb{K} \}$ and that $P[\varepsilon]$ is a locally $\mathbb{K}$-ringed space.
2. Show that $i^* \colon \mathbb{K}[\varepsilon] \to \mathbb{K}$, $a + b\varepsilon \mapsto a$, is a $\mathbb{K}$-algebra homomorphism that yields a morphism of locally $\mathbb{K}$-ringed spaces $i \colon P \to P[\varepsilon]$ (with P as in Problem 4.4).
3. Let $t \colon P[\varepsilon] \to X$ be a morphism of locally $\mathbb{K}$-ringed spaces, $x := t(*)$. Show that $t_x \colon \mathcal{O}_{X,x} \to \mathbb{K}[\varepsilon]$ induces a $\mathbb{K}$-linear map $\xi_t \colon \mathfrak{m}_x/\mathfrak{m}_x^2 \to \mathbb{K}\varepsilon \cong \mathbb{K}$ (where the isomorphism is the map $a\varepsilon \mapsto a$).
4. Show that the construction in 3 yields a bijection between the set of morphisms of locally $\mathbb{K}$-ringed spaces $P[\varepsilon] \to X$ and the set

$$\left\{ (x, \xi) \; ; \; x \in X \text{ with } \kappa(x) = \mathbb{K}, \xi \in (\mathfrak{m}_x/\mathfrak{m}_x^2)^{\vee} \right\}.$$

Problem 4.6. Let $(X, \mathcal{O}_X)$ be a locally ringed space.

1. Let $U \subseteq X$ be an open and closed subset. Show that there exists a unique section $e_U \in \Gamma(X, \mathcal{O}_X)$ such that $e_U|_V = 1$ for all open subsets V of U and $e_U|_V = 0$ for all open subsets V of $X \setminus U$. Show that $U \mapsto e_U$ yields a bijection

$$\mathrm{OC}(X) \leftrightarrow \mathrm{Idem}(\Gamma(X, \mathcal{O}_X)) \qquad (*)$$

   from the set of open and closed subsets of X to the set of idempotent elements of the ring $\Gamma(X, \mathcal{O}_X)$ ($e \in \Gamma(X, \mathcal{O}_X)$ is called *idempotent* if $e^2 = e$).
2. Show that $e_U e_{U'} = e_{U \cap U'}$ for $U, U' \in \mathrm{OC}(X)$.
3. Show that X is connected if and only if there exists no idempotent element $e \in \Gamma(X, \mathcal{O}_X)$ with $e \neq 0, 1$.

4. Now suppose that X is locally connected (Appendix Problem 12.22). Show that the bijection (*) induces a bijection between connected components of X and indecomposable idempotents (an idempotent is called *indecomposable* if it cannot be written as a sum of non-zero idempotents).

Problem 4.7. Let $n > 1$ be an odd integer and endow the topological space $\mathbb{R}$ via the two atlases $(\mathbb{R}, \mathrm{id}_{\mathbb{R}})$ and $(\mathbb{R}, \Phi)$ with $\Phi \colon \mathbb{R} \to \mathbb{R}$, $x \mapsto x^n$, in two ways with the structure of a smooth manifold. Show that these two structures are not equal but that the resulting manifolds are isomorphic.

Problem 4.8. Show that a premanifold is Hausdorff if and only if for all $p, q \in M$ with $p \neq q$ there exists $U \subseteq M$ open with $p, q \in U$ and $f \in \mathcal{O}_M(U)$ such that $f(p) \neq f(q)$.

Problem 4.9. Let M be a Hausdorff premanifold. Show that the following assertions are equivalent:

(i) The connected components of M are second countable.
(ii) The connected components of M are σ-compact.
(iii) M is paracompact.

Problem 4.10. Show that a Hausdorff premanifold is a manifold if and only if it is paracompact and has countably many connected components.
   Hint: Problem 4.9.

Problem 4.11. Let $U_1 = U_2 = \mathbb{R}$ (considered as 1-dimensional real analytic manifold) and glue U_1 and U_2 along $\mathrm{id}_{\mathbb{R} \setminus \{0\}}$ to a premanifold X.

1. Show that X is a second countable connected real analytic premanifold that is not Hausdorff. Let $\Phi_i := \mathrm{id}_{\mathbb{R}} \colon U_i \xrightarrow{\sim} \mathbb{R}$. Then $\mathcal{A} = (U_i, \Phi_i)_{i=1,2}$ is an atlas of X. Sketch X.
2. For $\alpha, \beta \in \widehat{\mathbb{N}}$ with $\alpha < \beta$ let $\varphi = \varphi_\alpha^\beta \colon \mathbb{R} \to \mathbb{R}$ be a C^α-diffeomorphism with $\varphi_\alpha(0) = 0$ and such that $\varphi_{|\mathbb{R} \setminus \{0\}}$ is a C^ω-diffeomorphism but that φ is not a C^β-map in 0. Show that (U_1, Φ_1) and $(U_2, \varphi \circ \Phi_2)$ form an atlas of the structure of a real analytic premanifold X^φ on the topological space X. Show that the underlying C^α-premanifolds X_α and $(X^\varphi)_\alpha$ are isomorphic but that X_β and $(X^\varphi)_\beta$ are not isomorphic.
3. Give an example of a map φ_α^β with the above properties.

Problem 4.12. Let $(\Omega, \leq)$ be an infinite well-ordered set and let $\bar{L} := \Omega \times [0, 1)$ endowed with the lexicographic order (i.e., $(\omega_1, x_1) \leq (\omega_2, x_2) \Leftrightarrow \omega_1 < \omega_2$ or $[\omega_1 = \omega_2$ and $x_1 \leq x_2]$). Endow $\bar{L}$ with the order topology (Appendix Problem 12.24). Let $\omega_0 \in \Omega$ be the smallest element and define the *open long ray* as the subspace $L := \bar{L} \setminus \{(\omega_0, 0)\}$.

1. Show that L is a connected normal Hausdorff space.
2. Let $\Omega := \mathbb{N}$. Show that L is homeomorphic to $\mathbb{R}_{>0}$.
3. Suppose that Ω is uncountable. Show that L is not second countable.
4. Endow L with the structure of a 1-dimensional real analytic premanifold.

Problem 4.13. Let M_1, M_2, and N be premanifolds. Show that a map $F \colon N \to M_1 \times M_2$ is a morphism of premanifolds if and only if each component $N \to M_i$ of F is a morphism of premanifolds.

Problem 4.14. Let M be a connected premanifold. Show that there exists a universal cover space $\pi \colon \tilde{M} \to M$ and that there exists a unique premanifold structure on $\tilde{M}$ such that π is a local isomorphism of premanifolds.
Hint: Problem 2.20.

Problem 4.15. Let M be a connected complex premanifold.

1. Show that every non-constant holomorphic function $f \colon M \to \mathbb{C}$ is open.
   Hint: Appendix Problem 16.4.
2. Let M be compact. Show that every holomorphic function $f \colon M \to \mathbb{C}$ is constant.

Problem 4.16. Let N be a premanifold, let $\pi \colon M' \to M$ be a surjective local isomorphism of premanifolds (for instance a covering map) and let $F \colon M \to N$ be a map. Show that F is a morphism of premanifolds if and only if $F \circ \pi$ is a morphism of premanifolds.

Linearization of Manifolds

5

It is a standard and very useful technique to approximate geometric objects and maps by linear objects and maps. A principal (trivial) example in basic analysis is approximating an open subspace U of $\mathbb{K}^n$ at a point by the linear space $\mathbb{K}^n$ itself by visualizing $\mathbb{K}^n$ as a tangent space to U at the given point. Another (less trivial) example is the derivative of a differentiable map f on U in a point that is the best approximation of f nearby this point by a linear map. In the first section we extend this technique to premanifolds and morphisms of premanifolds. Thus we first define the tangent space of a premanifold at a point, which will be a $\mathbb{K}$-vector space that we visualize as the "best linear approximation" of the premanifold at the given point. Then we define the derivative in a point x of a morphism f. This will be a linear map from tangent space at x to the tangent space at $f(x)$.

In the next section we study local properties of morphisms of premanifolds, in particular those given by properties of the derivative. The main result of Sect. 5.3 then will show that those morphisms whose derivative have locally constant rank as linear maps can themselves be linearized locally by choosing an appropriate chart.

These techniques allow us in the following sections to study submanifolds, fiber products of manifolds (in particular intersection of submanifolds), and quotient manifolds.

Notation: From now on a premanifold will always be either a real C^α-premanifold with $\alpha \in \widehat{\mathbb{N}}$ or a complex premanifold. Hence we no longer consider the case of a C^0-premanifold.

5.1 Tangent Spaces

We will now define the notion of the tangent space of a premanifold in a point. There are several ways to think of tangent vectors at a point p on the premanifolds M. We start by considering them as "derivatives $c'(0)$ of curves" $c: I \to M$ ($I \subseteq \mathbb{K}$ a "small" open neighborhood of 0) with $c(0) = p$. Of course the notion of the derivative of the curve c

© Springer Fachmedien Wiesbaden 2016
91
T. Wedhorn, *Manifolds, Sheaves, and Cohomology*, Springer Studium Mathematik – Master,
DOI 10.1007/978-3-658-10633-1_5

on M does not exist yet (it is one of the goals of this section to define this notion!). But if it existed, we would expect it to satisfy formal properties of the derivative such as the chain rule

$$\text{``} Df(p) \circ c'(0) = (f \circ c)'(0)\text{''} \tag{*}$$

for every $\mathbb{K}$-valued morphism f defined in some neighborhood of p. Also by taking for f the components of a chart at p we would expect that "$c'(0)$" is uniquely determined by "$Df(p) \circ c'(0)$" if f runs through all such functions.

Now, $c'(0)$ may not yet be defined, but the right-hand side of (*) perfectly makes sense as $f \circ c$ is a map from the open subset I of $\mathbb{K}$ to $\mathbb{K}$ for which a derivative is already defined. Hence we simply define a tangent vector as an equivalence class of curves $c : I \to M$ with $c(0) = p$, where such curves c_1 and c_2 are called equivalent if $(f \circ c_1)'(0) = (f \circ c_2)'(0)$ for all $\mathbb{K}$-valued morphisms f defined in an open neighborhood of p.

The set of these tangent vectors will be the tangent space of M at p. We then endow this set with the structure of a $\mathbb{K}$-vector space by "transport of structure" via a chart.

Hence let us first explain what we mean by "transport of structure". Given a set T and a bijective map $\alpha : T \to V$, where V is a $\mathbb{K}$-vector space, we define an addition and a scalar multiplication on T by $t_1 + t_2 := \alpha^{-1}(\alpha(t_1) + \alpha(t_2))$ and $\lambda \cdot t := \alpha^{-1}(\lambda \alpha(t))$. In other words, we endow T with the unique structure of a $\mathbb{K}$-vector space such that α becomes a $\mathbb{K}$-linear isomorphism. If $\varphi : V \xrightarrow{\sim} W$ is a $\mathbb{K}$-linear isomorphism, then the structures on T defined by α and by $\varphi \circ \alpha$ coincide.

After this preparation we can now define tangent spaces via the procedure motivated above.

Definition and Remark 5.1 (Tangent space). Let M be a premanifold, $p \in M$ a point, $m := \dim_p(M)$.

1. Define a set

$$T_p(M) := \{ (I, c) \mid I \subseteq \mathbb{K} \text{ open with } 0 \in I,$$
$$c : I \to M \text{ morphism with } c(0) = p \} / \sim$$

   where $(I_1, c_1) \sim (I_2, c_2)$ if and only if for every germ $f \in \mathcal{O}_{M,p}$ one has $(f \circ c_1)'(0) = (f \circ c_2)'(0)$.

2. Let (U, x) be a chart at p (i.e., $p \in U$), then the map

$$T_p(x) : T_p(M) \to \mathbb{K}^m, \qquad (I, c) \mapsto (x \circ c_{|c^{-1}(U)})'(0)$$

   yields a well-defined bijection (an inverse map is given by sending $a \in \mathbb{K}^m$ to the equivalence class of (I, c_a), where $c_a(t) := x^{-1}(x(p) + ta)$, restricted to a sufficiently small neighborhood I of 0). We use this bijection to endow $T_p(M)$ with the structure of a $\mathbb{K}$-vector space via transport of structure as explained above. This is

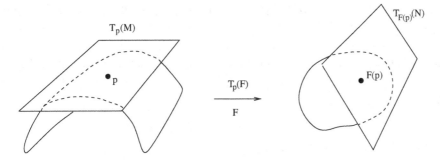

Figure 5.1 Tangent map in a point p

independent of the choice of the chart as every change of charts has as derivative a $\mathbb{K}$-linear isomorphism $\mathbb{K}^m \xrightarrow{\sim} \mathbb{K}^m$. As $m = \dim_p(M)$ we find

$$\dim_{\mathbb{K}}(T_p(M)) = \dim_p(M). \tag{5.1}$$

This definition of tangent vectors makes it easy to define the derivative of a morphism F: It is simply F applied to a tangent vector.

Remark and Definition 5.2 (Derivative). Let $F: M \to N$ be a morphism of premanifolds. Then we define the *derivative of F at p* or the *tangent map of F at p* (see Fig. 5.1) by

$$dF(p) := T_p(F): T_p(M) \to T_{F(p)}(N), \qquad [(I,c)] \mapsto [(I, F \circ c)].$$

To see that this is a well-defined $\mathbb{K}$-linear map let us first consider the case that M and N are open submanifolds of $\mathbb{K}^m$ and $\mathbb{K}^n$, respectively. As charts at p and $F(p)$ we choose the inclusions $M \hookrightarrow \mathbb{K}^m$ and $N \hookrightarrow \mathbb{K}^n$. We obtain isomorphisms

$$\mathbb{K}^m \xrightarrow{\sim} T_p(M), \qquad a \mapsto [(I, c_a)],$$
$$T_{F(p)}(N) \xrightarrow{\sim} \mathbb{K}^n, \qquad [(J,c)] \mapsto c'(0),$$

where $c_a(t) = p + ta$ and I is small enough such that $p + ta \in M$ for all $t \in I$. Hence

$$\mathbb{K}^m \xrightarrow{\sim} T_p(M) \xrightarrow{T_p(F)} T_{F(p)}(N) \xrightarrow{\sim} \mathbb{K}^n, \qquad a \mapsto \lim_{t \to 0} \frac{F(p + ta) - F(p)}{t},$$

in other words, this composition is the derivative of F at p. This shows that $T_p(F)$ is a well-defined $\mathbb{K}$-linear map in this case.

In general, for all charts (U, x) at p and (V, y) at $F(p)$ the chain rule (Appendix Proposition 16.5) shows

$$T_{F(p)}(y)[(I, F \circ c)] = (y \circ F \circ c)'(0) = D(y \circ F \circ x^{-1})(x(p))(x \circ c)'(0)$$
$$= D(y \circ F \circ x^{-1})T_{F(p)}(y)[(I,c)].$$

The arguments above show in particular that we have the following description of the tangent map.

Remark 5.3. Let $F: M \to N$ be a morphism of premanifolds. Given a chart $(U, x, V, y, \tilde{F})$ for F (Definition 4.22) we obtain for all $q \in U$ a commutative diagram of $\mathbb{K}$-linear maps

$$
\begin{array}{ccc}
T_q(M) & \xrightarrow{\;\;T_q(F)\;\;} & T_{F(q)}(N) \\
{\scriptstyle T_q(x)}\Big\downarrow{\scriptstyle \cong} & & {\scriptstyle \cong}\Big\downarrow{\scriptstyle T_{F(q)}(y)} \\
\mathbb{K}^m = T_{x(q)}(\tilde{U}) & \xrightarrow{\;\;T_{x(q)}(\tilde{F})\;\;} & T_{y(F(q))}(\tilde{V}) = \mathbb{K}^n,
\end{array}
\tag{5.2}
$$

and $T_{x(q)}(\tilde{F})$ is given by the Jacobi matrix of $\tilde{F}$ in $x(q)$ (Appendix Remark 16.18).

Remark 5.4 (Chain rule). Let $F: M \to N$ and $G: N \to Q$ be morphisms of premanifolds. Then the definition of the derivative implies immediately the following elegant version of the *chain rule*:

$$
T_p(G \circ F) = T_{F(p)}(G) \circ T_p(F) \qquad \text{``Chain rule''}.
\tag{5.3}
$$

It might be surprising that the chain rule seems to be a triviality. The difficulty is hidden in the fact that our definition of derivative of a morphism of premanifolds is well defined. This was proved in Remark 5.2 and used the classical chain rule.

We can express the chain rule also as follows. Let (PMfd*) be the category of premanifolds M together with a point $p \in M$. Morphisms $(M, p) \to (N, q)$ are morphisms of premanifolds $F: M \to N$ such that $F(p) = q$. Then

$$
(M, p) \mapsto T_p(M), \quad (F: (M, p) \to (N, q)) \mapsto (T_p(F): T_p(M) \to T_q(N))
$$

is a (covariant) functor (PMfd*) $\to$ ($\mathbb{K}$-Vec). In particular, if F is an isomorphism of premanifolds, then $T_p(F)$ is an isomorphism of $\mathbb{K}$-vector spaces.

Remark 5.5.

1. Let M be a premanifold, $U \subseteq M$ open (hence U is premanifold), $p \in U$. Then $T_p(U) = T_p(M)$.
2. Let M, N be premanifolds, $p \in M$, $q \in N$. If (U, x) is a chart of M at p and (V, y) is a chart of N at q, then $(U \times V, x \times y)$ is a chart of $M \times N$ at (p, q). Therefore Remark 5.1 2 shows

$$
T_{(p,q)}(M \times N) = T_p(M) \times T_q(N).
$$

A similar statement holds for the derivative of a product of morphisms of premanifolds.

Tangent vectors as derivations

We will now see that we can also consider tangent vectors more algebraically as derivations. Let us start with the general definition of a derivation.

Definition 5.6. Let A be a commutative $\mathbb{K}$-algebra and M an A-module. An element of

$$\mathrm{Der}_{\mathbb{K}}(A, M) := \{\xi\colon A \to M \mid \xi \ \mathbb{K}\text{-linear map},$$
$$\xi(ab) = a\xi(b) + b\xi(a) \text{ for all } a, b \in A\}$$

is called a $\mathbb{K}$-*derivation on A with values in M*. Then $\mathrm{Der}_{\mathbb{K}}(A, M)$ is a $\mathbb{K}$-subspace (even an A-submodule) of the A-module of all maps $A \to M$.

Of course the same definition works just fine if we replace $\mathbb{K}$ with an arbitrary commutative ring but we will not need this generality.

Remark 5.7 (Tangent vectors as derivations). Let M be a premanifold, $p \in M$ a point, $n := \dim_p(M)$. Then $\xi = [(I, c)] \in T_p(M)$ defines a map

$$\xi\colon \mathcal{O}_{M,p} \mapsto \mathbb{K}, \qquad f \mapsto (f \circ c)'(0).$$

Then ξ is $\mathbb{K}$-linear and $\xi(fg) = f(p)\xi(g) + g(p)\xi(f)$. In other words, ξ is a $\mathbb{K}$-derivation on $\mathcal{O}_{M,p}$ with values in $\mathbb{K}$, where we consider $\mathbb{K}$ as an $\mathcal{O}_{M,p}$-module via the scalar multiplication $f \cdot a = f(p)a$ for $f \in \mathcal{O}_{M,p}$ and $a \in \mathbb{K}$. We obtain a $\mathbb{K}$-linear map

$$\iota\colon T_p(M) \hookrightarrow \mathrm{Der}_{\mathbb{K}}(\mathcal{O}_{M,p}, \mathbb{K}), \tag{5.4}$$

which is injective by the definition of $T_p(M)$. We use this map to consider $T_p(M)$ as a $\mathbb{K}$-subspace of $\mathrm{Der}_{\mathbb{K}}(\mathcal{O}_{M,p}, \mathbb{K})$.

If $F\colon M \to N$ is a morphism of manifolds, then the $\mathbb{K}$-algebra homomorphism $F_p\colon \mathcal{O}_{N,F(p)} \to \mathcal{O}_{M,p}, g \mapsto g \circ F$, induces a $\mathbb{K}$-linear map

$$F_*\colon \mathrm{Der}_{\mathbb{K}}(\mathcal{O}_{M,p}, \mathbb{K}) \to \mathrm{Der}_{\mathbb{K}}(\mathcal{O}_{N,F(p)}, \mathbb{K}), \qquad \xi \mapsto \xi \circ F_p.$$

The diagram

$$
\begin{array}{ccc}
T_p(M) & \xrightarrow{\ T_p(F)\ } & T_{F(p)}(N) \\
\Big\downarrow{\scriptstyle\iota} & & \Big\downarrow{\scriptstyle\iota} \\
\mathrm{Der}_{\mathbb{K}}(\mathcal{O}_{M,p}, \mathbb{K}) & \xrightarrow{\ F_*\ } & \mathrm{Der}_{\mathbb{K}}(\mathcal{O}_{N,F(p)}, \mathbb{K})
\end{array}
$$

commutes. Indeed, let $\xi = [(I, c)] \in T_p(M)$ identified with the derivation $f \mapsto (f \circ c)'(0)$. Then

$$F_*(f \mapsto (f \circ c)'(0)) = (g \mapsto (g \circ F \circ c)'(0)) \in \mathrm{Der}_{\mathbb{K}}(\mathcal{O}_{N,F(p)}, \mathbb{K}),$$

which is identified with $[(I, F \circ c)] = T_p(F)(\xi)$.

If we choose a chart we obtain the following coordinate description of the tangent space.

Remark 5.8. Let $m = \dim_p(M)$ and (U, x) be a chart at p with coordinate function $x^i : U \to \mathbb{K}$. Then $T_p(x) : T_p(M) \xrightarrow{\sim} T_{x(p)}(x(U)) = \mathbb{K}^m$ is a $\mathbb{K}$-linear isomorphism with coordinates $dx^i = T_p(x^i) : T_p(M) \to \mathbb{K}$. In particular, $(dx^i)_{1 \le i \le m}$ is a $\mathbb{K}$-basis of the dual space $T_p(M)^\vee$.

On $\mathbb{K}^m$ we have the partial derivatives in direction of the standard basis vectors e_i. We use the chart to make these into derivations on $\mathcal{O}_{M,p}$. More precisely, we define for $i = 1, \ldots, m$ a derivation

$$\partial_i := \frac{\partial(\cdot)}{\partial x^i}(p) : \mathcal{O}_{M,p} \to \mathbb{K},$$

$$\partial_i(f) := \frac{\partial f}{\partial x^i}(p) := \frac{\partial}{\partial x_i}(f \circ x^{-1})(x(p)) = \frac{d}{dt}|_{t=0} f(x^{-1}(x(p) + te_i)).$$

In other words, let $c_{e_i} : I \to U$, $c_{e_i}(t) := x^{-1}(x(p) + te_i)$ (with $0 \in I \subseteq \mathbb{K}$ so small such that $x(p) + te_i \in x(U)$ for all $t \in I$). Then

$$\partial_i = [(I, c_{e_i})] \in T_p(M).$$

Clearly $dx^j(\partial_i) = \delta_{ij}$ (Kronecker delta). Hence $(\partial_i)_{1 \le i \le m}$ is linearly independent and therefore a basis of $T_p(M)$ because $\dim T_p(M) = m$. It is the basis dual to the basis $(dx^i)_i$ of $T_p(M)^\vee$.

For U open neighborhood of p and $f \in \mathcal{O}_M(U)$ we set $\frac{\partial f}{\partial x^i}(p) := \partial_i(f_p)$.

We give a further description of the space of derivations. This is the following purely algebraic fact applied to the $\mathbb{K}$-algebra $\mathcal{O}_{M,p}$ and its maximal ideal $\mathfrak{m}_p$.

Remark 5.9. Let k be a field, let A be k-algebra that is a local ring with maximal ideal $\mathfrak{m}$ and residue field k, i.e., the composition $k \to A \to A/\mathfrak{m}$ is an isomorphism that we use to identify k with $A/\mathfrak{m}$. We claim that $\xi \mapsto \xi_{|\mathfrak{m}}$ induces an isomorphism of k-vector spaces

$$\text{Der}_k(A, k) \xrightarrow{\sim} \text{Hom}_k(\mathfrak{m}/\mathfrak{m}^2, k) = (\mathfrak{m}/\mathfrak{m}^2)^\vee. \tag{5.5}$$

Indeed for $a, b \in \mathfrak{m}$ we have $\xi(ab) = a\xi(b) + b\xi(a) = 0$ and hence $\xi_{|\mathfrak{m}^2} = 0$. Hence ξ induces a homomorphism of abelian groups $\mathfrak{m}/\mathfrak{m}^2 \to k$, which is k-linear again because of the Leibniz rule.

To see that this map is bijective, observe that we have a k-linear isomorphism $A \to k \oplus \mathfrak{m}$, $a \mapsto (\bar{a}, a - \bar{a})$, where $\bar{a}$ is the image of a in $A/\mathfrak{m} = k$. In other words, the exact sequence $0 \to \mathfrak{m} \to A \to k \to 0$ splits via the given k-algebra structure $k \to A$

(Appendix Definition 14.5). If $t \colon \mathfrak{m}/\mathfrak{m}^2 \to k$ is a k-linear map we obtain a k-derivation $\xi \colon A = k \oplus \mathfrak{m} \to k$ by $\xi(\lambda, a) = t(a \bmod \mathfrak{m}^2)$. This defines an inverse map to (5.5).

Altogether we obtain $\mathbb{K}$-linear maps

$$T_p(M) \overset{\iota}{\longrightarrow} \mathrm{Der}_{\mathbb{K}}(\mathcal{O}_{M,p}, \mathbb{K}) \overset{\sim}{\to} (\mathfrak{m}_p/\mathfrak{m}_p^2)^{\vee} \tag{5.6}$$

for every premanifold M and every $p \in M$. The first map ι is injective and the second map is an isomorphism. After choosing a chart (U, x) at p, the image of ι can be described as the derivations that are linear combinations of $\partial_1, \ldots, \partial_m$, $m = \dim_p(M)$. In particular $\dim_{\mathbb{K}}(\mathfrak{m}_p/\mathfrak{m}_p^2) \geq \dim_p(M)$. For real C^α-premanifolds with $\alpha < \infty$, the map ι is not an isomorphism (in fact, $\mathfrak{m}_p/\mathfrak{m}_p^2$ is an infinite-dimensional $\mathbb{K}$-vector space in this case, see Problem 5.3). But for C^∞-premanifolds and (real or complex) analytic premanifolds, we will now show that $\mathfrak{m}_p/\mathfrak{m}_p^2$ has the same dimension as $T_p(M)$ and hence ι is an isomorphism. Indeed, by choosing a chart at p it suffices to prove the following lemma.

Lemma 5.10. *Let $X = \mathbb{K}^m$ with coordinate functions $x^i \colon X \to \mathbb{K}$. For $\mathbb{K} = \mathbb{R}$ let $\mathcal{O}_X$ be the sheaf of C^∞-functions or of analytic functions on X. For $\mathbb{K} = \mathbb{C}$ let $\mathcal{O}_X$ be the sheaf of holomorphic functions on X. Let $a = (a_1, \ldots, a_m) \in \mathbb{K}^m$ and let $\mathfrak{m}_a$ be the maximal ideal of $\mathcal{O}_{X,a}$. Then $\mathfrak{m}_a$ is generated by the germs of $x^i - a_i$, $i = 1, \ldots, m$, and their images in $\mathfrak{m}_a/\mathfrak{m}_a^2$ form a $\mathbb{K}$-basis.*

Proof. We may assume that $a = 0$. Let $f \in \mathfrak{m} := \mathfrak{m}_0$, i.e., f is C^α-function with $\alpha \in \{\infty, \omega\}$ defined on some neighborhood U of 0 with $f(0) = 0$. By shrinking U, we may assume that U is convex. Set

$$f_i(u) := \int_0^1 \partial_i f(tu)\, dt \qquad \text{for } u \in U.$$

Then the fundamental theorem of calculus implies that $f = \sum_{i=1}^n f_i x^i$ (here we use $f(0) = 0$). Moreover, as integration and partial differentiation commute, f_i is again a C^α-function (here we need $\alpha \geq \infty$; for $\alpha < \infty$, we could only conclude that f_i is a $C^{\alpha-1}$-function). Therefore we see that the ideal $\mathfrak{m}$ is generated by the coordinate functions $x^1, \ldots, x^m$. Hence their images in $\mathfrak{m}/\mathfrak{m}^2$ form a generating system. As we already know that $\dim_{\mathbb{K}}(\mathfrak{m}/\mathfrak{m}^2) \geq m$, these images form a $\mathbb{K}$-basis. $\qquad \square$

Summarizing we see that we have the following description of the tangent space.

Proposition 5.11. *For real C^α-premanifolds M with $\alpha \geq \infty$ or for complex premanifolds M and for all $p \in M$ one has $\mathbb{K}$-linear isomorphisms*

$$T_p(M) \overset{\sim}{\to} \mathrm{Der}_{\mathbb{K}}(\mathcal{O}_{M,p}, \mathbb{K}) \overset{\sim}{\to} (\mathfrak{m}_p/\mathfrak{m}_p^2)^{\vee}.$$

Finally, Problem 5.5 gives a further description of the tangent space of a premanifold M as in Proposition 5.11: Let $P[\varepsilon]$ be the locally $\mathbb{K}$-ringed space whose underlying topological space is a single point $*$ with $\mathcal{O}_{P[\varepsilon]}(P[\varepsilon]) = \mathbb{K}[\varepsilon] := \mathbb{K}[T]/(T^2)$, where ε denotes the image of T in $\mathbb{K}[T]/(T^2)$. Then one has for $p \in M$

$$T_p(M) = \left\{ \begin{matrix} t \colon P[\varepsilon] \to M \text{ morphism of locally} \\ \mathbb{K}\text{-ringed spaces with } t(*) = p \end{matrix} \right\}. \tag{5.7}$$

It is also possible to describe the $\mathbb{K}$-vector space structure on $T_p(M)$ via the right-hand side (Problem 5.5). Moreover, it is not difficult to see that the map induced by a morphism of premanifolds $F \colon M \to N$ on tangent spaces is given by

$$T_p(M) \ni (t \colon P[\varepsilon] \to M) \mapsto F \circ t \in T_{F(p)}(M).$$

We conclude the section by explaining that the tangent space does not change when the structure is weakened.

Remark 5.12. Let M be a real C^β-premanifold, let $\alpha \in \widehat{\mathbb{N}}$ with $\alpha \leq \beta$, and let $\iota_M \colon M \to M_\alpha$ be the canonical morphism of locally $\mathbb{R}$-ringed spaces, where M_α is the underlying C^α-premanifold (Remark 4.23). Then $\iota^\flat = \iota^\sharp \colon \mathcal{O}_M \to \mathcal{O}_{M_\alpha}$ is the inclusion and the restriction of derivations defines for all $p \in M$ a $\mathbb{K}$-linear map $\rho_M \colon \mathrm{Der}_{\mathbb{K}}(\mathcal{O}_{M_\alpha,p}, \mathbb{K}) \to \mathrm{Der}_{\mathbb{K}}(\mathcal{O}_{M,p}, \mathbb{K})$, which is functorial in M, i.e., for every morphism of C^β-premanifolds $F \colon M \to N$ the diagram

$$
\begin{array}{ccc}
\mathrm{Der}_{\mathbb{K}}(\mathcal{O}_{M_\alpha,p}, \mathbb{K}) & \xrightarrow{(F_\alpha)_*} & \mathrm{Der}_{\mathbb{K}}(\mathcal{O}_{N_\alpha,F(p)}, \mathbb{K}) \\
\rho_M \downarrow & & \downarrow \rho_N \\
\mathrm{Der}_{\mathbb{K}}(\mathcal{O}_{M,p}, \mathbb{K}) & \xrightarrow{F_*} & \mathrm{Der}_{\mathbb{K}}(\mathcal{O}_{N,F(p)}, \mathbb{K})
\end{array}
$$

commutes. The functoriality shows in particular that if we choose a chart (U, x) at p, then the i-th partial derivative $\frac{\partial}{\partial x^i}$ for M_α along the i-th coordinate ($i = 1, \ldots, \dim_p(M)$) is sent to the i-th partial derivative for M. As these derivations form a basis of the tangent spaces, one sees that ρ_M induces an isomorphism

$$T_p(M_\alpha) \xrightarrow{\sim} T_p(M), \tag{5.8}$$

which we use in the sequel to identify these tangent spaces. The functoriality of ρ_M also shows that for a morphism of C^β-premanifolds $F \colon M \to N$ the derivatives of F and of F_α are the same via the identification (5.8).

Remark 5.13. Let M be a complex premanifold, let $M_{\mathbb{R}}$ be the underlying real analytic manifold, and let $M_{\mathbb{R};\mathbb{C}} := (M, \mathcal{O}_{M_{\mathbb{R}};\mathbb{C}}) \to M_{\mathbb{R}}$ and $M_{\mathbb{R};\mathbb{C}} \to M$ be the canonical morphisms (Remark 4.24). Let $p \in M$ and let $\mathfrak{m}_p$, $\mathfrak{m}_{\mathbb{C},p}$ and $\mathfrak{m}_{\mathbb{R},p}$ be the maximal ideal of $\mathcal{O}_{M,p}$, $\mathcal{O}_{M_{\mathbb{R}};\mathbb{C},p}$ and $\mathcal{C}^{\omega}_{M_{\mathbb{R}},p}$, respectively. Then $\mathfrak{m}_{\mathbb{C},p} = \mathfrak{m}_{\mathbb{R},p} \otimes_{\mathbb{R}} \mathbb{C}$ and hence we obtain a functorial isomorphism of $\mathbb{C}$-vector spaces

$$T_p(M_{\mathbb{R};\mathbb{C}}) := (\mathfrak{m}_{\mathbb{C},p}/\mathfrak{m}^2_{\mathbb{C},p})^{\vee} = (\mathfrak{m}_{\mathbb{R},p}/\mathfrak{m}^2_{\mathbb{R},p})^{\vee} \otimes_{\mathbb{R}} \mathbb{C} = T_p(M_{\mathbb{R}}) \otimes_{\mathbb{R}} .\mathbb{C} \qquad (*)$$

On the other hand, the inclusion $\mathcal{O}_M \hookrightarrow \mathcal{O}_{M_{\mathbb{R}};\mathbb{C}}$ sends $\mathfrak{m}_p$ to $\mathfrak{m}_{\mathbb{C},p}$ and hence induces a functorial $\mathbb{C}$-linear map $\pi\colon T_p(M_{\mathbb{R};\mathbb{C}}) \to T_p(M)$. Let

$$\tau\colon T_p(M_{\mathbb{R}}) \to T_p(M)$$

be the composition of the $\mathbb{R}$-linear injective map $T_p(M_{\mathbb{R}}) \to T_p(M_{\mathbb{R};\mathbb{C}})$, $\xi \mapsto \xi \otimes 1$, followed by π. Then τ is a functorial $\mathbb{R}$-linear map. We claim that τ is an isomorphism.

Indeed, let $(U, z\colon U \to \mathbb{C}^m)$ be a chart of M and p with $z(p) = 0$. Let z^i_p be the image of the i-the coordinate function in $\mathfrak{m}_p/\mathfrak{m}^2_p$, and let x^i_p and y^i_p be the image of $\mathrm{Re}(z^i)$ and $\mathrm{Im}(z^i)$ in $\mathfrak{m}_{\mathbb{R},p}/\mathfrak{m}^2_{\mathbb{R},p}$. Lemma 5.10 shows that $(z^1_p, \ldots, z^m_p)$ is a $\mathbb{C}$-basis of $\mathfrak{m}_p/\mathfrak{m}^2_p$ and $(x^1_p, y^1_p, \ldots, x^m_p, y^m_p)$ is an $\mathbb{R}$-basis of $\mathfrak{m}_{\mathbb{R},p}/\mathfrak{m}^2_{\mathbb{R},p}$. Then $(z^1_p, \bar{z}^1_p, \ldots, z^m_p, \bar{z}^m_p)$ is an $\mathbb{R}$-basis of $\mathfrak{m}_p/\mathfrak{m}^2_p$. As the standard real analytic coordinate change $(z^i, \bar{z}^i) \mapsto (x^i, y^i)$ is an isomorphism, we see that the dual map of τ is an isomorphism. Hence τ is an isomorphism.

In the sequel we will use τ to identify $T_p(M_{\mathbb{R}})$ with the underlying real vector space of the $\mathbb{C}$-vector space $T_p(M)$. The functoriality of τ shows that if $F\colon M \to N$ is a holomorphic map, then via this identification of tangent spaces the derivatives of F and of $F_{\mathbb{R}}$ coincide.

5.2 Local Properties of Morphisms, Local Isomorphisms, Immersions, Submersions

Local properties may be checked via charts.

Remark 5.14 (Chart principle). If **P** is a property of a continuous map F between premanifolds M and N that is local on source and target (Appendix Sect. 12.5) and that is stable under composition with isomorphisms of manifolds, then a continuous map $F\colon M \to N$ has the property **P** if and only if for every $p \in M$ there exists a chart $(U, x, V, y, \tilde{F})$ of F at p such that $\tilde{F}$ has the property **P**.

Examples are the properties **P** of being open or being a morphism of premanifolds. Important applications of the chart principle are the following two results.

Proposition 5.15. *Let* $F: M \to N$ *be a morphism of premanifolds. Then* F *is locally constant if and only if* $T_p(F) = 0$ *for all* $p \in M$.

Proof. Both properties ("locally constant" and "$T_p(F) = 0$ for all p") are local on source and target and stable under composition with isomorphisms (for the second property this follows from the chain rule, Remark 5.4). Hence by choosing charts for F we may assume that M and N are open in some $\mathbb{K}$-vector space. Then the claim follows from the analogous result of local analysis (Appendix Proposition 16.6). $\square$

The chart principle also yields immediately the following theorem of inverse functions.

Theorem 5.16. *Let* $F: M \to N$ *be a morphism of premanifolds, let* $p \in M$. *Then* F *is a local isomorphism at* p *if and only if* $T_p(F)$ *is an isomorphism of* $\mathbb{K}$-*vector spaces.*

Proof. Again by choosing charts for F we may assume that M and N are open in some $\mathbb{K}$-vector space, and the claim follows from the analogous result of local analysis (Appendix Theorem 16.16). $\square$

Definition 5.17. Let $F: M \to N$ be a morphism of premanifolds, let $p \in M$ and let $T_p(F): T_p(M) \to T_{F(p)}(N)$ be the induced tangent map.

1. $\mathrm{rk}_p(F) := \mathrm{rk}(T_p(F))$ is called the *rank of F at p*.
2. F is called an *immersion at p* if $T_p(F)$ is injective. F is called an *immersion* if it is an immersion at every $p \in M$.
3. F is called an *submersion at p* if $T_p(F)$ is surjective. F is called an *submersion* if it is a submersion at every $p \in M$.

Remark 5.18.

1. The chain rule (5.3) implies that the composition of two submersions (respective of two immersions) is again a submersion (respectively an immersion). Moreover if $F \circ G$ is a submersion and G is surjective (respectively if $F \circ G$ is an immersion), then F is a submersion (respectively then G is an immersion).
2. By Theorem 5.16 a morphism of premanifolds is a local isomorphism if and only if it is a submersion and an immersion.

Proposition 5.19. *Let* $F: M \to N$ *be a morphism of premanifolds and fix* $r \in \mathbb{N}_0$. *Then* $\{ p \in M \; ; \; \mathrm{rk}_p(F) \geq r \}$ *is open in* M.

Proof. Again the chart principle allows us to assume that $M \subseteq \mathbb{K}^m$, $N \subseteq \mathbb{K}^n$ open. As F is a C^1-map, the map $M \ni p \mapsto T_p(F) \in \mathrm{Hom}_{\mathbb{K}}(\mathbb{K}^m, \mathbb{K}^n)$ is continuous, and $\{ A \in \mathrm{Hom}_{\mathbb{K}}(\mathbb{K}^m, \mathbb{K}^n) \; ; \; \mathrm{rk}(A) \geq n \}$ is open because via the identification $\mathrm{Hom}_{\mathbb{K}}(\mathbb{K}^m, \mathbb{K}^n) = M_{n \times m}(\mathbb{K})$ its complement is the closed set of all matrices whose all r-minors vanish. $\square$

Corollary 5.20. *Let $F: M \to N$ be a morphism of premanifolds.*

1. *The set $U := \{p \in M ; F$ is an immersion at $p\}$ is open in M, and $\mathrm{rk}_p(F) = \dim_p(M)$ for all $p \in U$.*
2. *The set $V := \{p \in M ; F$ is a submersion at $p\}$ is open in M, and $\mathrm{rk}_p(F) = \dim_{F(p)}(N)$ for all $p \in V$.*

5.3 Morphisms of Locally Constant Rank

In this section we will show that morphisms of locally constant rank are locally linear and deduce some properties of such morphisms.

Corollary 5.20 shows that for immersions and for submersions F the map $U \ni p \mapsto \mathrm{rk}_p(F)$ is locally constant. In other words, immersions and submersions are examples of morphisms of locally constant rank in the following sense.

Definition 5.21. Let $F: M \to N$ be a morphism of premanifolds and let $p \in M$. Then F is called *of locally constant rank at* p if there exists $p \in W \subseteq M$ open such that $W \to \mathbb{N}_0, q \mapsto \mathrm{rk}_q(F)$ is constant. The morphism F is called *of locally constant rank* if it is of locally constant rank at all $p \in M$.

Clearly, the set of points in M, where F is locally of constant rank, is open in M. One can show that it is also dense in M (Problem 5.6).

The main result about morphisms of locally constant rank shows that they have particularly simple charts.

Theorem 5.22 (Rank theorem). *Let $F: M \to N$ be a morphism of premanifolds and let $p \in M$, $m := \dim_p(M)$, $n := \dim_{F(p)}(N)$, $k \in \mathbb{N}_0$. Assume that there exists $p \in W \subseteq M$ open such that $\mathrm{rk}_q(F) = k \in \mathbb{N}_0$ for all $q \in W$. Then there exists a chart of F at p*

$$
\begin{array}{ccc}
M \supseteq U & \xrightarrow{\ F_{|U}\ } & V \subseteq N \\
{\scriptstyle x}\downarrow{\scriptstyle \cong} & & {\scriptstyle \cong}\downarrow{\scriptstyle y} \\
\mathbb{K}^m \supseteq \tilde{U} & \xrightarrow{\ \tilde{F}\ } & \tilde{V} \subseteq \mathbb{K}^n,
\end{array}
$$

such that $\tilde{F}(x_1, \ldots, x_m) = (x_1, \ldots, x_k, 0, \ldots, 0)$ for all $(x_1, \ldots, x_m) \in \tilde{U}$.

Proof. Let $(U, x, V, y, \tilde{F})$ be a chart of F at p. By replacing U by $U \cap W$ we can assume $U \subseteq W$. Due to Remark 5.3, we obtain a commutative diagram of $\mathbb{K}$-vector spaces

$$
\begin{array}{ccc}
T_q(M) & \xrightarrow{\ T_q(F)\ } & T_{F(q)}(N) \\
T_q(x) \Big\downarrow \cong & & \cong \Big\downarrow T_{F(q)}(y) \\
T_{x(q)}(\tilde{U}) & \xrightarrow{\ T_{x(q)}(\tilde{F})\ } & T_{y(F(q))}(\tilde{V}),
\end{array}
$$

for all $q \in U \subseteq W$. In particular, the rank of $\tilde{F}$ at $x(q) \in \tilde{U}$ has to be k for every $q \in U$ so that $\tilde{U} \ni \tilde{q} \mapsto \mathrm{rk}_{\tilde{q}}\, \tilde{F}$ is constant (with value k) because $x(U) = \tilde{U}$. Therefore we may assume $p \in M \subseteq \mathbb{K}^m$, $N \subseteq \mathbb{K}^n$ open and $F \colon M \to N$ is a morphism of manifolds of constant rank. Composing x with a permutation of coordinates we may also rearrange coordinates.

Denote all elements $q \in M \subseteq \mathbb{K}^m$ as $q = (x, y)$ where $x \in \mathbb{K}^k$, $y \in \mathbb{K}^{m-k}$ and set

$$
\frac{\partial F}{\partial x}(x, y) := \left(\frac{\partial F_i}{\partial x_j}(x, y) \right)_{1 \le i \le n, 1 \le j \le k} \in M_{n \times k}(\mathbb{K}),
$$

$$
\frac{\partial F}{\partial y}(x, y) := \left(\frac{\partial F_i}{\partial y_j}(x, y) \right)_{1 \le i \le n, 1 \le j \le m-k} \in M_{n \times (m-k)}(\mathbb{K})
$$

for all morphisms of manifolds F between open subsets of $\mathbb{K}^m$ and $\mathbb{K}^n$. Moreover, let $F' \colon M \to \mathbb{K}^k$ and $F'' \colon M \to \mathbb{K}^{n-k}$ be the morphisms of manifolds with $F(q) = (F'(q), F''(q))$ for $q \in M$. As $J_F(p)$ has rank k, we may assume that $\frac{\partial F'}{\partial x}(p)$ is invertible after rearranging the coordinates. With this notation we have for the Jacobi matrix

$$
J_F(x, y) = \begin{pmatrix} \frac{\partial F'}{\partial x}(x, y) & \frac{\partial F'}{\partial y}(x, y) \\ \frac{\partial F''}{\partial x}(x, y) & \frac{\partial F''}{\partial y}(x, y) \end{pmatrix} \in M_{n \times m}(\mathbb{K}),
$$

where the upper-left matrix is an invertible square matrix of size k for $(x, y) = p$.

Define $\tilde{\varphi} \colon M \to \mathbb{K}^m$, $(x, y) \mapsto (F'(x, y), y)$. Then the Jacobi matrix of $\tilde{\varphi}$ has for all $q \in M$ the form

$$
J_{\tilde{\varphi}}(q) = \begin{pmatrix} \frac{\partial F'}{\partial x}(q) & * \\ 0 & I_{m-k} \end{pmatrix}
$$

so that $J_{\tilde{\varphi}}(p) \in \mathrm{GL}_m(\mathbb{K})$. By the inverse function theorem (Appendix Theorem 16.16) there exists an open neighborhood U of p such that $\varphi := \tilde{\varphi}_{|U} \colon U \to \tilde{\varphi}(U) =: \tilde{U} \subseteq \mathbb{K}^m$ is an isomorphism of manifolds. By shrinking U we may assume that $\tilde{U}$ is connected and of the form $\tilde{U} = U' \times U''$, where $U' \subseteq \mathbb{K}^k$, $U'' \subseteq \mathbb{K}^{m-k}$ are open subsets. Then U' and U'' are connected (Appendix Proposition 12.41).

For $(x, y) \in \tilde{U}$ we have $F(\varphi^{-1}(x, y)) = (x, g(x, y))$ where $g \colon \tilde{U} \to \mathbb{K}^{n-k}$ is defined as $F'' \circ \varphi^{-1}$. Hence we find

$$J_{F \circ \varphi^{-1}}(x, y) = \begin{pmatrix} I_k & 0 \\ \frac{\partial g}{\partial x}(x, y) & \frac{\partial g}{\partial y}(x, y) \end{pmatrix}.$$

We have $J_{F \circ \varphi^{-1}}(x, y) = J_F(\varphi^{-1}(x, y)) J_{\varphi^{-1}}(x, y)$ by the chain rule. As $J_{\varphi^{-1}}(x, y)$ is invertible, $\mathrm{rk}(J_{F \circ \varphi^{-1}}(x, y)) = \mathrm{rk}(J_F(\varphi^{-1}(x, y)))$ is equal to k on $\tilde{U}$. Thus we must have $\frac{\partial g}{\partial y}(x, y) = 0$ for every $(x, y) \in U' \times U''$ and since U'' is connected, g does only depend on x (Appendix Proposition 16.6).

Finally define $V := U' \times \mathbb{K}^{n-k}$ and $\psi \colon V \to V$, $(u, v) \mapsto (u, v - g(u))$. This map is an isomorphism of manifolds and for $(x, y) \in \tilde{U}$ we find

$$\tilde{F}(x, y) := \psi(F(\varphi^{-1}(x, y))) = \psi(x, g(x)) = (x, g(x) - g(x))$$
$$= (x_1, \ldots, x_k, 0, \ldots, 0)$$

as desired. $\square$

Corollary 5.23. *Let $F \colon M \to N$ be a morphism of premanifolds, let $p \in M$, $m := \dim_p(M)$, $n := \dim_{F(p)}(N)$. Assume that F is an immersion (respectively a submersion) at p. Then $m \le n$ (respectively $m \ge n$) and there exists a chart $(U, x, V, y, \tilde{F})$ of F at p such that $\tilde{F}(x_1, \ldots, x_m) = (x_1, \ldots, x_m, 0, \ldots, 0)$ (respectively $\tilde{F}(x_1, \ldots, x_m) = (x_1, \ldots, x_n)$).*

Proof. If F is an immersion at p, then there exists an open neighborhood U of p such that F is an immersion at q for all $q \in U$ and such that $U \ni q \mapsto \mathrm{rk}_q(F)$ is constant (Corollary 5.20). Now we can apply the rank theorem.

The same argument shows the claim if F is a submersion at p. $\square$

Corollary 5.24. *Let $F \colon M \to N$ be a submersion of premanifolds. Then F is an open map.*

Proof. By the chart principle and by Corollary 5.23 we may assume that F is a projection map $(x_1, \ldots, x_m) \mapsto (x_1, \ldots, x_n)$ with $n \le m$. This is an open map. $\square$

Proposition 5.25. *Let $F \colon M \to N$ be a morphism of premanifolds and let Z be a third premanifold.*

1. *Suppose that F is an immersion. Then a continuous map $G \colon Z \to M$ is a morphism of premanifolds if and only if $F \circ G$ is a morphism of premanifolds.*
2. *Suppose that F is a surjective submersion. Then a map $G \colon N \to Z$ is a morphism of premanifolds if and only if $G \circ F$ is a morphism of premanifolds.*

The hypothesis in 1 that G is continuous cannot be omitted (Problem 5.11), see however Problem 5.12.

Proof. Let us show 1. The property of a continuous map being a morphism of premanifolds is local on source and target. Hence by the chart principle we may assume that $Z \subseteq \mathbb{K}^z$, $M \subseteq \mathbb{K}^m$ and $N \subseteq \mathbb{K}^n$ are open. As F is an immersion, we may also assume by Corollary 5.23, that F is of the form $(x_1, \ldots, x_m) \mapsto (x_1, \ldots, x_m, 0, \ldots, 0)$. Hence it remains to show that a continuous map $G: Z \to \mathbb{K}^m$ is a C^α-map if and and only if $Z \to \mathbb{K}^n$, $z \mapsto (G(z), 0, \ldots, 0)$ is a C^α-map. But this is clear.

A similar argument shows Assertion (2) once we have seen that G is continuous. But for $V \subseteq Z$ open, we have $G^{-1}(V) = F(F^{-1}(G^{-1}(V)))$ because F is surjective and this is open in N because F is open (Corollary 5.24). $\qquad\square$

5.4   Submanifolds

For every subspace S of a premanifold on M we define a function on S to be a map $S \to \mathbb{K}$ that is locally the restriction of a morphism $U \to \mathbb{K}$ for some $U \subseteq M$ open. We obtain the sheaf $\mathcal{O}_S$ of functions on S, and S is be defined to be subpremanifold if $(S, \mathcal{O}_S)$ is a premanifold. Then the main result (Theorem 5.34) will be that a subpremanifold is always isomorphic to an embedding in the following sense.

Definition 5.26. A morphism $\iota: M \to N$ of premanifolds is called an *embedding* if ι is a topological embedding (Appendix Definition 12.19) and an immersion.

Example 5.27. Any morphism $\iota: M \to N$ of premanifolds that is a section of a morphism $\pi: N \to M$ (i.e., $\pi \circ \iota = \mathrm{id}_M$) is an embedding: It is a topological embedding by Appendix Remark 12.20 4 and for any $p \in M$ one has $T_{\iota(p)}(\pi) \circ T_p(\iota) = T_p(\mathrm{id}_M) = \mathrm{id}_{T_p(M)}$, which shows that ι is an immersion.

In particular for every morphism of premanifolds $F: M \to N$ its *graph morphism* $\Gamma_F: M \to M \times N$, $m \mapsto (m, F(m))$, is an embedding. It is a section of the projection $M \times N \to M$.

Remark 5.28. An embedding is an injective immersion. The converse is not true. We give two examples:

1. $\varphi: (-\pi/4, \pi/2) \to \mathbb{R}^2$, $t \mapsto (\sin 2t)(-\sin(t), \cos(t))$. This is shown in Fig. 5.2.
2. Endow $\mathbb{Q}$ with the discrete topology (which makes $\mathbb{Q}$ into a 0-dimensional smooth manifold). Then the inclusion $\mathbb{Q} \hookrightarrow \mathbb{R}$ is an injective immersion (here $\mathbb{R}$ is endowed with the usual structure of a 1-dimensional smooth manifold).

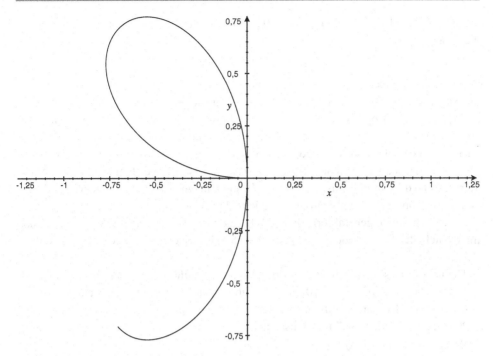

Figure 5.2 Locally around the origin the image of φ is not homeomorphic to an open interval

Remark 5.29. The composition $G \circ F$ of two embeddings F and G is again an embedding because $G \circ F$ is a topological embedding by Appendix Remark 12.20 3 and an immersion by the chain rule.

Definition and Remark 5.30 (Submanifolds). Let $(M, \mathcal{O}_M)$ be a premanifold.

1. Let $S \subseteq M$ be a subspace. A *function on S (C^α-function* if M is real C^α-manifold; *holomorphic function* if M is complex manifold) is a map $f \colon S \to \mathbb{K}$ such that for all $s \in S$ there exists $s \in U \subseteq M$ open and $g \in \mathcal{O}_M(U)$ such that $g_{|U \cap S} = f_{|U \cap S}$. Define a sheaf $\mathcal{O}_S$ of $\mathbb{K}$-valued functions by

$$\mathcal{O}_S(V) := \{ f \colon V \to \mathbb{K} \text{ function on } V \}, \qquad V \subseteq S \text{ open.}$$

   Then $(S, \mathcal{O}_S)$ is a $\mathbb{K}$-locally ringed space (all stalks $\mathcal{O}_{S,s}$ are local $\mathbb{K}$-algebras with residue field $\mathbb{K}$ by the same argument as in Remark 3.17).
2. A *(sub)premanifold of M* is a subspace S of M such that $(S, \mathcal{O}_S)$ is a (pre)manifold.

Remark 5.31. Let M be a premanifold, $S \subseteq M$ a closed subspace and $i \colon S \to M$ the inclusion. Let $\mathcal{I}_S$ be the sheaf of functions vanishing on S, i.e., for $U \subseteq M$ open we define $\mathcal{J}_S(U) := \{ f \in \mathcal{O}_M(U) \, ; \, f_{|S} = 0 \}$. Then we have an exact sequence of sheaves of abelian groups

$$0 \longrightarrow \mathcal{J}_S \overset{\iota}{\longrightarrow} \mathcal{O}_M \overset{\rho}{\longrightarrow} i_*\mathcal{O}_S \longrightarrow 0, \tag{5.9}$$

where ι is the inclusion and ρ is given by $\mathcal{O}_M(U) \ni g \mapsto g_{|U \cap S} \in i_*\mathcal{O}_S(U)$. To show the exactness of (5.9) we have to show that the induced sequence of stalks is exact. Fix $x \in X$. As $X \setminus S$ is open in X, $(i_*\mathcal{O}_S)_x = 0$ and $(\mathcal{J}_S)_x = \mathcal{O}_{X,x}$ for $x \in X \setminus S$. Hence we may assume that $x \in S$. Then $f \in (i_*\mathcal{O}_S)_x = \mathcal{O}_{S,x}$ is represented by a function (U, f), where U is a neighborhood of x and where $f : U \cap S \to \mathbb{K}$ is the restriction of a morphism $g : U \to \mathbb{K}$. Hence we see that $\rho_x : \mathcal{O}_{M,x} \to (i_*\mathcal{O}_S)_x$ is surjective. Elements of its kernel are represented by morphisms $g : U \to \mathbb{K}$ (U some neighborhood of x depending on g) whose restriction to $U \cap S$ is zero, i.e., the kernel of ρ_x is $\mathcal{J}_{S,x}$.

Once we have the general formalism of $\mathcal{O}_M$-modules available (Sect. 8.3), one checks immediately that $\mathcal{J}_S$ is an ideal of $\mathcal{O}_M$ and that (5.9) is an exact sequence of $\mathcal{O}_M$-modules.

The next goal is to show that a subspace of a premanifold is a subpremanifold if and only if it is the image of an embedding. Such subpremanifolds are also often called *regular submanifolds*. There are useful more general classes of submanifolds (see Problem 5.16 or Problem 5.18), which will not be used here.

We start with some easy remarks.

Remark 5.32. Let M be a premanifold, $S \subseteq M$ a subspace.

1. If S is open in M, then $(S, \mathcal{O}_S)$ is the open subpremanifold of M defined in Definition 4.16.
2. If S is a discrete subspace of M, then S is a 0-dimensional subpremanifold of M.
3. Let $\Phi : M \overset{\sim}{\to} N$ be an isomorphism of premanifolds. Then S is a subpremanifold of M if and only if $\Phi(S)$ is a subpremanifold of N.
4. Let $U_i \subseteq M$ be open for $i \in I$ such that $S \subseteq \bigcup_i U_i$. Then S is a subpremanifold of M if and only if $S \cap U_i$ is a subpremanifold of U_i for all $i \in I$.

As an immersion looks locally as the inclusion of a subspace into a vector space, we study this case first.

Example 5.33. Let $M \subseteq \mathbb{K}^m$ be open and $W \subseteq \mathbb{K}^m$ be a $\mathbb{K}$-subspace. Let us show that $S := W \cap M$ is a submanifold of M of dimension $\dim(W)$.

Clearly, $S \subseteq W$ is open. We may assume that $S \neq \emptyset$. After applying a $\mathbb{K}$-linear automorphism of $\mathbb{K}^m$ (which is a bianalytic isomorphism) we may also assume that $W = \{ (x_1, \ldots, x_k, 0, \ldots, 0) \in \mathbb{K}^m \, ; \, x_i \in \mathbb{K} \}$ for some $0 \le k \le m$.

Now there are two structures of a $\mathbb{K}$-ringed space on S. One as an open submanifold of W, call its structure sheaf $\mathcal{C}_S^\alpha$. Then $(S, \mathcal{C}_S^\alpha)$ is a C^α-manifold. The other is the sheaf $\mathcal{O}_S$ defined in Definition 5.30.

We claim that $\mathcal{O}_S = \mathcal{C}_S^\alpha$. Let $V \subseteq S$ be open. We have to show that a function $f : V \to \mathbb{K}$ is C^α if and only if for every $p \in V$ there exists $p \in U \subseteq \mathbb{R}^m$ open and a C^α-function $g : U \to \mathbb{K}$ such that $g_{|U \cap V} = f_{|U \cap V}$.

Clearly we have $\mathcal{O}_S(V) \subseteq \mathcal{C}_S^\alpha(V)$. Conversely, let $f : V \to \mathbb{K}$ be a C^α-function. Let

$$U := \{ (x_1, \ldots, x_k, t_{k+1}, \ldots, t_m) \in \mathbb{K}^m \; ; \; x = (x_1, \ldots, x_k) \in V \} \cap M.$$

Then $U \subseteq M$ is open and $U \cap S = V$. Define

$$g : U \to \mathbb{K}, \qquad g(x_1, \ldots, x_k, t_{k+1}, \ldots, t_m) := f(x_1, \ldots, x_k).$$

Then $g \in C_M^\alpha(U)$ with $g_{|V} = f$. Therefore $f \in \mathcal{O}_S(V)$.

Theorem 5.34. *Let M be a premanifold.*

1. *Let $S \subseteq M$ be a subpremanifold. Then $S \subseteq M$ is locally closed and the inclusion $\iota : S \hookrightarrow M$ is an embedding.*
2. *Conversely, let $i : N \to M$ be an embedding. Then $S := i(N)$ is locally closed in M, S is a subpremanifold and $i : N \to S$ is an isomorphism of premanifolds.*

Proof. 2. By Corollary 5.20 we may assume that $p \mapsto \mathrm{rk}_p(i)$ is constant and that N is of constant dimension n. Let $p \in N$. By the rank theorem there exists $p \in V = V_p \subseteq N$ open and a commutative diagram

$$
\begin{array}{ccc}
N \supseteq V & \xrightarrow{\quad i_{|V} \quad} & U \subseteq M \\
{\scriptstyle y} \downarrow {\scriptstyle \cong} & & {\scriptstyle \cong} \downarrow {\scriptstyle x} \\
\mathbb{K}^n \supseteq \tilde{V} & \xrightarrow{\tilde{i}:(x_1,\ldots,x_n) \mapsto (x_1,\ldots,x_n,0,\ldots,0)} & \tilde{U} \subseteq \mathbb{K}^m,
\end{array}
$$

where $U = U_p$ open in M.

As i is a homeomorphism $N \to S$, $i(V)$ is open in S, say $i(V) = U' \cap S$ with $U' \subseteq M$ open. Replacing U by $U \cap U'$ and $\tilde{U}$ by $x(U \cap U')$, we can assume that $U \cap S = i(V)$. Similarly we may assume $\tilde{i}(\tilde{V}) = \tilde{U} \cap (\mathbb{K}^n \times \{0\})$, hence $\tilde{i}(\tilde{V}) \subseteq \tilde{U}$ closed and therefore $i(V)$ closed in U. Hence U is an open neighborhood of $i(p)$ such that $U \cap S$ is locally closed in U. As p was arbitrary, this shows that S is locally closed.

Moreover, $i(V)$ is a submanifold of U because $\tilde{i}(\tilde{V})$ is a submanifold of $\tilde{U}$. As the U_p for $p \in N$ cover S, this implies that S is a subpremanifold of M.

Finally, the chart shows that for each U_p the map $i : i^{-1}(U_p \cap S) = V_p \to U_p \cap S$ is an isomorphism of manifolds. Hence $i : N \to S$ is an isomorphism of premanifolds because being an isomorphism can be checked locally on the target.

1. The map ι is a morphism of premanifolds: Let $U \subseteq M$ be open, $f \in \mathcal{O}_M(U)$. We have to show that $f \circ (\iota_{|\iota^{-1}(U)}) = f_{|U \cap S} \in \mathcal{O}_S(U \cap S)$. This holds by definition of $\mathcal{O}_S$.

The map ι is an embedding: we have to show that ι is an immersion. Let $p \in S$. By definition of $\mathcal{O}_S$ the homomorphism of $\mathbb{K}$-algebras

$$\iota_p^\sharp \colon \mathcal{O}_{M,p} \to \mathcal{O}_{S,p}, \qquad [(f, U)] \mapsto [(f_{|U \cap S}, U \cap S)]$$

is surjective. Hence the $\mathbb{K}$-linear map

$$\iota_* \colon \operatorname{Der}_{\mathbb{K}}(\mathcal{O}_{S,p}, \mathbb{K}) \to \operatorname{Der}_{\mathbb{K}}(\mathcal{O}_{M,p}, \mathbb{K})$$

is injective. Therefore $T_p(\iota)$ is injective by Remark 5.7.

Hence ι is an embedding and hence S is locally closed in M by 2. $\square$

The proof of Theorem 5.34 2 shows that there is also the following more classical characterization of a subpremanifold.

Remark 5.35. A subspace S of M is a subpremanifold if and only if for all $p \in S$ there exists a chart $x \colon U \to \mathbb{K}^m$ of M at p such that $x(U \cap S) = x(U) \cap L$, where L is a subspace of $\mathbb{K}^m$ (here x can be chosen such that $L = \mathbb{K}^d \times \{0\}$ with $d = \dim_p(S)$).

This description in particular shows:

Corollary 5.36. *Let M be a premanifold and let S be a subpremanifold. Then S is open in M if and only if $\dim_p(S) = \dim_p(M)$ for all $p \in S$.*

Corollary 5.37. *Let M be a (pre)manifold, S a sub(pre)manifold of M, and let T be a sub(pre)manifold of S. Then T is a sub(pre)manifold of M.*

Proof. Let $i \colon S \to M$ and $j \colon T \to S$ be the inclusions. By Theorem 5.34 1 these are embeddings. Hence $i \circ j$ is an embedding and therefore T is a subpremanifold of M by Theorem 5.34 2. $\square$

Proposition 5.38. *Let $F \colon M \to N$ be a morphism of premanifolds, let $S \subseteq M$ and $T \subseteq N$ be subpremanifolds such that $F(S) \subseteq T$. Then the induced map $F_{|S} \colon S \to T$ is a morphism of premanifolds.*

Proof. Let $i \colon S \to M$ and $j \colon T \to N$ be the inclusions. Then $F \circ i = j \circ F_{|S}$. Hence $j \circ F_{|S}$ is a morphism. As $F_{|S}$ is continuous and j is an immersion, this implies that $F_{|S}$ is a morphism by Proposition 5.25. $\square$

If $F \colon M \to N$ is a morphism of premanifolds, then a fiber of F is in general not a subpremanifold of M. For instance the union of the coordinate axes in $\mathbb{K}^2$ is the fiber $F^{-1}(0)$ of the analytic map $\mathbb{K}^2 \to \mathbb{K}$, $(x, y) \mapsto xy$, and this is not a submanifold of $\mathbb{K}^2$ (see Problem 5.13). The following result and its corollaries give criteria under which a fiber of a morphism of premanifolds is a subpremanifold.

Proposition 5.39. *Let $F: M \to N$ be a morphism of premanifolds, let $q \in N$, and assume that the rank of F is locally constant on an open neighborhood of $F^{-1}(q)$. Then $F^{-1}(q)$ is a subpremanifold of M. For $p \in F^{-1}(q)$ one has*

$$T_p(F^{-1}(q)) = \ker(T_p(F): T_p(M) \to T_q(N)), \qquad (5.10)$$

in particular $\dim_p(F^{-1}(q)) = \dim_p(M) - \mathrm{rk}_p(F)$.

Proof. By the rank theorem 5.22 we can choose for all $p \in F^{-1}(q)$ a chart $(U, x, V, y, \tilde{F})$ of F at p such that $x(p) = 0$, the rank of F on U is constant, and $\tilde{F}$ is given by

$$\mathbb{K}^m \supseteq \tilde{U} \ni (x_1, \dots, x_m) \mapsto (x_1, \dots, x_k, 0 \dots, 0) \in \tilde{V} \subseteq \mathbb{K}^n.$$

By Remark 5.32 4 it suffices to show that $U \cap F^{-1}(q)$ is a submanifold. By Remark 5.32 3 we may assume that $F = \tilde{F}$. But then $F^{-1}(q) = \tilde{F}^{-1}(0) = \tilde{U} \cap W$ with $W = \{(0, \dots, 0, x_{k+1}, \dots, x_m) \in \mathbb{K}^m\}$ and the claim follows from Example 5.33. $\qquad\square$

Corollary 5.40. *Let $F: M \to N$ be a morphism of premanifolds, let $q \in N$, and assume that F is a submersion at p for all $p \in F^{-1}(q)$. Then $F^{-1}(q)$ is a submanifold of M with $\dim_p(F^{-1}(q)) = \dim_p(M) - \dim_q(N)$ for all $p \in F^{-1}(q)$.*

A point $q \in N$ such that F is a submersion at p for all $p \in F^{-1}(q)$ is called a *regular value*. Note that all $q \in N \setminus f(M)$ are regular values.

Proof. By Corollary 5.20 there exists an open neighborhood U of $F^{-1}(q)$ such that $F_{|U}$ is a submersion and in particular $F_{|U}$ has locally constant rank. Hence we may apply Proposition 5.39. $\qquad\square$

Corollary 5.41. *Let $U \subseteq \mathbb{K}^m$ be open and let $F: U \to \mathbb{K}$ be a C^α-function. Let $q \in \mathbb{K}$ such that for all $p \in F^{-1}(q)$ there exists $1 \le i \le m$ with $\frac{\partial F}{\partial x_i}(p) \ne 0$. Then $M := F^{-1}(q)$ is a C^α-submanifold of U (and hence of $\mathbb{K}^m$) of dimension $m - 1$ with*

$$T_p(M) = \left\{ y \in \mathbb{K}^m ; \sum_{i=1}^m y_i \frac{\partial F}{\partial x_i}(p) = 0 \right\}.$$

Example 5.42. Let Q be a non-degenerate quadratic form on $\mathbb{R}^m$. Then

$$S := \{ x \in \mathbb{R}^m ; Q(x) = 1 \}$$

is an analytic submanifold of $\mathbb{R}^m$: By the classification of real quadratic forms we may assume that $Q(x) = x_1^2 + \cdots + x_k^2 - x_{k+1}^2 - \cdots - x_m^2$ after a linear change of coordinates.

Figure 5.3 Quadrik

As a polynomial function, Q is real analytic. For all $p = (p_1, \ldots, p_m) \in \mathbb{R}^m$ we have $\partial_i Q(p) \neq 0$ if $p_i \neq 0$. Hence we can apply Corollary 5.41 to $F = Q$ and $q = 1$. Fig. 5.3 shows the set

$$\left\{ x \in \mathbb{R}^3 \; ; \; x_1^2 + x_2^2 - x_3^2 = 1 \right\}.$$

In particular we see that the *n-sphere*

$$S^n := \left\{ (x_1, \ldots, x_{n+1}) \in \mathbb{R}^{n+1} \; ; \; x_1^2 + \cdots + x_{n+1}^2 = 1 \right\}$$

is a real analytic submanifold of $\mathbb{R}^{n+1}$ with

$$T_p(S^n) = \left\{ (y_1, \ldots, y_{n+1}) \in \mathbb{R}^{n+1} \; ; \; \sum_{i=1}^{n+1} y_i x_i = 0 \right\}, \qquad p = (x_1, \ldots, x_{n+1}) \in S^n.$$

We conclude this section by stating several important results that we will neither prove nor use in the sequel.

Remark and Definition 5.43. The first is Sard's Theorem, which is the fact that for a morphism $F \colon M \to N$ of manifolds "most" values are regular. More precisely, let N be a second countable premanifold. Then every atlas has a countable subatlas (Proposition 1.4 1). We say that a subset $Z \subseteq N$ is a *zero set* if there exists a countable atlas $(U_n, \Phi_n)_{n \in \mathbb{N}}$ of N such that $\Phi_n(Z \cap U_n)$ is contained in a subset of $\mathbb{K}^m$ that has Lebesgue measure 0. Then Z has this property for all countable atlases (as the image of a set of measure zero under a C^1-diffeomorphism is again a set of measure zero by the transformation formula).

We can now state the theorem of Sard (see [Ste] II, §3 for a proof for real premanifolds, the case of complex premanifolds can be reduced to the real analytic case by Remark 5.13).

Theorem 5.44 (Theorem of Sard). *Let M and N be second countable premanifolds and let $F: M \to N$ be a morphism of premanifolds. Suppose that M and N are either complex premanifolds or that M and N are real C^α-premanifolds such that $\alpha \geq \max\{\dim_{F(p)} N - \dim_p M + 1, 1\}$ for all $p \in M$. Then the set of non-regular values of F is a subset of measure zero in N.*

Remark 5.45. Sometimes it can be useful to know whether a premanifold M can be embedded into the standard affine space $\mathbb{K}^N$ for some N. As subspaces of Hausdorff (respectively second countable) spaces are again Hausdorff (respectively second countable) a necessary criterion certainly is that M is a manifold. We list the following results.

Suppose that M is a connected real C^α-manifold of dimension $n \in \mathbb{N}_0$ with $\alpha \in \widehat{\mathbb{N}}$. Then there exists a closed embedding $i: M \hookrightarrow \mathbb{R}^N$ ([Whi1] IV, Theorem 1A for $1 \leq \alpha \leq \infty$ with $N = 2n$; the cases $2 \leq \alpha \leq \infty$ with $N = 2n + 1$ follow quite easily from Sard's Theorem, see [Ste] II Theorem 4.4; for $\alpha = \omega$ this is shown in [Gra] Theorem 3).

Assume that M is a complex manifold. Then there do not exist holomorphic embeddings into $\mathbb{C}^N$ in general. For instance, if M is compact and connected, then for all N every holomorphic map $M \to \mathbb{C}^N$ is constant (Problem 4.15). Complex manifolds M such that there exists $N \in \mathbb{N}$ and a closed holomorphic embedding $i: M \to \mathbb{C}^N$ are called *Stein manifolds*[1]. Every non-compact connected complex manifold of dimension 1 is a Stein manifold ([GuRo] Chap. IX, Sect. B, Theorem 10).

If M is a compact complex manifold, one can ask whether there exists an embedding of M into some other nice "standard space", for instance an embedding $M \hookrightarrow \mathbb{P}^N(\mathbb{C})$ (automatically closed) for some $N \in \mathbb{N}$. But a result of Chow ([GuRo] Chap. V, Sec. D, Theorem 7) shows that if there exists such an embedding, then M is *projective*, i.e., one can find homogeneous polynomials $f_1, \ldots, f_r \in \mathbb{C}[T_0, \ldots, T_N]$ such that

$$M = \left\{ z = (z_0 : \cdots : z_N) \in \mathbb{P}^N(\mathbb{C}) \, ; \, f_1(z) = \cdots = f_r(z) = 0 \right\}.$$

This shows that compact manifolds that are embeddable in projective space are rather special. For instance complex tori T (i.e., $T = V/\Lambda$, where V is a finite-dimensional $\mathbb{C}$-vector space of dimension g and $\Lambda = \mathbb{Z}v_1 \oplus \cdots \oplus \mathbb{Z}v_{2g}$ for $\mathbb{R}$-linearly independent vectors $v_1, \ldots, v_{2g} \in V$) are almost never embeddable into projective space if $\dim(T) > 1$ (more

---

[1] This is not the usual definition of a Stein manifold, and the fact that this property is equivalent to the usual definition is a deep theorem by Remmert ([GuRo] Chap. VII, C, Theorem 13); it also shows that one can always choose $N = 2n + 1$ if M is of (complex) dimension $n \in \mathbb{N}$.

precisely the locus of projective complex tori in the g^2-dimensional space classifying all complex tori has dimension $g(g + 1)/2$.

5.5 Fiber Products of Manifolds

In Corollary 5.40 we saw a criterion for the fiber of a submersion to be a submanifold. This is a special case of the following existence criterion of fiber products of premanifolds. Let $F: M \to S$ and $G: N \to S$ be morphisms of premanifolds and define a set

$$M \times_S N := \{ (p, q) \in M \times N \; ; \; F(p) = G(q) \} . \qquad (5.11)$$

We endow $M \times_S N$ with the subspace topology of $M \times N$. In other words, $M \times_S N$ is the fiber product of F and G in the category of topological spaces (Appendix Example 13.46). If G is the inclusion of a subspace T of S, then $M \times_S T = F^{-1}(T)$ is the inverse image. In particular, if T consists of a single point $s \in S$, then $M \times_S \{s\} = F^{-1}(s)$ is the fiber in s.

The goal in this section is to give a criterion when $M \times_S N$ is a subpremanifold of $M \times N$. In this case we call $M \times_S N$ the *fiber product premanifold of F and G*. It follows then from Proposition 5.38 that $M \times_S N$ is in fact a categorical fiber product (Appendix Definition 13.43) in the category of premanifolds. It is even a fiber product in the category of all $\mathbb{K}$-ringed spaces (Problem 5.20).

Definition 5.46. Let $F: M \to S$ and $G: N \to S$ be morphisms of premanifolds and let $z = (p, q) \in M \times N$ with $F(p) = G(q) =: s$. Then F and G are called *transversal at z* if $T_p(F) + T_q(G): T_p(M) \oplus T_q(N) \to T_s(S)$ is surjective.

Theorem 5.47. *Let $F: M \to S$ and $G: N \to S$ be morphisms of premanifolds, let $Z := M \times_S N$ and suppose that F and G are transversal at all points z of Z. Then Z is a subpremanifold of $M \times N$ and for all $z = (p, q) \in Z$ we have*

$$T_z Z = \{ (\xi, \eta) \in T_p M \times T_q N \; ; \; T_p(F)(\xi) = T_q(G)(\eta) \} . \qquad (5.12)$$

If G is the inclusion of a point $s \in S$, then F is transversal to G in all points of $Z = F^{-1}(s)$ if and only if F is a submersion in all points of Z. Hence in this special case the theorem is just Corollary 5.40. In the proof we will reduce to this case.

If G is the inclusion of a submanifold of S, then we identify Z with $F^{-1}(N)$ and (5.12) can be written as

$$T_p(F^{-1}(N)) = T_p(F)^{-1}(T_{F(p)}(N)), \qquad p \in F^{-1}(N). \qquad (5.13)$$

Proof. *(i).* We start with the following remark. Let $(U_i)_i$ be an open covering of S and set $Z_i := F^{-1}(U_i) \times_{U_i} G^{-1}(U_i)$. Then $(Z_i)_i$ is an open covering of Z and Z is a subpremanifold of $M \times N$ if and only if Z_i is a subpremanifold of $F^{-1}(U_i) \times G^{-1}(U_i)$. As also the transversality condition is local on S, we may prove the theorem locally on S.

(ii). Now we show that we may assume that G is the embedding of a submanifold of S. Indeed, we may write $Z = (F \times G)^{-1}(\Delta_S)$, where $\Delta_S \subseteq S \times S$ is the diagonal (a submanifold by Example 5.27). Moreover, we claim that F and G are transversal in $(p, q) \in Z$ if and only if $F \times G$ is transversal to the diagonal of $S \times S$ in (p, q). Indeed, set $P := T_p(F)(T_p M)$, $Q := T_q(G)(T_q N)$, $T := T_s S$, where $s = F(p) = G(q)$. Then the claim follows from the easy linear algebra fact that $P + Q = T$ if and only if $(P \oplus Q) + \Delta_T = T \oplus T$, where Δ_T is the diagonal in $T \oplus T$. Finally

$$\{ (\xi, \eta) \in T_p M \times T_q N \; ; \; T_p(F)(\xi) = T_q(G)(\eta) \}$$
$$\cong \{ (\xi, \eta, \sigma) \in (T_p M \times T_q N) \times T_s S \; ; \; (T_p(F)(\xi), T_q(G)(\eta)) = (\sigma, \sigma) \},$$

which shows that (5.12) holds if and only if the corresponding description of the tangent space of the fiber product of $F \times G$ and Δ_S holds. Hence from now on we may assume that G is a submanifold of S.

(iii). We now reduce to the case that G is the embedding of a single point of s. By (i) we may work locally on S. Hence by Remark 5.35 we may assume there exists an open embedding $\Phi \colon S \to \mathbb{K}^s$ and a subspace L of $\mathbb{K}^s$ such that $\Phi(N) = \Phi(S) \cap L$. Let $f \colon S \to W := \mathbb{K}^s / L$ be the composition of Φ with the projection $\mathbb{K}^s \to W$. Then $f^{-1}(0) = N$ and hence $F^{-1}(N) = (f \circ F)^{-1}(0)$. For $p \in F^{-1}(N)$ the morphism F is transversal to $N \to G$ if and only if $T_p(\Phi \circ F)(T_p M) + L = \mathbb{K}^s$, i.e., if and only if $f \circ F$ is a submersion in p. Hence we may replace F by f, S by W, and N by $\{0\}$.

(iv). Now we apply Corollary 5.40. We obtain that $F^{-1}(N)$ is a subpremanifold of M and hence that the graph of $F_{|F^{-1}(N)}$ is a subpremanifold of $M \times N$. By (5.10) we find

$$T_p(F^{-1}(N)) = \ker(T_p(f) \colon T_p(M) \to W)$$
$$= T_p(\Phi \circ F)^{-1}(L) = T_p(F)^{-1}(T_{F(p)}(N)). \qquad \square$$

Remark 5.48. Let $F \colon M \to S$ and $G \colon N \to S$ be morphisms of premanifolds and let $Z := M \times_S N$. The reduction steps in the proof of Theorem 5.47 show that the openness of a set of points where a morphism is a submersion (Corollary 5.20) implies that $\{ z \in Z \; ; \; F \text{ and } G \text{ are transversal in } z \}$ is open in Z.

Remark 5.49. Let $F \colon M \to S$ and $G \colon N \to S$ be morphisms of premanifolds, let $Z := M \times_S N$, let $\pi \colon Z \to M$ and $\varpi \colon Z \to N$ be the projections, and suppose that F and G are transversal at all points z of Z. Then (5.12) implies that for $z = (p, q) \in Z$,

$s := F(p) = G(q)$ the commutative diagram

$$
\begin{array}{ccc}
T_z Z & \xrightarrow{\ T_z(\pi)\ } & T_p M \\
{\scriptstyle T_z(\varpi)}\downarrow & & \downarrow{\scriptstyle T_p(F)} \\
T_q(N) & \xrightarrow{\ T_q(G)\ } & T_s S
\end{array}
$$

induces isomorphisms

$$
\mathrm{Ker}(T_z(\pi)) \xrightarrow{\sim} \mathrm{Ker}(T_q(G)),
$$
$$
\mathrm{Coker}(T_z(\pi)) \xrightarrow{\sim} \mathrm{Coker}(T_q(G))
$$

(and similarly for $T_z(\varphi)$ and $T_p(F)$).

As submersions are clearly transversal to all other morphisms, Theorem 5.47 and Remark 5.49 imply Corollary 5.50.

Corollary 5.50. *Let $F\colon M \to S$ and $G\colon N \to S$ be morphisms of premanifolds and suppose that F is a submersion. Then $M \times_S N$ is a subpremanifold of $M \times N$ and the projection $M \times_S N \to N$ is a submersion.*

5.6   Quotients of Manifolds

Let M be a premanifold and let $R \subseteq M \times M$ be an equivalence relation, i.e., R is a subset of $M \times M$ such that for $m, m', m'' \in R$ one has

(a) $(m, m') \in R$, $(m', m'') \in R \Rightarrow (m, m'') \in R$,
(b) $(m, m) \in R$,
(c) $(m, m') \in R \Rightarrow (m', m) \in R$.

We denote by M/R the set of equivalence classes. Let $p\colon M \to M/R$ be the projection and endow M/R with the quotient topology, i.e., $V \subseteq M/R$ is open if and only if $p^{-1}(V)$ is open in M.

We study the question when M/R can be endowed with the structure of a premanifold.

Proposition 5.51. *Let N be a topological space, M a premanifold and $p\colon M \to N$ a surjective map. Then there exists at most one structure of a premanifold on N such that p is a submersion.*

Proof. Let N_1 and N_2 be premanifolds with underlying topological space N such that the morphisms $p_1: M \to N_1$ and $p_2: M \to N_2$ with underlying map p are submersions. Now id: $N_1 \to N_2$ is a continuous map and we clearly have id $\circ p_1 = p_2$ as maps of topological spaces. By Proposition 5.25 2 id: $N_1 \to N_2$ is already a morphism of premanifolds. By the same argument id: $N_2 \to N_1$ is a morphism of premanifolds. Therefore N_1 and N_2 carry the same premanifold structure. $\qquad\square$

If there exists on N the structure of a premanifold such that p is a submersion, then Proposition 5.25 2 shows that the structure sheaf of N can be described by

$$\mathcal{O}_N(V) := \left\{ f: V \to \mathbb{K} \text{ map} ;\; f \circ p_{|p^{-1}(V)} \in \mathcal{O}_M(p^{-1}(V)) \right\}.$$

Remark and Definition 5.52 (Quotient manifold). Let M be a premanifold and let $R \subseteq M \times M$ be an equivalence relation. Then Proposition 5.51 shows that there exists at most one structure of a premanifold on M/R such that the projection $p: M \to M/R$ is a submersion. In this case the premanifold M/R is called the *quotient premanifold of M by R* and the equivalence relation R is called *regular*.

Theorem 5.53. *Let M be a premanifold and let $R \subseteq M \times M$ be an equivalence relation. Then the following assertions are equivalent:*

(i) *There exists the structure of a premanifold on M/R such that the projection $p: M \to M/R$ is a submersion.*

(ii) *R is a submanifold of $M \times M$ and the restriction $\mathrm{pr}_1: R \to M$ of the first projection $M \times M \to M$ is a submersion.*

If these assertions hold, M/R is Hausdorff if and only if R is closed in $M \times M$.

Proof. "(i) $\Rightarrow$ (ii)". Assume that M/R is a premanifold in such a way that the projection $p: M \to M/R$ is a submersion. Then we have $R = M \times_{M/R} M$ as topological spaces. Hence, Corollary 5.50 implies (ii).

"(ii) $\Rightarrow$ (i)". Conversely, let R be a submanifold of $M \times M$ such that $\mathrm{pr}_1: R \to M$ is a submersion. Notice that this implies that $\mathrm{pr}_2: R \to M$ is also a submersion because $\tau: R \to R, (x, y) \mapsto (y, x)$ is an isomorphism with $\mathrm{pr}_2 = \mathrm{pr}_1 \circ \tau$.

Idea of proof. The idea to construct the premanifold structure on M/R is rather simple: If we already had constructed this structure and knew that p is a submersion, then locally on an open subspace U of M the morphism p would have a section s (because locally a submersion looks like a surjective linear map that clearly has a section). Then the image of s is a submanifold N of U isomorphic to $p(U)$ and every point of U is equivalent to precisely one point of N. Hence we will construct U and N and then endow $p(U)$ with the premanifold structure of N.

To do this, we again think backwards: If we had already constructed M/R, then for $x_0 \in M$ the tangent map $T_{x_0}(p)$ would be surjective and hence the restriction to a subspace

L' of $T_{x_0}(M)$ is an isomorphism. This would be a complement to the subspace L that we expect to become the kernel of $T_{x_0}(p)$ (see (*) below). If N then is a subpremanifold of M in a neighborhood of x_0 with $T_{x_0}(N) = L'$, then N is locally isomorphic to M/R by the theorem of inverse functions 5.16.

Step 1: p is open. For $U \subseteq M$ open, we have

$$p^{-1}(p(U)) = \{ x \in M \; ; \; \exists \, y \in U : (x, y) \in R \} = \mathrm{pr}_1((M \times U) \cap R).$$

Since the submersion $\mathrm{pr}_1 : R \to M$ is open (Corollary 5.24), $p^{-1}(p(U)) \subseteq M$ is open so that $p(U) \subseteq X/R$ is open by definition of the quotient topology.

Step 2: Locality on M. Next we show that we can define the premanifold structure on M/R locally on M. Assume that there exists an open covering $(U_i)_{i \in I}$ of M and the structure of a premanifold on $p(U_i)$ such that the projection $p_{|U_i} : U_i \to p(U_i)$ is a submersion for $i \in I$. Then $p(U_i \cap U_j)$ is an open subset of $p(U_i)$ and of $p(U_j)$ for $i, j \in I$ so that there are two structures of a premanifold on $p(U_i \cap U_j)$ making it into an open submanifold of $p(U_i)$ or $p(U_j)$ respectively. By assumption we find the map $p_{|U_i \cap U_j} : U_i \cap U_j \to p(U_i \cap U_j)$ to be a submersion for both structures on $p(U_i \cap U_j)$ so that both premanifolds structures on $p(U_i \cap U_j)$ agree by Proposition 5.51. Hence, there exists a unique structure on M/R compatible with the given structure on $p(U_i)$ for $i \in I$ because $(p(U_i))_{i \in I}$ is an open covering of M/R.

After these preparations the main point of the proof is the following.

Main construction. For all $x_0 \in M$ there exists an open neighborhood $U \subseteq M$ of x_0, a submanifold N of U containing x_0 and a morphism $r : U \to N$ of premanifolds such that for all $x \in U$, $r(x)$ is the unique point of N equivalent to x.

Step 3: "Main construction $\Rightarrow$ (i)". Let us first show that the construction yields a premanifold structure on $p(U)$ such that $p : U \to p(U)$ is a submersion (then we have shown (i) by Step 2). As the inclusion of N into U is a right inverse of $r : U \to N$, we find that r is a submersion. The continuous map $p : N \to p(U)$ is bijective and open, hence a homeomorphism. Thus, transporting the premanifold structure of N to $p(U)$ via $p : N \to p(U)$ makes $p : U \to p(U)$ a submersion because we have $p_{|U} = p_{|N} \circ r$.

Step 4: Construction of U, N, and r. Define

$$L := \{ \xi \in T_{x_0}(M) \; ; \; (\xi, 0) \in T_{x_0, x_0}(R) \} \tag{*}$$

and choose a submanifold $\tilde{N}$ of M such that $x_0 \in \tilde{N}$ and $L' := T_{x_0}(\tilde{N})$ is a complementary vector space of L in $T_{x_0}(M)$. Because $\mathrm{pr}_1 : R \to M$ is a submersion, we find $\tilde{N} \times M$ and R to be transversal in $M \times M$ and therefore

$$\Sigma := (\tilde{N} \times M) \cap R$$

is a submanifold of $M \times M$ by Theorem 5.47.

Next we claim that $\mathrm{pr}_2 : \Sigma \to M$ is a local isomorphism at $(x_0, x_0) \in \Sigma$. It suffices to show that $T_{(x_0, x_0)}(\mathrm{pr}_2) : T_{(x_0, x_0)}(\Sigma) \to T_{x_0}(M)$ is an isomorphism (Theorem 5.16). For the surjectivity let $\eta \in T_{x_0}(M)$. There exists $\xi \in T_{x_0}(M)$ such that $(\xi, \eta) \in T_{x_0, x_0}(R)$ because $\mathrm{pr}_2 \colon R \to M$ is a submersion. Choose $\xi_1 \in L, \xi_2 \in L'$ with $\xi_1 + \xi_2 = \xi$ so that

$$(\xi_2, \eta) = (\xi, \eta) - (\xi_1, 0) \in T_{x_0, x_0}(R) \qquad \text{and} \qquad (\xi_2, \eta) \in T_{x_0}(\tilde{N}) \times T_{x_0}(M).$$

Thus $(\xi_2, \eta) \in T_{x_0, x_0}(\Sigma)$ so that $T_{(x_0, x_0)}(\mathrm{pr}_2)$ is surjective. For $(\xi, \eta) \in \ker(T_{(x_0, x_0)}(\mathrm{pr}_2))$ we have $\eta = 0$ and $(\xi, 0) \in T_{x_0, x_0}(\Sigma)$. Hence, we must have $(\xi, 0) \in T_{x_0, x_0}(R)$ and $\xi \in T_{x_0}(\tilde{N})$ and therefore $\xi \in L \cap L' = 0$ so that $T_{(x_0, x_0)}(\mathrm{pr}_2)$ is injective.

By definition there exist open neighborhoods $W, V \subseteq M$ of x_0 such that $\mathrm{pr}_2 \colon \Sigma \cap (W \times W) \to V$ is an isomorphism of premanifolds. In particular, its inverse has to be of the form

$$V \to \Sigma \cap (W \times W), \qquad x \mapsto (r(x), x)$$

for a morphism $r \colon V \to \tilde{N} \cap W$ and we must have $V \subseteq W$. Moreover, for $x \in \tilde{N} \cap V$ we have $(x, x) \in \Sigma \cap (W \times W)$ so that $r(x) = x$ because $\mathrm{pr}_2 \colon \Sigma \cap (W \times W) \to V$ is bijective.

Set $U := V \cap r^{-1}(\tilde{N} \cap V)$ and $N := \tilde{N} \cap U$. Then $\tilde{N} \cap V \subseteq U$ because $r_{|\tilde{N} \cap V} = \mathrm{id}$. Hence $r(U) \subseteq U$ and we obtain a morphism $r \colon U \to N$. We have $x_0 \in U$ and hence $x_0 = r(x_0) \in N$. For $x \in U$ we have $(r(x), x) \in R$ so that $r(x)$ is equivalent to x. Conversely, suppose that $x \in U$ and $y \in N$ are equivalent. We have $y, x \in V \subseteq W$, $(y, x) \in R$ and $y \in \tilde{N}$ so that $(y, x) \in \Sigma \cap (W \times W)$. Because $\mathrm{pr}_2 \colon \Sigma \cap (W \times W) \to V$ is bijective, we must have $(y, x) = (r(x), x)$.

Step 5: "M/R Hausdorff $\Leftrightarrow R \subseteq M \times M$ closed". By Step 1, $p \times p$ is open and surjective. Hence the diagonal $\Delta_{M/R}$ is closed in $M/R \times M/R$ if and only if $(p \times p)^{-1}(\Delta_{M/R}) = R$ is closed in $M \times M$. $\square$

Proposition 5.54. *Let $F \colon M \to N$ be a morphism of premanifolds. Let R and S be regular equivalence relations on M and N respectively. Suppose that F maps equivalence classes with respect to R to equivalence classes with respect to S. Then the induced map $\bar{F} \colon M/R \to N/S$ is a morphism of premanifolds.*

Proof. Let $\pi_M \colon M \to M/R$ and $\pi_N \colon N \to N/S$ the projections. Then $\bar{F} \circ \pi_M = \pi_N \circ F$. Hence $\bar{F} \circ \pi_M$ is a morphism. As π_M is a surjective submersion, this implies that $\bar{F}$ is a morphism (Proposition 5.25). $\square$

5.7   Problems

Problem 5.1. Let M be a real C^α-premanifold with $\alpha \geq \infty$ or a complex manifold, let $p \in M$, and $f \in \mathcal{O}_{M,p}$.

1. Choose a representative $(U, \tilde{f})$ of the germ f and a chart (U, Φ) at p and define the *vanishing order of f at p* as

$$\operatorname{ord}_p(f) := \inf \left\{ r \in \mathbb{N}_0 \, ; \, D^r(f \circ \Phi^{-1})(\Phi(p)) \neq 0 \right\}.$$

   Show that $\operatorname{ord}_p(f)$ does not depend on the choice of U, $\tilde{f}$, or Φ and that

$$\operatorname{ord}_p(f) = \sup \left\{ r \in \mathbb{N}_0 \, ; \, f \in \mathfrak{m}_p^r \right\}.$$

2. Let $(V, \| \, \|)$ be a normed $\mathbb{K}$-vector space of finite dimension and suppose that $M \subseteq V$ is open. Show that

$$\operatorname{ord}_p(f) = \sup \left\{ r \in \mathbb{R}_{\geq 0} \, ; \, \lim_{M \ni x \to p} \|x - p\|^{-r} |f(x)| = 0 \right\}$$

   and that $\operatorname{ord}_p(f)$ is the supremum of all $r \in \mathbb{N}_0$ such that the Taylor expansion up to degree r of f at p is zero.
3. Let M be a real analytic or a complex premanifold. Show that $\operatorname{ord}_p(f) = \infty$ if and only if $f = 0$ and hence $\bigcap_{r \geq 0} \mathfrak{m}_p^r = 0$.

Problem 5.2. Let $p \in \mathbb{R}^n$. Show that attaching to $f \in C^\infty_{\mathbb{R}^n, p}$ its Taylor series in p (in variables $t_1, \ldots, t_n$) induces an isomorphism of $\mathbb{R}$-algebras

$$C^\infty_{\mathbb{R}^n, p} \bigg/ \left(\bigcap_{r \geq 0} \mathfrak{m}_p^r \right) \xrightarrow{\sim} \mathbb{R}[\![t_1, \ldots, t_n]\!],$$

where the right-hand side denotes the $\mathbb{R}$-algebra of formal power series in $t_1, \ldots, t_n$ with values in $\mathbb{R}$.
Hint: The injectivity is Problem 5.1.

Problem 5.3. Let M be a real C^α-premanifold with $1 \leq \alpha < \infty$, let $p \in M$ and assume that $\dim_p(M) > 0$. We wish to show that $\mathfrak{m}_p/\mathfrak{m}_p^2$ is an infinite-dimensional $\mathbb{R}$-vector space.

1. Show that it suffices to consider the case $M = \mathbb{R}$ and $p = 0$, which we assume from now on.

2. For $f \in \mathfrak{m} := \mathfrak{m}_0$ define

$$o(f) := \mathrm{ord}_0(f) := \sup \left\{ \beta \geq 0 \; ; \; \lim_{x \to 0} |x|^{-\beta} f(x) = 0 \right\}.$$

Show that $o(f) > \alpha + 1$ or $o(f) \in \mathbb{N}$.

3. Show that $\left\{ |x|^\beta \; ; \; \alpha < \beta < \alpha + 1 \right\}$ is linearly independent in $\mathfrak{m}/\mathfrak{m}^2$.

Problem 5.4. Let M and N be real C^α-premanifolds with $\alpha \geq \infty$ or let M and N be complex premanifolds. Let $f: M \to N$ be a morphism of premanifolds. Show that f is a local isomorphism if and only if $f_x: \mathcal{O}_{Y,f(x)} \to \mathcal{O}_{X,x}$ is an isomorphism for all $x \in X$.

Problem 5.5. For every $\mathbb{K}$-vector space V define the $\mathbb{K}$-algebra $\mathbb{K}[V]$ as follows. As a $\mathbb{K}$-vector space we set $\mathbb{K}[V] := \mathbb{K} \oplus V$. Define a multiplication on $\mathbb{K}[V]$ by

$$(\lambda, v) \cdot (\lambda', v') := (\lambda \lambda', \lambda v' + \lambda' v).$$

1. Show that these definitions make $\mathbb{K}[V]$ into a commutative local $\mathbb{K}$-algebra whose maximal ideal is V. Show that $V^2 = 0$ and that $\mathbb{K}[\mathbb{K}] \cong \mathbb{K}[T]/(T^2) =: \mathbb{K}[\varepsilon]$.
2. Show that for every $\mathbb{K}$-linear map $u: V \to W$ of $\mathbb{K}$-vector spaces the map $\mathbb{K}[V] \to \mathbb{K}[W]$, $(\lambda, v) \mapsto (\lambda, u(v))$, is a homomorphism of $\mathbb{K}$-algebras and that one obtains a functor $V \mapsto \mathbb{K}[V]$ from the category of $\mathbb{K}$-vector spaces to the category of $\mathbb{K}$-algebras.
3. Let $P[V]$ be the locally $\mathbb{K}$-ringed space consisting of a singe point $*$ and with $\mathcal{O}_{P[V]}(P[V]) = \mathbb{K}[V]$. Let M be a real C^α-premanifold with $\alpha \geq \infty$ or a complex manifold. Show that the identification (5.7) yields for all $r \in \mathbb{N}$ an identification

$$\mathrm{Hom}_{\mathbb{K}}(V, T_p(M)) = \left\{ \begin{array}{l} t: P[V] \to M \text{ morphism of locally} \\ \mathbb{K}\text{-ringed spaces with } t(*) = p \end{array} \right\}. \qquad (*)$$

Show that addition $\mathbb{K} \times \mathbb{K} \to \mathbb{K}$ and for $\lambda \in \mathbb{K}$ scalar multiplication $\mathbb{K} \to \mathbb{K}$, $\alpha \mapsto \lambda \alpha$ induce via (*) the $\mathbb{K}$-vector space structure on $T_p(M)$ defined in Definition 5.1.

Problem 5.6. Let X be a topological space and let $\rho: X \to \mathbb{N}_0$ be a map such that $\{x \in X \; ; \; \rho(x) \geq r\}$ is open in X for all $r \in \mathbb{N}_0$ and such that there exists an open covering $(U_i)_i$ such that $\rho_{|U_i}$ is bounded for all i. Show that

$$\left\{ x \in X \; ; \; \exists \text{ neighborhood } W \text{ of } x \text{ such that } \rho_{|W} \text{ is constant} \right\}$$

is open and dense in X.

Deduce that if $F: M \to N$ is a morphism of premanifolds, then the set of points of M where F is of locally constant rank is open and dense in M.

Problem 5.7. Show that the underlying real analytic manifold of $\mathbb{P}^1(\mathbb{C})$ is isomorphic to the 2-sphere S^2.

Problem 5.8. Let $X := \{ (x, |x|) \in \mathbb{R}^2 \, ; \, x \in \mathbb{R} \}$. Is X a C^1-submanifold of $\mathbb{R}^2$?

Problem 5.9. Let M be a compact premanifold of constant dimension n. Show that M cannot be embedded into $\mathbb{K}^n$.

Problem 5.10. Let $n \in \mathbb{N}$ and consider det: $M_n(\mathbb{K}) \to \mathbb{K}$. Show that det is a morphism of real analytic (for $\mathbb{K} = \mathbb{R}$) manifolds respectively of complex manifolds (for $\mathbb{K} = \mathbb{C}$). Determine $\mathrm{rk}_p(\det)$ for all $p \in M_n(\mathbb{K})$.

Problem 5.11. Show that $F \colon (0, 2\pi) \to \mathbb{R}^2$, $t \mapsto (\sin^3 t, \sin t \cos t)$ is an injective immersion of real manifolds. Define a map

$$G \colon (0, 2\pi) \to (0, 2\pi), \qquad t \mapsto \begin{cases} t + \pi, & \text{if } t < \pi; \\ 0, & \text{if } t = \pi; \\ t - \pi, & \text{if } t > \pi. \end{cases}$$

Show that $F \circ G$ is a morphism of real manifolds but that G is not continuous.

Problem 5.12. Let $F \colon M \to N$ be an embedding of premanifolds, let Z be a premanifold, and let $G \colon Z \to M$ be a map. Show that G is a morphism of premanifolds if and only if $F \circ G$ is a morphism of premanifolds.

Problem 5.13. Show that $S := \{ (x, y) \in \mathbb{R}^2 \, ; \, xy = 0 \}$ is not a real C^1-submanifold of $\mathbb{R}^2$.
Hint: What would be $T_0(S)$?

Problem 5.14. Let F be an injective (respectively bijective) morphism of premanifolds that is locally of constant rank. Show that F is an immersion (respectively an isomorphism).

Problem 5.15. Let V and W be finite-dimensional $\mathbb{K}$-vector spaces, $n := \dim_{\mathbb{K}}(V)$ and $m := \dim_{\mathbb{K}}(W)$ and let r be an integer with $0 \leq r \leq \min\{m, n\}$. Show that $\{ f \in \mathrm{Hom}_{\mathbb{K}}(V, W) \, ; \, \mathrm{rk}(f) = r \}$ is a analytic submanifold of $\mathrm{Hom}_{\mathbb{K}}(V, W)$ of dimension $r(m + n - r)$.

Problem 5.16. Let M be a premanifold. A pair (N, i), where N is a premanifold and $i \colon N \to M$ is an injective immersion is called *immersed subpremanifold of M*.

1. Show that every submanifold together with the inclusion is an immersed subpremanifold but that the converse does not hold in general.
2. Let (N, i) be an immersed subpremanifold and suppose that N is compact and M is Hausdorff. Show that $i(N)$ is a closed subpremanifold of M.

Problem 5.17. Consider the real analytic map $\gamma \colon \mathbb{R} \to \mathbb{R}^2$, $t \mapsto (\sin^3 t, \sin t \cos t)$ and let $N_1 := (-\pi, \pi)$ and $N_2 := (0, 2\pi)$ considered as open submanifolds of $\mathbb{R}$. Show that $(N_1, \gamma_{|N_1})$ and $(N_2, \gamma_{|N_2})$ are immersed subpremanifolds of M, that $\gamma(N_1) = \gamma(N_2)$ but that there exists no isomorphism of manifolds $\varphi \colon N_1 \xrightarrow{\sim} N_2$ such that $\gamma_{|N_1} = \gamma_{|N_2} \circ \varphi$.

Problem 5.18. Let M be a premanifold.

1. Show that for an immersed premanifold (N, i) (Problem 5.16) the following assertions are equivalent:
    (i) For each premanifold Z a map $Z \to N$ is a morphism of premanifolds if and only if $i \circ f \colon Z \to M$ is a morphism of premanifolds.
    (ii) For each premanifold Z a map $Z \to N$ is continuous if and only if $i \circ f \colon Z \to M$ is a morphism of premanifolds.
    In this case (N, i) is called *initial subpremanifold of M*.
2. Show that every subpremanifold together with the inclusion is an initial subpremanifold.
    Hint: Problem 5.12.
3. Let (N_1, i_1) and (N_2, i_2) be initial subpremanifolds of M with $i_1(N_1) = i_2(N_2)$. Show that there exists a unique isomorphism $F \colon N_1 \xrightarrow{\sim} N_2$ such that $i_1 = i_2 \circ F$ (compare to Problem 5.17).

Problem 5.19. Let $r \in \mathbb{R} \setminus \mathbb{Q}$ and $\mathbb{R} \to \mathbb{R}/\mathbb{Z}$, $x \mapsto \bar{x}$ the canonical map. Show that $i \colon \mathbb{R} \to \mathbb{R}/\mathbb{Z} \times \mathbb{R}/\mathbb{Z}$, $x \mapsto (\bar{x}, \overline{rx})$ is an injective immersion with dense image. Deduce that i is not an embedding of manifolds.

Problem 5.20. Show that the fiber product constructed in Theorem 5.47 is a fiber product in the category of all $\mathbb{K}$-ringed spaces.

Problem 5.21. Let G be a finite group of automorphisms of a premanifold M and let $M^G := \{\, p \in M \; ; \; g(p) = p \text{ for all } g \in G \}$.

1. Show that for all $p \in M^G$ there exists a system of local coordinate at p with respect to which G acts linearly.
2. Show that M^G is a submanifold of M and that $T_p(M^G) = T_p(M)^G$ for $p \in M^G$.

Problem 5.22. Let $F \colon M \to N$ be a submersion of premanifolds. Show that the equivalence relation $p \sim q :\Leftrightarrow F(p) = F(q)$ is regular and that $M/\!\sim \,\cong F(M)$ as open premanifolds of N.

Lie Groups

6

A Lie group is a premanifold together with a group structure such that multiplication and inversion are morphisms of premanifolds. They play an important role as symmetry groups of premanifolds and this is also the main focus of this chapter. Hence after defining Lie groups and Lie subgroups and studying their topology we will consider Lie groups acting on premanifolds in Sect. 6.2. The main result is the existence of quotients of premanifolds by a proper and free Lie group action in the last section.

6.1 Definition and Examples of Lie Groups

Definition 6.1. A *real Lie group* (respectively a *complex Lie group*) is a real analytic premanifold (respectively a complex premanifold) G endowed with a group structure such that

$$m_G: G \times G \to G, \qquad (g, h) \mapsto gh,$$
$$i_G: G \to G, \qquad g \mapsto g^{-1}$$

are real analytic maps (respectively holomorphic maps). As usual, we speak simply of a *Lie group* if we consider the real and the complex case simultaneously.

A *homomorphism of Lie groups* is a morphism of premanifolds that is a group homomorphism.

We obtain the categories of real and of complex Lie groups.

In the real case we could also have defined Lie groups and their homomorphisms via C^α-premanifolds with some fixed $\alpha \in \mathbb{N}_0 \cup \{\infty\}$. But one would obtain the same notion: One can show that weakening the structure from analytic to C^α (Remark 4.23) yields an equivalence from the category of real analytic Lie groups to the category of real C^α-Lie groups ([BouLie1] III, §8.1).

© Springer Fachmedien Wiesbaden 2016

T. Wedhorn, *Manifolds, Sheaves, and Cohomology*, Springer Studium Mathematik – Master,

DOI 10.1007/978-3-658-10633-1_6

Remark and Definition 6.2. The group law on a complex Lie group G defines on the underlying real analytic premanifold $G_{\mathbb{R}}$ the structure of a real Lie group. We obtain a functor $G \mapsto G_{\mathbb{R}}$ from the category of complex Lie groups to the category of real Lie groups.

One can show that this functor has a left adjoint functor $H \mapsto H_{\mathbb{C}}$, and $H_{\mathbb{C}}$ is called the *complexification of the real Lie group* H ([HiNe] 15.1.4).

The following result in particular shows that the analyticity of the multiplication already implies that the inversion is analytic.

Proposition 6.3. *Let G be a real analytic premanifold or a complex premanifold endowed with a group structure such that $m_G: G \times G \to G$, $(g, h) \mapsto gh$ is a morphism.*

1. *Then G is a Lie group.*
2. *For $g \in G$ the maps $l_g: G \to G$, $h \mapsto gh$ and $r_g: G \to G$, $h \mapsto hg$ are isomorphisms of premanifolds.*
3. *The map $i_G: G \to G$, $g \mapsto g^{-1}$, is an isomorphism of premanifolds.*

Proof. 2. Both maps l_g and r_g are bijective morphisms and their inverse, namely $l_{g^{-1}}$ and $r_{g^{-1}}$ respectively, is again a morphism.

1. Let $\sigma: G \times G \to G \times G$, $(g, h) \mapsto (g, gh)$. For all $(g, h) \in G \times G$, $T_{(g,h)}(\sigma)$ has the form

$$\begin{pmatrix} \mathrm{id}_{T_g G} & 0 \\ T^1_{(g,h)}(m_G) & T^2_{(g,h)}(m_G) \end{pmatrix},$$

where $T^i m_G$ denotes the i-th partial derivative of m_G. As $T^2_{(g,h)}(m_G) = T_h(l_g)$ is an isomorphism by 2, $T_{(g,h)}(\sigma)$ is an isomorphism. Therefore σ is a local isomorphism (Theorem 5.16) and hence an isomorphism because σ is bijective. For $g \in G$ we have $\sigma^{-1}(g, e) = (g, g^{-1})$. Hence i_G is a morphism.

3. The morphism i_G has as inverse again a morphism of premanifolds, namely itself. $\square$

Example 6.4. Let $n \in \mathbb{N}_0$.

1. Let V be a finite-dimensional $\mathbb{R}$- (respectively $\mathbb{C}$-) vector space. Then $(V, +)$ is a real (respectively complex) Lie group.
2. $GL_n(\mathbb{R})$ is real Lie group, $GL_n(\mathbb{C})$ is a complex Lie group: multiplication of matrices is polynomial and in particular analytic, we conclude by Proposition 6.3 (it also follows directly from Cramer's rule, that inversion of matrices is also polynomial).
   More generally, if V is a finite-dimensional $\mathbb{R}$- (respectively $\mathbb{C}$-) vector space, then $GL(V)$ is a real (respectively complex) Lie group.

A Lie group has the following topological properties.

Proposition 6.5. *Let G be a Lie group.*

1. *G is Hausdorff and paracompact. All connected components of G are open and closed in G and have the same dimension.*
2. *The underlying premanifold of G is a manifold if and only if G has countably many connected components.*

Proof. Every premanifold is a T_1-space. In particular $\{e\}$ is closed in G, where $e \in G$ is the neutral element. Therefore G is Hausdorff by Appendix Proposition 12.58.

2. If G has uncountably many connected components, G cannot be second countable. Conversely, suppose that G has countably many connected components. As countable unions of second countable spaces are again second countable, it suffices to show that a connected Lie group is second countable. This follows from Corollary 1.6.

1. As G is locally connected, its connected components are open and closed (Proposition 2.13). If G^1 is a connected component, then left multiplication with any $g \in G^1$ yields an isomorphism $G^0 \xrightarrow{\sim} G^1$, where G^0 is the identity component of G. In particular $\dim(G^1) = \dim(G^0)$. By 2, G^0 is a manifold and hence paracompact (Theorem 4.28). Therefore G is the topological sum of its paracompact connected components and hence paracompact. $\qquad\square$

Definition 6.6. Let G be a Lie group. A *Lie subgroup of G* is a subgroup of G that is a subpremanifold.

Similarly, as there are more general notions of submanifolds, there are also more general notions of Lie subgroups.

Proposition 6.7. *Let H be a Lie subgroup of G.*

1. *H is a Lie group.*
2. *H is closed in G.*

Proof. 1. By Proposition 6.3 it suffices to show that the multiplication $H \times H \to H$ is a morphism. This follows from Proposition 5.38 applied to the multiplication morphism m_G.

2. As a submanifold, H is locally closed in G. Every locally closed topological subgroup is closed (Appendix Proposition 12.58). $\qquad\square$

It can be shown that if G is a real Lie group and H is any closed subgroup, then H is a real Lie subgroup of G (e.g., [BouLie1] III, §8.2, Theorem 2). The analogous result does not hold for complex Lie groups (for instance, $\mathbb{R} \subseteq \mathbb{C}$ is a closed subgroup of the complex Lie group $(\mathbb{C}, +)$ but it is not a complex Lie group).

6.2   Lie Group Actions

We now consider the important case that a Lie group acts on a premanifold. Let us first recall some notions of a group G acting on a set X. Then X is called a *G-set* and the *action* is a map $G \times X \to X$, usually simply denoted by $(g, x) \mapsto gx$, that satisfies $1x = x$ and $g(hx) = (gh)x$ for all $g, h \in G$ and $x \in X$.

1. The action is called *transitive* (respectively *simply transitive*) if for all $x, y \in X$ there exists a (respectively there exists a unique) $g \in G$ with $gx = y$.
2. For $x \in X$ the subgroup $G_x := \mathrm{Stab}_G(x) := \{ g \in G \; ; \; gx = x \}$ of G is called the *stabilizer of x in G*. The subset $Gx := \{ gx \; ; \; g \in G \}$ is called the *G-orbit of x*. The map $G \mapsto X$, $g \mapsto gx$, induces a bijection $G/G_x \xrightarrow{\sim} Gx$ of sets. The set of G-orbits in X is denoted by X/G.
3. The action is called *free* if for all $x \in X$ its stabilizer G_x is trivial.
4. The action is called *faithful* if $1 \in G$ is the only element $g \in G$ such that $x \mapsto gx$ is the identity for all $x \in X$ (in other words $\bigcap_{x \in X} G_x = \{1\}$).

If X and Y are G-sets, a map $f \colon X \to Y$ is called *G-equivariant* of *morphism of G-sets*, if $f(gx) = gf(x)$ for all $x \in X$ and $g \in G$. Then G-sets and G-equivariant maps together with the usual composition form a category.

Definition 6.8. Let G be a real (respectively complex) Lie group. A *C^α-G-premanifold* (respectively a *complex G-premanifold*) is a real C^α-premanifold (respectively a complex premanifold) M together with a C^α-map (respectively a holomorphic map) $G \times M \to M$, which defines an action of the group G on the set M. In the real case we consider the analytic premanifold G as a C^α-premanifold (Remark 4.23).

If we consider the cases of a real Lie group and a real C^α-premanifold ($\alpha \in \widehat{\mathbb{N}}$ fixed) and the complex case simultaneously, we simply speak of a *G-premanifold* or of the Lie group G *acting on the premanifold M*.

A *morphism of G-premanifolds* is a G-equivariant morphism of premanifolds.

We obtain the category of G-premanifolds.

Proposition 6.9. *Let G be a Lie group, let M and N be G-premanifolds, and let $F \colon M \to N$ be a morphism of G-premanifolds. For all $p \in M$ and $g \in G$ one has*

$$\dim_p(M) = \dim_{gp}(M), \qquad \mathrm{rk}_p(F) = \mathrm{rk}_{gp}(F).$$

Moreover, F is a submersion (respectively immersion) at p if and only if F is a submersion (respectively immersion) at gp.

Proof. The map $g_M: M \to M$, $m \mapsto gm$ is an isomorphism of premanifolds with inverse $(g^{-1})_M$. Set $q := F(p) \in N$. Then $F(gp) = gq$ and all assertions follow from the commutative diagram

$$
\begin{array}{ccc}
T_p(M) & \xrightarrow{\ T_p(F)\ } & T_q(N) \\
{\scriptstyle T_p(g_M)}\Big\downarrow & & \Big\downarrow{\scriptstyle T_q(g_N)} \\
T_{gp}(M) & \xrightarrow{\ T_{gp}(F)\ } & T_{gq}(N),
\end{array}
$$

where the vertical arrows are isomorphisms. $\square$

Corollary 6.10. *Notation as in Proposition 6.9. Assume that the action of G on M is transitive. Then F is of constant rank and for all $q \in N$ the fiber $F^{-1}(q)$ is a subpremanifold of M.*

Proof. By Proposition 6.9 the map $M \ni p \mapsto \mathrm{rk}_p(F)$ is constant. The second assertion follows from Proposition 5.39. $\square$

Example 6.11. Let $\beta: \mathbb{K}^n \times \mathbb{K}^n \to \mathbb{K}$ be a bilinear form with matrix $B \in M_n(\mathbb{K})$ with respect to the standard basis of $\mathbb{K}^n$. Let

$$
\begin{aligned}
G &:= \{\, g \in \mathrm{GL}_n(\mathbb{K}) \, ; \, \beta(gv, gw) = \beta(v, w) \ \forall\, v, w \in \mathbb{K}^n \,\} \\
&= \{\, g \in \mathrm{GL}_n(\mathbb{K}) \, ; \, {}^t g B g = B \,\}.
\end{aligned}
$$

This is a closed subgroup of $\mathrm{GL}_n(\mathbb{K})$. For instance $G = O(p, q)$ for $B = \begin{pmatrix} I_p & 0 \\ 0 & -I_q \end{pmatrix}$ and $\mathbb{K} = \mathbb{R}$, or $B = \mathrm{Sp}_{2m}(\mathbb{K})$ for $n = 2m$ and $B = \begin{pmatrix} 0 & I_m \\ -I_m & 0 \end{pmatrix}$.

Then $G = F^{-1}(B)$, where

$$
F: \mathrm{GL}_n(\mathbb{K}) \to M_n(\mathbb{K}), \qquad F(A) = {}^t A B A.
$$

Let $\mathrm{GL}_n(\mathbb{K})$ act transitively on $\mathrm{GL}_n(\mathbb{K})$ from the left via $(g, A) \mapsto Ag^{-1}$ and let it act on $M_n(\mathbb{K})$ by $(g, X) \mapsto {}^t g^{-1} X g^{-1}$ for $g \in \mathrm{GL}_n(\mathbb{K})$, $X \in M_n(\mathbb{K})$. Then F is $\mathrm{GL}_n(\mathbb{K})$-equivariant:

$$
F(g \cdot A) = F(Ag^{-1}) = {}^t g^{-1\, t} A B A g^{-1} = g \cdot F(A).
$$

Hence G is a submanifold of $\mathrm{GL}_n(\mathbb{K})$ by Corollary 6.10 and hence a Lie subgroup of $\mathrm{GL}_n(\mathbb{K})$.

Corollary 6.12. *Let G be a Lie group and let M be a G-premanifold. Then for all $p \in M$ the stabilizer G_p of p is a Lie subgroup of G.*

Proof. The morphism $a: G \to M$, $g \mapsto gp$, is G-equivariant if we let G act on itself via left multiplication. Hence $G_p = a^{-1}(p)$ is a Lie subgroup by Corollary 6.10. $\square$

6.3   Quotients by Lie Group Actions

Our next goal is a criterion when the quotient of a premanifold by a Lie group action is again a premanifold. For this we first define the notion of a proper action.

Proper Actions

Definition 6.13. Let G be a topological group, X a topological space and suppose that G acts on X via a continuous map $G \times X \to X$. Then we say that G *acts properly on X* or that the action of G on X is *proper* if the map

$$\theta: G \times X \to X \times X, \qquad (g, x) \mapsto (x, gx)$$

is a proper map (Definition 1.30).

Note that the condition does *not* say that the action map $G \times X \to X$ is proper.

Remark 6.14. The topological group G acts properly on X if and only if θ is closed and for all $x \in X$ the stabilizer $G_x := \{\, g \in G \; ; \; gx = x \,\}$ is a compact subgroup of G.

Indeed, by definition of properness, G acts properly on X if and only if θ is closed and for all $(x, y) \in X \times X$ the fiber $\theta^{-1}(x, y) = G_{x,y} \times \{x\}$ is compact, where $G_{x,y} = \{\, g \in G \; ; \; gx = y \,\}$. But if $G_{x,y} \neq \emptyset$ we can choose $h \in G_{x,y}$ and $G_x \to G_{x,y}, g \mapsto hg$, is a homeomorphism with inverse $k \mapsto h^{-1}k$.

We refer to Problem 6.2 for other characterizations of proper action.

Example 6.15. Let G be a topological group and $H \subseteq G$ be a closed subgroup. Suppose that G acts properly on a topological space X. Then Proposition 1.32 shows that H acts properly on X because the closed embedding $H \times X \hookrightarrow G \times X$ is proper (Example 1.31) and separated.

Consider in particular the case that G acts on itself by left multiplication. Then $G \times G \to G \times G, (g, g') \mapsto (g', gg')$ is a homeomorphism and in particular proper. Hence the action of H on G by left multiplication is proper.

Actions of compact groups are always proper, more precisely:

Proposition 6.16. *Let G be a compact Hausdorff topological group acting on a Hausdorff space X. Then this action is proper and the canonical map $\pi: X \to X/G$ is proper (X/G endowed with the quotient topology of X).*

Proof. As G is compact, the projection $\mathrm{pr}_2: G \times X \to X$ is proper (Proposition 1.29). The map $a: G \times X \to X, (g, x) \mapsto gx$ is the composition of the homeomorphism $G \times X \to$

$G \times X$, $(g, x) \mapsto (g, gx)$ (an inverse is given by $(g, x) \mapsto (g, g^{-1}x)$) followed by pr_2. Hence a is proper. The map $\theta: (g, x) \mapsto (x, gx)$ is the composition of the diagonal $\Delta: G \times X \to (G \times X) \times (G \times X)$ followed by the product of pr_2 and a. The diagonal is a closed embedding and in particular proper because G and X are Hausdorff. Hence we see that θ is proper because products and compositions of proper maps are again proper (Proposition 1.34 and Proposition 1.32).

It remains to show that π is proper. Let $A \subseteq X$ be closed, then $\pi^{-1}(\pi(A)) = GA$ is the image of $G \times A$ under the action map a. As we have seen that a is proper, GA is closed in X and hence $\pi(A)$ is closed in X/G by the definition of the quotient topology. Hence we see that π is closed. The fibers of π are the G-orbits in X, which are compact by Lemma 6.17. Hence π is proper (Theorem 1.30). $\square$

Lemma 6.17. *Let G be a compact topological group acting on a Hausdorff topological space. Then for all $x \in X$ the G-orbit Gx is compact and closed in X. The map $a: G \to X$, $g \mapsto gx$ induces a homeomorphism $G/G_x \xrightarrow{\sim} Gx$.*

Proof. The stabilizer $G_x = a^{-1}(x)$ is closed in G. Hence the quotient G/G_x is Hausdorff (Appendix Proposition 12.58). As image of the compact space G it is also compact. Therefore the bijective continuous map $G/G_x \to Gx$ is a homeomorphism (Appendix Corollary 12.53). Hence Gx is compact and thus closed in X. $\square$

Remark 6.18. Let G be a topological group acting continuously on a topological space X. Then the action of G on X is proper (respectively free) if and only if the map $\theta: G \times X \to X \times X$, $(g, x) \mapsto (x, gx)$ is proper (respectively injective). Therefore the action is proper and free if and only if θ is closed topological embedding.

Existence of Quotients by Lie Group Actions
We can now formulate the main result of this section.

Theorem 6.19. *Let G be a Lie group and let M be a G-premanifold. Suppose that the action is free and proper. Then there is a unique structure of a premanifold on the set of orbits M/G such that the projection $\pi: M \to M/G$ is a submersion. Its topology is the quotient topology of M. It is Hausdorff.*

Proof. The uniqueness follows from Proposition 5.51. If such a premanifold structure exists on M/G, then π is surjective and open (Corollary 5.24) and hence the topology of M/G is the quotient topology. It remains to show the existence of the premanifold structure. We use Theorem 5.53. Hence we have to show:

1. $R := \{(p, gp) \, ; \, p \in M, g \in G\}$ is a closed submanifold of $M \times M$.
2. $\mathrm{pr}_1: R \to M$ is a submersion.

Now R is the image of $\theta: G \times M \to M \times M$, $(g, p) \mapsto (p, gp)$ and θ is a closed topological embedding by Remark 6.18. Hence for (1) it suffices to show that θ is an immersion. For $g \in G$ and $p \in M$ the derivative $T_{(g,p)}(\theta)$ is of the form

$$\begin{pmatrix} 0 & \mathrm{id}_{T_p(M)} \\ T_g(a_p) & * \end{pmatrix},$$

where $a_p: G \to M$, $g \mapsto gp$. As the action is free, a_p is injective and by equivariance, a_p is constant rank. Hence a_p is an immersion and therefore $T_{(g,p)}(\theta)$ is injective.

Let us show (2). As θ yields an isomorphism $G \times M \xrightarrow{\sim} R$, we have a commutative diagram

$$ \begin{CD} G \times M \end{CD} \tag{6.1} $$

where the square is a fiber product diagram and where θ is an isomorphism. As $\mathrm{pr}_2: G \times M \to M$ is a submersion, $\mathrm{pr}_1: R \to M$ is a submersion. $\square$

The proof shows that to get a premanifold structure on M/G such that $M \to M/G$ is a submersion it suffices to assume that $\theta: G \times M \to M \times M$, $(g, p) \mapsto (p, gp)$ is an embedding (instead of the properness of the action). In this case the premanifold M/G is Hausdorff if and only if θ is a closed embedding.

Example 6.20. Let V be a finite-dimensional $\mathbb{K}$-vector space. The action of $\mathbb{K}^\times$ on V by scalar multiplication is not proper as the stabilizer of $0 \in V$ is not compact. But its restriction to $V \setminus \{0\}$ is a proper and free action. Its quotient is the projective space $\mathbb{P}(V)$ of lines in V.

Remark 6.21. We keep the notation and hypotheses of Theorem 6.19. Let $p \in M$ and $a_p: G \to M$, $g \mapsto gp$. Then the sequence

$$ 0 \longrightarrow T_e(G) \xrightarrow{T_e(a_p)} T_p M \xrightarrow{T_p(\pi)} T_{\pi(p)}(M/G) \longrightarrow 0 \tag{6.2} $$

is exact. In particular one has $\dim_{\pi(p)}(M/G) = \dim_p(M) - \dim G$. If $\dim(G) = 0$, then π is a local isomorphism.

Indeed, in the proof of Theorem 6.19 we have seen that a_p is an immersion, hence $T_e(a_p)$ is injective. As π is a submersion, $T_p(\pi)$ is surjective. To show exactness in the middle, we apply Remark 5.49 to (6.1) and obtain

$$ \ker(T_p(\pi)) \cong \ker(T_{(p,ep)}(\mathrm{pr}_1)) \cong \ker(T_{(e,p)}(\mathrm{pr}_2: G \times M \to M)) \cong T_e(G). $$

Figure 6.1 Klein bottle

Example 6.22. Let $M = S^1 \times S^1$ and consider the action by $G = \mathbb{Z}/2\mathbb{Z}$ for which the non-trivial element acts by $(z, w) \mapsto (1/z, -w)$. This is a free action and it is proper by Proposition 6.16. The quotient manifold M/G is a 2-dimensional real analytic manifold called the *Klein bottle* (see Fig. 6.1, note that there is a self intersection as the Klein bottle cannot be embedded smoothly into $\mathbb{R}^3$).

Corollary 6.23. *Let G be a Lie group and let H be a Lie subgroup.*

1. *There exists on G/H a unique structure of an analytic Hausdorff premanifold such that the projection $\pi\colon G \to G/H$ is a submersion.*
2. *If G has only countably many connected components, then G/H is an analytic manifold.*
3. *The dimension of G/H is constant and $\dim(G/H) = \dim(G) - \dim(H)$.*
4. *If H is normal in G, then the induced group law on G/H makes G/H into a Lie group.*

Proof. The Lie group H acts properly (Example 6.15) and freely on G. Hence 1 follows from Theorem 6.19. Remark 6.21 implies 3, and 4 follows from Proposition 5.54 applied to the multiplication morphism of G.

Finally, if G has countably many connected components, then G is second countable (Proposition 6.5). Therefore G/H is second countable because $G \to G/H$ is surjective and open (Remark 1.3 2). Hence G/H is a manifold. $\square$

Corollary 6.24. *Let $\varphi\colon G \to G'$ be a morphism of Lie groups and let $H' \subseteq G'$ be a Lie subgroup. Then $\varphi^{-1}(H')$ is a Lie subgroup of G. In particular $\mathrm{Ker}(\varphi)$ is a Lie subgroup of G.*

Proof. Let F be the composition of φ followed by the projection $G' \to G'/H'$ and let $\bar{e}' \in G'/H'$ be the image in the neutral element of G'. Let $g \in G$ act on G by left multiplication and on G'/H' by left multiplication with $\varphi(g)$. Then F is G-equivariant and the action of G on G is transitive. Hence we can apply Corollary 6.10 to see that the subgroup $F^{-1}(e') = \varphi^{-1}(H')$ is a subpremanifold. $\square$

Example 6.25. Let $n \geq 1$. Then $\mathrm{SL}_n(\mathbb{K}) = \mathrm{Ker}(\det\colon \mathrm{GL}_n(\mathbb{K}) \to \mathbb{K}^\times)$ is a Lie subgroup of $\mathrm{GL}_n(\mathbb{K})$.

Example 6.26 (Flag manifold). Let $E = \mathbb{K}^n$ and let $\underline{n} = (n_1, \dots, n_r)$ be a tuple of integers $n_i \in \mathbb{N}$ with $\sum_i n_i = n$. For $j = 0, \dots, r$ set $m_j := n_1 + \cdots + n_j$ and let $\mathrm{Flag}_{\underline{n}}(E)$ be the set of flags of type $\underline{n}$ of E, i.e., flags of subspaces

$$0 = W_0 \subset W_1 \subset \cdots \subset W_r = E$$

with $\dim W_i = m_i$. Then $\mathrm{GL}_n(\mathbb{K})$ acts on $\mathrm{Flag}_{\underline{n}}(E)$ via $(g, (W_i)_i) \mapsto (g(W_i))_i$. By basis extension this action is transitive. Let E_i be the subspace generated by the standard vectors $e_1, \dots, e_{m_i}$. Then the stabilizer $L_{\underline{n}}$ of the flag $(E_i)_i$ in $\mathrm{GL}_n(\mathbb{K})$ is the Lie subgroup of matrices of the form

$$\begin{pmatrix} A_1 & * & \cdots & * \\ 0 & A_2 & \ddots & \vdots \\ \vdots & \ddots & \ddots & * \\ 0 & \cdots & 0 & A_r \end{pmatrix}$$

with $A_i \in \mathrm{GL}_{n_i}(\mathbb{K})$. Hence we may apply Corollary 6.23 to endow $\mathrm{Flag}_{\underline{n}}(E) = \mathrm{GL}_n(\mathbb{K})/L_{\underline{n}}$ with the structure of an analytic manifold.

Let $0 \leq d \leq n$, then $\mathrm{Grass}_d(E) := \mathrm{Flag}_{(d, n-d)}(E)$ is the *Grassmann manifold* of d-dimensional subspaces of E.

In Example 4.26 we endowed projective space $\mathbb{P}^n(\mathbb{K})$ with the structure of an analytic manifold. As a set we have $\mathbb{P}^n(\mathbb{K}) = \mathrm{Grass}_1(\mathbb{K}^{n+1})$ and it is not difficult to check directly the manifold structure defined in Example 4.26 and in Example 6.26 are the same. This can also be deduced from the following result.

Proposition 6.27. *Let G be a Lie group, let M be a G-premanifold of type C^α with $\alpha \in \widehat{\mathbb{N}}$. Suppose that the G-action on M is transitive and choose $p \in M$. Let $H := \mathrm{Stab}_G(p)$ be its stabilizer. Suppose that G has only countably many connected components. Then $a\colon G \to M$, $g \mapsto gp$ induces an isomorphism of manifolds*

$$\bar{a}\colon G/H \xrightarrow{\sim} M, \qquad gH \mapsto gp.$$

Proof. The group H is a Lie subgroup by Corollary 6.12. Clearly, $\bar{a}$ is bijective. It is a morphism of premanifolds by Proposition 5.54. It suffices to show that $\bar{a}$ is a submersion (looking at charts [Corollary 5.23] we see that every injective submersion is a local isomorphism and hence every bijective submersion is an isomorphism).

As G acts transitively on G/H and on M, there exist $m, k, r \in \mathbb{N}_0$ such that $\dim_p(M) = m$, $\dim_q(G/H) = k$ and $\mathrm{rk}_q(\bar{a}) = r$ for all $p \in M$, $q \in G/H$. Assume that $r < m$ (in particular $m > 0$). As G/H is a second countable (Corollary 6.23 2), G/H is Lindelöf (Proposition 1.4). Hence there exist countably many charts $(U_n, x_n, V_n, y_n, \tilde{a}_n)$, $n \in \mathbb{N}$, of $\bar{a}$ such that $\bigcup_n U_n = G/H$ and such that $\tilde{a}_n$ is given by $(z_1, \dots, z_k) \mapsto (z_1, \dots, z_r, 0, \dots, 0)$. Hence $\mathrm{Im}(\tilde{a}_n)$ is contained in a zero set (Remark 5.43) of $\mathbb{K}^m$ and hence $\mathrm{Im}(\bar{a})$ has measure zero. But this is a contradiction to the surjectivity of $\bar{a}$. □

The hypothesis that G has only countably many connected components is necessary. Let $G = \mathbb{R}$ be endowed with the discrete topology making G into a 0-dimensional real Lie group. Let G act on $M = \mathbb{R}$ (with the usual structure of a 1-dimensional real manifold) via $(g, p) \mapsto g + p$. Then $G/G_p \cong G$ is discrete for $p \in M$ but M is not.

If the underlying topological space of G is second countable, then G has only countably many connected components. This is for instance the case if G is a closed subgroup of $\mathrm{GL}_n(\mathbb{K})$.

Example 6.28. Consider $\mathbb{H} := \{ z \in \mathbb{C} \; ; \; \mathrm{Im}(z) > 0 \}$ as an open real analytic submanifold of $\mathbb{C} = \mathbb{R}^2$. Let the real Lie group $G := \mathrm{SL}_2(\mathbb{R})$ act on $\mathbb{H}$ by *Möbius transformation*

$$\left(\begin{pmatrix} a & b \\ c & d \end{pmatrix}, z \right) \mapsto \frac{az + b}{cz + d}.$$

This makes $\mathbb{H}$ into a $\mathrm{SL}_2(\mathbb{R})$-manifold. The action is transitive and

$$\mathrm{Stab}_G(i) = \left\{ \begin{pmatrix} a & b \\ c & d \end{pmatrix} \in \mathrm{SL}_2(\mathbb{R}) \; ; \; \frac{ai + b}{ci + d} = i \right\}$$

$$= \left\{ \begin{pmatrix} a & b \\ c & d \end{pmatrix} \in \mathrm{SL}_2(\mathbb{R}) \; ; \; a = d, b = -c \right\}$$

$$= \mathrm{SO}(2).$$

Hence Proposition 6.27 shows that $\mathbb{H} = \mathrm{SL}_2(\mathbb{R}) / \mathrm{SO}(2)$.

To see how many connected components a Lie group precisely has, the following remark is often helpful.

Remark 6.29. Let G be a Lie group. Its identity component G^0 is an open and closed normal subgroup, in particular it is a Lie subgroup with $\dim(G^0) = \dim(G)$. Hence if we consider the abstract group $\pi_0(G)$ as a 0-dimensional Lie group, then the canonical map $\varpi_G \colon G \to G/G^0 = \pi_0(G)$ is a surjective submersion.

Any homomorphism of Lie groups $\varphi \colon H \to G$ induces a unique group homomorphism $\pi_0(\varphi) \colon \pi_0(H) \to \pi_0(G)$ making the diagram of Lie group homomorphisms

$$
\begin{array}{ccc}
H & \xrightarrow{\;\;\varphi\;\;} & G \\
{\scriptstyle \varpi_H}\downarrow & & \downarrow{\scriptstyle \varpi_G} \\
\pi_0(H) & \xrightarrow{\;\pi_0(\varphi)\;} & \pi_0(G)
\end{array}
$$

commutative. We obtain a functor π_0 from the category of Lie groups to the category of groups (identified with the category of 0-dimensional Lie groups).

Let H be a Lie subgroup of G and let Γ be the image of $\pi_0(H) \to \pi_0(G)$. As $G \to G/H$ is open and surjective, it induces a bijective map

$$
\pi_0(G)/\Gamma \xrightarrow{\;\sim\;} \pi_0(G/H). \tag{6.3}
$$

In particular we see that G is connected if G/H and H are connected.

Example 6.30. The action of the real Lie group $SO(n)$ on S^{n-1} by matrix multiplication is analytic. The stabilizer of $(1, 0, \ldots, 0) \in S^{n-1}$ is $SO(n-1)$ and we deduce $S^{n-1} \cong SO(n)/SO(n-1)$. Arguing by induction on n starting with $SO(1) = 1$ we deduce via Remark 6.29 that $SO(n)$ is connected for all n. As $SO(n)$ is of index 2 in $O(n)$ we also see that $O(n)$ has two connected components and that $O(n)^0 = SO(n)$.

Let $B \subseteq GL_n(\mathbb{R})$ be the subgroup of upper triangular matrices. As an analytic manifold, B is isomorphic to $\mathbb{R}^{n(n-1)/2}$. In particular, B is connected. By Gram-Schmidt orthogonalization, multiplication of matrices yields an isomorphism $B \times O(n) \xrightarrow{\sim} GL_n(\mathbb{R})$ of real analytic manifolds. Hence we deduce that $GL_n(\mathbb{R})$ has two connected components. Its identity component is the subgroup of matrices with positive determinant.

6.4　Problems

Problem 6.1. Let G be a topological group that acts properly on a topological space X. Let Y be a G-invariant subspace. Show that the action of G on Y is proper.

Problem 6.2. Let G be a topological group acting continuously on a topological space X.

1. Suppose that G acts freely on X and let $R := \{ (gx, x) \, ; \, g \in G, x \in X \} \subseteq X \times X$. Show that $\varpi \colon R \to G$, $(gx, x) \mapsto g$ is well defined. Show that G acts properly on X if and only if R is closed in $X \times X$ and ϖ is continuous.

2. Suppose that G and X are Hausdorff and that G is locally compact. Show that G acts properly on X if and only if for all $x, y \in X$ there exist neighborhoods U of x and V of y such that $\{g \in G \; ; \; gU \cap V \neq \emptyset\}$ is relatively compact in G (Appendix Problem 12.31).

Problem 6.3. Let G be a finite group and let M be a G-manifold such that the action of G on M is free. Show that there exists on M/G a unique structure of a premanifold such that $\pi: M \to M/G$ is a submersion. Show that π is a local isomorphism.

Problem 6.4. Let G and H be Lie groups acting properly and freely on premanifolds M and N, respectively. Let $\varphi: G \to H$ be a homomorphism of Lie groups and $F: M \to N$ a morphism of premanifolds such that $F(gm) = \varphi(g)F(m)$. Show that there exists a unique morphism $\bar{F}: M/G \to N/H$ such that the following diagram commutes:

$$
\begin{array}{ccc}
M & \xrightarrow{\ F\ } & N \\
\downarrow & & \downarrow \\
M/G & \xrightarrow{\ \bar{F}\ } & N/H \ .
\end{array}
$$

Problem 6.5. Let $M = (-1, 1) \times S^1$ (considered as a real analytic manifold) and let $G = \mathbb{Z}/2\mathbb{Z}$ act on M such that the non-trivial element acts by $(t, z) \mapsto (-t, -z)$. Show that this action is free and proper and that M/G is isomorphic to the Möbius band.

Problem 6.6. Show that the quotient manifold $\mathbb{R}/\mathbb{Z}$ and the submanifold S^1 of $\mathbb{R}^2$ are isomorphic real analytic manifolds.

Problem 6.7. Let $\varphi: G \to G'$ be a homomorphism of Lie groups. Show that the following assertions are equivalent:

(i) φ is a submersion (respectively a local isomorphism).
(ii) $T_1(\varphi): T_1(G) \to T_1(G')$ is surjective (respectively bijective).
(iii) $\varphi_{|G^0}: G^0 \to G'^0$ is surjective (respectively a covering map), where H^0 denotes the connected component of 1 in a topological group H.

Problem 6.8. Let G be a connected Lie group. Show that there exists a universal cover space $\pi: \tilde{G} \to G$ and that for all $\tilde{e} \in \pi^{-1}(e)$ there exists a unique Lie group structure on $\tilde{G}$ such that $\tilde{e}$ is the neutral element of $\tilde{G}$ and such that π is a local isomorphism of premanifolds and a group homomorphism. Show that $(\tilde{G}, \pi)$ is unique up to a unique isomorphism of Lie groups.
Hint: Problem 4.14.

Problem 6.9. Let G be a Hausdorff locally compact second countable topologically group that acts continuously and transitively on a Hausdorff Baire space X (Problem 1.9). Let $x \in X$ and $G_x = \{ g \in G \; ; \; gx = x \}$. Show that the map $G \times X$, $g \mapsto gx$, is open and that the induced map $G/G_x \to X$ is a homeomorphism.
Hint: Problem 1.11.

Problem 6.10. Let G be a Lie group with countable many connected components, let M be a G-premanifold, and let $p \in M$. Let $Gp \subseteq M$ be the orbit of p.

1. Show that the following assertions are equivalent:
   (i) $Gp \subseteq M$ is locally closed.
   (ii) Gp is a submanifold of M.
   Hint: If $Gp \subseteq M$ is locally closed, use Problem 6.9 and Problem 1.10 to see that $G \to Gp$, $g \mapsto gp$, is open.
2. Suppose that Gp is locally closed in M. Show that $G \to M$, $g \mapsto gp$, induces a G-equivariant isomorphism of premanifolds $G/\operatorname{Stab}_G(p) \xrightarrow{\sim} Gp$.
3. Suppose that the action of G on M is proper. Show that 1 and 2 hold for each orbit.
4. Let $r \in \mathbb{R}$. Show that $\mathbb{R} \times S^1 \times S^1$, $(x, (z_1, z_2)) \mapsto (e^{2\pi i x} z_1, e^{2\pi i r x} z_2)$, defines an analytic action of $(\mathbb{R}, +)$ on $S^1 \times S^1$. Show that the equivalent assertions of 1 hold if and only if $r \in \mathbb{Q}$.
   Hint: Problem 5.19

Problem 6.11. For $n \geq 2$ show that there are isomorphisms of real analytic manifolds $S^{2n+1} \cong \mathrm{U}(n)/\mathrm{U}(n-1) \cong \mathrm{SU}(n)/\mathrm{SU}(n-1)$. Deduce that $\mathrm{U}(n)$ and $\mathrm{SU}(n)$ are connected for all $n \geq 1$. Show that $\mathrm{GL}_n(\mathbb{C})$ is connected.

Problem 6.12. Let E and F be finite-dimensional $\mathbb{K}$-vector spaces with $d := \dim(F) \leq \dim(E)$. Let $I(F, E)$ be the set of injective $\mathbb{K}$-linear maps $F \to E$ and define on $I(E, F)$ an equivalence relation by $u \sim u' :\Leftrightarrow \exists v \in \mathrm{GL}(F): u' = u \circ v$. Show that $I(F, E)$ is an open submanifold of $\operatorname{Hom}_{\mathbb{K}}(F, E)$, that $\sim$ is a regular equivalence relation and that $I(F, E)/\sim \; \cong \operatorname{Grass}_d(F)$.

Problem 6.13. Use the notations of Example 6.26.

1. Show that the *Plücker embedding*

$$\operatorname{Grass}_d(E) \to \mathbb{P}(\Lambda^d(E)) := \operatorname{Grass}_1(\Lambda^d(E)), \qquad F \mapsto \Lambda^d(F)$$

   is a closed embedding of analytic manifolds.
2. Show that

$$\operatorname{Flag}_{\underline{n}}(E) \to \prod_{i=1}^{r-1} \operatorname{Grass}_{m_i}(E), \quad (W_1 \subset \cdots \subset W_{r-1}) \mapsto (W_i)_{1 \leq i \leq r-1}$$

is a closed embedding of analytic manifolds.
3. Deduce that $\mathrm{Grass}_d(E)$ and, more generally, $\mathrm{Flag}_{\underline{n}}(E)$ are compact.

Problem 6.14. Let $(V, \langle\,,\,\rangle)$ be a symplectic $\mathbb{K}$-space, i.e., V is a finite-dimensional $\mathbb{K}$-vector space with a non-degenerate alternating bilinear form $\langle\,,\,\rangle$. Show that the *symplectic group*

$$\mathrm{Sp}(V, \langle\,,\,\rangle) := \{\, g \in \mathrm{GL}(V)\,;\, \langle gv, gw \rangle = \langle v, w \rangle \text{ for all } v, w \in V \,\}$$

is a closed Lie subgroup of $\mathrm{GL}(V)$ and endow the set of *Lagrangians of* $(V, \langle\,,\,\rangle)$ (i.e., the set of $\mathbb{K}$-subspaces U of V with $U^\perp = U$) with the structure of an analytic quotient manifold of $\mathrm{Sp}(V, \langle\,,\,\rangle)$.

Torsors and Non-abelian Čech Cohomology

7

Very often one is confronted with construction problems where it is easier to make the construction locally. Then the question arises whether these local constructions "glue" to a global object. For instance, on premanifolds it is often easier to construct objects if the premanifold is isomorphic to an open ball in $\mathbb{K}^m$, which is locally always the case. In this chapter we study a general technique to deal with such gluing problems – at least if the difference of two possible local objects is given by a sheaf of groups. For instance two primitives of a $\mathbb{K}$-valued function always differ by a locally constant $\mathbb{K}$-valued function and these form a sheaf of groups.

7.1 Torsors

The precise mathematical notion that captures "local objects always differing by a sheaf of groups" is that of a torsor. The idea is to encode all local objects into a sheaf on which a sheaf of groups then acts simply transitively.

Definition 7.1 (Torsor). Let X be a topological space, let $\mathcal{F}$ and $\mathcal{G}$ be sheaves of sets on X and let $\mathcal{A}$ be a sheaf of groups on X.

1. An $\mathcal{A}$-*sheaf on* X is a sheaf of sets $\mathcal{F}$ on X together with a morphisms of sheaves of sets

$$a : \mathcal{A} \times \mathcal{F} \to \mathcal{F}$$

such that for every open subset $U \subseteq X$ the map

$$a_U : \mathcal{A}(U) \times \mathcal{F}(U) \to \mathcal{F}(U)$$

is an action of the group $\mathcal{A}(U)$ on the set $\mathcal{F}(U)$.

© Springer Fachmedien Wiesbaden 2016
T. Wedhorn, *Manifolds, Sheaves, and Cohomology*, Springer Studium Mathematik – Master,
DOI 10.1007/978-3-658-10633-1_7

A *morphism* $\mathcal{F} \to \mathcal{G}$ *of* $\mathcal{A}$-*sheaves* is a morphism $\varphi \colon \mathcal{F} \to \mathcal{G}$ of sheaves such that for all $U \subseteq X$ the map

$$\varphi_U \colon \mathcal{F}(U) \to \mathcal{G}(U)$$

is $\mathcal{A}(U)$-equivariant. We obtain the category of $\mathcal{A}$-sheaves on X.

2. An $\mathcal{A}$-sheaf T on X is called an $\mathcal{A}$-*pseudotorsor* if for all $U \subseteq X$ open the action $\mathcal{A}(U) \times T(U) \to \mathcal{F}(U)$ is simply transitive.

3. An $\mathcal{A}$-pseudotorsor T on X is called an $\mathcal{A}$-*torsor* if there exists an open covering $X = \bigcup_i U_i$ such that $T(U_i) \neq \emptyset$ for all i.

   We obtain the full subcategory of $\mathcal{A}$-torsors of the category of $\mathcal{A}$-sheaves. We denote this category by $(\mathrm{Tors}(\mathcal{A}))$.

Example 7.2. Let $M \subseteq \mathbb{C}$ be open and let $f \colon M \to \mathbb{C}$ be a holomorphic function. Define a sheaf Prim_f on M by

$$\mathrm{Prim}_f(U) := \left\{ F \colon U \to \mathbb{C} \text{ holomorphic} ; \; F' = f_{|U} \right\}$$

for $U \subseteq M$ open. Then any two primitives of $f_{|U}$ differ by a locally constant function. In other words, the constant sheaf $\mathbb{C}_M$ acts on it (recall: $\mathbb{C}_M(U) = \{ h \colon U \to \mathbb{C} ; \; h \text{ locally constant} \}$) by addition:

$$\mathbb{C}_M(U) \times \mathrm{Prim}_f(U) \to \mathrm{Prim}_f(U), \qquad (h, F) \mapsto h + F.$$

This makes Prim_f into a $\mathbb{C}_M$-pseudotorsor because two primitives of a holomorphic functions only differ by a locally constant function. Now the local existence of primitives just means that the $\mathbb{C}_M$-pseudotorsor Prim_f is in fact a torsor. Here this is the case because we may find a covering of M by convex open subsets U_i and $\mathrm{Prim}_f(U_i) \neq \emptyset$ for all $i \in I$.

Next we come to the question of global existence of objects on a space X. If the local objects are encoded by a torsor T on X, then this is equivalent to the question of whether $T(X)$ is non-empty.

Recall that a *pointed set* is a pair (H, x) consisting of a set H and an element $x \in H$. A *morphism of pointed sets* $(H, x) \to (H', x')$ is a map $\varphi \colon H \to H'$ with $\varphi(x) = x'$. We obtain the category of pointed sets. Every group G gives rise to a pointed set (G, e), where e is the neutral element.

Definition 7.3. Let X be a topological space and let $\mathcal{A}$ be a sheaf of groups on X.

1. Let $\mathcal{A}$ act on itself by left multiplication. This makes $\mathcal{A}$ into an $\mathcal{A}$-torsor. An $\mathcal{A}$-torsor that is isomorphic to $\mathcal{A}$ is called a *trivial* $\mathcal{A}$-*torsor*.

2. Define $H^1(X, \mathcal{A})$ as the set of isomorphism classes of $\mathcal{A}$-torsors on X. It is a pointed set; the distinguished element is the isomorphism class of trivial torsors. It is called the *(first) cohomology set of* $\mathcal{A}$.

3. Let T be an $\mathcal{A}$-torsor over X, and let $\mathcal{U} = (U_i)_{i \in I}$ be an open covering of X. Then we say that $\mathcal{U}$ trivializes T if $T(U_i) \neq \emptyset$ for all $i \in I$. We denote by $H^1(\mathcal{U}, \mathcal{A})$ the isomorphism classes of $\mathcal{A}$-torsors over X trivialized by $\mathcal{U}$. Then

$$H^1(X, \mathcal{A}) = \bigcup_{\mathcal{U}} H^1(\mathcal{U}, \mathcal{A}).$$

Proposition 7.4. *Let X be a topological space and let $\mathcal{A}$ be a sheaf of groups on X.*

1. *Every morphism in the category* $(\mathrm{Tors}(\mathcal{A}))$ *is an isomorphism.*
2. *An $\mathcal{A}$-torsor T over X is trivial if and only if $\mathcal{T}(X) \neq \emptyset$.*

Proof. (1). Let T and S be $\mathcal{A}$-torsors on X and let $\varphi \colon T \to S$ be an $\mathcal{A}$-equivariant morphism of sheaves on X. If $X = \bigcup_i U_i$ is an open covering and $\varphi_{|U_i}$ is an isomorphism for all $i \in I$, then φ is an isomorphism (Proposition 3.18). Choosing a covering $(U_i)_i$ that trivializes S and T and replacing X by one of the U_i we may assume that $T(X) \neq \emptyset \neq S(X)$. Then $T(U) \neq \emptyset \neq S(U)$ for all $U \subseteq X$ open. Hence it suffices to show: Let A be a group acting simply transitively on two non-empty sets X and Y. Then every A-equivariant map $\varphi \colon X \to Y$ is bijective. Let $x \in X$, $y := \varphi(x)$. Then we have a diagram with bijective diagonal maps

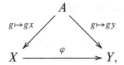

which commutes because φ is G-equivariant. Hence φ is bijective.

(2). If $T \cong \mathcal{A}$, then $T(X) \neq \emptyset$ because $\mathcal{A}(X)$ is a group, in particular it is non-empty. Conversely, assume there exists $s \in T(X)$. Then

$$\mathcal{A} \to \mathcal{F}, \quad \mathcal{A}(U) \ni g \mapsto g \cdot (s_{|U}) \in T(U), \qquad U \subseteq X \text{ open}$$

is an isomorphism of $\mathcal{A}$-torsors. $\square$

The first important result says that for constant group sheaves on simply connected spaces H^1 is trivial:

Proposition 7.5. *Let X be a topological space. Let $\mathcal{A}$ be a locally constant sheaf of groups on X. Suppose that X is simply connected and locally path connected. Then $\mathcal{A}$ is a constant sheaf of groups and $H^1(X, \mathcal{A}) = 0$.*

Proof. As any $\mathcal{A}$-torsor $\mathcal{T}$ is a locally isomorphic to $\mathcal{A}$ (Proposition 7.4), it is a locally constant sheaf on X. Hence it suffices to show that any locally constant sheaf $\mathcal{F}$ on X is constant. Let $p\colon E \to X$ be the étalé space corresponding to $\mathcal{F}$. Then p is a covering map (Example 3.41). As X is simply connected and locally path connected, this covering map has to be trivial (Proposition 2.41), i.e., $E = X \times F$, where F is some fiber of p, considered as a discrete topological space. This shows that $\mathcal{F}$ is a constant sheaf. $\square$

Using Example 7.2 we obtain the following criterion for the existence of global primitives.

Example 7.6. Let $M \subseteq \mathbb{C}$ be open and let $f\colon M \to \mathbb{C}$ be a holomorphic function. Then the torsor Prim_f (Example 7.2) is trivial if and only if there exists $F\colon M \to \mathbb{C}$ such that $F' = f$.

If $H^1(M, \mathbb{C}_M) = 0$, then every holomorphic function on M has a primitive. Hence we can deduce from Proposition 7.5 the classical result that if M is simply connected, then every holomorphic function on M has a primitive.

Note that to get this result we used that Prim_f is a torsor (and not only a pseudotorsor), i.e., that locally a primitive always exists. Hence the cohomological formalism allowed us to extend a local existence result to a global one.

7.2   Non-abelian Čech Cohomology

Notation: Let X be a topological space and $\mathcal{A}$ a sheaf of groups on X.

We may also give a more concrete description of $H^1(X, \mathcal{A})$ in terms of cocycles, which is often advantageous for concrete calculations. To ease the notation, for two sections $s \in \Gamma(U, \mathcal{A})$ and $t \in \Gamma(V, \mathcal{A})$ we will often write $st \in \Gamma(U \cap V, \mathcal{A})$ instead of $s_{|U \cap V} t_{|U \cap V}$, and $s = t$ instead of $s_{|U \cap V} = t_{|U \cap V}$.

Definition and Remark 7.7 (Čech cohomology). Fix an open covering $\mathcal{U} = (U_i)_{i \in I}$ of X. For $i_1, \ldots, i_r \in I$ define $U_{i_1 \ldots i_r} := U_{i_1} \cap \cdots \cap U_{i_r}$.

1. A *Čech 1-cocycle of $\mathcal{A}$ on $\mathcal{U}$* is a tuple $\theta = (g_{ij})_{i,j \in I}$, where $g_{ij} \in \mathcal{A}(U_{ij})$ such that the cocycle condition

$$g_{kj} g_{ji} = g_{ki} \in \mathcal{A}(U_{ijk}) \tag{7.1}$$

   holds for all i, j, k. Choosing $i = j = k$ we see that $g_{ii} = 1$; choosing $k = i$ we see that $g_{ij} = g_{ji}^{-1}$ for all $i, j \in I$. The set of all Čech 1-cocycles of $\mathcal{A}$ on $\mathcal{U}$ is denoted by $\check{Z}^1(\mathcal{U}, \mathcal{A})$.
2. Two Čech 1-cocycles θ and θ' on $\mathcal{U}$ are called *cohomologous* if there exist $h_i \in \mathcal{A}(U_i)$ for all i such that we have

$$h_i g_{ij} = g'_{ij} h_j \qquad \in \mathcal{A}(U_{ij})$$

for all $i, j \in I$. This is easily checked to be an equivalence relation on $\check{Z}^1(\mathcal{U}, \mathcal{A})$. The equivalence classes are called *Čech cohomology classes*, and the set of cohomology classes of Čech 1-cocycles on $\mathcal{U}$ is called the *(first) Čech cohomology of $\mathcal{A}$ on $\mathcal{U}$* and is denoted by $\check{H}^1(\mathcal{U}, \mathcal{A})$. This is a pointed set in which the distinguished element is the cohomology class of the cocycle (e_{ij}) with $e_{ij} = 1$ for all i, j. It is denoted by 1.

Then a Čech 1-cocycle $(g_{ij}) \in \check{Z}^1(\mathcal{U}, \mathcal{A})$ is cohomologous to 1 if there exist $h_i \in \mathcal{A}(U_i)$ such that $g_{ij} = h_i^{-1} h_j$.

Definition and Remark 7.8 (Refinement of Čech cohomology). Let $\mathcal{V} = (V_j)_{j \in J}$ be a second open covering of X. A morphism of open coverings $\mathcal{V} \to \mathcal{U}$ is a map $\tau \colon J \to I$ such that $V_j \subseteq U_{\tau(j)}$ for all $j \in J$ (then $\mathcal{V}$ is a refinement of $\mathcal{U}$). We obtain the category of open coverings $(\mathrm{Cov}(X))$ of X.

If $(g_{ii'})$ is a Čech 1-cocycle on $\mathcal{U}$, then the tuple $\tau^*(g)_{jj'} = g_{\tau(j)\tau(j')}|_{V_j \cap V_{j'}}$ is a Čech 1-cocycle on $\mathcal{V}$. It is easy to see that this construction induces a map

$$\tau^* \colon \check{H}^1(\mathcal{U}, \mathcal{A}) \to \check{H}^1(\mathcal{V}, \mathcal{A}).$$

Therefore we obtain a functor $\mathcal{U} \to \check{H}^1(\mathcal{U}, \mathcal{A})$ from $(\mathrm{Cov}(X))^{\mathrm{opp}}$ to the category of pointed sets. We leave it to the reader to make the (easy) check that the category $(\mathrm{Cov}(X))^{\mathrm{opp}}$ is filtered (Problem 7.1). The pointed set

$$\check{H}^1(X, \mathcal{A}) := \operatorname*{colim}_{\mathcal{U}} \check{H}^1(\mathcal{U}, \mathcal{A}),$$

is called the *(first) Čech cohomology of $\mathcal{A}$ on X*. Note that the explicit description of a filtered colimit of sets in Appendix Example 13.38 shows that the map from $\check{H}^1(\mathcal{U}, \mathcal{A})$ to the colimit in the category of sets the distinguished points are all sent to the same point e. Hence the colimit of the $\check{H}^1(\mathcal{U}, \mathcal{A})$ in the category of pointed sets is the colimit in the category of sets endowed with the point e.

Remark 7.9 (Abelian Čech cohomology). If $\mathcal{A}$ is a sheaf of abelian groups, the set of Čech 1-cocycles of $\mathcal{A}$ on $\mathcal{U}$ forms an abelian group with respect to componentwise multiplication $(g_{ij})(g'_{ij}) := (g_{ij} g'_{ij})$. The equivalence relation of being cohomologous is compatible with the group structure and therefore $\check{H}^1(\mathcal{U}, \mathcal{A})$ is an abelian group. For every refinement $\mathcal{V}$ of $\mathcal{U}$ the map $\check{H}^1(\mathcal{U}, \mathcal{A}) \to \check{H}^1(\mathcal{V}, \mathcal{A})$ is a homomorphism of abelian groups. Therefore $\check{H}^1(X, \mathcal{A})$ is an abelian group.

Example 7.10 (Čech cohomology of the circle). Let $\mathcal{U} = (U_0, U_1)$ be an open covering of X consisting of two open subsets. Then

$$\check{Z}^1(\mathcal{U}, \mathcal{A}) \xrightarrow{\sim} \mathcal{A}(U_0 \cap U_1),$$

and a Čech 1-cocycle of $\mathcal{A}$ on $\mathcal{U}$ is given by an element $g = g_{01} \in \mathcal{A}(U_0 \cap U_1)$. Moreover, $g, g' \in \mathcal{A}(U_0 \cap U_1)$ are cohomologous if and and only if there exist $h_i \in \mathcal{A}(U_i)$, $i = 0, 1$, such that $g' = h_0 g h_1$.

Let $X = S^1 = \{z \in \mathbb{C} \; ; \; |z| = 1\}$, A be an abelian group, $\mathcal{A} = A_{S^1}$ the associated constant sheaf, $U_0 = S^1 \setminus \{1\}$, $U_1 = S^1 \setminus \{-1\}$. Then

$$\mathcal{A}(U_0) = \mathcal{A}(U_1) = A$$

because U_0 and U_1 are connected. Moreover $U_0 \cap U_1 = W_+ \sqcup W_-$ with $W_\pm :=$ $\{z \in S^1 \; ; \; \pm \mathrm{Im}(z) > 0\}$. Hence

$$\mathcal{A}(U_0 \cap U_1) \xrightarrow{\sim} A \times A, \qquad f \mapsto (f_{|W_+}, f_{|W_-}).$$

The restriction map $\mathcal{A}(U_i) \mapsto \mathcal{A}(U_0 \cap U_1)$ is given by $A \ni a \mapsto (a, a) \in A^2$. Hence we see:

$$g = (a_+, a_-) \in A^2 \text{ and } g' = (a'_+, a'_-) \text{ cohomologous}$$
$$\Leftrightarrow \exists b_0, b_1 \in A \colon a'_+ = b_0 + a_+ + b_1, \qquad a'_- = b_0 + a_- + b_1$$
$$\Leftrightarrow \exists b \in A \colon a'_+ = a_+ + b, \qquad a'_- = a_- + b$$
$$\Leftrightarrow a'_+ - a'_- = a_+ - a_-.$$

In other words $H^1(\mathcal{U}, A_{S^1}) \cong A$.

Remark 7.11. Let $X = \coprod_i X_i$. Then $\check{H}^1(X, \mathcal{A}) = \prod_{i \in I} \check{H}^1(X_i, \mathcal{A}_{|X_i})$.

Remark 7.12 (Čech cohomology and torsors). We will now construct an isomorphism of pointed sets

$$H^1(X, \mathcal{A}) \cong \check{H}^1(X, \mathcal{A}). \tag{7.2}$$

Let T be an $\mathcal{A}$-torsor and let $\mathcal{U} = (U_i)_{i \in I}$ be an open covering of X that trivializes T, i.e., $T(U_i) \neq \emptyset$ for all i. Set $U_{ij} = U_i \cap U_j$ for all $i, j \in I$. Choose elements $t_i \in T(U_i)$. As $\mathcal{A}$ acts simply transitively, there exists a unique element $g_{ij} \in \mathcal{A}(U_{ij})$ such that $g_{ij} t_j = t_i$. We have $g_{kj} g_{ji} t_i = t_k = g_{ki} t_i$ and thus $g_{kj} g_{ji} = g_{ki}$. Hence $(g_{ij})_{ij}$ is a Čech 1-cocycle. For a different choice of elements t_i we obtain a cohomologous 1-cocycle. Hence this construction yields a morphism of pointed sets

$$c_{\mathcal{A}, \mathcal{U}} \colon H^1(\mathcal{U}, \mathcal{A}) \to \check{H}^1(\mathcal{U}, \mathcal{A}). \tag{7.3}$$

If $\tau \colon \mathcal{V} \to \mathcal{U}$ is a morphism of open coverings, one easily checks that

$$
\begin{array}{ccc}
H^1(\mathcal{U}, \mathcal{A}) & \xrightarrow{\ c_{\mathcal{A}, \mathcal{U}}\ } & \check{H}^1(\mathcal{U}, \mathcal{A}) \\
\cap \downarrow & & \downarrow \tau^* \\
H^1(\mathcal{V}, \mathcal{A}) & \xrightarrow{\ c_{\mathcal{A}, \mathcal{V}}\ } & \check{H}^1(\mathcal{V}, \mathcal{A})
\end{array}
\tag{7.4}
$$

is a commutative diagram. By taking inductive limits we obtain a map of pointed sets

$$
c_{\mathcal{A}} \colon H^1(X, \mathcal{A}) \to \check{H}^1(X, \mathcal{A}).
\tag{7.5}
$$

Proposition 7.13. *The maps $c_{\mathcal{A}, \mathcal{U}}$ are isomorphisms of pointed sets. In particular, $c_{\mathcal{A}}$ is an isomorphism.*

Proof. We define an inverse of $c_{\mathcal{A}, \mathcal{U}}$ as follows. Let (g_{ij}) be a representative of a class in $\check{H}^1(\mathcal{U}, \mathcal{A})$. For $V \subseteq X$ open we set

$$
T(V) = \left\{ (t_i) \in \prod_i \mathcal{A}(U_i \cap V) \; ; \; t_i t_j^{-1} = g_{ij} \right\}.
\tag{7.6}
$$

Endowed with the obvious restriction maps, T is a sheaf. We define an $\mathcal{A}$-action on T via $g \cdot (t_i)_i = (t_i g^{-1})_i$. For a fixed $k \in I$ and for $V \subseteq U_k$ the map $\mathcal{A}(V) \to T(V)$, $g \mapsto (g_{ik} g^{-1})_i$, defines an isomorphism of $\mathcal{A}_{|U_k}$-sheaves $\mathcal{A}_{|U_k} \to T_{|U_k}$ whose inverse is given by $(t_i) \mapsto t_k^{-1}$. Thus T is an $\mathcal{A}$-torsor that is trivialized by $\mathcal{U}$. If (g_{ij}) is replaced by a cohomologous cocycle $(g'_{ij}) = (h_i g_{ij} h_j^{-1})$ with an associated $\mathcal{A}$-torsor T', then $(t_i) \mapsto (h_i t_i)_i$ defines an isomorphism $T \xrightarrow{\sim} T'$ of $\mathcal{A}$-torsors. $\qquad\square$

Remark 7.14. The commutative diagram (7.4) then also shows that the maps $\tau^* \colon \check{H}^1 (\mathcal{U}, \mathcal{A}) \to \check{H}^1(\mathcal{V}, \mathcal{A})$ are injective and depend only on $\mathcal{U}$ and $\mathcal{V}$ (but not on the choice of a morphism $\tau \colon \mathcal{V} \to \mathcal{U}$). In particular $\check{H}^1(\mathcal{U}, \mathcal{A}) \to \check{H}^1(X, \mathcal{A})$ is injective.

Corollary 7.15. *If $\mathcal{A}$ and $\mathcal{A}'$ are two sheaves of groups on a topological space X, then $H^1(X, \mathcal{A} \times \mathcal{A}') = H^1(X, \mathcal{A}) \times H^1(X, \mathcal{A}')$.*

Proof. This is clear for Čech cohomology. $\qquad\square$

For concrete calculations of $\check{H}^1(X, \mathcal{A})$ it suffices very often to calculate $\check{H}^1(\mathcal{U}, \mathcal{A})$ for a sufficiently nice open covering $\mathcal{U}$ by the following result.

Proposition 7.16. *Let $\mathcal{U} = (U_i)_{i \in I}$ be an open covering of X such that $\check{H}^1(U_i, \mathcal{A}) = 0$ for all $i \in I$. Then $\check{H}^1(\mathcal{U}, \mathcal{A}) \to \check{H}^1(X, \mathcal{A})$ is an isomorphism.*

Proof. It suffices to show: Let $\mathcal{V} = (V_j)_{j \in J}$ be a refinement of $\mathcal{U}$ and let $\tau \colon J \to I$ be a morphism of open coverings. Then $\tau^* \colon \check{H}^1(\mathcal{U}, \mathcal{A}) \to \check{H}^1(\mathcal{V}, \mathcal{A})$ is bijective.

It is injective by Remark (7.14). Hence it remains to show that τ^* is surjective. Let $U_i \cap \mathcal{V} := (U_i \cap V_j)_{j \in J}$. This is an open covering of U_i. As $\check{H}^1(U_i \cap \mathcal{V}, \mathcal{A}) \to \check{H}^1(U_i, \mathcal{A})$ is injective (Remark 7.14), we find

$$\check{H}^1(U_i \cap \mathcal{V}, \mathcal{A}) = 1. \tag{*}$$

Let $\eta = (\eta_{jj'})_{j,j' \in J} \in \check{Z}^1(\mathcal{V}, \mathcal{A})$ and define $\eta^{(i)} \in \check{Z}^1(U_i \cap \mathcal{V}, \mathcal{A})$ by

$$\eta^{(i)}_{jj'} := \eta_{jj'}|_{U_i \cap V_j \cap V_{j'}}.$$

By (*) there exist $g_j^{(i)} \in \mathcal{A}(U_i \cap V_j)$ such that

$$\eta_{jj'} = \left(g_j^{(i)}\right)^{-1} g_{j'}^{(i)}$$

and hence for all $i, i' \in I$

$$g_j^{(i')}\left(g_j^{(i)}\right)^{-1} = g_{j'}^{(i')}\left(g_{j'}^{(i)}\right)^{-1} \in \mathcal{A}(U_i \cap U_{i'} \cap V_j \cap V_{j'}). \tag{**}$$

Now define $\theta \in \check{Z}^1(\mathcal{U}, \mathcal{A})$ by

$$\theta_{i'i}|_{V_j} := g_j^{(i')}\left(g_j^{(i)}\right)^{-1}.$$

This is possible because of (**). Moreover set $h_j := \left(g_j^{(\tau(j))}\right)^{-1} \in \mathcal{A}(U_{\tau(j)} \cap V_j)$. Then we have on $V_j \cap V_{j'} = U_{\tau(j)} \cap U_{\tau(j')} \cap V_j \cap V_{j'}$:

$$
\begin{aligned}
h_j(\tau^*(\theta))_{jj'} = h_j \theta_{\tau(j)\tau(j')} &= h_j g_j^{(\tau(j))}\left(g_j^{(\tau(j'))}\right)^{-1} \\
&= \left(g_j^{(\tau(j'))}\right)^{-1} = \left(g_j^{(\tau(j'))}\right)^{-1} g_{j'}^{(\tau(j'))} h_{j'} \\
&= \eta_{jj'} h_{j'},
\end{aligned}
$$

and hence $\tau^*(\theta)$ is cohomologous to η. $\qquad\qquad\square$

Corollary 7.17. *Let X be a locally path connected space and let G be a group. Let $\mathcal{U} = (U_i)_i$ be an open covering of X such that U_i is simply connected for all i. Then $H^1(\mathcal{U}, G_X) = H^1(X, G_X)$.*

Proof. We can apply Proposition 7.16 by Proposition 7.5. $\qquad\qquad\square$

Example 7.18. Let $X = S^1$, let A be an abelian group, and let $\mathcal{U} = (U_0, U_1) := (S^1 \setminus \{1\}, S^1 \setminus \{-1\})$. Then U_i is homeomorphic to the open interval $(0, 1)$ and hence simply connected. Hence Corollary 7.17 implies that $\check{H}^1(S^1, A_{S^1}) = \check{H}^1(\mathcal{U}, A_{S^1}) \cong A$ by Example 7.10.

7.3   First Term Sequence of Cohomology

An important tool to calculate cohomology pointed sets is by relating them to already-known cohomology pointed sets using long exact cohomology sequences. This is particularly useful if the cohomology sets carry a group structure but can be also used in general. Hence we explain first what is meant by an exact sequence of pointed sets.

Definition 7.19. A sequence of morphisms of pointed sets

$$\ldots \xrightarrow{\varphi_{i-1}} (H_{i-1}, e_{i-1}) \xrightarrow{\varphi_i} (H_i, e_i) \xrightarrow{\varphi_{i+1}} (H_{i+1}, e_{i+1}) \xrightarrow{\ \ \ldots\ \ }$$

is called *exact* if for all i

$$\mathrm{Ker}(\varphi_{i+1}) := \{\, h \in H_i \ ;\ \varphi_{i+1}(h) = e_{i+1} \,\} = \mathrm{Im}(\varphi_i).$$

Note that if $\varphi \colon (H, e) \to (H', e')$ is a morphism of pointed sets, then $\mathrm{Ker}(\varphi) = \{e\}$ does not imply that φ is injective in general.

Remark 7.20. Let $\varphi \colon \mathcal{A} \to \mathcal{A}'$ be a homomorphism of sheaves of groups. Sending a Čech cocycle (g_{ij}) of $\mathcal{A}$ to $(\varphi(g_{ij}))$ defines a homomorphism of pointed sets

$$\check{H}^1(\varphi) \colon \check{H}^1(X, \mathcal{A}) \to \check{H}^1(X, \mathcal{A}'). \tag{7.7}$$

We obtain a functor $\check{H}^1(X, \cdot)$ from the category of sheaves of groups on X to the category of pointed sets.

We can also define $H^1(\varphi) \colon H^1(X, \mathcal{A}) \to H^1(X, \mathcal{A}')$ via torsors as follows. Let T be an $\mathcal{A}$-torsor. We let $\mathcal{A}$ act on $\mathcal{A}' \times T$ via

$$(a, (a', t)) \mapsto (a'\varphi(a^{-1}), at)$$

for $a \in \mathcal{A}(U)$, $a' \in \mathcal{A}'(U)$, $t \in T(U)$, $U \subseteq X$ open. Let $\mathcal{A}' \times^{\mathcal{A}} T$ be the sheafification of the presheaf that sends $U \subseteq X$ open to the set of $\mathcal{A}(U)$-orbits in $\mathcal{A}' \times T$. We let $\mathcal{A}'$ act on $\mathcal{A}' \times^{\mathcal{A}} T$ from the left by multiplication on the first factor. This makes $\mathcal{A}' \times^{\mathcal{A}} T$ into an $\mathcal{A}'$-torsor (see Problem 7.3) and $T \mapsto \mathcal{A}' \times^{\mathcal{A}} T$ defines the map $H^1(\varphi)$ of pointed sets.

Remark 7.21. Let X be a topological space and let

$$1 \to \mathcal{A}' \xrightarrow{\iota} \mathcal{A} \xrightarrow{\pi} \mathcal{A}'' \to 1 \tag{7.8}$$

be an exact sequence of sheaves of groups on X. Define a *connecting map*

$$\delta \colon \mathcal{A}''(X) \to \check{H}^1(X, \mathcal{A}') \tag{7.9}$$

as follows. For $g'' \in \mathcal{A}''(X)$ let $\mathcal{U} = (U_i)_i$ be an open covering of X such that there exist $g_i \in \mathcal{A}(U_i)$ whose image in $\mathcal{A}''(U_i)$ is $g''_{|U_i}$. For all i, j let $g'_{ij} \in \mathcal{A}'(U_{ij})$ be the unique element that is mapped to $g_i g_j^{-1} \in \mathcal{A}(U_i \cap U_j)$. Then (g'_{ij}) is a Čech cocycle on $\mathcal{U}$. A different choice of elements g_i yields a cohomologous cocycle. Therefore its class $\delta(g'')$ in $\check{H}^1(X, \mathcal{A})$ is well defined. It is clear that δ is a morphism of pointed sets.

One can define δ also via torsors as a map $\delta \colon \mathcal{A}''(X) \to H^1(X, \mathcal{A}')$ as follows. Let $g'' \in \mathcal{A}(X'')$. Define a sheaf of sets $T_{g''}$ on X by

$$T_{g''}(U) := \left\{ t \in \mathcal{A}(U) \, ; \, \pi(t) = g''_{|U} \right\}. \tag{7.10}$$

Then $\mathcal{A}'(U)$ acts on $T_{g''}(U)$ by $(g', t) \mapsto \iota(g')t$ and this makes $T_{g''}$ into an $\mathcal{A}'$-torsor. Then $\delta(g'')$ is the isomorphism class of $T_{g''}$.

Proposition 7.22. *The following sequence of pointed sets is exact:*

$$\begin{aligned} 1 \longrightarrow \mathcal{A}'(X) &\xrightarrow{\iota_X} \mathcal{A}(X) \xrightarrow{\pi_X} \mathcal{A}''(X) \\ &\xrightarrow{\delta} \check{H}^1(X, \mathcal{A}') \xrightarrow{\check{H}^1(\iota)} \check{H}^1(X, \mathcal{A}) \xrightarrow{\check{H}^1(\pi)} \check{H}^1(X, \mathcal{A}''). \end{aligned} \tag{7.11}$$

Moreover, one has the following assertions.

1. *Assume that $\mathcal{A}'$ is a subgroup sheaf of the center of $\mathcal{A}$ (in particular it is a sheaf of abelian groups). Then δ is a homomorphism of groups, and $\check{H}^1(\iota)$ induces an injection of the group $\mathrm{Coker}(\delta)$ into the pointed set $\check{H}^1(X, \mathcal{A})$.*
2. *Assume that $\mathcal{A}'$, $\mathcal{A}$, and $\mathcal{A}''$ are abelian sheaves. Then the sequence (7.11) is an exact sequence of abelian groups.*

The same sequence exists with H^1 instead of $\check{H}^1$ and it is an instructive (and not too difficult) exercise to check all of the above assertions in the language of torsors.

Proof. ι_X injective. Proposition 3.18.

$\mathrm{Im}(\iota_X) = \mathrm{Ker}(\pi_X)$. "$\subseteq$" is clear. Let $g \in \mathrm{Ker}(\pi_X)$. Then $g_x \in \mathrm{Ker}(\pi_x) = \mathrm{Im}(\iota_x)$ for all $x \in X$. Hence there exists $x \in U^x \subseteq X$ open and $g'^x \in \mathcal{A}'(U^x)$ such that $\iota(g'^x) = g_{|U^x}$. For $x, y \in X$ one has

$$\iota(g'^x_{|U^x \cap U^y}) = g_{|U^x \cap U^y} = \iota(g'^y_{|U^x \cap U^y})$$

and hence $g'^x|_{U^x \cap U^y} = g'^y|_{U^x \cap U^y}$ because $\iota \colon \mathcal{A}'(U^x \cap U^y) \to \mathcal{A}(U^x \cap U^y)$ is injective. Hence there exists $g' \in \mathcal{A}'(X)$ such that $\iota(g') = g$.

$\mathrm{Im}(\pi_X) \subseteq \mathrm{Ker}(\delta)$. Let $g'' = \pi_X(g) \in \mathrm{Im}(\pi_X)$. Then we may choose $g_i = g_{|U_i}$ in the definition of δ and hence $g'_{ij} = 1$ and therefore $g'' \in \mathrm{Ker}(\delta)$.

$\mathrm{Im}(\pi_X) \supseteq \mathrm{Ker}(\delta)$. Let $g'' \in \mathrm{Ker}(\delta)$, i.e., we find an open covering $(U_i)_i$ of X, $g_i \in \mathcal{A}(U_i)$ and $h'_i \in \mathcal{A}'(U_i)$ such that

$$\pi(g_i) = g''_{|U_i} \qquad \text{and} \qquad \iota(h_i'^{-1} h'_j) = g_i g_j^{-1} \Leftrightarrow 1 = \iota(h'_i) g_i g_j^{-1} \iota(h'_j)^{-1}$$

for all $i, j \in I$. Replacing g_i by $\iota(h'_i) g_i$ we may assume that $g_i = g_j \in \mathcal{A}(U_i \cap U_j)$. Hence there exists $g \in \mathcal{A}(X)$ such that $g_{|U_i} = g_i$. Then $\pi(g)_{|U_i} = g''_{|U_i}$ and hence $\pi(g) = g''$.

$\mathrm{Im}(\delta) = \mathrm{Ker}(\check{H}^1(\iota))$. Let $(U_i)_i$ be an open covering and let $\theta = (g'_{ij})_{i,j} \in \check{H}^1(\mathcal{U}, \mathcal{A}')$. Then:

$$\theta \in \mathrm{Im}(\delta) \Leftrightarrow \check{H}^1(\iota)(\theta) = (g_i g_j^{-1})_{i,j} \text{ with } g_i \in \mathcal{A}(U_i)$$
$$\Leftrightarrow \iota^1(\theta) \text{ is cohomologous to } 1.$$

$\mathrm{Im}(\check{H}^1(\iota)) = \mathrm{Ker}(\check{H}^1(\pi))$. This follows from $\mathrm{Im}(\iota_{U_i}) = \mathrm{Ker}(\pi_{U_i})$ for every open subset U_i of X.

Assertions (1) and (2) are easy to check. $\qquad\qquad\qquad\qquad\qquad\qquad\square$

Remark 7.23. One can define the first terms of the cohomology sequence (7.11) in the following more general situation. Let $\mathcal{A}'$ be a sheaf of subgroups of a sheaf of groups $\mathcal{A}$ and let $\mathcal{A}/\mathcal{A}'$ be the sheafification of the presheaf of left cosets $U \mapsto \mathcal{A}(U)/\mathcal{A}'(U)$. Then $(\mathcal{A}/\mathcal{A}')(X)$ is a pointed set (the distinguished element is the image of $1 \in \mathcal{A}(X)$ in $(\mathcal{A}/\mathcal{A}')(X)$). The same construction of δ and the same arguments as above show that one still has an exact sequence of pointed sets

$$1 \longrightarrow \mathcal{A}'(X) \longrightarrow \mathcal{A}(X) \longrightarrow (\mathcal{A}/\mathcal{A}')(X)$$
$$\overset{\delta}{\longrightarrow} \check{H}^1(X, \mathcal{A}') \longrightarrow \check{H}^1(X, \mathcal{A}). \qquad (7.12)$$

7.4 Problems

Problem 7.1. Let X be a topological space. Show that the category $(\mathrm{Cov}(X))^{\mathrm{opp}}$ is filtered.

Problem 7.2. Let X be a topological space, G a sheaf of groups acting on a sheaf of sets T. The action is called *transitive* (respectively *free*) if the morphism $G \times T \to T \times T$, $(g, t) \mapsto (t, gt)$ for $g \in G(U)$, $t \in T(U)$, $U \subseteq X$ open, is a surjective (respectively injective) morphism of sheaves.

1. Show that G acts freely on T if and only if for all $U \subseteq X$ open and for all $t \in T(U)$ the stabilizer of t in $G(U)$ is trivial.

2. Show that T is a pseudotorsor if and only if the G-action is free and transitive.

3. Let e be the final sheaf, i.e., $e(U)$ is the set consisting of a single element for all $U \subseteq X$ open. Show that e is a final object in $(\mathrm{Sh}(X))$ and that a G-pseudotorsor T is a torsor if and only if $T \to e$ is an epimorphism.

Problem 7.3. Let X be a topological space and G be a sheaf of groups on X. Let T (respectively T') be a sheaf of sets on X that is endowed with a right action $T \times G \to T$ (respectively a left action $G \times T' \to T'$) of G. Define a left G-action of $T \times T'$ by $g \cdot (t, t') := (tg^{-1}, gt')$ for $g \in G(U)$, $t \in T(U)$, $t' \in T'(U)$, $U \subseteq X$ open. The *contracted product* $T \times^G T'$ is defined as the sheaf associated with the presheaf whose value on U is the set of $G(U)$-orbits of $T(U) \times T'(U)$.

Let $\varphi \colon G \to G'$ be a homomorphism of sheaves of groups on X and let G'_r be the sheaf G' with the right G-action given by $g' \cdot g := g'\varphi(g)$ on local sections. Let G' act on $G'_r \times^G T$ from the left on $G'_r \times^G T$ via the first factor.

1. Show that for every sheaf E on X one has $\mathrm{Hom}_{(\mathrm{Sh}(X))}(E, T \times^G T') = \mathrm{Hom}_{(\mathrm{Sh}(X))}(E, T \times T')^G$, where $(\ )^G$ denotes G-invariants.

2. Show that $T \mapsto G'_r \times^G T$ defines a functor from the category of sheaves with G-action to the category of sheaves with G'-action. Show that this functor is left adjoint to the functor $T' \mapsto T'^\varphi$, where T'^φ is the sheaf T' with G-action $(g, t') \mapsto \varphi(g)t'$.

3. Suppose that G acts transitively (respectively freely) on T (Problem 7.2). Show that G' acts transitively (respectively freely) on $G'_r \times^G T$. Deduce that if T is a G-torsor, then $G'_r \times^G T$ is a G'-torsor defining a map $H^1(\varphi) \colon H^1(X, G) \to H^1(X, G')$ of pointed sets.

4. Show that the following diagram is commutative:

$$
\begin{array}{ccc}
H^1(X, G) & \xrightarrow{\ H^1(\varphi)\ } & H^1(X, G') \\
{\scriptstyle c_G} \big\downarrow & & \big\downarrow {\scriptstyle c_{G'}} \\
\check{H}^1(X, G) & \xrightarrow{\ \check{H}^1(\varphi)\ } & \check{H}^1(X, G').
\end{array}
$$

Problem 7.4. Let $M \subseteq \mathbb{C}$ be open, $f \colon M \to \mathbb{C}$ be a holomorphic map and let $\mathcal{L}_f(U)$ be the set of logarithms of f, i.e., the set of holomorphic maps $g \colon U \to \mathbb{C}$ such that $\exp \circ g = f_{|U}$ for $U \subseteq M$ open. Endow $\mathcal{L}_f$ with the structure of a $\mathbb{Z}_M$-pseudotorsor, which is a torsor if and only if $f(z) \neq 0$ for all $z \in M$. Show that a holomorphic map $f \colon M \to \mathbb{C}^\times$ has a logarithm if M is simply connected.

Problem 7.5. Let X be a topological space, $x \in X$, G a group, and let $\mathcal{G}$ be the skyscraper sheaf in x with values G (Problem 3.6). Show that $H^1(X, \mathcal{G}) = 1$.

Problem 7.6. Let $n \geq 1$ and let A be an abelian group.

1. Show $H^1(\mathbb{P}^1(\mathbb{R}), A_{\mathbb{P}^1(\mathbb{R})}) = A$ and $H^1(\mathbb{P}^n(\mathbb{R}), A_{\mathbb{P}^n(\mathbb{R})}) = \{a \in A \; ; \; 2a = 0\}$ for $n \geq 2$.
2. Show that $H^1(\mathbb{P}^n(\mathbb{C}), A_{\mathbb{P}^n(\mathbb{C})}) = 0$.

Problem 7.7. Let $X = \mathbb{P}^1(\mathbb{C})$, $\mathcal{U} = (U_0, U_1)$ with $U_0 = \mathbb{P}^1(\mathbb{C}) \setminus \{(1 : 0)\}$ and $U_1 = \mathbb{P}^1(\mathbb{C}) \setminus \{(0 : 1)\}$, and let $\mathcal{O}_{\mathbb{P}^1(\mathbb{C})}$ the sheaf of holomorphic functions on $\mathbb{P}^1(\mathbb{C})$. Show that $H^1(\mathcal{U}, \mathcal{O}_{\mathbb{P}^1(\mathbb{C})}) = 0$. In Theorem 10.22 we will see that $H^1(\mathbb{C}, \mathcal{O}_{\mathbb{C}}) = 0$. Deduce that $H^1(\mathbb{P}^1(\mathbb{C}), \mathcal{O}_{\mathbb{P}^1(\mathbb{C})}) = 0$.
Hint: Use the theory of Laurent series.

Problem 7.8. Let $X \subseteq \mathbb{C}$ be open and let $\mathcal{O}_X$ be the sheaf of holomorphic functions. Let $f : X \to \mathbb{C}$ be a holomorphic function.

1. Show that the following sequence of sheaves of abelian groups is exact:
$$1 \longrightarrow (\mathbb{C}^\times)_X \longrightarrow \mathcal{O}_X^\times \xrightarrow{\mathrm{dlog}} \mathcal{O}_X \to 0,$$
where $\mathrm{dlog}(g) := g'/g$.
2. Show that if X is simply connected, then there exists a holomorphic map $g : X \to \mathbb{C}^\times$ with $g'/g = f$.
3. Show that the sheaf of holomorphic solutions of the differential equation $u' = fu$ is a locally constant sheaf of 1-dimensional $\mathbb{C}$-vector spaces. Show that the differential equation has a solution without zeros if X is simply connected.

Bundles

8

We will now study morphisms of premanifolds $\pi\colon X \to B$ that look locally on B like a fixed morphism $p\colon Z \to B$. We then call π a twist of p. Important special cases are fiber bundles, where $p\colon B \times F \to B$ is a projection. Moreover, one often endows twists with an additional datum that restricts the changes between different local isomorphisms of p and π to a fixed subsheaf of the sheaf all automorphisms of p. For fiber bundles this subsheaf will usually be given by a faithful action of a Lie group G on the fiber F. An important special case of a fiber bundle is the notion of a principal bundle for a Lie group G. They are in a way the universal fiber bundles for a given structure group G (Remark 8.22).

Most of the results in Sect. 8.1 hold verbatim if one replaces "premanifold" with "-topological space" and "Lie group" with "topological group". Notable exceptions are Remark 8.10 and Example 8.21 as for topological groups the projection $G \to G/H$ does not have local sections in general.

In the next section we study a very important special case of fiber bundles: vector bundles. This is the main topic of this chapter. There are two ways to look at vector bundles. The first point of view is that of a fiber bundle where the typical fiber is a vector space and where the local coordinate changes are linear. This is explained in Sect. 8.2. The second point of view is to consider vector bundles as locally free modules over the structure sheaf. Hence we introduce such modules in Sect. 8.3 and prove in Sect. 8.4 that both point of views are equivalent.

Among the most important vector bundles is the tangent bundle whose fiber at a point p of a premanifold M is the tangent space $T_p(M)$. This will be described as a geometric vector bundle as well as a locally free module. The dual of the tangent bundle is the vector bundle of differential 1-forms and by taking exterior powers we obtain differential forms of arbitrary degree. We conclude the chapter by defining the de Rham complex of a premanifold.

© Springer Fachmedien Wiesbaden 2016
T. Wedhorn, *Manifolds, Sheaves, and Cohomology*, Springer Studium Mathematik – Master,
DOI 10.1007/978-3-658-10633-1_8

8.1 Twists, Fiber Bundles, and Principal Bundles

Twists will be premanifolds endowed with a morphism to a fixed premanifold B, the "base premanifold". Hence let us introduce the following general terminology.

Definition 8.1. Let C be a category and let B be an object of C. Define the *category of B-objects of C*, denoted by $C_{/B}$:

(a) Objects in $C_{/B}$ are morphisms $p: X \to B$ in C, usually denoted by (X, p).
(b) A morphism $(X, p) \to (Y, q)$ in $C_{/B}$ is a morphism $\phi: X \to Y$ in C such that $q \circ f = p$. It is called a *B-morphism*.
    Composition is given by composition in C.

In the sequel we will take for C the category of premanifolds (real C^α for a fixed α or complex). We fix a premanifold B. Then objects in $C_{/B}$ are called *B-premanifolds*.

Twists
Let $p: Z \to B$ be a morphism of premanifolds, in other words, (Z, p) is a B-premanifold. Let

$$\mathcal{A}ut(p) = \mathcal{A}ut_B(Z)$$

be the sheaf of groups on B that attaches to $U \subseteq B$ the group of automorphism of U-premanifolds $\alpha: p^{-1}(U) \overset{\sim}{\to} p^{-1}(U)$.

Definition and Remark 8.2 (Twists). We fix a morphism $p: Z \to B$ and a sheaf $\mathcal{G}$ of subgroups of $\mathcal{A}ut_B(Z)$.

1. For a morphism of premanifolds $\pi: X \to B$ a $\mathcal{G}$-*twisting atlas of p for π* is a family $(U_i, h_i)_{i \in I}$ where $(U_i)_i$ is an open covering of B and where h_i is an isomorphism of U_i-premanifolds

$$h_i: \pi^{-1}(U_i) \overset{\sim}{\to} p^{-1}(U_i) \tag{8.1}$$

such that for all $i, j \in I$ we have

$$g_{ij} := h_{i|\pi^{-1}(U_{ij})} \circ h_j^{-1}{}_{|p^{-1}(U_{ij})} \in \mathcal{G}(U_{ij}), \tag{8.2}$$

where as usual $U_{ij} := U_i \cap U_j$.
   Note the similarity to the notion of an atlas of a premanifold, where one also has a family of charts covering the premanifold such that the chart changes have certain properties (namely being C^α-maps).
2. Given such a $\mathcal{G}$-twisting atlas $(U_i, h_i)_{i \in I}$ and a refinement $(V_j)_{j \in J}$ of $(U_i)_i$ defined by $\tau: J \to I$ such that $V_j \subseteq U_{\tau(j)}$ we let $k_j: \pi^{-1}(V_j) \overset{\sim}{\to} p^{-1}(V_j)$ be the restriction of

$h_{\tau(j)}$. Then $(V_j, k_j)_{j \in J}$ is again a $\mathcal{G}$-twisting atlas of p for π, called a *refinement of* $(U_i, h_i)_{i \in I}$.

Two $\mathcal{G}$-twisting atlases $(U_i, h_i)_{i \in I}$ and $(U_i', h_i')_{i \in I'}$ are called *equivalent* if there exists refinements $(V_j, k_j)_{j \in J}$ and $(V_j, k_j')_{j \in J}$ of $(U_i, h_i)_{i \in I}$ and $(U_i', h_i')_{i \in I'}$ respectively (with the same underlying open covering $(V_j)_j$) such that $k_j \circ (k_j')^{-1} \in \mathcal{G}(V_j)$.

3. A *twist of p with structure sheaf $\mathcal{G}$* is a morphism of premanifold $\pi: X \to B$ together with an equivalence class of $\mathcal{G}$-twisting atlases of p for π.

4. Let $\xi = (\pi: X \to B, [U_i, h_i]_{i \in I})$ and $\xi' = (\pi': X' \to B, [U_i, h_i']_{i \in I})$ be twists of p with structure sheaf $\mathcal{G}$ (after replacing $\mathcal{G}$-twisting atlases by equivalent atlases we can assume that the underlying open covering of both atlases is the same). An *isomorphism* $\xi \xrightarrow{\sim} \xi'$ *of twists of p with structure sheaf $\mathcal{G}$* is an isomorphism of B-manifolds $F: X \xrightarrow{\sim} X'$ such that for all $i \in I$ there exists $g_i \in \mathcal{G}(U_i)$ making the following diagram commutative:

$$
\begin{array}{ccc}
\pi^{-1}(U_i) & \xrightarrow{F} & \pi'^{-1}(U_i) \\
\downarrow{h_i} & & \downarrow{h_i'} \\
p^{-1}(U_i) & \xrightarrow{g_i} & p^{-1}(U_i).
\end{array}
$$

Remark and Definition 8.3 (Twists with full structure sheaf). If $\mathcal{G} = \mathcal{A}ut(p)$, then any two $\mathcal{G}$-twisting atlases are equivalent and it suffices to ask only for the existence of an $\mathcal{A}ut(p)$-twisting atlas. In other words, a twist of p with $\mathcal{A}ut(p)$-structure is simply a morphism $\pi: X \to B$ of premanifolds such that there exists an open covering $(U_i)_i$ of B and isomorphism of U_i-premanifolds $\pi^{-1}(U_i) \xrightarrow{\sim} p^{-1}(U_i)$. In this case we simply speak of a *twist of p*.

In general the equivalence class of a $\mathcal{G}$-twisting atlas is a datum containing additional information: There exist non-isomorphic twists of p with structure sheaf $\mathcal{G}$ such that the underlying B-manifolds are isomorphic (Problem 8.1).

Remark 8.4 (Twists and Čech cohomology). Let $p: Z \to B$ be a morphism of premanifolds and $\mathcal{G}$ a sheaf of subgroups of $\mathcal{A}ut_B(Z)$. Let $\xi = (\pi: X \to B, [U_i, h_i]_{i \in I})$ be a twist of p with structure sheaf $\mathcal{G}$. Then the g_{ij} defined in (8.2). Satisfy for all $i, j, k \in I$ the cocycle condition

$$g_{kj} g_{ji} = g_{ki} \in \mathcal{G}(U_{ijk}). \tag{8.3}$$

We obtain an element $(g_{ij})_{ij} \in \check{Z}^1(\mathcal{U}, \mathcal{G})$, where $\mathcal{U} := (U_i)_i$. Let $\gamma_{\mathcal{U}}(\xi)$ be its class in $\check{H}^1(\mathcal{U}, \mathcal{G})$. Refining $(U_i, h_i)_i$ by a refinement $\mathcal{V}$ replaces $\gamma_{\mathcal{U}}(\xi)$ by its image in $\check{H}^1(\mathcal{V}, \mathcal{G})$. If one chooses an equivalent $\mathcal{G}$-twisting atlas, then $(g_{ij})_{ij}$ is replaced by a cohomologous cocycle. Similarly, any isomorphism of twists replaces $(g_{ij})_{ij}$ by a cohomologous cocycle. Hence the image $\gamma(\xi)$ of $\gamma_{\mathcal{U}}(\xi)$ in $\check{H}^1(B, \mathcal{G})$ depends only on the isomorphism class of the twist ξ of p with structure sheaf $\mathcal{G}$.

Conversely, the twist π (or rather a B-premanifold isomorphic to $\pi\colon X \to B$) can be glued from the family $(p^{-1}(U_i))_i$ along the $p^{-1}(U_{ij})$ using the automorphisms g_{ij}, see Proposition 4.11. Using the notation there, the isomorphisms h_i are then given by the inverse of the maps ψ_i considered as isomorphisms onto their image.

Altogether, we obtain the following classification result of twists.

Theorem 8.5. *The map γ constructed in Remark 8.4 yields a bijection of pointed sets*

$$\gamma\colon \left\{ \begin{array}{c} \text{isomorphism classes of twists of } p \\ \text{with structure sheaf } \mathcal{G} \end{array} \right\} \xrightarrow{\ \sim\ } \check{H}^1(B, \mathcal{G}). \tag{8.4}$$

The trivial class in $\check{H}^1(B, \mathcal{G})$ corresponds to the isomorphism class of the trivial twist $p\colon Z \to B$ endowed with the equivalence class of the trivial $\mathcal{G}$-twisting atlas (B, id_Z).

Via the isomorphism $\check{H}^1(B, \mathcal{G}) \cong H^1(B, \mathcal{G})$ (Proposition 7.13), every isomorphism class of a twist with structure sheaf $\mathcal{G}$ also corresponds to the isomorphism class of a $\mathcal{G}$-torsor. This can even be made into an equivalence of categories; see Problem 8.2 for details.

Remark 8.6. One can work more generally with a not necessarily injective homomorphism $\alpha\colon \mathcal{G} \to \mathcal{A}ut(p)$. In this case one should add the $(g_{ij})_{i,j}$ as an additional datum to a $\mathcal{G}$-twisting atlas. Using the obvious notion of "refinement" for such a $\mathcal{G}$-twisting atlas one obtains the notion of equivalence for such twisting atlases and hence a generalization of a twist. The constructions in Remark 8.4 then still yield a bijective map

$$\left\{ \begin{array}{c} \text{isomorphism classes of twists of } p \\ \text{with structure sheaf } \mathcal{G} \end{array} \right\} \xrightarrow{\ \sim\ } \check{H}^1(B, \mathcal{G}).$$

We leave details to the reader.

Remark 8.7 (Pullback of twists). Let $p\colon Z \to B$ be a submersion and let $r\colon B' \to B$ be a morphism of premanifolds. We set $Z' := B' \times_B Z$. As p is a submersion, we obtain by Corollary 5.50 a commutative diagram

$$\begin{array}{ccc} Z' & \longrightarrow & Z \\ {\scriptstyle p'} \downarrow & & \downarrow {\scriptstyle p} \\ B' & \xrightarrow{\ r\ } & B \end{array}$$

of premanifolds, where $r^*(p) := p'$ is the projection. Moreover p' is again a submersion.

Let $\pi\colon X \to B$ be a twist of p. As π looks locally on B like p, π is also a submersion, $X' := B' \times_B X$ is a premanifold, and the projection $r^*(\pi) := \pi'\colon X' \to B'$ is a submersion. If $(U_i, h_i)_i$ is a twisting atlas of p for π, then $(f^{-1}(U_i), \mathrm{id}_{B'} \times h_i)_i$ is a twisting atlas of p' for π'. In particular, π' is a twist of p', called the *pullback of π via r*.

Sheaves of Groups defined by Lie Groups

The most important special case of a twist is a fiber bundle where the structure sheaf is given by the action of a Lie group on the fiber. Hence let us first study how Lie groups give rise to sheaves of groups on a premanifold B.

Definition and Remark 8.8. Let M be a premanifold. Let $\mathcal{O}_{B;M}$ be the sheaf of morphisms from B to M, i.e., for $U \subseteq B$ open we let $\mathcal{O}_{B;M}(U)$ be the set of morphisms of premanifolds $U \to M$. Then $\mathcal{O}_{B;M}$ is a sheaf of M-valued functions on B. If $\varphi\colon M \to N$ is a morphism of premanifolds, composition with φ yields a morphism of sheaves $\varphi_B\colon \mathcal{O}_{B;M} \to \mathcal{O}_{B;N}$. We obtain a functor from the category of premanifolds to the category of sheaves of sets on B.

If $M = G$ is a Lie group, then the group structure on G endows $\mathcal{O}_{B;G}(U)$ with a group structure by pointwise multiplication making $\mathcal{O}_{B;G}$ into a sheaf of groups. For a homomorphism $\varphi\colon G \to H$ of Lie groups, φ_B is a morphism of sheaves of groups $\mathcal{O}_{B;G} \to \mathcal{O}_{B;H}$. We obtain a functor from the category of Lie groups to the category of sheaves of groups on B.

Example 8.9.

1. If G is an abstract group considered as a 0-dimensional Lie group, then $\mathcal{O}_{B;G}$ is the constant group sheaf G_B on B attached to G.
2. Let $G = (\mathbb{K}, +)$ considered as a 1-dimensional Lie group. Then $\mathcal{O}_{B;\mathbb{K}}$ is the underlying sheaf of abelian groups of the structure sheaf $\mathcal{O}_B$ on B.
3. Let $G = \mathrm{GL}_n(\mathbb{K})$. Then $\mathrm{GL}_{n,B} := \mathcal{O}_{B;\mathrm{GL}_n(\mathbb{K})}$ is the group sheaf that attaches to $U \subseteq B$ open the group $\mathrm{GL}_n(\mathcal{O}_B(U))$. In particular $\mathcal{O}_B^\times := \mathcal{O}_{B;\mathbb{K}^\times}$ is the sheaf of $\mathbb{K}$-valued nowhere vanishing morphisms.

To calculate $\check{H}^1(B, \mathcal{O}_{B;G})$ it is useful to have exact cohomology sequences. To use this tool, we show that every short exact sequence of Lie groups yields an exact sequence of sheaves on any premanifold.

Remark 8.10. Let $\iota\colon H \hookrightarrow G$ be the inclusion of a Lie subgroup into a Lie group G and let G/H be the quotient manifold (Corollary 6.23). Then ι_B identifies $\mathcal{O}_{B;H}$ with a sheaf of subgroups of $\mathcal{O}_{B;G}$. We claim that $\mathcal{O}_{B;G/H}$ is the sheaf $\mathcal{O}_{B;G}/\mathcal{O}_{B;H}$ of left cosets (Remark 7.23).

Indeed, composition with the projection $\pi\colon G \to G/H$ yields a morphism of sheaves $\pi_B\colon \mathcal{O}_{B;G} \to \mathcal{O}_{B;G/H}$ and for two morphisms $g, g'\colon U \to G$, $U \subseteq B$ open, one has

$\pi \circ g = \pi \circ g'$ if and only if $(g^{-1}g')(b) \in H$ for all $b \in U$. Hence π_B induces an injective morphism of sheaves $\mathcal{O}_{B;G}/\mathcal{O}_{B;H} \to \mathcal{O}_{B;G/H}$. As π is a surjective submersion, it has locally on G/H sections that are morphisms of premanifolds. This shows that π_B is surjective. Hence $\mathcal{O}_{B;G}/\mathcal{O}_{B;H} \to \mathcal{O}_{B;G/H}$ is also surjective.

We obtain by Remark 7.23 a cohomology sequence

$$
\begin{aligned}
1 &\longrightarrow \mathrm{Hom}_{(\mathrm{PMfd})}(B, H) \longrightarrow \mathrm{Hom}_{(\mathrm{PMfd})}(B, G) \longrightarrow \mathrm{Hom}_{(\mathrm{PMfd})}(B, G/H) \\
&\overset{\delta}{\longrightarrow} \check{H}^1(B, \mathcal{O}_{B;H}) \longrightarrow \check{H}^1(B, \mathcal{O}_{B;G}).
\end{aligned}
\tag{8.5}
$$

If H is normal in G, then G/H is a Lie group and (8.5) can be extended by $\check{H}^1(X, \mathcal{O}_{B;G/H})$ on the right-hand side (Proposition 7.22). If G is an abelian Lie group, this sequence is an exact sequence of abelian groups.

Example 8.11. We consider the complex exponential sequence

$$
0 \longrightarrow \mathbb{Z} \longrightarrow \mathbb{C} \overset{\exp(2\pi i \cdot)}{\longrightarrow} \mathbb{C}^\times \longrightarrow 1,
$$

which we consider as an exact sequence of complex Lie groups. Hence we obtain by Remark 8.10 for every complex premanifold B an exact sequence of sheaves of abelian groups

$$
0 \longrightarrow \mathbb{Z}_B \longrightarrow \mathcal{O}_B \overset{f \mapsto \exp(2\pi i f)}{\longrightarrow} \mathcal{O}_B^\times \longrightarrow 1,
$$

yielding a cohomology sequence

$$
\begin{aligned}
0 &\longrightarrow \mathbb{Z}^{\pi_0(B)} \longrightarrow \mathcal{O}_B(B) \longrightarrow \mathcal{O}_B^\times(B) \\
&\longrightarrow \check{H}^1(B, \mathbb{Z}_B) \longrightarrow \check{H}^1(B, \mathcal{O}_B) \longrightarrow \check{H}^1(B, \mathcal{O}_B^\times).
\end{aligned}
\tag{8.6}
$$

Hence we deduce that every non-vanishing holomorphic function on B has a logarithm if and only if $H^1(B, \mathbb{Z}_B) \to H^1(B, \mathcal{O}_B)$ is injective. This is for instance the case if $H^1(B, \mathbb{Z}_B) = 0$ and hence if B is simply connected (Proposition 7.5).

Moreover, by (8.15) below we will see that $\check{H}^1(B, \mathcal{O}_B^\times)$ classifies (complex holomorphic) line bundles on B. To use the cohomological formalism to study $\check{H}^1(B, \mathcal{O}_B^\times)$ it would be desirable to extend (8.6) to the right by defining higher cohomology groups $H^p(B, \cdot)$ for $p \geq 2$. This will be the topic of Chap. 10.

Fiber Bundles

We now define fiber bundles over a fixed premanifold B as twists of the projection $B \times F \to B$, where F is some fixed premanifold. The structure sheaf will be given by a faithful action of a Lie group G on F.

Remark and Definition 8.12 (Fiber bundles). Let us fix a premanifold F, a Lie group G, and a faithful action of G on F by a morphism of premanifolds $G \times F \to F$. Let $p: B \times F \to B$ be the projection. Then

$$\mathcal{O}_{B;G} \to \mathcal{A}ut(p), \qquad g \mapsto \big((b, f) \mapsto (b, g(b)f)\big) \tag{8.7}$$

$(g \in \mathcal{O}_{B;G}(U), (b, f) \in U \times F, U \subseteq B$ open$)$ is a homomorphism of sheaves of groups over B. It is injective because the G-action on F is faithful. We use it to consider $\mathcal{O}_{B;G}$ as a sheaf of subgroups of $\mathcal{A}ut(p)$.

A twist $\pi: X \to B$ of $p: B \times F \to B$ (without specifying any structure group, Remark 8.3) is called a *fiber bundle with basis B and typical fiber F*. A twist $(\pi: X \to B, [U_i, h_i]_{i \in I})$ of $p: B \times F \to B$ with structure group sheaf $\mathcal{O}_{B;G}$ is called a *fiber bundle with basis B, typical fiber F, and structure group G*. A $\mathcal{O}_{B;G}$-twisting atlas for a fiber bundle with basis B and typical fiber F is called a *G-bundle atlas*. An equivalence class of G-bundle atlases is called a *G-bundle structure*. In other words, a fiber bundle with basis B, typical fiber F, and structure group G is the same as a fiber bundle with basis B and typical fiber F together with the datum of a G-bundle structure.

An isomorphism of twists of p with structure group sheaf $\mathcal{O}_{B;G}$ is called an *isomorphism of fiber bundles with basis B, typical fiber F, and structure group G*. We obtain a category in which all morphisms are isomorphisms, denoted by $(\mathrm{FIB}_{F,G}(B))$. We denote by $\mathrm{Fib}_{F,G}(B)$ the set of isomorphisms classes in $(\mathrm{FIB}_{F,G}(B))$.

By Theorem 8.5 we have a bijection

$$\mathrm{Fib}_{F,G}(B) \xrightarrow{\sim} \check{H}^1(B, \mathcal{O}_{B;G}). \tag{8.8}$$

The trivial cocycle in $\check{H}^1(B, \mathcal{O}_{B;G})$ corresponds to the isomorphism class of the trivial fiber bundle $p: B \times F \to B$ endowed with the equivalence class of the trivial G-bundle atlas $(B, \mathrm{id}_{B \times F})$.

Very often, one writes also simply $\check{H}^1(B, G)$ instead of $\check{H}^1(B, \mathcal{O}_{B;G})$.

Remark 8.13. If the action of G on F is not faithful, the homomorphism (8.7) is not injective and one has to fix the cocycle defining the fiber bundle as an extra datum (Remark 8.6).

Example 8.14. Let F be a discrete space, considered as a 0-dimensional premanifold, and consider the group $\mathrm{Aut}(F)$ of all permutations of F as a 0-dimensional Lie group. Then the action of $\mathrm{Aut}(F)$ on F is faithful and a fiber bundle with basis B, typical fiber F, and structure group $\mathrm{Aut}(F)$ is the same as a covering space $p: \tilde{B} \to B$ of B such that $p^{-1}(b) \cong F$ for all $b \in B$ (Remark 8.3).

Example 8.15 (Möbius band). Let $B = S^1 = \{z \in \mathbb{C} \; ; \; |z| = 1\}$, $F = (-1, 1)$ and let $G := \{\pm 1\}$, considered as a 0-dimensional real Lie group, act on F via multiplication.

By Example 7.18 we have $\check{H}^1(S^1, G_{S^1}) \cong \check{H}^1(\mathcal{U}, G_{S^1}) = \check{Z}^1(\mathcal{U}, G_{S^1})/\sim$ where $\mathcal{U} = (U_0, U_1) := S^1 \setminus \{1\}, S^1 \setminus \{-1\})$. Moreover $\check{Z}^1(\mathcal{U}, G_{S^1}) = G_{S^1}(U_0 \cap U_1) = G \times G$ and $(\varepsilon_+, \varepsilon_-) \sim (\varepsilon'_+, \varepsilon'_-)$ if $\varepsilon_+ \varepsilon_- = \varepsilon'_+ \varepsilon'_-$.

Therefore there are precisely two isomorphism classes of fiber bundles with typical fiber $(-1, 1)$ and structure group $\{\pm 1\}$ over S^1. The fiber bundle corresponding to the trivial class is $B \times F$.

The non-trivial isomorphism class of the fiber bundle is given by the cocycle $(+1, -1) \in \check{Z}^1(\mathcal{U}, G_{S^1})$ (or by the cohomologous cocycle $(-1, +1)$). If we write $U_0 \cap U_1 = W_+ \cup W_-$ with $W_{\pm} := \{z \in S^1 \ ; \ \pm \mathrm{Im}(z) > 0\}$, then it is obtained from gluing $U_0 \times F$ and $U_1 \times F$ along $(U_0 \cap U_1) \times F$ via the automorphism

$$(U_0 \cap U_1) \times F \ni (b, x) \mapsto \begin{cases} (b, x), & \text{if } x \in W_+; \\ (b, -x), & \text{if } x \in W_-. \end{cases}$$

We obtain the Möbius band (Example 4.30).

Fiber bundles "inherit" the following properties from the trivial bundle $B \times F \to B$.

Proposition 8.16. *Let $F \neq \emptyset$ be a premanifold and let $\pi\colon X \to B$ be a fiber bundle with basis B and typical fiber F.*

1. *Then π is a surjective submersion.*
2. *If F is Hausdorff, then π is separated. If B and F are manifolds, then X is a manifold.*

Proof. 1.   Working locally on B we can assume that $X = B \times F$ and that π is the projection. This shows that π is surjective submersion.

2.   To see that π is separated, we can again assume the π is the projection $B \times F \to B$. This is separated because F is Hausdorff. If B is also Hausdorff, then X is Hausdorff by Corollary 1.27. If B is second countable, then B is a Lindelöf space (Theorem 4.28). Hence we can assume there exists a countable open covering $(U_n)_n$ of B such that for all n one has isomorphisms $h_n\colon \pi^{-1}(U_n) \xrightarrow{\sim} U_n \times F$. As $U_n \times F$ is second countable, X has a countable covering by second countable open subspaces. Hence X is second countable. $\qquad \square$

Remark and Definition 8.17 (Pullback of fiber bundles). Let $F \neq \emptyset$ be a premanifold and let G be a Lie group acting faithfully on F by a morphism of premanifolds $G \times F \to F$. Let $r\colon B' \to B$ be a morphism of premanifolds. Then the projection $B' \times_B (B \times F) \to B' \times F$ is an isomorphism with inverse $(b', f) \mapsto (b', r(b'), f)$. Hence if $\pi\colon X \to B$ is a fiber bundle with typical fiber F, then its pullback $\pi' := r^*(\pi)\colon X' := B' \times_B X \to B'$ is a fiber bundle with typical fiber F over B' (Remark 8.7). Note that the projection $X' \to X$ induces isomorphisms of the fibers $\pi'^{-1}(b') \xrightarrow{\sim} \pi^{-1}(r(b'))$.

Now let $(U_i, h_i)_i$ be a G-bundle atlas. We set

$$h'_i := \mathrm{id}_{r^{-1}(U_i)} \times h_i : r^{-1}(U_i) \times_{U_i} \pi^{-1}(U_i) = \pi'^{-1}\left(r^{-1}(U_i)\right)$$
$$\xrightarrow{\sim} r^{-1}(U_i) \times_{U_i} (U_i \times F) = r^{-1}(U_i) \times F.$$

Then $(r^{-1}(U_i), h'_i)_i$ is a G-bundle atlas for π'. It is immediate that this construction sends equivalent G-bundle atlases for π to equivalent G-bundle atlases for π'. Therefore we have constructed for every fiber bundle $\xi = (\pi, [U_i, h_i]_i)$ with typical fiber F and structure group G over B a fiber bundle $r^*(\xi)$ with typical fiber F and structure group G over B'.

If $f : \xi_1 \xrightarrow{\sim} \xi_2$ is an isomorphism of fiber bundles with typical fiber F and structure group G over B, then $r^*(f) := \mathrm{id}_{B'} \times f : B' \times_B X_1 \xrightarrow{\sim} B' \times_B X_2$ is an isomorphism of fiber bundles with typical fiber F and structure group G over B'. We obtain a functor

$$r^* : (\mathrm{FIB}_{F,G}(B)) \longrightarrow (\mathrm{FIB}_{F,G}(B')) \tag{8.9}$$

called *pullback under r*.

If $(g_{ij} : U_i \cap U_j \to G)_{ij}$ is the cocycle of $\mathcal{O}_{B;G}$ given by a G-bundle atlas $(U_i, h_i)_i$ of ξ, then $(g_{ij} \circ r : r^{-1}(U_i \cap U_j) \to G)_{ij}$ is the cocycle of $\mathcal{O}_{B';G}$ given by the G-bundle atlas $(r^{-1}(U_i), h'_i)_i$ of $r^*(\xi)$. The map $(g_{ij})_{ij} \mapsto (g_{ij} \circ r)_{ij}$ yields a morphism of pointed sets $r^* : \check{H}^1(B, \mathcal{O}_{B;G}) \to \check{H}^1(B', \mathcal{O}_{B';G})$. It is a homomorphism of abelian groups if G is abelian. We obtain a commutative diagram

$$
\begin{array}{ccc}
\mathrm{Fib}_{F,G}(B) & \xrightarrow{\sim} & \check{H}^1(B, \mathcal{O}_{B;G}) \\
{\scriptstyle r^*}\downarrow & & \downarrow{\scriptstyle r^*} \\
\mathrm{Fib}_{F,G}(B') & \xrightarrow{\sim} & \check{H}^1(B', \mathcal{O}_{B';G}).
\end{array}
\tag{8.10}
$$

Definition and Remark 8.18 (Morphisms of fiber bundles). Let $F \neq \emptyset$ be a premanifold and let G be a Lie group acting faithfully on F by a morphism of premanifolds $G \times F \to F$. Let $r : B' \to B$ be a morphism of premanifolds and let $\xi = (X \to B, [U_i, h_i]_i)$ (respectively $\xi' = (X' \to B', [U'_i, h'_i]_i)$) be fiber bundles with typical fiber F and structure group G over B (respectively B'). A *morphism $\xi' \to \xi$ of fiber bundles with typical fiber F and structure group G along r* is an isomorphism $R : \xi' \xrightarrow{\sim} r^*(\xi)$ of fiber bundles with typical fiber F and structure group G over B'. Sometimes one writes $(\hat{r}, r)$ for such a morphism, where $\hat{r}$ is the composition $X' \xrightarrow{R} B' \times_B X \xrightarrow{\mathrm{pr}_2} X$.

Principal Bundles

An important special case of a fiber bundle is a principal bundle.

Definition 8.19 (Principal bundle). Let G be a Lie group and consider the faithful action on itself by left multiplication. A fiber bundle with basis B, typical fiber G, and structure group G is called a *principal G-bundle over B*.

We denote by $\mathrm{Princ}_G(B)$ the set of isomorphism classes of principal G-bundles. By (8.8) we have a bijection

$$\mathrm{Princ}_G(B) \xrightarrow{\sim} \check{H}^1(B, \mathcal{O}_{B;G}). \tag{8.11}$$

There is a different way to describe the G-bundle structure for a principal G-bundle.

Proposition 8.20. *Let G be a Lie group. The following data are equivalent:*

(a) *A principal G-bundle $(\pi\colon X \to B, [U_i, h_i]_{i \in I})$ over B.*
(b) *A morphism of premanifolds $\pi\colon X \to B$ and a right action of G on X by a morphism of premanifolds such that there exists an open covering $(U_i)_{i \in I}$ of B and for all $i \in I$ an isomorphism of U_i-premanifolds $h_i\colon \pi^{-1}(U_i) \xrightarrow{\sim} U_i \times G$ such that*

$$h_i^{-1}(b, gg') = h_i^{-1}(b, g)g', \qquad b \in U_i, g, g' \in G. \tag{8.12}$$

Moreover, if these data are given, then the right action of G on X preserves the fibers $X(b) := \pi^{-1}(b)$ and the induced action of G on $X(b)$ is simply transitive for all $b \in B$.

Proof. Let there be given a datum as in (b). Then $\pi\colon X \to B$ is a fiber bundle with basis B and typical fiber G. Let us show the last assertion. The G-right action preserves the fibers because of (8.12) and because the h_i preserve fibers. For $x \in X$ and $g' \in G$ we choose $i \in I$ with $b := \pi(x) \in U_i$. Let $h_i(x) = (b, g)$. Then $xg' = h_i^{-1}(b, gg')$ by (8.12) and hence $\pi(xg') = b$. Moreover, the h_i induce for all b an isomorphism

$$h_{i,b}\colon X(b) \xrightarrow{\sim} \{b\} \times G = G$$

and (8.12) shows that $h_{i,b}$ is G-equivariant if we endow the right-hand side with the (simply transitive) action of G by right multiplication. Hence the right action of G on $X(b)$ is simply transitive.

Let us now construct a G-bundle structure from a datum as in (b). For $i, j \in I$, $b \in U_i \cap U_j$ the map $h_{i,b} \circ h_{j,b}^{-1}\colon G \to G$ is a G-equivariant isomorphism. Hence it is necessarily of the from $g \mapsto g_{i,j,b}g$ for some $g_{i,j,b} \in G$. Therefore $b \mapsto h_{i,b} \circ h_{j,b}^{-1}$ defines a morphism $U_i \cap U_j \to G$, $b \mapsto g_{i,j,b}$, for $i, j \in I$, and $(U_i, h_i)_i$ is a G-bundle atlas for π. A different choice of $(U_i, h_i)_i$ as in (b) yields equivalent G-bundle atlases. We obtain a well-defined G-bundle structure on π.

Conversely, let $(\pi\colon X \to B, [U_i, h_i]_{i \in I})$ be a principal G-bundle. For $b \in B$ choose $i \in I$ with $b \in U_i$. Then h_i induces an isomorphism of the fibers $h_{i,b}\colon X(b) \xrightarrow{\sim} \{b\} \times G =$

G, which we use to define a right action on $X(b)$ by transport of structure, i.e., the right action is given by

$$(x, g) \mapsto h_{i,b}^{-1}(h_{i,b}(x)g), \qquad x \in X(b), g \in G.$$

For every $j \in I$ with $b \in U_j$ there exists $g_{i,j,b} \in G$ such that $h_{i,b}(x) = g_{i,j,b}h_{j,b}(x)$ and hence we obtain the same right G-action on $X(b)$ if we use h_j for the definition. Therefore the G-action is independent of the choice of i and we obtain a right G-action on X. The same argument shows that passing to an equivalent $\mathcal{O}_{B;G}$-twisting atlas does not change the G-action on X. To see that this action is a morphism of premanifolds we may work locally on B. Then over U_i this follows because h_i is an isomorphism of premanifolds. $\qquad\square$

Example 8.21. Let G be a Lie group, let H be a Lie subgroup of G, let G/H be the quotient (Corollary 6.23), and let $\pi: G \to G/H$ be the projection. Then the action of H on G by right multiplication makes π into a principal H-bundle over G/H.

Indeed, as π is a surjective submersion, it looks locally as a projection (Corollary 5.23) and we find an open covering $(U_i)_i$ of G/H and for all i sections $s_i: G/H \to G$ of π. Then

$$h_i: \pi^{-1}(U_i) \xrightarrow{\sim} U_i \times H, \qquad g \mapsto \big(\pi(g), s_i(\pi(g))^{-1}g\big)$$

is an U_i-isomorphism with inverse morphism $(b, h) \mapsto s_i(b)h$. Clearly, h_i^{-1} satisfies (8.12).

Remark 8.22 (Bundles associated with principal bundles). Let G be a Lie group and let F be a G-premanifold.

Suppose that G acts faithfully on F. Then we obtain from (8.11) and (8.8) isomorphisms of pointed sets

$$\mathrm{Princ}_G(B) \xrightarrow{\sim} \check{H}^1(B, \mathcal{O}_{B;G}) \xrightarrow{\sim} \mathrm{Fib}_{F,G}(B). \qquad (8.13)$$

This isomorphism can be described by the following construction of fiber bundles from a principal bundle. Let $\pi: X \to B$ be a principal G-bundle over B whose G-bundle structure is given by a right action $X \times G \to X$ (Proposition 8.20). Then

$$(x, f) \cdot g := (xg, g^{-1}f)$$

defines a right G-action on $X \times F$ by a morphism of premanifolds. Let $X \times^G F$ be the set of G-orbits in $X \times F$ endowed with the quotient topology. Then

$$\tilde{\pi}: X \times^G F \to B, \qquad [x, f] \mapsto \pi(x)$$

is a well-defined continuous map.

Let us define the structure of a fiber bundle with typical fiber F and structure group G on $\tilde{\pi}$. Let $(U_i, h_i: \pi^{-1}(U_i) \xrightarrow{\sim} U_i \times G)$ be a G-bundle atlas for the principal bundle π and let (g_{ij}) be the attached cocycle. Consider the continuous maps

$$\tilde{k}_i: U_i \times F \longrightarrow \tilde{\pi}^{-1}(U_i), \qquad (b, f) \mapsto \left[h_i^{-1}(b, 1), f\right].$$

By definition of the quotient topology, $\tilde{\pi}^{-1}(U_i)$ is open in X. Using that G acts simply transitively on the fibers of π and that the h_i are G-equivariant, one checks that $\tilde{k}_i$ is bijective. We have

$$\left(\tilde{k}_j^{-1} \circ \tilde{k}_i\right)(b, f) = (b, g_{ji}f), \qquad b \in U_i \cap U_j, f \in F.$$

Therefore $\tilde{k}_j^{-1} \circ \tilde{k}_i$ is over $U_i \cap U_j$ an isomorphism of premanifolds. In particular, $\tilde{k}_i$ is a local isomorphism and hence an isomorphism because $\tilde{k}_i$ is bijective. As $\tilde{k}_j^{-1} \circ \tilde{k}_i$ is an isomorphism of premanifolds over $U_i \cap U_j$, there is a unique structure of a premanifold on $X \times^G F$ such that $\tilde{k}_i$ is an isomorphism onto the open submanifold $\tilde{\pi}^{-1}(U_i)$. Hence $\tilde{\pi}: X \to B$ is a fiber bundle over B with typical fiber F. Setting $\tilde{h}_i := \tilde{k}_i^{-1}$, $(U_i, \tilde{h}_i)$ is a G-bundle atlas for $\tilde{\pi}$, which is given by the cocycle $(g_{ij})_{i,j}$.

Note that for the construction of $X \times^G F$ it was not necessary to assume that G acts faithfully on F (Remark 8.13). For instance we could have taken G acting trivially on the one point manifold $F = \{*\}$. Then $X \times^G F = X/G$ is the set of G-orbits in X. Moreover, $\tilde{\pi}: X/G \to B$ is a fiber bundle with typical fiber $\{*\}$ over B, hence $\tilde{\pi}$ is an isomorphism of premanifolds.

The construction above yields in particular the following functoriality in G for G-principal bundles.

Remark 8.23 (Functoriality for principal bundles). Let $\varphi: G \to H$ be a homomorphism of Lie groups. Then G acts on H from the left by $(g, h) \mapsto \varphi(g)h$. This action is faithful if and only if φ is injective. Let $\xi = (\pi: X \to B, X \times G \to X)$ be a principal G-bundle over B and set $\varphi(\xi) := X \times^G H$. By Remark 8.22 this is a fiber bundle over B with typical fiber H. Then H acts on it from the right by the second factor $[x, h] \cdot h' \mapsto [x, hh']$ making $\varphi(\xi)$ into a principal H-bundle. We obtain a commutative diagram

$$\begin{array}{ccc}
\mathrm{Princ}_G(B) & \xrightarrow{\sim} & \check{H}^1(B, \mathcal{O}_{B;G}) \\
{\scriptstyle \xi \mapsto \varphi(\xi)} \downarrow & & \downarrow {\scriptstyle \check{H}^1(\varphi_B)} \\
\mathrm{Princ}_H(B) & \xrightarrow{\sim} & \check{H}^1(B, \mathcal{O}_{B;H}).
\end{array}$$

8.2 Vector Bundles

Among the most important examples of fiber bundles are vector bundles. We continue to denote by B a premanifold.

Definition 8.24 (Vector bundle). Let V be a finite-dimensional $\mathbb{K}$-vector space. The Lie group $GL(V)$ acts faithfully on V by left multiplication. A fiber bundle ξ with basis B, with typical fiber V, and structure group $GL(V)$ is called a *vector bundle with typical fiber V over B*. We say that ξ has *rank n* if $\dim_{\mathbb{K}}(V) = n$. A vector bundle of rank 1 is called a *line bundle*.

If B is a real C^{α}-premanifold we may also choose for V to be a finite-dimensional complex vector space V and consider $GL_{\mathbb{C}}(V)$ as a real Lie group. In this case we speak of a *complex C^{α}-vector bundle on B of complex rank* $\dim_{\mathbb{C}}(V)$.

If B is a complex premanifold, we may also consider a real vector bundle in the category of C^{α}-premanifolds ($\alpha \in \widehat{\mathbb{N}}$) on the underlying real C^{α} premanifold of B. In this case we speak of a *C^{α}-vector bundle on B*.

In the sequel, if we speak simply of a vector bundle over a premanifold B, we will either mean

(a) vector bundles with a real vector space as fiber over a real C^{α}-premanifold B (also called *real C^{α}-vector bundle*), or

(b) vector bundles with a complex vector space as fiber over a complex premanifold B (also called *holomorphic vector bundles*).

Remark 8.25. By (8.8) we have a bijection

$$\left\{ \begin{array}{c} \text{isomorphism classes of vector bundles} \\ \text{of rank } n \text{ with basis } B \end{array} \right\} \xrightarrow{\sim} \check{H}^{1}(B, GL_{n,B}) \qquad (8.14)$$

and in particular

$$\mathrm{Pic}(B) := \left\{ \begin{array}{c} \text{isomorphism classes of line bundles} \\ \text{with basis } B \end{array} \right\} \xrightarrow{\sim} \check{H}^{1}\left(B, \mathcal{O}_{B}^{\times}\right). \qquad (8.15)$$

As $\mathcal{O}_{B}^{\times} = GL_{1,B}$ is a sheaf of abelian groups, $\mathrm{Pic}(B)$ has the structure of an abelian group. It is called the *Picard group of B*.

Example 8.26. The map $e \colon \mathbb{R} \times \{\pm 1\} \xrightarrow{\sim} \mathbb{R}^{\times}$, $(x, \varepsilon) \mapsto \varepsilon \exp(x)$ is an isomorphism of real Lie groups. Its inverse is given by $z \mapsto (\log(|z|), \mathrm{sgn}(z))$. Let B be a real premanifold. Applying the functor $G \mapsto \mathcal{O}_{B;G}$ to the isomorphism e (Remark 8.8) we obtain an isomorphism $\mathcal{O}_{B} \times (\mathbb{Z}/2\mathbb{Z})_{B} \xrightarrow{\sim} \mathcal{O}_{B}^{\times}$ of sheaves of abelian groups on B (identifying the groups $\{\pm 1\}$ and $\mathbb{Z}/2\mathbb{Z}$). Therefore by (8.15) we obtain an isomorphism of abelian groups

$$\mathrm{Pic}(B) \xrightarrow{\sim} H^1(B, \mathcal{O}_B) \times H^1(B, (\mathbb{Z}/2\mathbb{Z})_B). \qquad (8.16)$$

We will see that for real C^α-manifolds B one has $H^1(B, \mathcal{O}_B) = 0$: For $\alpha \leq \infty$ we will prove this in Corollary 9.15; for real analytic manifolds see Remark 10.21. Therefore we see

$$\mathrm{Pic}(B) \cong H^1(B, (\mathbb{Z}/2\mathbb{Z})_B) \qquad (8.17)$$

for real manifolds B.

The vector bundle structure can also be described as follows.

Proposition 8.27. *Let V be a finite-dimensional $\mathbb{K}$-vector space. The following data are equivalent:*

(a) *A vector bundle $(\pi\colon E \to B, [U_i, h_i]_{i \in I})$ with typical fiber V over B.*
(b) *A morphism $\pi\colon E \to B$ of premanifolds and for each $b \in B$ the structure of a $\mathbb{K}$-vector space on $E(b) := \pi^{-1}(b)$ such that there exists an open covering $(U_i)_i$ of B and isomorphisms of U_i-premanifolds $h_i\colon \pi^{-1}(U_i) \xrightarrow{\sim} U_i \times V$ such that*

$$h_{i,b} := h_{i \mid E(b)}\colon E(b) \xrightarrow{\sim} \{b\} \times V = V$$

is an isomorphism of $\mathbb{K}$-vector spaces.

In the sequel we will denote vector bundles usually by (E, π) and it is understood that they are endowed with both of the above structures. In particular the fibers $E(b)$ will always be endowed with the structure of a $\mathbb{K}$-vector space.

Proof. The proof is similar to the proof of Proposition 8.20. Given a datum as in (b), π is a fiber bundle over B with typical fiber F. Moreover, we find that for $i, j \in I$ the map $b \mapsto h_{i,b} \circ h_{j,b}^{-1}$ is a morphism $U_i \cap U_j \to \mathrm{GL}(V)$. Hence $(U_i, h_i)_i$ is a $\mathcal{O}_{B;\mathrm{GL}(V)}$-bundle atlas for π. Any two choices of $(U_i, h_i)_{i \in I}$ as in (b) define equivalent $\mathcal{O}_{B;\mathrm{GL}(V)}$-bundle atlases for π and hence a well-defined $\mathcal{O}_{B;\mathrm{GL}(V)}$-bundle structure on π.

Conversely, let $(U_i, h_i)_{i \in I}$ be a $\mathcal{O}_{B;\mathrm{GL}(V)}$-bundle atlas for π. For $b \in B$ choose $i \in I$ with $b \in U_i$. Then h_i induces a bijection $h_{i,b}\colon E(b) \xrightarrow{\sim} \{b\} \times V = V$, which we use to define a structure of $\mathbb{K}$-vector space on $E(b)$ by transport of structure. For any $i' \in I$ with $b \in U_{i'}$, $h_{i,b}$ and $h_{i',b}$ differ only by a linear automorphism of V and hence define the same vector space structure on E_b. Therefore the vector space structure is independent of the choice of i. The same argument shows that passing to an equivalent $\mathcal{O}_{B;\mathrm{GL}(V)}$-twisting atlas does not change the vector space structure on $E(b)$. $\qquad\qquad \square$

Although for fiber bundles (over a fixed basis) we defined only the notion of an isomorphism, the alternative description in Proposition 8.27 allows us to define a more general notion of morphism for vector bundles. First let us agree that we mean by a *vector bundle over B* (without specifying a typical fiber or a rank) a decomposition $B = \coprod_i B_i$ of B into open and closed subpremanifolds and for all i a finite-dimensional $\mathbb{K}$-vector space V_i and a vector bundle over B_i with typical fiber V_i. Then for a vector bundle (E, π) over B its *rank* is a locally constant function $B \to \mathbb{N}_0$, $b \mapsto \mathrm{rk}_b(E)$.

Definition 8.28. Let (E, π) and (E', π') be vector bundles over B (possibly of different rank and in particular with different typical fiber). A *morphism of vector bundles* $(E, \pi) \to (E', \pi')$ is a B-morphism $\Phi \colon E \to E'$ of premanifolds such that for all $p \in B$ the maps $\Phi(p) := \Phi_{|E(b)} \colon E(b) \to E'(b)$ are $\mathbb{K}$-linear.

The composition of two morphisms of vector bundles is again a morphism. We obtain the *category of vector bundles* over B, which we denote by $(\mathrm{Vec}(B))$.

8.3 $\mathcal{O}_X$-Modules

Vector bundles over a premanifold B can also be described as finite locally free modules over the structure sheaf $\mathcal{O}_B$. Let us first explain what we mean by this. In this section let $(X, \mathcal{O}_X)$ be an arbitrary ringed space.

The Category of $\mathcal{O}_X$-Modules

Definition 8.29. An $\mathcal{O}_X$-*(pre-)module* is a (pre-)sheaf $\mathcal{F}$ on X together with two morphisms (addition and scalar multiplication) of sheaves

$$\mathcal{F} \times \mathcal{F} \to \mathcal{F}, \quad (s, s') \mapsto s + s' \qquad \text{for } s, s' \in \mathcal{F}(U),\, U \subseteq X \text{ open,}$$
$$\mathcal{O}_X \times \mathcal{F} \to \mathcal{F}, \quad (a, s) \mapsto as \qquad \text{for } a \in \mathcal{O}_X(U),\, s \in \mathcal{F}(U),\, U \subseteq X \text{ open}$$

such that addition and scalar multiplication by $\mathcal{O}_X(U)$ define on $\mathcal{F}(U)$ the structure of an $\mathcal{O}_X(U)$-module.

If $\mathcal{F}_1$ and $\mathcal{F}_2$ are $\mathcal{O}_X$-premodules, a morphism of sheaves $w \colon \mathcal{F}_1 \to \mathcal{F}_2$ is called a *homomorphism of $\mathcal{O}_X$-premodules*, if for each open subset $U \subseteq X$ the map $\mathcal{F}_1(U) \to \mathcal{F}_2(U)$ is an $\mathcal{O}_X(U)$-module homomorphism, i.e.,

$$w_U(s + s') = w_U(s) + w_U(s'),$$
$$w_U(as) = a w_U(s)$$

for all $s, s' \in \mathcal{F}(U)$ and $a \in \mathcal{O}_X(U)$. A *homomorphism of $\mathcal{O}_X$-modules* is a homomorphism of $\mathcal{O}_X$-premodules.

The composition of two homomorphisms of $\mathcal{O}_X$-modules is again a homomorphism of $\mathcal{O}_X$-modules. We obtain the category of $\mathcal{O}_X$-modules, which we denote by $(\mathcal{O}_X\text{-Mod})$. The $\mathcal{O}_X$-module $\mathcal{F}$ such that $\mathcal{F}(U) = \{0\}$ for all open sets $U \subseteq X$ is called the zero module and simply denoted by 0.

Example 8.30.

1. Let X be a topological space and let $\mathbb{Z}_X$ be the constant sheaf of rings on X with value $\mathbb{Z}$. As $\mathbb{Z}$-modules are simply abelian groups, a $\mathbb{Z}_X$-module is simply a sheaf of abelian groups on X.
2. Let A be a commutative ring. Let X be a space that consists of a single point and let $\mathcal{O}_X$ be the sheaf of rings with $\mathcal{O}_X(X) = A$. Then an $\mathcal{O}_X$-module $\mathcal{F}$ is just an A-module M (by attaching $M = \mathcal{F}(X)$ to $\mathcal{F}$).

Remark 8.31. Let $\mathcal{F}$ be an $\mathcal{O}_X$-module and $x \in X$. The $\mathcal{O}_X(U)$-module structures on $\mathcal{F}(U)$, where U is an open neighborhood of x, induce on the stalk $\mathcal{F}_x$ an $\mathcal{O}_{X,x}$-module structure. If $\mathcal{O}_{X,x}$ is a local ring with maximal ideal $\mathfrak{m}_x$ and residue field $\kappa(x) := \mathcal{O}_{X,x}/\mathfrak{m}_x$, then we call the $\kappa(x)$-vector space $\mathcal{F}(x) := \mathcal{F}_x/\mathfrak{m}_x\mathcal{F}_x$ the *fiber of $\mathcal{F}$ in x*.

If $w\colon \mathcal{F} \to \mathcal{F}'$ is a homomorphism of $\mathcal{O}_X$-modules, the induced map on stalks $w_x\colon \mathcal{F}_x \to \mathcal{F}'_x$ is a homomorphism of $\mathcal{O}_{X,x}$-modules. If $\mathcal{O}_{X,x}$ is local, it also induces a $\kappa(x)$-linear map $w(x)\colon \mathcal{F}(x) \to \mathcal{F}'(x)$ on fibers.

Remark 8.32. As the sheafification commutes with formation of finite products, addition and scalar multiplication on an $\mathcal{O}_X$-premodule $\mathcal{F}$ induce by functoriality an addition and a scalar multiplication on the sheafification $\mathcal{F}^a$ of $\mathcal{F}$. Hence $\mathcal{F}^a$ is an $\mathcal{O}_X$-module.

Definition 8.33. An $\mathcal{O}_X$-*submodule of* $\mathcal{F}$ is an $\mathcal{O}_X$-module $\mathcal{G}$ such that $\mathcal{G}(U)$ is a subset of $\mathcal{F}(U)$ for all open sets $U \subseteq X$ and such that the inclusions $\iota_U\colon \mathcal{G}(U) \hookrightarrow \mathcal{F}(U)$ form a homomorphism $\iota\colon \mathcal{G} \to \mathcal{F}$ of $\mathcal{O}_X$-modules.

The $\mathcal{O}_X$-submodules of $\mathcal{O}_X$ are called *ideals of $\mathcal{O}_X$*.

Example 8.34. Let M be a premanifold and let $S \subseteq M$ be a subset. Then the sheaf $\mathcal{I}_S$ of functions vanishing on S (Remark 5.31) is an ideal of $\mathcal{O}_M$.

Remark 8.35. Let $\mathcal{F}$ and $\mathcal{F}'$ be $\mathcal{O}_X$-modules. If w_1 and w_2 are two homomorphisms $\mathcal{F} \to \mathcal{F}'$, we can define their sum in the obvious way:

$$(w_1 + w_2)_U := w_{1,U} + w_{2,U}\colon \mathcal{F}(U) \to \mathcal{F}'(U).$$

Let $a \in \Gamma(X, \mathcal{O}_X)$ and $w\colon \mathcal{F} \to \mathcal{F}'$ be a homomorphism. Then we define a homomorphism $aw\colon \mathcal{F} \to \mathcal{F}'$ by $(aw)_U := (a_{|U})w_U$. In this way we endow the set of homomorphisms $\mathcal{F} \to \mathcal{F}'$ of $\mathcal{O}_X$-modules with the structure of a $\Gamma(X, \mathcal{O}_X)$-module. We denote this $\Gamma(X, \mathcal{O}_X)$-module by $\mathrm{Hom}_{\mathcal{O}_X}(\mathcal{F}, \mathcal{F}')$. The composition of homomorphisms is $\Gamma(X, \mathcal{O}_X)$-bilinear. In particular, the additive group structures on the sets $\mathrm{Hom}_{\mathcal{O}_X}(\mathcal{F}, \mathcal{F}')$ make $(\mathcal{O}_X\text{-Mod})$ into a preadditive category (Appendix Definition 15.21).

Remark 8.36 (Construction with $\mathcal{O}_X$-modules). As a general principle, functorial construction for A-modules, where A is a commutative ring, can be generalized to constructions for $\mathcal{O}_X$-modules. Given a functor on usual A-modules, we apply it to all sets of sections $\mathcal{F}(U)$. We obtain an $\mathcal{O}_X$-premodule. If this is not already a sheaf, we sheafify to obtain an $\mathcal{O}_X$-module.

For instance let us consider limits and colimits of $\mathcal{O}_M$-modules. Let $\mathcal{I}$ be a small category and let $\mathcal{F}: \mathcal{I} \to (\mathcal{O}_X\text{-Mod})$, $i \mapsto \mathcal{F}_i$ be an $\mathcal{I}$-diagram of $\mathcal{O}_X$-modules.

Remark 8.37 (Limits of $\mathcal{O}_X$-modules). The explicit description of limits of R-modules shows that the $\mathcal{O}_X$-premodule $U \mapsto \lim_{\mathcal{I}} \mathcal{F}_i(U)$ (taking the limit in the category of $\mathcal{O}_X(U)$-modules) is in fact an $\mathcal{O}_X$-module denoted by $\lim_{\mathcal{I}} \mathcal{F}_i$ and called the *limit of the diagram* $\mathcal{F}$. As for sheaves (Remark 3.61) one obtains for all $x \in X$ a functorial morphism

$$(\lim_{\mathcal{I}} \mathcal{F}_i)_x \longrightarrow \lim_{\mathcal{I}}(\mathcal{F}_i)_x, \tag{8.18}$$

which is an isomorphism if $\mathcal{I}$ is finite. This is a homomorphism of $\mathcal{O}_{X,x}$-modules.

Taking for $\mathcal{I}$ a category whose only morphisms are the identities, we obtain as a special case the *product of the family* $(\mathcal{F}_i)_{i \in I}$ with $I := \mathrm{Ob}(\mathcal{I})$. It is denoted by $\prod_{i \in I} \mathcal{F}_i$. The universal property for products implies that composition with projections $\mathrm{pr}_j: \prod_i \mathcal{F}_i \to \mathcal{F}_j$ yields for every $\mathcal{O}_X$-module $\mathcal{G}$ a $\Gamma(X, \mathcal{O}_X)$-linear bijection

$$\mathrm{Hom}_{(\mathcal{O}_X\text{-Mod})}\left(\mathcal{G}, \prod_{i \in I} \mathcal{F}_i\right) \xrightarrow{\sim} \prod_{i \in I} \mathrm{Hom}_{(\mathcal{O}_X\text{-Mod})}(\mathcal{G}, \mathcal{F}_i). \tag{8.19}$$

Another special case is the *kernel* of a homomorphism $w: \mathcal{F} \to \mathcal{G}$ of $\mathcal{O}_X$-modules as the limit of the diagram

$$\mathcal{F} \underset{0}{\overset{w}{\rightrightarrows}} \mathcal{G}. \tag{8.20}$$

It is denoted by $\mathrm{Ker}(w)$ and it is the sheaf $U \mapsto \mathrm{Ker}(w_U: \mathcal{F}(U) \to \mathcal{G}(U))$. By (8.18) we have $\mathrm{Ker}(w)_x = \mathrm{Ker}(w_x)$. In particular, w is an injective homomorphism of sheaves if and only if $\mathrm{Ker}(w) = 0$.

Remark 8.38 (Colimits of $\mathcal{O}_X$-modules). Consider the $\mathcal{O}_X$-premodule $U \mapsto \mathrm{colim}_{\mathcal{I}} \mathcal{F}_i(U)$, where we take the colimit in the category of $\mathcal{O}_X(U)$-modules. It is in general not a sheaf. Its sheafification is an $\mathcal{O}_X$-module denoted by $\mathrm{colim}_{\mathcal{I}} \mathcal{F}_i$ and called the *colimit* *of the diagram* $\mathcal{F}$. The same argument as for sheaves (Remark 3.63) shows that one obtains for all $x \in X$ a functorial isomorphism of $\mathcal{O}_{X,x}$-modules

$$(\operatorname*{colim}_{\mathcal{I}} \mathcal{F}_i)_x \xrightarrow{\sim} \operatorname*{colim}_{\mathcal{I}}(\mathcal{F}_i)_x. \tag{8.21}$$

Again taking for $\mathcal{I}$ a category whose only morphisms are the identities, we obtain as a special case the *direct sum of the family* $(\mathcal{F}_i)_{i \in I}$ with $I := \mathrm{Ob}(\mathcal{I})$. It is denoted by $\bigoplus_{i \in I} \mathcal{F}_i$. If $\mathcal{F}_i = \mathcal{F}$ for all i, we also write $\mathcal{F}^{(I)}$. The universal property for direct sums of modules (Appendix Sect. (14.3)) and the universal property of the sheafification yield for every $\mathcal{O}_X$-module $\mathcal{G}$ an isomorphism of $\Gamma(X, \mathcal{O}_X)$-modules

$$\mathrm{Hom}_{(\mathcal{O}_X\text{-Mod})}\left(\bigoplus_{i \in I} \mathcal{F}_i, \mathcal{G}\right) \xrightarrow{\sim} \prod_{i \in I} \mathrm{Hom}_{(\mathcal{O}_X\text{-Mod})}(\mathcal{F}_i, \mathcal{G}). \tag{8.22}$$

If I is finite, then $\bigoplus_{i \in I} \mathcal{F}_i = \prod_{i \in I} \mathcal{F}_i$.

Another special case is the *cokernel* of a homomorphism $w : \mathcal{F} \to \mathcal{G}$ of $\mathcal{O}_X$-modules as the colimit of the diagram (8.20). It is denoted by $\mathrm{Coker}(w)$ and it is the sheaf associated with the presheaf $U \mapsto \mathrm{Coker}(w_U : \mathcal{F}(U) \to \mathcal{G}(U))$. By (8.21) we have $\mathrm{Coker}(w)_x = \mathrm{Coker}(w_x)$. In particular, w is a surjective homomorphism of sheaves if and only if $\mathrm{Coker}(w) = 0$.

Let $\mathcal{F}$ be an $\mathcal{O}_X$-submodule of $\mathcal{G}$ and let $i : \mathcal{F} \to \mathcal{G}$ be the inclusion. Then

$$\mathcal{G}/\mathcal{F} := \mathrm{Coker}(i)$$

is called the *quotient of $\mathcal{G}$ by $\mathcal{F}$*. It is the $\mathcal{O}_X$-module associated with the $\mathcal{O}_X$-premodule $U \mapsto \mathcal{G}(U)/\mathcal{F}(U)$.

For a homomorphism $w : \mathcal{F} \to \mathcal{G}$ of $\mathcal{O}_X$-modules we can also define its *image*

$$w(\mathcal{F}) := \mathrm{Im}(w) := \mathrm{Ker}(\mathcal{G} \to \mathrm{Coker}(w))$$

and its *coimage*

$$\mathrm{Coim}(w) := \mathrm{Coker}(\mathrm{Ker}(w) \to \mathcal{F}).$$

It is the sheaf associated with the presheaf $U \mapsto \mathrm{Im}(w_U : \mathcal{F}(U) \to \mathcal{G}(U))$ respective to $U \mapsto \mathrm{Coim}(w_U)$. As kernel and cokernel are compatible with the formation of stalks, we also have $\mathrm{Im}(w)_x = \mathrm{Im}(w_x)$ and $\mathrm{Coim}(w)_x = \mathrm{Coim}(w_x)$. As w_x induces for all $x \in X$ an isomorphism $\mathrm{Coim}(w_x) \xrightarrow{\sim} \mathrm{Im}(w_x)$ of $\mathcal{O}_{X,x}$-modules, we deduce that w induces an isomorphism $\mathrm{Coim}(w) \xrightarrow{\sim} \mathrm{Im}(w)$. In particular we see:

Proposition 8.39. *The category of $\mathcal{O}_X$-modules is an abelian category.*

Therefore, we also have the notion of exact sequences of $\mathcal{O}_X$-modules: A sequence $\cdots \to \mathcal{F}_{i-1} \xrightarrow{w_{i-1}} \mathcal{F}_i \xrightarrow{w_i} \mathcal{F}_{i+1} \to \cdots$ of homomorphisms of $\mathcal{O}_X$-modules is called *exact* if $\mathrm{Ker}(w_i) = \mathrm{Im}(w_{i-1})$ for all i.

Finite Locally Free $\mathcal{O}_X$-modules

As for modules over rings, every family of global sections of an $\mathcal{O}_X$-module $\mathcal{F}$ yields a homomorphism of a free $\mathcal{O}_X$-module to $\mathcal{F}$. More precisely:

Remark and Definition 8.40 (Bases of $\mathcal{O}_X$-modules). Let $\mathcal{F}$ be an $\mathcal{O}_X$-module and let $\underline{s} := (s_i)_{i \in I}$ be a family of sections $s_i \in \mathcal{F}(X)$. Then one obtains a homomorphism of $\mathcal{O}_X$-modules

$$w_{\underline{s}} \colon \mathcal{O}_X^{(I)} \longrightarrow \mathcal{F}, \qquad (f_i)_{i \in I} \mapsto \sum_{i \in I} f_i(s_{i|U}), \tag{8.23}$$

for $U \subseteq X$ open and $f_i \in \mathcal{O}_X(U)$. As for modules over rings, $\underline{s}$ is called a *generating system* (respectively a *basis*) if $w_{\underline{s}}$ is surjective (respectively bijective).

Definition 8.41. An $\mathcal{O}_X$-module $\mathcal{F}$ is called *finite locally free* if there exists an open covering $(U_i)_i$ of X such that for all i the $\mathcal{O}_{U_i}$-module $\mathcal{F}_{|U_i}$ is isomorphic to $\mathcal{O}_{U_i}^n$ for some $n \in \mathbb{N}_0$ (depending on i).

The full subcategory of $(\mathcal{O}_X\text{-Mod})$ whose objects are the finite locally free $\mathcal{O}_X$-modules is denoted by $(\mathrm{FLF}(X))$.

If $\mathcal{F}$ is a finite locally free $\mathcal{O}_X$-module, its stalk $\mathcal{F}_x$ is a finitely generated free $\mathcal{O}_{X,x}$-module for all $x \in X$. We set $\mathrm{rk}_x(\mathcal{F}) := \mathrm{rk}_{\mathcal{O}_{X,x}}(\mathcal{F}_x)$ and obtain a locally constant function

$$X \to \mathbb{N}_0, \qquad x \mapsto \mathrm{rk}_x(\mathcal{F}),$$

called the *rank of $\mathcal{F}$*.

Next we define the $\mathcal{O}_X$-module of all homomorphisms between two $\mathcal{O}_X$-modules. In particular we obtain the dual of an $\mathcal{O}_X$-module. The general principle is again the one sketched in Remark 8.36.

Definition and Remark 8.42. Let $(X, \mathcal{O}_X)$ be a ringed space and let $\mathcal{F}$ and $\mathcal{G}$ be two $\mathcal{O}_X$-modules. The presheaf

$$U \mapsto \mathrm{Hom}_{\mathcal{O}_{X|U}}(\mathcal{F}_{|U}, \mathcal{G}_{|U})$$

with the obvious restriction maps is a sheaf. The right-hand side is a $\Gamma(U, \mathcal{O}_X)$-module. Therefore this sheaf has the structure of an $\mathcal{O}_X$-module, and we denote this $\mathcal{O}_X$-module by $\mathcal{H}om_{\mathcal{O}_X}(\mathcal{F}, \mathcal{G})$. Moreover:

1. By (8.19) and (8.22) one has for every family $(\mathcal{F}_i)_{i \in I}$ of $\mathcal{O}_X$-modules and every $\mathcal{O}_X$-module $\mathcal{G}$ isomorphisms of $\mathcal{O}_X$-modules

$$\mathcal{H}om_{\mathcal{O}_X}\left(\mathcal{G}, \prod_{i \in I} \mathcal{F}_i\right) \xrightarrow{\sim} \prod_{i \in I} \mathcal{H}om_{\mathcal{O}_X}(\mathcal{G}, \mathcal{F}_i),$$

$$\mathcal{H}om_{\mathcal{O}_X}\left(\bigoplus_{i \in I} \mathcal{F}_i, \mathcal{G}\right) \xrightarrow{\sim} \prod_{i \in I} \mathcal{H}om_{\mathcal{O}_X}(\mathcal{F}_i, \mathcal{G}). \tag{8.24}$$

2. As for modules (Appendix Example 14.7), evaluation on 1 yields an identification

$$\mathcal{H}om_{\mathcal{O}_X}(\mathcal{O}_X, \mathcal{F}) = \mathcal{F}. \tag{8.25}$$

3. If $\mathcal{F}$ and $\mathcal{G}$ are finite locally free, then $\mathcal{H}om_{\mathcal{O}_X}(\mathcal{F}, \mathcal{G})$ is finite locally free and $\mathrm{rk}_x(\mathcal{H}om_{\mathcal{O}_X}(\mathcal{F}, \mathcal{G})) = \mathrm{rk}_x(\mathcal{F})\,\mathrm{rk}_x(\mathcal{G})$ for all $x \in X$.

Indeed, let $(U_i)_i$ be an open covering such that $\mathcal{F}_{|U_i} \cong \mathcal{O}_{U_i}^n$ and $\mathcal{G}_{|U_i} \cong \mathcal{O}_{U_i}^m$ for all i. Then

$$
\begin{aligned}
\mathcal{H}om_{\mathcal{O}_X}(\mathcal{F}, \mathcal{G})_{|U_i} &= \mathcal{H}om_{\mathcal{O}_{U_i}}(\mathcal{F}_{|U_i}, \mathcal{G}_{|U_i}) \\
&\cong \mathcal{H}om_{\mathcal{O}_{U_i}}(\mathcal{O}_{U_i}^n, \mathcal{O}_{U_i}^m) \\
&\overset{(8.24)}{=} \mathcal{H}om_{\mathcal{O}_{U_i}}(\mathcal{O}_{U_i}, \mathcal{O}_{U_i})^{nm} \\
&\overset{(8.25)}{=} \mathcal{O}_{U_i}^{nm}.
\end{aligned}
$$

4. For all $x \in X$, sending a homomorphism of $\mathcal{O}_X$-modules to its germ in x yields a homomorphism of $\mathcal{O}_{X,x}$-modules

$$\mathcal{H}om_{\mathcal{O}_X}(\mathcal{F}, \mathcal{G})_x \to \mathrm{Hom}_{\mathcal{O}_{X,x}}(\mathcal{F}_x, \mathcal{G}_x). \tag{8.26}$$

This is an isomorphism if $\mathcal{F}$ and $\mathcal{G}$ are finite locally free (see also Problem 8.5). Indeed, by replacing X by a sufficiently small neighborhood of x we may assume that $\mathcal{F}$ and $\mathcal{G}$ are free. Then by (8.24) we may assume that $\mathcal{F} = \mathcal{G} = \mathcal{O}_X$. Then this is clear by (8.25).

The surjectivity of (8.26) means that every $\mathcal{O}_{X,x}$-linear map $w_0 : \mathcal{F}_x \to \mathcal{G}_x$ is induced by a homomorphism of $\mathcal{O}_U$-modules $w_U : \mathcal{F}_{|U} \to \mathcal{G}_{|U}$ for some open neighborhood U of x. The injectivity of (8.26) means that any two such homomorphisms w_U and $w_{U'}$ become equal after restricting to an open neighborhood of x contained in $U \cap U'$.

Definition 8.43 (Dual $\mathcal{O}_X$-module). Let $(X, \mathcal{O}_X)$ be a ringed space and let $\mathcal{F}$ be an $\mathcal{O}_X$-module. The $\mathcal{O}_X$-module

$$\mathcal{F}^\vee := \mathcal{H}om_{\mathcal{O}_X}(\mathcal{F}, \mathcal{O}_X) \tag{8.27}$$

is called the *dual $\mathcal{O}_X$-module of $\mathcal{F}$*.

Remark 8.44. If $\mathcal{F} = \mathcal{O}_X^n$, then $\mathcal{F}^\vee \cong \mathcal{O}_X^n$ by Remark 8.42 2 and 1. Hence, if $\mathcal{F}$ is a finite locally free $\mathcal{O}_X$-module, then its dual $\mathcal{F}^\vee$ is again locally free, and $\mathrm{rk}_x(\mathcal{F}^\vee) = \mathrm{rk}_x(\mathcal{F})$ for all $x \in X$.

8.4   Vector Bundles and Finite Locally Free Modules

Now let M be a premanifold. We will show that the category of vector bundles over M and the category of finite locally free $\mathcal{O}_M$-modules are equivalent. We start by defining a functor

$$\text{Mod: } (\text{Vec}(M)) \to (\text{FLF}(M)). \tag{8.28}$$

Let $\pi\colon E \to M$ be a vector bundle. In particular, all fibers $E(p) := \pi^{-1}(p)$ for $p \in M$ are endowed with the structure of a $\mathbb{K}$-vector space. We attach an $\mathcal{O}_M$-module $\mathcal{E} := \text{Mod}(E, \pi)$ to (E, π) as follows. For $U \subseteq M$ open set

$$\mathcal{E}(U) := \{\, s\colon U \to E \text{ morphism of premanifold} ;\ \pi \circ s = \text{id}_U \,\}.$$

For $s, s' \in \mathcal{E}(U)$ define $s + s' \in \mathcal{E}(U)$ by $(s + s')(p) := s(p) + s'(p)$ (addition inside the $\mathbb{K}$-vector space $E(p)$). For $f \in \mathcal{O}_M(U)$, $s \in \mathcal{E}(U)$ define $fs \in \mathcal{E}(U)$ by $(fs)(p) := f(p)s(p)$ (scalar multiplication inside $E(p)$). This makes $\mathcal{E}$ into an $\mathcal{O}_M$-module.

Let $\Phi\colon (E, \pi) \to (E', \pi')$ be a morphism of vector bundles. Set $\mathcal{E} := \text{Mod}(E, \pi)$, $\mathcal{E}' := \text{Mod}(E', \pi')$. For $U \subseteq M$ open define

$$\varphi_U\colon \mathcal{E}(U) \to \mathcal{E}'(U), \qquad s \mapsto \Phi_{|\pi^{-1}(U)} \circ s.$$

As Φ is linear on fibers, φ_U is a homomorphism of $\mathcal{O}_M(U)$-modules. Therefore $\text{Mod}(\Phi) := (\varphi_U)_U$ is a homomorphism of $\mathcal{O}_M$-modules. We obtain a functor Mod from the category of vector bundles over M to the category of $\mathcal{O}_M$-modules.

Let $\mathcal{E} = \text{Mod}(E, \pi)$ for a vector bundle (E, π). Let us show that $\mathcal{E}$ is finite locally free. Let $W \subseteq M$ be open such that $\pi^{-1}(W) = W \times \mathbb{K}^n$, for some $n \in \mathbb{N}$. For $U \subseteq W$ one then has

$$
\begin{aligned}
\mathcal{E}(U) &\cong \{\, s\colon U \to U \times \mathbb{K}^n \text{ morphism} ;\ s(p) = (p, v) \text{ for some } v \in \mathbb{K}^n \} \\
&= \text{Hom}_{(\text{PMfd})}(U, \mathbb{K}^n) \\
&= \mathcal{O}_M(U)^n
\end{aligned}
$$

and hence $\mathcal{E}_{|W} \cong (\mathcal{O}_{M|W})^n$. This concludes the construction of the functor (8.28).

Note that we have $\text{rk}_p(E) = \text{rk}_p(\mathcal{E})$ for $p \in M$.

Proposition 8.45. *The functor* Mod *from the category of vector bundles over M to the category of finite locally free $\mathcal{O}_M$-modules is an equivalence of categories.*

Proof. Construction of a functor $G\colon (\text{FLF}(M)) \to (\text{Vec}(M))$. Let $\mathcal{E}$ be a finite locally free $\mathcal{O}_M$-module. Let $p \in M$. Recall that $\mathcal{O}_{M,p}$ is a local $\mathbb{K}$-algebra with maximal ideal $\mathfrak{m}_p$ and residue class field $\mathcal{O}_{M,p}/\mathfrak{m}_p = \mathbb{K}$. The stalk $\mathcal{E}_p$ is a free $\mathcal{O}_{M,p}$-module of rank $\text{rk}_p(\mathcal{E})$. Let $\mathcal{E}(p) := \mathcal{E}_p/\mathfrak{m}_p\mathcal{E}_p$ be the fiber of $\mathcal{E}$. It is a finite-dimensional $\mathbb{K}$-vector space of dimension $\mathrm{rk}_p(\mathcal{E})$. For $p \in U \subseteq M$ open and $s \in \mathcal{E}(U)$ we denote by $s(p)$ the image of $s_p \in \mathcal{E}_p$ in $\mathcal{E}(p)$. We define the set

$$E := \coprod_{p \in M} \mathcal{E}(p)$$

and the map $\pi\colon E \to M$ such that $\pi^{-1}(p) = \mathcal{E}(p)$ for $p \in M$.

Let $U \subseteq M$ be open such that $\mathcal{E}_{|U} \cong (\mathcal{O}_{M|U})^n$. In other words, there exists a basis $e = (e_1, \ldots, e_n)$ of $\mathcal{E}_{|U}$ (Remark 8.40). In particular $(e_1(p), \ldots, e_n(p))$ is a $\mathbb{K}$-basis of $\mathcal{E}(p)$. Hence we get a bijection

$$\Phi_{U,e}\colon U \times \mathbb{K}^n \xrightarrow{\sim} \pi^{-1}(U) = \coprod_{p \in U} \mathcal{E}(p), \qquad (p, \lambda) \mapsto \sum_{i=1}^{n} \lambda_i e_i(p).$$

This bijection endows $\pi^{-1}(U)$ with the structure of a premanifold. For a different choice $\tilde{U}$ and $\tilde{e}$ the "change of charts" $\Phi_{U,e}^{-1} \circ \Phi_{\tilde{U},\tilde{e}}$ is an automorphism of $(U \cap \tilde{U}) \times \mathbb{K}^n$ given by a matrix in $\mathrm{GL}_n(\mathcal{O}_M(U \cap \tilde{U}))$. This makes (E, π) into a vector bundle over M.

A morphism $w\colon \mathcal{E} \to \mathcal{E}'$ between finite locally free $\mathcal{O}_M$-modules induces $\mathbb{K}$-linear maps $w(p)\colon \mathcal{E}(p) \to \mathcal{E}'(p)$ for all $p \in M$ (Remark 8.31) and hence a map $E \to E'$ on the attached vector bundles. Covering M by open subsets U as above, one sees that this map is a morphism of vector bundles. Hence we obtain a functor G from the category of finite locally free $\mathcal{O}_M$-modules to the category of vector bundles over M.

G and Mod are quasi-inverse. Let (E, π) be a vector bundle over M and let $\mathcal{E} = \mathrm{Mod}(E, \pi)$. For $p \in M$ let $s \in \mathcal{E}_p$, i.e., s is the equivalence class of a pair $(U, \tilde{s})$, where $p \in U \subseteq M$ and $\tilde{s}\colon U \to E$ is a section of π. The homomorphism of $\mathcal{O}_{M,p}$-modules

$$\mathcal{E}_p \to E(p), \qquad s \mapsto s(p)$$

is surjective and its kernel is $\mathfrak{m}_p \mathcal{E}_p$. Hence we obtain an isomorphism of $\mathbb{K}$-vector spaces $E(p) \cong \mathcal{E}(p) = (G \circ \mathrm{Mod})(E, \pi)(p)$ and hence a bijective map $E \to (G \circ \mathrm{Mod})(E, \pi)$. Looking at charts $\Phi_{U,e}$ as above one sees that this is locally an isomorphism of vector bundles. Hence it is an isomorphism of vector bundles. It is clear that this isomorphism is functorial in (E, π).

Conversely, let $\mathcal{E}$ be a finite locally free $\mathcal{O}_M$-module and let $(E, \pi) = G(\mathcal{E})$. Let $U \subseteq M$ be open, $s \in \mathcal{E}(U)$. Then $s(p) \in \mathcal{E}(p) = E(p)$ and we obtain a map $s\colon U \to E$, $p \mapsto s(p)$ such that $\pi \circ s = \mathrm{id}_U$. Looking at charts, it is easy to see that s is a morphism of manifolds and hence an element of $\mathrm{Mod}(E, \pi)(U)$. We obtain maps $\mathcal{E}(U) \to (\mathrm{Mod} \circ G)(\mathcal{E})(U)$ and it is immediate that these maps define an isomorphism of $\mathcal{O}_M$-modules, which is functorial in $\mathcal{E}$. $\square$

Let $\mathcal{E}$ be a finite locally free $\mathcal{O}_M$-module and let (E, π) be the corresponding vector bundle. Then the proof shows that the $\mathbb{K}$-vector spaces $E(p)$ and $\mathcal{E}/\mathfrak{m}_p \mathcal{E}_p$ are identified.

Definition and Remark 8.46 (Dual and direct sum of vector bundles). Let M be a premanifold, let (E, π) and (E', π') be vector bundles over M and let $\mathcal{E}$ and $\mathcal{E}'$, respectively, be the corresponding finite locally free $\mathcal{O}_M$-modules.

1. The vector bundle $(E^\vee, \pi^\vee)$ corresponding to the dual $\mathcal{O}_M$-module $\mathcal{E}^\vee$ is called the *dual of* (E, π). For each point $p \in M$ the fiber of $E^\vee$ in p is given by the dual space of $E(p)$:
$$E^\vee(p) = \mathcal{E}_p^\vee / \mathfrak{m}_p \mathcal{E}_p^\vee \overset{(8.26)}{=} (\mathcal{E}_p / \mathfrak{m}_p \mathcal{E}_p)^\vee = E(p)^\vee.$$

2. The vector bundle corresponding to $\mathcal{E} \oplus \mathcal{E}'$ is called the *direct sum of* (E, π) *and* (E', π'). It is denoted by $(E \oplus E', \pi \oplus \pi')$. For $p \in M$ one has $(E \oplus E')(p) = E(p) \oplus E'(p)$.

Definition 8.47. Let $(X, \mathcal{O}_X)$ be a ringed space, let $\mathcal{E}$ be an $\mathcal{O}_X$-module, and let $\mathcal{F}$ be an $\mathcal{O}_X$-submodule of $\mathcal{E}$. Then $\mathcal{F}$ is called a *direct summand of* $\mathcal{E}$ if there exists a submodule $\mathcal{G}$ of $\mathcal{E}$ such that $\mathcal{F} \oplus \mathcal{G} = \mathcal{E}$. The submodule $\mathcal{F}$ is called a *local direct summand* if there exists an open covering $(U_i)_i$ of X such that $\mathcal{F}_{|U_i}$ is a direct summand of $\mathcal{E}_{|U_i}$ for all $i \in I$.

Remark 8.48. Let $\mathcal{F}$ be a submodule of an $\mathcal{O}_X$-module $\mathcal{E}$. Let $i : \mathcal{F} \to \mathcal{E}$ be the inclusion and let $p : \mathcal{E} \to \mathcal{E}/\mathcal{F}$ be the projection. As for modules over a ring (Appendix Remark 14.5) one sees that the following assertions are equivalent:

(i) The submodule $\mathcal{F}$ is a direct summand of $\mathcal{E}$.
(ii) There exists a homomorphism of $\mathcal{O}_X$-modules $r : \mathcal{E} \to \mathcal{F}$ such that $r \circ i = \mathrm{id}_\mathcal{E}$.
(iii) There exists a homomorphism of $\mathcal{O}_X$-modules $s : \mathcal{E}/\mathcal{F} \to \mathcal{E}$ such that $p \circ s = \mathrm{id}_{\mathcal{E}/\mathcal{F}}$.

Local direct summands of a finite locally free module $\mathcal{E}$ correspond to subbundles of the vector bundle attached to $\mathcal{E}$:

Proposition and Definition 8.49. *Let M be a premanifold, $\mathcal{E}$ and $\mathcal{F}$ finite locally free $\mathcal{O}_M$-modules and let E and F be the corresponding vector bundles over M. Let $w : \mathcal{E} \to \mathcal{F}$ be a homomorphism of $\mathcal{O}_M$-modules and let $\varpi : E \to F$ be the corresponding morphism of vector bundles.*

1. *The map ϖ is surjective if and only if the morphism of sheaves w is surjective. In this case we say that F is a* quotient bundle *of E.*
2. *The map ϖ is injective if and only if the morphism of sheaves w is injective and $w(\mathcal{E})$ is a local direct summand of $\mathcal{F}$. In this case we say that E is a* subbundle *of F.*

Proof. The map ϖ is surjective (respectively injective) if and only if for all $p \in M$ the induced linear map on the fiber $\varpi(p): E(p) \to F(p)$ is surjective (respectively injective).

We show (1). Let $\mathfrak{m}_p$ be the maximal ideal of $\mathcal{O}_{M,p}$ and let $w(p)$ be the $\mathbb{K}$-linear map $\mathcal{E}(p) := \mathcal{E}_p/\mathfrak{m}_p\mathcal{E}_p \to \mathcal{F}(p) := \mathcal{F}_p/\mathfrak{m}_p\mathcal{F}_p$ induced by w. In the proof of Proposition 8.45 we have seen that $E(p) = \mathcal{E}(p)$, $F(p) = \mathcal{F}(p)$ and $w(p) = \varpi(p)$. The stalk $\mathcal{E}_p$ is a finitely generated free $\mathcal{O}_{M,p}$-module. Hence the first assertion follows immediately from Nakayama's lemma (Appendix 14.38).

Let us show (2). Suppose that w is injective and that $w(\mathcal{E})$ is a local direct summand of $\mathcal{F}$. Let $p \in M$. To see that $\varpi(p)$ is injective we may pass to a sufficiently small neighborhood of p and can assume that there exists a homomorphism of $\mathcal{O}_X$-modules $r: \mathcal{E} \to \mathcal{F}$ such that $r \circ w = \mathrm{id}_{\mathcal{E}}$ (Remark 8.48). Then r induces a $\mathbb{K}$-linear map $r(p): E(p) \to F(p)$ with $r(p) \circ w(p) = \mathrm{id}$. As $w(p) = \varpi(p)$, we see that $\varpi(p)$ is injective.

Conversely, assume that $\varpi(p) = w(p)$ is injective for all $p \in M$. Let $\bar{r}_0: \mathcal{F}(p) \to \mathcal{E}(p)$ be a $\mathbb{K}$-linear map with $\bar{r}_0 \circ w(p) = \mathrm{id}_{\mathcal{E}(p)}$. Let $(f_1, \ldots, f_n)$ be an $\mathcal{O}_{M,p}$-basis of $\mathcal{F}_p$. For all i let $\bar{e}_i$ be the image of f_i under $\mathcal{F}_p \longrightarrow \mathcal{F}(p) \overset{\bar{r}_0}{\longrightarrow} \mathcal{E}(p)$ and let $e_i \in \mathcal{E}_p$ be an element whose image in $\mathcal{E}(p)$ is $\bar{e}_i$. Sending f_i to e_i defines an $\mathcal{O}_{M,p}$-linear map $r': \mathcal{F}_p \to \mathcal{E}_p$. Then $r' \circ w_p$ is an endomorphism of the finitely generated free $\mathcal{O}_{M,p}$-module $\mathcal{E}_p$, which is modulo $\mathfrak{m}_p$ the identity. Hence it is bijective (Appendix Proposition 14.39). Hence if we set $r_0 := (r' \circ w_p)^{-1} \circ r'$ we have $r_0 \circ w_p = \mathrm{id}_{\mathcal{E}_p}$.

As (8.26) is surjective for homomorphisms $\mathcal{F} \to \mathcal{E}$, we find an open neighborhood U of p and a homomorphism $r: \mathcal{F}_{|U} \to \mathcal{E}_{|U}$ of $\mathcal{O}_U$-modules such that $r_p = r_0$. As (8.26) is injective for homomorphisms $\mathcal{E} \to \mathcal{E}$, we have $r \circ w_{|U} = \mathrm{id}_{\mathcal{E}_{|U}}$ after possibly shrinking U. As p was arbitrary, this shows that w is injective and that $w(\mathcal{E})$ is a local direct summand of $\mathcal{F}$ by Remark 8.48. $\square$

The proof shows that if $\varpi(p)$ is injective for a single point p, then there exists an open neighborhood U of p such that $E_{|U}$ is a subbundle of $F_{|U}$.

The local-global principle explained in Chap. 7 yields a cohomological criterion for a local direct summand to be a direct summand and hence for subbundles to be global direct summands:

Remark 8.50 (Splitting of submodules). Let $(X, \mathcal{O}_X)$ be a ringed space. Let $\mathcal{E}$ be a local direct summand of an $\mathcal{O}_X$-module $\mathcal{F}$. Let $p: \mathcal{F} \to \mathcal{F}/\mathcal{E}$ be the projection and let $\mathcal{S}$ be the sheaf of splittings of p, i.e., for $U \subseteq X$ open, $\mathcal{S}(U)$ is the set of homomorphisms of $\mathcal{O}_U$-modules $r: (\mathcal{F}/\mathcal{E})_{|U} \to \mathcal{F}_{|U}$ with $p_{|U} \circ r = \mathrm{id}$. For any $r, r' \in \mathcal{S}(U)$ one has $p_{|U} \circ (r - r') = 0$, i.e., $r - r'$ takes values in $\mathcal{E}$. Hence the sheaf

$$\mathcal{A} := \mathcal{H}om_{\mathcal{O}_M}(\mathcal{F}/\mathcal{E}, \mathcal{E}),$$

considered as a sheaf of abelian groups, acts simply transitively on $\mathcal{S}$ by $(w, s) \mapsto w + s$, $w \in \mathcal{A}(U)$, $s \in \mathcal{S}(U)$, $U \subseteq X$ open. This makes $\mathcal{S}$ into a pseudo-$\mathcal{A}$-torsor. By hypothesis there exists on open covering $(U_i)_i$ with $\mathcal{S}(U_i) \neq \emptyset$ for all i. Hence $\mathcal{S}$ is an $\mathcal{A}$-torsor. Its class

$$\sigma(\mathcal{E}) \in H^1(X, \mathcal{H}om_{\mathcal{O}_M}(\mathcal{F}/\mathcal{E}, \mathcal{E}))$$

is trivial if and only if $\mathcal{S}(X) \neq \emptyset$, i.e., if and only if $\mathcal{E}$ is a (global) direct summand of $\mathcal{F}$.

Definition and Remark 8.51 (Split subbundles). We keep the notation of Definition 8.49. We call the vector bundle E a *split subbundle of F* if w is injective and $w(\mathcal{E})$ is a direct summand of $\mathcal{F}$, i.e., if F is the direct sum of the image of ϖ and a subbundle of F.

If E is a subbundle of F, then Remark 8.50 shows that there exists a class $\sigma(E) \in H^1(M, \mathcal{H}om_{\mathcal{O}_M}(\mathcal{F}/\mathcal{E}, \mathcal{E}))$ that vanishes if and only if E is a split subbundle of F.

8.5 Tangent Bundle

In this and the following section we will assume that we are either in the case of real C^∞-(pre)manifolds (real smooth case) or in the case of real C^ω-(pre)manifolds with $\mathbb{K} = \mathbb{R}$ (real analytic case) or in the case of complex (pre)manifolds with $\mathbb{K} = \mathbb{C}$ (complex case).

Definition and Remark 8.52. Let M be a premanifold. We define the *tangent bundle* T_M as a set

$$T_M := \coprod_{p \in M} T_p(M).$$

Let $\pi = \pi_M \colon T_M \to M$ be the map such that $\pi^{-1}(p) = T_p(M)$ for all p.

Definition of an atlas of T_M: Let $x \colon U \to \tilde{U} \subseteq \mathbb{K}^m$ be a chart of M. For $p \in U$ we obtain an isomorphism $T_p(x) \colon T_p(M) \to \mathbb{K}^m$. Let $V := \pi^{-1}(U) \subseteq T_M$ and define as a chart for T_M

$$y \colon V \to \tilde{V} := \tilde{U} \times \mathbb{K}^m \subseteq \mathbb{K}^{2m}, \qquad T_p(M) \ni \xi \mapsto (x(p), T_p(x)(\xi)).$$

If we define for a second chart (U', x') of M a chart (V', y') of T_M in the same way, the change of charts $y' \circ y^{-1}$ is given by

$$\begin{aligned} y(V \cap V') = x(U \cap U') \times \mathbb{K}^m &\longrightarrow x'(U \cap U') \times \mathbb{K}^m = y'(V \cap V'), \\ (\tilde{p}, v) &\longmapsto \left((x' \circ x^{-1})(\tilde{p}), J_{x' \circ x^{-1}}(\tilde{p})(v)\right), \end{aligned} \tag{*}$$

which is again an isomorphism of premanifolds. Hence if (U, x) runs through an atlas of M the (V, y) form an atlas of T_M. We obtain on T_M the structure of a premanifold[1].

---

[1] If M was a C^α-premanifold with $\alpha < \infty$, then the second component of (*) would be only $C^{\alpha-1}$ and hence T_M would be only $C^{\alpha-1}$-premanifold. Taking this added complication into account, most results of Sect. 8.5 also hold for real C^α-premanifolds with $\alpha < \infty$.

Clearly, (T_M, π) is a vector bundle over M and we have $\mathrm{rk}_p(T_M) = \dim_p(M)$. We denote by $\mathcal{T}_M$ the finite locally free $\mathcal{O}_M$-module corresponding to T_M (Proposition 8.45).

If M is a manifold, then T_M is a manifold (Proposition 8.16).

Definition and Remark 8.53. Let $F: M \to N$ be a morphism of premanifolds. Define

$$T_F: T_M \to T_N, \qquad T_p(M) \ni \xi \mapsto T_p(F)(\xi) \in T_{F(p)}(N).$$

Then the diagram

$$\begin{CD} T_M @>T_F>> T_N \\ @V\pi_M VV @VV\pi_N V \\ M @>F>> N \end{CD}$$

is commutative, and T_F is $\mathbb{K}$-linear on fibers. We claim that T_F is a morphism of premanifolds. Indeed, this can be checked locally on M and N and we may assume that $M \subseteq \mathbb{K}^m$ and $N \subseteq \mathbb{K}^n$ are open. Then T_F is given by

$$T_F: T_M = M \times \mathbb{K}^m \longrightarrow N \times \mathbb{K}^n = T_N, \qquad (p, x) \mapsto (F(p), DF(p)),$$

which is a morphism of premanifolds[2].

For morphisms $F: M \to N$ and $G: N \to P$ the chain rule takes the following elegant form:

$$T_{G \circ F} = T_G \circ T_F. \tag{8.29}$$

Sections of the $\mathcal{O}_M$-module $\mathcal{T}_M$ are called vector fields:

Definition 8.54. Let M be a premanifold. For $U \subseteq M$ open, elements of

$$\mathcal{T}_M(U) = \{\, s: U \to T_M \; ; \; \pi \circ s = \mathrm{id}_U \,\}$$

are called *vector fields over U*. For a vector field X we sometimes write X_p instead of $X(p)$ for $p \in U$.

Example 8.55. Let V be a finite-dimensional $\mathbb{K}$-vector space and let $M \subseteq V$ be a submanifold. For every $p \in M$ we have $T_p(M) \subseteq T_p(V) = V$ and hence we may view the tangent bundle T_M as the submanifold of $V \times V$

$$T_M = \{\, (p, v) \in M \times V \; ; \; v \in T_p(M) \,\}.$$

Hence a vector field X is the same as a morphism $F: M \to V$ of manifolds with $F(p) \in T_p(M)$. The correspondence is given by $X(p) = (p, F(p))$ for all $p \in M$.

---

[2] Again, T_F would be only a morphism of $C^{\alpha-1}$-premanifolds for $\alpha < \infty$.

As a special case let $M = U \subseteq V = \mathbb{K}^n$ be open. Then $T_U = U \times \mathbb{K}^n$ and a vector field X corresponds to a morphism $F: U \to \mathbb{K}^n$.

Our next goal is to give a concrete description of $\mathcal{T}_M$ by derivations generalizing the description of a single tangent space by derivations (Proposition 5.11).

Definition and Remark 8.56. Let $(X, \mathcal{O}_X)$ be a $\mathbb{K}$-ringed space.

1. A $\mathbb{K}$-derivation of $\mathcal{O}_X$ is a $\mathbb{K}$-linear homomorphism $D: \mathcal{O}_X \to \mathcal{O}_X$ of sheaves such that $D_U(fg) = f D_U(g) + g D_U(f)$ for all $U \subseteq X$ open, $f, g \in \mathcal{O}_X(U)$. Denote by $\mathrm{Der}_{\mathbb{K}}(\mathcal{O}_X)$ the $\mathbb{K}$-vector space of $\mathbb{K}$-derivations of $\mathcal{O}_X$. It is an $\mathcal{O}_X(X)$-module via
$$(g \cdot D)_U(f) := g_{|U} D_U(f) \qquad \text{for } D \in \mathrm{Der}_{\mathbb{K}}(\mathcal{O}_X),\, g \in \mathcal{O}_X(X),\, f \in \mathcal{O}_X(U).$$

2. Define an $\mathcal{O}_X$-module via
$$\mathcal{D}er_{\mathbb{K}}(\mathcal{O}_X)(U) := \mathrm{Der}_{\mathbb{K}}(\mathcal{O}_{X|U}).$$

Proposition 8.57. *Let M be a premanifold.*

1. *Let X be a vector field over U. Then the* Lie *derivative of X*
$$\mathcal{L}_X: \mathcal{O}_M(U) \to \mathcal{O}_M(U), \qquad f \mapsto (p \mapsto T_p(f)(X_p))$$
is a $\mathbb{K}$-derivation of $\mathcal{O}_M(U)$.

2. *The morphism of sheaves*
$$\mathcal{L}: \mathcal{T}_M \to \mathcal{D}er_{\mathbb{K}}(\mathcal{O}_M),$$
$$\mathcal{T}_M(U) \ni X \mapsto (\mathcal{L}_{X_{|V}})_{V \subseteq U \text{ open}} \in \mathcal{D}er_{\mathbb{K}}(\mathcal{O}_M)(U) \tag{8.30}$$
is an isomorphism of $\mathcal{O}_M$-modules.

Proof. Assertion (1) follows from the product rule.

Let us show that $\mathcal{L}$ is $\mathcal{O}_M$-linear. For $X, Y \in \mathcal{T}_M(U)$ and $g \in \mathcal{O}_M(U)$ we have
$$\mathcal{L}_{gX+Y}(f)(p) = T_p(f)((gX + Y)(p)) = T_p(f)(g(p)X(p) + Y(p))$$
$$= g(p)T_p(f)(X(p)) + T_p(f)(Y(p)) = (g\mathcal{L}_X + \mathcal{L}_Y)(f)(p).$$

Let us construct an inverse map. Let $(D_V)_{V \subseteq U \text{ open}} \in \mathcal{D}er_{\mathbb{K}}(\mathcal{O}_M)(U)$. For $p \in U$ recall that $T_p(M) = \mathrm{Der}_{\mathbb{K}}(\mathcal{O}_{M,p}, \mathbb{K})$ (Proposition 5.11). Define
$$X_p \in \mathrm{Der}_{\mathbb{K}}(\mathcal{O}_{M,p}, \mathbb{K}) \qquad \text{by} \qquad X_p([V, f]) := D_V(f)(p).$$

Using a chart it is easy to check that $U \ni p \mapsto X_p \in T_p(M)$ is a vector field on U and that this defines an inverse map to $\mathcal{L}$. $\qquad\square$

Remark 8.58. Let M be a premanifold. Let (U, x) be a chart with coordinate functions $x^1, \ldots, x^m$. Then

$$\frac{\partial}{\partial x^i} \colon \mathcal{O}_{M|U} \to \mathcal{O}_{M|U}, \qquad \mathcal{O}_M(V) \ni f \mapsto \frac{\partial f}{\partial x^i} \in \mathcal{O}_M(V), \qquad V \subseteq U \text{ open}$$

is an element of $\mathcal{D}er_{\mathbb{K}}(\mathcal{O}_M)(U)$ and $(\frac{\partial}{\partial x^1}, \ldots, \frac{\partial}{\partial x^m})$ is a basis of the free $\mathcal{O}_U$-module $\mathcal{D}er_{\mathbb{K}}(\mathcal{O}_M)_{|U}$ by Remark 5.8.

8.6 Differential Forms and De Rham Complex

Recall that in this section we only consider real C^α-premanifolds with $\alpha \geq \infty$ and complex premanifolds. Differential forms are by definition sections of exterior powers of the dual of the $\mathcal{O}_M$-module $\mathcal{T}_M$. Hence we introduce first the notion of "exterior powers" for general $\mathcal{O}_X$-modules using the construction principle of Remark 8.36. We refer to Appendix Sect. 14.3 for the notion of the exterior power of a module over a commutative ring.

Definition and Remark 8.59 (Exterior powers of $\mathcal{O}_X$-modules). Let $(X, \mathcal{O}_X)$ be a ringed space, let $\mathcal{F}$ be an $\mathcal{O}_X$-module, and let $r \geq 0$ be an integer. The $\mathcal{O}_X$-module attached to the $\mathcal{O}_X$-premodule

$$U \mapsto \bigwedge_{\Gamma(U, \mathcal{O}_X)}^{r} \mathcal{F}(U)$$

is denoted by $\Lambda_{\mathcal{O}_X}^r \mathcal{F}$ or simply $\Lambda^r \mathcal{F}$. It is called the *r-th exterior power of $\mathcal{F}$*.

As the formation of exterior powers is functorial for modules over a ring, we see that $\Lambda^r \mathcal{F}$ is a covariant functor in $\mathcal{F}$.

Remark and Definition 8.60 (Exterior powers of finite locally free modules). Let $\mathcal{F}$ be a finite locally free $\mathcal{O}_X$-module of constant rank $n \in \mathbb{N}_0$. Then $\Lambda^r \mathcal{F}$ is a locally free $\mathcal{O}_X$-module of rank $\binom{n}{r}$ (Appendix Proposition 14.22).

In particular, $\det(\mathcal{F}) := \Lambda^n \mathcal{F}$ is a locally free $\mathcal{O}_X$-module of rank 1 that we call the *determinant of $\mathcal{F}$*. If u is an endomorphism of the $\mathcal{O}_X$-module $\mathcal{O}_X^n$, then u is given by a matrix $A \in M_n(\mathcal{O}_X(X))$. By Appendix Proposition 14.22, $\Lambda^n(u)$ is given by the multiplication with $\det(A)$.

The construction in Appendix Remark 14.24 for modules yields a homomorphism of $\mathcal{O}_X$-modules

$$\Lambda^r(\mathcal{F}^\vee) \longrightarrow (\Lambda^r \mathcal{F})^\vee. \tag{8.31}$$

We claim that this is an isomorphism if $\mathcal{F}$ is finite locally free. Indeed, this can be checked locally on X. Hence we can assume $\mathcal{F} = \mathcal{O}_X^n$ and the claim follows because (14.24) is an isomorphism for finitely generated free modules (Appendix Remark 14.24).

Definition 8.61. Let M be a premanifold. Define locally free $\mathcal{O}_M$-modules

$$\Omega_M^1 := (\mathcal{T}_M)^\vee, \qquad \Omega_M^i := \bigwedge^i (\Omega_M^1), \qquad i \geq 0.$$

In particular $\Omega_M^0 = \mathcal{O}_M$. The $\mathcal{O}_M$-module Ω_M^i is called the *sheaf of i-differential forms*. For every $U \subseteq M$ open we call an element of $\Omega_M^i(U)$ an *i-form on U*. If M is a real C^∞-premanifold (respectively a real analytic premanifold, respectively a complex premanifold), then elements of $\Omega_M^i(U)$ are called *smooth i-forms* (respectively *real analytic i-forms*, respectively *holomorphic i-forms*).

Remark 8.62. For every $i \geq 0$ we have $\Omega_M^i = \mathcal{H}om(\bigwedge^i \mathcal{T}_M, \mathcal{O}_M)$ by Remark 8.60 and hence for $U \subseteq M$ open

$$\Omega_M^i(U) = \{\alpha \colon \mathcal{T}_U \times \cdots \times \mathcal{T}_U \to \mathcal{O}_U \; ; \; \alpha \text{ alternating } i\text{-multilinear form}\} .$$

If M is a premanifold of dimension $d \in \mathbb{N}_0$, then $\operatorname{rk} \Omega_M^1 = \operatorname{rk} \mathcal{T}_M = d$ and hence

$$\operatorname{rk} \Omega_M^r = \binom{d}{r} \tag{8.32}$$

by Remark 8.60. In particular, $\Omega_M^r = 0$ for $r > d$.

Remark and Definition 8.63. Let M be a premanifold. We define

$$d \colon \mathcal{O}_M \to \Omega_M^1 \tag{8.33}$$

as follows. For $f \in \mathcal{O}_M(U)$ let $df \colon \mathcal{T}_{M|U} = \mathcal{D}er_{\mathbb{K}}(\mathcal{O}_U, \mathcal{O}_U) \to \mathcal{O}_U$ be the morphism that sends $\partial \in \operatorname{Der}_{\mathbb{K}}(\mathcal{O}_V, \mathcal{O}_V)$ to $\partial(f_{|V})$ for $V \subseteq U$ open. Then d is a morphism of sheaves of $\mathbb{K}$-vector spaces (but not of $\mathcal{O}_M$-modules!) and

$$d(fg) = f\,dg + g\,df, \qquad f, g \in \mathcal{O}_M(U), U \subseteq M \text{ open}.$$

For $p \in M$ one can think of $(df)(p) \in \Omega_M^1(p) = T_p(M)^\vee$ as the infinitesimal variation of f at p in an unspecified direction, which gives after evaluation on a tangent vector the derivative of f at p in that tangent direction.

Remark 8.64 (Local description of differential forms). Let M be a premanifold, let (U, x) be a chart with coordinate functions $x^1, \ldots, x^m \colon U \to \mathbb{K}$ (and hence $\dim_p(M) = m$ for $p \in M$). Then $dx^i \in \Omega^1_M(U)$ and $(dx^1, \ldots, dx^m)$ is a basis of $\Omega^1_{M|U}$. This basis is dual to the basis $(\frac{\partial}{\partial x^i})_i$ of $\mathcal{T}_M(U)$ (Remark 8.58). For $f \in \mathcal{O}_M(U)$ one has

$$df = \sum_{i=1}^m \frac{\partial f}{\partial x^i} dx^i$$

(evaluate both sides on $\frac{\partial}{\partial x^j}$ to see the identity).

Let $r \geq 1$ be an integer. Then $\Omega^r_{M|U}$ is a free $\mathcal{O}_U$-module and the

$$dx^{i_1} \wedge \cdots \wedge dx^{i_r}, \qquad 1 \leq i_1 < \cdots < i_r \leq m$$

form a basis (Appendix Proposition 14.22).

Proposition 8.65. *Let M be a premanifold. There exists a unique family of morphisms $(d \colon \Omega^i_M \to \Omega^{i+1}_M)_{i \geq 0}$ such that:*

(a) *d is a morphism of sheaves of $\mathbb{K}$-vector spaces.*
(b) *$d \colon \Omega^0_M = \mathcal{O}_M \to \Omega^1_M$ is the morphism defined in Remark 8.63.*
(c) *For $U \subseteq M$ open, $\omega \in \Omega^r_M(U)$, $\eta \in \Omega^s_M(U)$ one has*

$$d(\omega \wedge \eta) = d\omega \wedge \eta + (-1)^r \omega \wedge d\eta.$$

(d) *$d \circ d = 0 \colon \Omega^{i-1}_M \to \Omega^{i+1}_M$ for all $i \geq 1$.*

Proof. As Ω^1_M generates $\bigwedge \Omega^1_M$ as $\mathcal{O}_M$-algebra, there exists at most one such family. Because of this uniqueness, it suffices to prove the existence locally. Hence we may assume that $M \subseteq \mathbb{K}^m$ is open. Let $x^1, \ldots, x^m \colon M \to \mathbb{K}$ be the coordinate functions. Then Ω^r_M is a free $\mathcal{O}_M$-module and the $dx^{i_1} \wedge \cdots \wedge dx^{i_r}$ for $1 \leq i_1 < \cdots < i_r \leq m$ form a basis of Ω^r_M. Conditions (a)–(d) imply that we must define

$$d(f dx^{i_1} \wedge \cdots \wedge dx^{i_r}) = df \wedge dx^{i_1} \wedge \cdots \wedge dx^{i_r}. \tag{8.34}$$

Then (a) and (b) are clearly satisfied. To check (c) and (d) is a straightforward computation (for (d) one uses that taking partial derivations in different directions commute with each other by Appendix Proposition 16.19). $\qquad \square$

Example 8.66 (Gradient, curl, and divergence). Gradient, curl, and divergence are all special cases of the above construction. Let $U \subseteq \mathbb{K}^m$ be open. Then a section of $\mathcal{T}_U$ over an $V \subseteq U$, i.e., a vector field X over V, can be considered as a morphism of manifolds $X : V \to \mathbb{K}^m$ (Example 8.55). Moreover, Ω_U^1 is a free $\mathcal{O}_U$-module with basis $(dx^1, \ldots, dx^m)$, where $x^i : U \to \mathbb{K}$ is the i-th coordinate function. Hence the choice of coordinates allows us to identify

$$\mathcal{O}_U = \Omega_U^0 \xrightarrow{\sim} \Omega_U^m,$$
$$f \mapsto f(dx^1 \wedge \cdots \wedge dx^m),$$

and

$$\mathcal{T}_U \xrightarrow{\sim} \Omega_U^1 \xrightarrow{\sim} \Omega_U^{m-1},$$
$$X = (f_1, \ldots, f_m) \mapsto \sum_{i=1}^m f_i dx^i \mapsto \sum_{i=1}^m f_i (-1)^{i-1} (dx^1 \wedge \cdots \wedge \widehat{dx^i} \wedge \cdots \wedge dx^m),$$

where as usual $\widehat{(\,)}$ denotes an omission of that term.

The identification of 1-forms and vector fields identifies $d : \Omega_M^0 \to \Omega_M^1$, $f \mapsto \sum_{i=1}^m \frac{\partial f}{\partial x^i} dx^i$ with the gradient $f \mapsto (\frac{\partial f}{\partial x^1}, \ldots, \frac{\partial f}{\partial x^m})$.

The identification of vector fields and $(m-1)$-forms and of m-forms and functions identifies $d : \Omega_U^{m-1} \to \Omega_U^m$ given by

$$\sum_{i=1}^m f_i (-1)^{i-1} \left(dx^1 \wedge \cdots \wedge \widehat{dx^i} \wedge \cdots \wedge dx^m \right) \mapsto \left(\sum_{i=1}^m \frac{\partial f}{\partial x^i} \right) (dx^1 \wedge \cdots \wedge dx^m)$$

with the divergence $X = (f_1, \ldots, f_m) \mapsto \sum_{i=1}^m \frac{\partial f}{\partial x^i}$.

Finally, if $m = 3$, then $d : \Omega_U^1 \to \Omega_U^2$ is given by sending the 1-form $f_1 dx^1 + f_2 dx^2 + f_3 dx^3$ to

$$\left(\frac{\partial f_2}{\partial x^1} - \frac{\partial f_1}{\partial x^2} \right) dx^1 \wedge dx^2 - \left(\frac{\partial f_1}{\partial x^3} - \frac{\partial f_3}{\partial x^1} \right) dx^1 \wedge dx^3 + \left(\frac{\partial f_3}{\partial x^2} - \frac{\partial f_2}{\partial x^3} \right) dx^2 \wedge dx^3.$$

Hence it is identified with the curl.

Definition 8.67. Let M be a premanifold. The complex

$$\mathcal{O}_M \xrightarrow{d} \Omega_M^1 \xrightarrow{d} \Omega_M^2 \xrightarrow{d} \cdots \tag{8.35}$$

is called the *de Rham complex of M*. Note that although the Ω_M^r are $\mathcal{O}_M$-modules, the maps d are only morphism of sheaves of $\mathbb{K}$-vector spaces.

For $U \subseteq M$ an r-form $\omega \in \Omega_M^r(U)$ is called *closed* (respectively *exact*) if $d\omega = 0$ (respectively if there exists $\eta \in \Omega_M^{r-1}(U)$ such that $d\eta = \omega$). As $d \circ d = 0$, every exact form is closed.

Theorem 8.68. (Poincaré lemma) *Let* $M := \{z \in \mathbb{K}^m \,;\, \forall i = 1, \ldots, m : |z_i| < 1\} \subseteq$ $\mathbb{K}^m$. *Then for every* $r \in \mathbb{N}$ *every closed form* $\omega \in \Omega_M^r(M)$ *is exact.*

We do not give a proof of this result here but refer to [AmEs3] XI Corollary 3.12 for the real C^∞-case. The explicit formula given there also shows the Poincare lemma in the real analytic case. For the complex case we refer to [Wel] II, Example 2.13.

Corollary 8.69. *Let M be a premanifold. Then the de Rham complex is an exact sequence of sheaves of $\mathbb{K}$-vector spaces.*

Proof. The exactness of the de Rham complex means that locally every closed form is exact. Hence we may assume that M is as in the Poincaré lemma 8.68. $\qquad\square$

As the kernel of $d : \mathcal{O}_M \to \Omega_M^1$ is the sheaf of locally constant $\mathbb{K}$-valued functions, Corollary 8.69 means (by Appendix Example 15.3) that we have a quasi-isomorphism of $\mathbb{K}_M$-modules

$$\mathbb{K}_M \xrightarrow{qis} \Omega_M^\bullet. \tag{8.36}$$

In Chap. 10 we will define the cohomology of a complex of sheaves and then (8.36) will imply that $\mathbb{K}_M$ and $\Omega_M^\bullet$ have the same cohomology.

Example 8.70. Let us give an immediate application of the Poincaré lemma to vector fields. Let $U \subseteq \mathbb{K}^m$ be open. Identifying morphisms of manifolds $X : U \to \mathbb{K}^m$, $X = (f_1, \ldots, f_m)$ with 1-forms ω_X over U, we see by Example 8.66 that $d(\omega_X) = 0$ if and only if

$$\frac{\partial f_i}{\partial x^j} = \frac{\partial f_j}{\partial x^i} \quad \forall\, i, j = 1, \ldots, m. \tag{*}$$

Hence from the exact de Rham complex we obtain an exact sequence of sheaves of $\mathbb{K}$-vector spaces

$$0 \longrightarrow \mathbb{K}_U \longrightarrow \mathcal{O}_U \xrightarrow{\text{grad}} \mathfrak{I} \to 0, \tag{**}$$

where for $V \subseteq U$ open $\mathfrak{I}(V)$ denotes the $\mathbb{K}$-vector space of those vector fields $(f_1, \ldots, f_m) : V \to \mathbb{K}^m$ satisfying (*) and where grad is the morphism sending a function on V to its gradient.

Hence the cohomology sequence obtained from (**) shows that every vector field $(f_1, \ldots, f_m)$ on U satisfying (*) is the gradient of a morphism $U \to \mathbb{K}$ if and only if $H^1(U, \mathbb{K}_U) \to H^1(U, \mathcal{O}_U)$ is injective. This is for instance the case if U is simply connected because then $H^1(U, \mathbb{K}_U) = 0$ by Proposition 7.5.

8.7 Problems

Problem 8.1. Let the group $G := \{\pm 1\}$ act on S^1 by a rotation by π, i.e., $(-1) \cdot e^{i\theta} := e^{i(\theta+\pi)} = -e^{i\theta}$. Show that there are precisely two isomorphism classes X and Y of fiber bundles with basis $B := S^1$, typical fiber S^1, and structure group $\{\pm 1\}$. Show that X and Y are isomorphic as B-manifolds.

Problem 8.2. Let $p: Z \to B$ be a morphism of premanifolds, $\mathcal{G} \subseteq \mathcal{A}ut(p)$ a sheaf of subgroups. Let $\xi = (\pi: X \to B, [U_i, h_i]_i)$ be a twist of p with structure sheaf $\mathcal{G}$. Denote the trivial twist $(p, [B, \mathrm{id}_Z])$ by ξ_0.

1. Show that for $U \subseteq B$ open $\xi_{|U} := (\pi_{|\pi^{-1}(U)}, [U_i \cap U, h_{i|\pi^{-1}(U_i \cap U)}])$ is a twist of $p_{|p^{-1}(U)}: p^{-1}(U) \to U$ with structure sheaf $\mathcal{G}_{|U}$.
2. For $U \subseteq B$ open let $T_\xi(U)$ be the set of isomorphisms $\xi_{|U} \xrightarrow{\sim} \xi_{0|U}$ of twists of $p_{|p^{-1}(U)}$ with structure sheaf $\mathcal{G}_{|U}$. Let $\mathcal{G}$ act on $\mathcal{T}_\xi$ via composition. Show that T_ξ is a $\mathcal{G}$-torsor.
3. Show that $\xi \mapsto T_\xi$ yields an equivalence of categories from the category of twists of p with structure sheaf $\mathcal{G}$ (morphisms in this category are isomorphisms of twists) and the category $(\mathrm{Tors}(\mathcal{G}))$. Show that this equivalence induces on isomorphism classes the composition of (8.4) with the isomorphism $\check{H}^1(B, \mathcal{G}) \xrightarrow{\sim} H^1(B, \mathcal{G})$ (7.5).

Problem 8.3. Let B and Z be premanifolds, let A be the group of automorphisms of the premanifold Z, and let $p: Z \to B$ be the constant morphism with image $\{b\}$ for some $b \in B$. Show that $\mathcal{A}ut(p)$ is the skyscraper sheaf in b with value A and deduce that there are no non-trivial twists of p.
Hint: Problem 7.5.

Problem 8.4. Show that for a sequence of $\mathcal{O}_X$-modules $0 \to \mathcal{F}' \to \mathcal{F} \to \mathcal{F}''$ (respectively $\mathcal{F}' \to \mathcal{F} \to \mathcal{F}'' \to 0$) the following assertions are equivalent:

(i) The sequence is exact.
(ii) The sequence $0 \to \mathrm{Hom}_{\mathcal{O}_X}(\mathcal{G}, \mathcal{F}') \to \mathrm{Hom}_{\mathcal{O}_X}(\mathcal{G}, \mathcal{F}) \to \mathrm{Hom}_{\mathcal{O}_X}(\mathcal{G}, \mathcal{F}'')$ (respectively $0 \to \mathrm{Hom}_{\mathcal{O}_X}(\mathcal{F}'', \mathcal{G}) \to \mathrm{Hom}_{\mathcal{O}_X}(\mathcal{F}, \mathcal{G}) \to \mathrm{Hom}_{\mathcal{O}_X}(\mathcal{F}', \mathcal{G}))$ of $\Gamma(X, \mathcal{O}_X)$-modules is exact for every $\mathcal{O}_X$-module $\mathcal{G}$.
(iii) The sequence $0 \to \mathcal{H}om_{\mathcal{O}_X}(\mathcal{G}, \mathcal{F}') \to \mathcal{H}om_{\mathcal{O}_X}(\mathcal{G}, \mathcal{F}) \to \mathcal{H}om_{\mathcal{O}_X}(\mathcal{G}, \mathcal{F}'')$ (respectively $0 \to \mathcal{H}om_{\mathcal{O}_X}(\mathcal{F}'', \mathcal{G}) \to \mathcal{H}om_{\mathcal{O}_X}(\mathcal{F}, \mathcal{G}) \to \mathcal{H}om_{\mathcal{O}_X}(\mathcal{F}', \mathcal{G}))$ of $\mathcal{O}_X$-modules is exact for every $\mathcal{O}_X$-module $\mathcal{G}$.

Problem 8.5. Let $(X, \mathcal{O}_X)$ be a ringed space. An $\mathcal{O}_X$-module $\mathcal{F}$ is called *of finite type* (respectively *of finite presentation*) if there exists an open covering $(U_i)_i$ of X and for all i an exact sequence of $\mathcal{O}_{U_i}$-modules $\mathcal{O}_{U_i}^n \to \mathcal{F}_{|U_i} \to 0$ for some $n \in \mathbb{N}_0$ (respectively $\mathcal{O}_{U_i}^m \to \mathcal{O}_{U_i}^n \to \mathcal{F}_{|U_i} \to 0$ for some $m, n \in \mathbb{N}_0$).

1. Show that every finite locally free $\mathcal{O}_X$-module is of finite presentation.
2. Show that (8.26) is an isomorphism for every $\mathcal{O}_X$-module $\mathcal{G}$ and for every $\mathcal{O}_X$-module $\mathcal{F}$ of finite presentation.
   Hint: Use Problem 8.4 and the five lemma to reduce to $\mathcal{F} = \mathcal{O}_X$.
3. Let $\mathcal{F}$ and $\mathcal{G}$ be $\mathcal{O}_X$-modules of finite presentation. Let $x \in X$ and let $w_0 \colon \mathcal{F}_x \xrightarrow{\sim} \mathcal{G}_x$ be an isomorphism of $\mathcal{O}_{X,x}$-modules. Show that there exists an open neighborhood U of x and an isomorphism $w \colon \mathcal{F}_{|U} \xrightarrow{\sim} \mathcal{G}_{|U}$ of $\mathcal{O}_U$-modules such that $w_x = w_0$.

Problem 8.6. Let $(X, \mathcal{O}_X)$ be a ringed space and let $\mathcal{F}$ be an $\mathcal{O}_X$-module of finite type (Problem 8.5).

1. Let $x \in X$ and $s_1, \ldots, s_n \in \mathcal{F}(X)$ such that the germs $(s_1)_x, \ldots, (s_n)_x$ generate the $\mathcal{O}_{X,x}$-module $\mathcal{F}_x$. Show that there exists an open neighborhood U of x such that $(s_1)_y, \ldots, (s_n)_y$ generate the $\mathcal{O}_{X,y}$-module $\mathcal{F}_y$ for all $y \in U$.
2. Show that for every $r \in \mathbb{N}_0$ the subset

$$\{ x \in X \; ; \; \mathcal{F}_x \text{ can be generated as } \mathcal{O}_{X,x}\text{-module by } r \text{ elements}\}$$

   is open in X.
3. Deduce that $\mathrm{Supp}(\mathcal{F}) := \{ x \in X \; ; \; \mathcal{F}_x \neq 0 \}$ is closed in X.

Problem 8.7. Let $\mathcal{F}$ and $\mathcal{G}$ be two $\mathcal{O}_X$-modules.

1. Show that

$$U \mapsto \mathcal{F}(U) \otimes_{\mathcal{O}_X(U)} \mathcal{G}(U)$$

   defines an $\mathcal{O}_X$-premodule. Its sheafification is an $\mathcal{O}_X$-module, which is called the *tensor product of $\mathcal{F}$ and $\mathcal{G}$* and denoted by $\mathcal{F} \otimes_{\mathcal{O}_X} \mathcal{G}$.
2. Let $\tau \colon \mathcal{F} \times \mathcal{G} \to \mathcal{F} \otimes_{\mathcal{O}_X} \mathcal{G}$ be the sheafification of the morphism of presheaves $\mathcal{F}(U) \times \mathcal{G}(U) \to \mathcal{F}(U) \otimes_{\mathcal{O}_X(U)} \mathcal{G}(U)$, $(f, g) \mapsto f \otimes g$. Show that for an $\mathcal{O}_X$-module $\mathcal{H}$ composition with τ yields a $\mathcal{O}_X(X)$-linear bijection from the set of $\mathcal{O}_X$-linear morphisms $\mathcal{F} \otimes_{\mathcal{O}_X} \mathcal{G} \to \mathcal{H}$ with the set of $\mathcal{O}_X$-bilinear maps $\mathcal{F} \times \mathcal{G} \to \mathcal{H}$.
3. Show that for all $x \in X$ there exists a functorial isomorphism of $\mathcal{O}_{X,x}$-modules $(\mathcal{F} \otimes_{\mathcal{O}_X} \mathcal{G})_x \to \mathcal{F}_x \otimes_{\mathcal{O}_{X,x}} \mathcal{G}_x$.
4. Show that if $\mathcal{F}$ and $\mathcal{G}$ are finite locally free $\mathcal{O}_X$-modules, then $\mathcal{F} \otimes_{\mathcal{O}_X} \mathcal{G}$ is finite locally free and $\mathrm{rk}_x(\mathcal{F} \otimes_{\mathcal{O}_X} \mathcal{G}) = \mathrm{rk}_x(\mathcal{F}) \, \mathrm{rk}_x(\mathcal{G})$ for all $x \in X$.
5. Suppose that $(X, \mathcal{O}_X)$ is a premanifold M and that $\mathcal{F}$ and $\mathcal{G}$ are finite locally free $\mathcal{O}_M$-modules. Let F and G be the corresponding vector bundles and define $F \otimes G$ to be the vector bundle corresponding to $\mathcal{F} \otimes_{\mathcal{O}_X} \mathcal{G}$. Show that for all $p \in M$ one has $(F \otimes G)(p) = F(p) \otimes_{\mathbb{K}} G(p)$ for the fibers.

Problem 8.8. Let $\mathcal{F}$, $\mathcal{G}$ and $\mathcal{H}$ be $\mathcal{O}_X$-modules. See Problem 8.7 for the definition of tensor products.

1. Show that there exists a functorial isomorphism of $\mathcal{O}_X$-modules

$$\mathcal{H}om_{\mathcal{O}_X}(\mathcal{F} \otimes_{\mathcal{O}_X} \mathcal{G}, \mathcal{H}) \xrightarrow{\sim} \mathcal{H}om_{\mathcal{O}_X}(\mathcal{F}, \mathcal{H}om_{\mathcal{O}_X}(\mathcal{G}, \mathcal{H})). \tag{8.37}$$

2. Show that there exists a functorial homomorphism of $\mathcal{O}_X$-modules

$$\mathcal{H}om_{\mathcal{O}_X}(\mathcal{F}, \mathcal{G}) \otimes_{\mathcal{O}_X} \mathcal{H} \longrightarrow \mathcal{H}om_{\mathcal{O}_X}(\mathcal{F}, \mathcal{G} \otimes_{\mathcal{O}_X} \mathcal{H}), \tag{8.38}$$

which is an isomorphism if $\mathcal{F}$ or $\mathcal{H}$ is finite locally free.

Problem 8.9. Let $(X, \mathcal{O}_X)$ be a ringed space. Denote by $\mathrm{Pic}(X)$ the set of isomorphism classes of finite locally free $\mathcal{O}_X$-modules of rank 1. Show that $([\mathcal{L}], [\mathcal{M}]) \mapsto [\mathcal{L} \otimes_{\mathcal{O}_X} \mathcal{M}]$ defines the structure of an abelian group on $\mathrm{Pic}(X)$ such that for $[\mathcal{L}] \in \mathrm{Pic}(X)$ one has $[\mathcal{L}]^{-1} = [\mathcal{L}^\vee]$.

Show that if $(X, \mathcal{O}_X) = B$ is a premanifold, then this group structure on $\mathrm{Pic}(B)$ coincides with the group structure defined in Remark 8.25.
Hint: Problem 8.8.

Problem 8.10. Let $(X, \mathcal{O}_X)$ be a ringed space, let $\mathcal{E}$, $\mathcal{F}$, $\mathcal{H}$ be $\mathcal{O}_X$-modules. Assume that $\mathcal{H}$ is finite locally free.

1. Show that there are functorial isomorphisms of $\Gamma(X, \mathcal{O}_X)$-modules

$$\mathrm{Hom}_{\mathcal{O}_X}(\mathcal{E} \otimes_{\mathcal{O}_X} \mathcal{F}, \mathcal{H}) \xrightarrow{\sim} \mathrm{Hom}_{\mathcal{O}_X}\left(\mathcal{E}, \mathcal{F}^\vee \otimes_{\mathcal{O}_X} \mathcal{H}\right) \xrightarrow{\sim} \mathrm{Hom}_{\mathcal{O}_X}\left(\mathcal{F}, \mathcal{E}^\vee \otimes_{\mathcal{O}_X} \mathcal{H}\right).$$

An $\mathcal{O}_X$-linear homomorphism $\beta : \mathcal{E} \otimes_{\mathcal{O}_X} \mathcal{F} \to \mathcal{H}$ is called a *pairing*. If $\mathcal{H} = \mathcal{O}_X$, then β is called a *bilinear form*. A pairing is called *perfect* if the corresponding $\mathcal{O}_X$-module homomorphisms $s_\beta : \mathcal{E} \to \mathcal{F}^\vee \otimes_{\mathcal{O}_X} \mathcal{H}$ and $r_\beta : \mathcal{F} \to \mathcal{E}^\vee \otimes_{\mathcal{O}_X} \mathcal{H}$ are both isomorphisms.

2. Show that if $\mathcal{E}$ and $\mathcal{F}$ are finite locally free, then r_β is an isomorphism if and only if s_β is an isomorphism.

Problem 8.11. Let $(X, \mathcal{O}_X)$ be a ringed space, $n \geq 1$ an integer, and let $\mathcal{F}$ be a locally free $\mathcal{O}_X$-module of rank n. Show that the wedge product yields for all $1 \leq r \leq n$ a perfect pairing (Problem 8.10)

$$\Lambda^r \mathcal{F} \otimes_{\mathcal{O}_X} \Lambda^{n-r} \mathcal{F} \longrightarrow \Lambda^n \mathcal{F} = \det(\mathcal{F}).$$

In particular we obtain an isomorphism of $\mathcal{O}_X$-modules

$$\Lambda^r \mathcal{F} \xrightarrow{\sim} (\Lambda^{n-r} \mathcal{F})^\vee \otimes_{\mathcal{O}_X} \det(\mathcal{F}).$$

Problem 8.12.

1. Let A be a commutative ring, M a finitely generated free A-module and let $\omega\colon M \times M \to A$ be a symplectic pairing (i.e., ω is an A-bilinear map, $\omega(m, m) = 0$ for all $m \in M$, and the induced map $M \to M^\vee$ is an isomorphism of A-modules). Show that the rank of M is even, say $2n$, and that there exists a basis of M such that ω is given with respect to the basis by the matrix

$$J := \begin{pmatrix} 0 & I_n \\ -I_n & 0 \end{pmatrix} \in M_{2n}(A),$$

   where I_n is the $(n \times n)$-identity matrix.

2. Let $(X, \mathcal{O}_X)$ be a ringed space, let $\mathcal{E}$ be a finite locally free $\mathcal{O}_X$-module, and let $\omega\colon \mathcal{E} \otimes_{\mathcal{O}_X} \mathcal{E} \to \mathcal{O}_X$ be an alternating bilinear form (Problem 8.10). It is called *symplectic* if ω is perfect.

   Show that if ω is a symplectic form on $\mathcal{E}$, then there exists for all $x \in X$ an open neighborhood U of x and integer $n \geq 0$ and an isomorphism of $\mathcal{O}_U$-modules $\mathcal{E}_{|U} \xrightarrow{\sim} \mathcal{O}_U^{2n}$, which identifies ω with the alternating form on the free module $\mathcal{O}_U^{2n}$ given by the matrix J.

   Remark: If $(X, \mathcal{O}_X)$ is a real manifold and $\mathcal{E}$ is the tangent bundle of X, then this result is called the *Theorem of Darboux*.

Problem 8.13. Let $V = \mathbb{K}^{2n}$ and ω_0 be the symplectic form on V given by the matrix J (notation as in Problem 8.12). Let $\mathrm{Sp}(V, \omega_0)$ be the Lie group of $\mathbb{K}$-linear automorphisms of V preserving ω_0 and let $\iota\colon \mathrm{Sp}(V, \omega_0) \to \mathrm{GL}(V)$ be the inclusion. Let M be a premanifold. Show that there exists an equivalence of the category of pairs (E, ω), where E is a vector bundle of rank $2n$ and ω is a symplectic form on E and where morphisms are isomorphisms of vector bundles respecting the symplectic forms, and the category $(\mathrm{Tors}(\mathcal{O}_{M;\mathrm{Sp}(V,\omega_0)}))$. In particular one obtains a bijection between isomorphism classes of pairs (E, ω) as above and $H^1(M, \mathcal{O}_{M;\mathrm{Sp}(V,\omega_0)}) = H^1(M, \mathrm{Sp}(V, \omega_0))$.

   Show that $H^1(\iota)$ corresponds to the map that sends the isomorphism class of (E, ω) to the isomorphism class of E.

Hint: Problem 8.12.

Problem 8.14. Let $n \in \mathbb{N}$. Show that for every premanifold M the category $(\mathrm{Tors}(\mathcal{O}_{M;\mathrm{SL}_n(\mathbb{K})}))$ is equivalent to the category of pairs $(\mathcal{E}, \delta)$, where $\mathcal{E}$ is a finite locally free $\mathcal{O}_M$-module of rank n and $\delta\colon \Lambda^n \mathcal{E} \xrightarrow{\sim} \mathcal{O}_M$ is an isomorphism of $\mathcal{O}_M$-modules and where morphisms are isomorphisms of finite locally free modules preserving δ. In particular one obtains a bijection between isomorphism classes of pairs $(\mathcal{E}, \delta)$ as above and $H^1(M, \mathcal{O}_{M;\mathrm{SL}_n(\mathbb{K})})$.

   Let L be the line bundle corresponding to $\Lambda^n \mathcal{E}$. Show that the datum of an isomorphism δ as above is equivalent to the datum of a global section $s\colon M \to L$ such that $s(p) \neq 0$ for all $p \in M$.

Problem 8.15. Let G be a Lie group, let H be a Lie subgroup. For every premanifold B let $\iota_B \colon \mathrm{Princ}_H(B) \to \mathrm{Princ}_G(B)$ be the map of pointed sets deduced by functoriality from the inclusion $H \to G$. Let $\xi \in \mathrm{Princ}_G(B)$ be given by $\pi \colon X \to B$ together with a right G-action.

1. Consider G/H as a G-premanifold and denote by ξ/H the fiber bundle over B with typical fiber G/H and structure group G associated with ξ via (8.13). Show that the quotient of X/H (with respect to the induced right action of H) exists, that π induces a morphism of premanifolds $\pi_{/H} \colon X/H \to B$, and that the underlying B-premanifold of ξ/H is given by $\pi_{/H}$.
2. Let $s \in \mathrm{Hom}_B(B, X/H)$, i.e., $s \colon B \to X/H$ is a morphism of premanifolds such that $\pi_{/H} \circ s = \mathrm{id}_B$. Let $\zeta_s \in \mathrm{Princ}_H(B)$ be the pullback of the principal H-bundle $X \to X/H$ by s. Show that $s \mapsto \zeta_s$ yields a bijection between $\mathrm{Hom}_B(B, X/H)$ and $\iota_B^{-1}(\xi)$.

Problem 8.16. Notation of Problem 8.15. Let $\xi \in \mathrm{Princ}_H(G/H)$ be the class of the principal H-bundle $G \to G/H$. Show that $\iota_{G/H}(\xi)$ is trivial.

Problem 8.17. Let $r \colon B' \to B$ be a morphism of premanifolds.

1. Show that the pullback of fiber bundles yields a functor r^* from the category of vector bundles over B to the category of vector bundles over B'. Show that for $b' \in B'$ and for a vector bundle E over B one has $\mathrm{rk}_{b'}(r^*(E)) = \mathrm{rk}_{r(b')}(E)$.
2. Show that if F is a subbundle (respectively a quotient bundle) of a vector bundle E on B, then $r^*(F)$ is a subbundle (respectively a quotient bundle) of $r^*(E)$.

Problem 8.18. Let $(f, f^\sharp) \colon (X, \mathcal{O}_X) \to (Y, \mathcal{O}_Y)$ be a morphism of ringed spaces and let $\mathcal{E}$ be an $\mathcal{O}_Y$-module.

1. Consider $\mathcal{O}_X$ as a $f^{-1}\mathcal{O}_Y$-module via $f^\sharp$. Show that $f^{-1}\mathcal{E}$ is a $f^{-1}\mathcal{O}_Y$-module and define $f^*(\mathcal{E}) := \mathcal{O}_X \otimes_{f^{-1}\mathcal{O}_Y} f^{-1}\mathcal{E}$ endowed with scalar multiplication by $\mathcal{O}_X$ via the first factor. Show that $\mathcal{E} \mapsto f^*\mathcal{E}$ defines a functor from the category of $\mathcal{O}_Y$-modules to the category of $\mathcal{O}_X$-modules, called *pullback of $\mathcal{O}_Y$-modules*.
2. Show that $f^*(\mathcal{E})$ is a finite locally free $\mathcal{O}_X$-module if $\mathcal{E}$ is a finite locally free $\mathcal{O}_Y$-module. In this case one has $\mathrm{rk}_x(f^*\mathcal{E}) = \mathrm{rk}_{f(x)}(\mathcal{E})$ for all $x \in X$.
3. Let $f \colon M \to N$ be a morphism of premanifolds. Show that the pullback of finite locally free $\mathcal{O}_N$-modules corresponds to the pullback of vector bundles (Problem 8.17) via the equivalence (8.28).

Problem 8.19. Recall that if A is a commutative ring and L is an A-module, then a *Lie bracket on L* is an A-bilinear map $L \times L \to L$, $(m, n) \mapsto [m, n]$ such that $[m, m] = 0$ for all $m \in L$ and such that the *Jacobi identity* $[[m, n], \ell] + [[n, \ell], m] + [[\ell, m], n] = 0$ holds for all $m, n, \ell \in L$. A pair $(L, [,])$ consisting of an A-module L and a Lie bracket on L is called a *Lie algebra over A*. A *homomorphism* $(L_1, [,]) \to (L_2, [,])$ *of Lie algebras over A* is an A-linear map $u : L_1 \to L_2$ such that $[u(m), u(n)] = u([m, n])$ for all $m, n \in L_1$.

Let $(X, \mathcal{O}_X)$ be a $\mathbb{K}$-ringed space. Show that $[D_1, D_2] := D_1 \circ D_2 - D_2 \circ D_1$ defines on $\mathrm{Der}_{\mathbb{K}}(\mathcal{O}_X)$ the structure of a Lie algebra over the ring $\mathcal{O}_X(X)$.

In particular if M is a real C^α-premanifold with $\alpha \geq \infty$ or a complex premanifold, then via (8.30), the $\mathcal{O}_M(M)$-module of vector fields over M is endowed with the structure of a Lie algebra.

Problem 8.20. Let M be a real C^α-premanifold with $\alpha \geq \infty$ or a complex premanifold. Show that there exists a functorial bijection between the set of morphisms of locally $\mathbb{K}$-ringed spaces $P[\varepsilon] \to M$ (Problem 4.5) and the tangent bundle T_M.

Problem 8.21. Let M and N be real C^α-premanifolds with $\alpha \geq \infty$ or complex premanifolds. Show that $T_{M \times N} = T_M \times T_N$.

Problem 8.22. Let G be a (real or complex) Lie group with multiplication $m_G : G \times G \to G$. Show that $T_{m_G} : T_{G \times G} = T_G \times T_G \to T_G$ defines on T_G the structure of a Lie group with identity element $0 \in T_e(G)$. Show that the canonical projection $T_G \to G$ is a homomorphism of Lie groups with kernel $T_e(G)$.

Problem 8.23. Let G be a (real or complex) Lie group. For $g \in G$ let $\ell_g : G \to G$, $h \mapsto gh$. Show that

$$\Phi : G \times T_e(G) \to T_G, \qquad (g, x) \mapsto T_e(\ell_g)(x) \tag{8.39}$$

is an isomorphism of vector bundles. Deduce that $\mathcal{T}_G$ and Ω_G^p, $p \geq 0$, are free $\mathcal{O}_G$-modules. Is Φ a homomorphism of Lie groups if we endow T_G with the Lie group structure defined in Problem 8.22?

Problem 8.24. Notation as in Problem 8.23. Let $\mathcal{T}_G(G)$ be the $\mathcal{O}_G(G)$-module of global vector fields on G.

1. Show that $G \times \mathcal{T}_G(G) \to \mathcal{T}_G(G)$, $(g, X) \mapsto T(\ell_g) \circ X \circ \ell_g^{-1}$, defines an action of the group G on $\mathcal{T}_G(G)$ by $\mathbb{K}$-linear automorphisms. Let $\mathrm{Lie}(G)$ be the $\mathbb{K}$-subspace of $\mathcal{T}_G(G)$ consisting of $X \in \mathcal{T}_G(G)$ such that $X = T(\ell_g) \circ X \circ \ell_g^{-1}$ for all $g \in G$.
2. Show that $\mathrm{Lie}(G)$ is $\mathbb{K}$-Lie subalgebra of the Lie algebra $\mathcal{T}_G(G)$ (Problem 8.19) and that this construction yields a functor Lie from the category of real (respectively complex) Lie groups to the category of Lie algebras over $\mathbb{R}$ (respectively over $\mathbb{C}$).

3. Show that the restriction of Φ^{-1} (Problem 8.23) to Lie(G) followed by the projection to $T_e(G)$ yields an isomorphism of $\mathbb{K}$-vector spaces Lie(G) $\tilde{\to}$ $T_e(G)$, which is functorial in G.
4. Let G be a Lie group and let G^0 be its identity component. Show that Lie(G^0) $\cong$ Lie(G).
5. Deduce from Problem 6.7 that a homomorphism of connected Lie groups $\varphi\colon G \to G'$ is a covering map if and only if Lie(φ): Lie(G) $\to$ Lie(G') is an isomorphism.

Remark: One can show that the functor Lie induces an equivalence between the category of simply connected Lie groups and the category of finite-dimensional $\mathbb{K}$-Lie algebras (e.g., [HiNe] 9.4.11 and 9.5.9) and deduce that two Lie groups have isomorphic Lie algebras if and only if their universal covers (Problem 6.8) are isomorphic.

Problem 8.25. Let $(X, \mathcal{O}_X)$ be a locally ringed space and let $\mathcal{E}$ be a finite locally free $\mathcal{O}_X$-module. Show that every local direct summand of $\mathcal{E}$ is finite locally free.
Hint: Use Problem 8.5 and Appendix Problem 14.31.

Problem 8.26. Let $D := (n_z)_{z \in \mathbb{C}} \in \mathbb{Z}^{\mathbb{C}}$ such that $\{ z \in \mathbb{C} \; ; \; n_z \neq 0 \}$ is discrete and closed in $\mathbb{C}$ (such a tuple is called a *divisor on* $\mathbb{C}$). For $U \subseteq \mathbb{C}$ open let $\mathcal{L}_D(U)$ be the set of meromorphic functions f on U such that $\operatorname{ord}_z(f) \geq n_z$ for all $z \in U$. Show that $\mathcal{L}_D$ is via restriction of meromorphic functions a sheaf on $\mathbb{C}$. Show that addition of meromorphic functions and multiplication with holomorphic functions makes $\mathcal{L}_D$ into a locally free $\mathcal{O}_{\mathbb{C}}$-module of rank 1.

Problem 8.27. Notation of Definition 8.49.

1. Suppose that F is a quotient bundle of E. Show that $w\colon \mathcal{E} \to \mathcal{F}$ has locally on M a section and deduce that Ker(w) is a local direct summand of $\mathcal{E}$ and hence corresponds to a subbundle of $\mathcal{E}$, denoted by Ker(ϖ) and called the *Kernel of* ϖ. Show that for the fibers one has Ker(ϖ)(p) = Ker($\varpi(p)$) for $p \in M$.
   Hint: Problem 8.25.
2. Let E be a subbundle of F. Show that $\mathcal{F}/\mathcal{E}$ is a finite locally free $\mathcal{O}_M$-module. The corresponding vector bundle is denoted by F/E and called the *quotient of F by the subbundle E*. Show that for the fibers one has $(F/E)(p) = F(p)/E(p)$ for $p \in M$.

Problem 8.28. Let E be a finite-dimensional $\mathbb{K}$-vector space and d an integer with $0 \leq d \leq \dim_{\mathbb{K}}(E)$. Let $M := \operatorname{Grass}_d(E)$. For a point $p \in M$ let $U_p \subseteq E$ be the corresponding $\mathbb{K}$-subspace of dimension d. Show that $U := \coprod_{p \in M} U_p$ is a subbundle of rank d of the trivial vector bundle $M \times E$ called the *tautological vector bundle on* $\operatorname{Grass}_d(E)$.

Problem 8.29. Notation as in Problem 8.28. Let $\mathcal{U}$ be the $\mathcal{O}_{\mathrm{Grass}_d(E)}$-module corresponding to the tautological vector bundle on $\mathrm{Grass}_d(E)$. It is a local direct summand of $\mathcal{E} := \mathcal{O}_{\mathrm{Grass}_d(E)} \otimes_{\mathbb{K}} E$. Show that $\mathcal{T}_{\mathrm{Grass}_d(E)} \cong \mathcal{H}om_{\mathcal{O}_{\mathrm{Grass}_d(E)}}(\mathcal{E}/\mathcal{U}, \mathcal{U})$. In particular $T_p(\mathrm{Grass}_d(E)) = \mathrm{Hom}_{\mathbb{K}}(E/U_p, U_p)$ for all $p \in \mathrm{Grass}_d(E)$.

Problem 8.30. Let M be a real C^α-premanifold with $\alpha \geq \infty$ or a complex premanifold. For $p \geq 0$ and $U \subseteq M$ open identify $\Omega_M^p(U)$ with the set of p-multilinear alternating morphisms $\mathcal{T}_U^p \to \mathcal{O}_U$.

1. Let $U \subseteq M$ be open, X a vector field over U, and $\omega \in \Omega_M^p(U)$, $p \geq 1$. Define $i_X(\omega) \in \Omega_M^{p-1}(U)$ by

$$i_X(\omega)(\xi_1, \dots, \xi_{p-1}) := \omega(X, \xi_1, \dots, \xi_{p-1})$$

    for $\xi_i \in \mathcal{T}_M(V)$, $V \subseteq U$ open. For $f \in \Omega_M^0(U) = \mathcal{O}_M(U)$ define $i_X(f) := 0$. Show that this defines a homomorphism $i_X \colon \Omega_M^p \to \Omega_M^{p-1}$ of $\mathcal{O}_M$-modules.
2. Show for $p, q \geq 0$ and for $\omega \in \Omega_M^p(U)$, $\eta \in \Omega_M^q(U)$ one has

$$i_X(\omega \wedge \eta) = i_X(\omega) \wedge \eta + (-1)^p \omega \wedge i_X(\eta).$$

3. For $p \geq 0$ define the *Lie derivative*

$$\mathcal{L}_X \colon \Omega_M^p \to \Omega_M^p, \qquad \mathcal{L}_X := d \circ i_X + i_X \circ d.$$

    Show that for $p = 0$ this is the map defined in Proposition 8.57.
4. Show that for differential forms ω and η one has

$$d \circ \mathcal{L}_X = \mathcal{L}_X \circ d,$$
$$\mathcal{L}_X(\omega \wedge \eta) = \mathcal{L}_X(\omega) \wedge \eta + \omega \wedge \mathcal{L}_X(\eta).$$

Soft Sheaves

9

In this chapter we study soft sheaves. The class of soft sheaves is on one hand large enough to include all kinds of interesting sheaves on real C^α-manifolds if $\alpha \leq \infty$, for instance the structure sheaf. On the other hand the cohomology of soft sheaves vanishes, and this yields immediately several local-global principles. The chapter starts with the definition and general properties of soft sheaves. In particular we will see that every global section has an arbitrary fine partition (Proposition 9.9). In Sect. 9.2 we will show that the structure sheaf of a real C^α-manifold for $\alpha \leq \infty$ is a soft sheaf. In the last section we show that the first cohomology vanishes for any soft sheaf of groups on a paracompact Hausdorff space. We obtain several immediate corollaries for real C^α-manifolds with $\alpha \leq \infty$: The fact that sections of vector bundles over closed sets can always be extended to a global section (and in particular a smooth version of Tietze's extension theorem), a description of the Picard group, or the fact that every subbundle of a vector bundle is split.

9.1 Definition and Examples of Soft Sheaves

Let X be a topological space, let $Z \subseteq X$ be a subspace, and let $i \colon Z \to X$ be the inclusion. Let $\mathcal{F}$ be a sheaf on X. Recall that we defined $\mathcal{F}(Z)$ as $(i^{-1}\mathcal{F})(Z)$, in other words

$$
\mathcal{F}(Z) = \Big\{ (s_i, U_i)_{i \in I} \, ; \, U_i \subseteq X \text{ open with } Z \subseteq \bigcup_i U_i,
$$
$$
s_i \in \mathcal{F}(U_i) \text{ with } (s_i)_z = (s_{i'})_z \tag{9.1}
$$
$$
\text{for all } i, i' \text{ and all } z \in Z \cap U_i \cap U_{i'} \Big\} / \sim
$$

with $(U_i, s_i)_{i \in I} \sim (V_j, t_j)_{j \in J}$ if $(s_i)_z = (t_j)_z$ for all $i \in I$, $j \in J$ and for all $z \in U_i \cap V_j \cap Z$.

© Springer Fachmedien Wiesbaden 2016   193
T. Wedhorn, *Manifolds, Sheaves, and Cohomology*, Springer Studium Mathematik – Master,
DOI 10.1007/978-3-658-10633-1_9

Our first goal is to show that under certain hypotheses, $\mathcal{F}(Z)$ has a simpler description. For every open neighborhood U of Z we have a restriction map

$$\mathcal{F}(U) \to \mathcal{F}(Z), \qquad s \mapsto s_{|Z} := [(U, s)]$$

and hence obtain a map

$$r_Z := r_Z^X := \operatorname*{colim}_{U} r_Z^U : \operatorname*{colim}_{Z \subseteq U} \mathcal{F}(U) \to \mathcal{F}(Z). \tag{9.2}$$

Here the left-hand side has the simple description

$$\operatorname*{colim}_{Z \subseteq U} \mathcal{F}(U) = \{ (U, s) \; ; \; U \subseteq X \text{ open with } Z \subseteq U, s \in \mathcal{F}(U) \} / \sim \tag{9.3}$$

with $(U, s) \sim (V, t)$ if there exists $Z \subseteq W \subseteq U \cap V$ open such that $s_{|W} = t_{|W}$.

The map r_Z is always injective. Let (U, s) and (V, t) represent elements in $\mathcal{F}'(Z)$ with $s_z = t_z$ for all $z \in Z$. Then there exists $Z \subseteq W \subseteq X$ open with $s_{|W} = t_{|W}$. We have the following criteria for r_Z to be bijective.

Proposition 9.1. *The map r_Z (9.2) is an isomorphism for every sheaf $\mathcal{F}$ on X if one of the following hypotheses is satisfied:*

1. *The space X is Hausdorff paracompact and $Z \subseteq X$ is closed.*
2. *The subspace Z has a fundamental system of Hausdorff paracompact open neighborhoods.*
3. *The subspace Z is compact and a relatively Hausdorff subspace of X (Definition 1.22).*

Hypothesis 2 is in particular satisfied for every subspace Z if X is Hausdorff and hereditarily paracompact, for instance if X is a paracompact Hausdorff premanifold or if X is metrizable (Proposition 1.13).

Proof. Let $s \in \mathcal{F}(Z)$ be represented by $(U_i, s_i)_{i \in I}$.

(i). Let us assume that Hypothesis 1 or Hypothesis 2 is satisfied. In this case we first show that we may assume that $(U_i)_i$ is a locally finite open covering of X and that X is paracompact Hausdorff. If 1 is satisfied, we first add $U_0 := X \setminus Z$ to the covering (with $s_0 \in \mathcal{F}(U_0)$ arbitrary) to obtain a covering of X and then replace it by a locally finite refinement. If 2 is satisfied, we replace X by an open Hausdorff paracompact neighborhood X' of Z contained in $\bigcup_i U_i$ and replace $(U_i \cap X')_i$ by a locally finite refinement. In both cases there exists an open covering $(V_i)_i$ of X with $\overline{V_i} \subseteq U_i$ by Corollary 1.21. Clearly $(V_i)_i$ is again locally finite.

(ii). We now conclude the proof if Hypothesis 1 or Hypothesis 2 is satisfied. Note that for all $i, j \in I$ the set $W_{ij} := \{x \in U_i \cap U_j \; ; \; (s_i)_x = (s_j)_x\}$ is open in X. Define

$$W := \{x \in X \; ; \; \forall\, i, j \in I : x \in \overline{V}_i \cap \overline{V}_j \Rightarrow (s_i)_x = (s_j)_x\}$$
$$= \bigcap_{i,j \in I} (\overline{V}_i^{\,c} \cup \overline{V}_j^{\,c} \cup W_{ij}),$$

where $(\ )^c$ denotes the complement in X. Then $Z \subseteq W$. It suffices to show that W is open in X. Indeed, then $s_{i\,|\,V_i \cap V_j \cap W} = s_{j\,|\,V_i \cap V_j \cap W}$ for all $i, j \in I$ implies that there exists $\tilde{s} \in \mathcal{F}(W)$ such that $\tilde{s}_{|\,V_i \cap W} = s_{i\,|\,V_i \cap W}$ for all i; hence $(W, \tilde{s})$ represents a preimage of s.

To see that W is open we choose for all $y \in X$ an open neighborhood W_y of y such that $I(y) := \{i \in I \; ; \; W_y \cap V_i \neq \emptyset\}$ is finite. For $y \in W$ one has

$$W \cap W_y = \{x \in W_y \; ; \; \forall\, i, j \in I(y) : x \in \overline{V}_i \cap \overline{V}_j \Rightarrow (s_i)_x = (s_j)_x\},$$

which is open in X. Hence every point y of W contains an open neighborhood contained in W, namely $W \cap W_y$. This shows that W is open in X.

(iii). Now suppose that Hypothesis 3 is satisfied. By Appendix Proposition 12.54, every $z \in Z$ has a compact neighborhood K_z of z in Z contained in U_{i_z} for some $i_z \in I$. Since Z is compact, finitely many $K_{z_1}, \ldots, K_{z_n}$ cover K. Set $K_r := K_{z_r}$, $U_r := U_{i_{z_r}}$ and $s_r := s_{i_{z_r}}$ for $r = 1, \ldots, n$.

For $n = 1$ we are done. Now $s_{1\,|\,U_1 \cap U_2}$ and $s_{2\,|\,U_1 \cap U_2}$ are two sections whose restriction to $K_1 \cap K_2$ is equal to $s_{|\,K_1 \cap K_2}$. Hence the injectivity of the map $r_{K_1 \cap K_2}^{U_1 \cap U_2}$ (9.2) yields an open neighborhood V of $K_1 \cap K_2$ in $U_1 \cap U_2$ with $s_{1\,|\,V} = s_{2\,|\,V}$. As Z is relatively Hausdorff in X and $K_i \setminus V$ is compact we find by Proposition 1.23 disjoint open neighborhoods $U_i' \subseteq U_i$ of $K_i \setminus V$ for $i = 1, 2$. The three sections $s_{1\,|\,U_1'}$, $s_{2\,|\,U_2'}$, and $s_{1\,|\,V} = s_{2\,|\,V}$ glue to a section on $U_1' \cup U_2' \cup V$, which extends $s_{|\,K_1 \cup K_2}$. We conclude by induction. $\qquad\square$

Definition 9.2. Let X be a paracompact Hausdorff space. A sheaf $\mathcal{F}$ on X is called *soft* if for every closed subspace A of X the restriction map $\mathcal{F}(X) \to \mathcal{F}(A)$ is surjective.

Even if one of the hypotheses of Proposition 9.1 is satisfied, the notion of sections over a subspace is somewhat subtle. Consider the following example.

Example 9.3. Let X be a paracompact Hausdorff space and let $\mathcal{C}_X$ be the sheaf of continuous $\mathbb{R}$-valued functions on X. We claim that $\mathcal{C}_X$ is soft.

Let $A \subseteq X$ be closed and let $s \in \mathcal{C}_X(A)$. We want to extend s to a section over X. As paracompact Hausdorff spaces are normal (Proposition 1.18), it would be tempting simply to apply Tietze's extension theorem (Theorem 1.15 (iii)) to conclude that s can be extended from A to X. But there is the following technical point to observe: The restriction $\mathcal{C}_{X|A}$ is usually *not* the sheaf of continuous functions on A. For instance if $A = \{x\}$ for some $x \in X$, then $\mathcal{C}_{X|A}(A) = \mathcal{C}_{X,x}$ and $\mathcal{C}_{X|A}$ is not the sheaf of continuous functions

$\{x\} \to \mathbb{R}$. More generally $\mathcal{C}_X(A)$ consists by Proposition 9.1 of germs of continuous functions defined in some open neighborhood of A as in (9.3). Hence there is only a map $\rho \colon \mathcal{C}_X(A) \to \mathcal{C}_A(A)$, which sends a representative (U, s) (U an open neighborhood of A, s a continuous function on A) to the continuous function $s_{|A}$ on A. The composition $\mathcal{C}_X(X) \longrightarrow \mathcal{C}_X(A) \xrightarrow{\rho} \mathcal{C}_A(A)$ is surjective by Tietze's extension theorem but it is not immediate that $\mathcal{C}_X(X) \longrightarrow \mathcal{C}_X(A)$ is surjective.

Nevertheless we can use Theorem 1.15 as follows. Let U be an open neighborhood of A, $\tilde{s}' \in \mathcal{C}_X(U)$ extending s. Applying Theorem 1.15 (iii) twice we find a closed neighborhood B of A and a closed neighborhood C of B with $C \subseteq U$. By Tietze's extension theorem (Theorem 1.15 (iv)) we find a continuous function $f \colon X \to \mathbb{R}$ with $f(b) = 1$ for all $b \in B$ and $f(x) = 0$ for all $x \in X \setminus C^\circ$. Then the germ of f in every point of A is 1 and hence $\tilde{s} := f_{|U} \tilde{s}'$ also extends s. It can be extended to a continuous function on X by 0.

Example 9.4. Let X be a connected paracompact Hausdorff space consisting of more than one point. Then the constant sheaf $\mathbb{Z}_X$ is not soft. Take $x, y \in X$, $x \neq y$, $A := \{x, y\}$. As X is Hausdorff, there exist open neighborhoods $U \ni x$ and $V \ni y$ with $U \cap V = \emptyset$. The locally constant function f with value 0 on U and value 1 on V defines a section of $\mathbb{Z}_X(A)$ that cannot be extended to X because every locally constant function on X is constant.

Remark 9.5. Let X be a paracompact Hausdorff space, let $A \subseteq X$ be a closed subspace, and let $\mathcal{F}$ be a soft sheaf on X. Then $\mathcal{F}_{|A}$ is a soft sheaf on A: If A is closed and $B \subseteq A$ is closed, then B is closed in X. Therefore the composition of restrictions $\mathcal{F}(X) \to \mathcal{F}(A) \to \mathcal{F}(B)$ is surjective and hence $\mathcal{F}(A) \to \mathcal{F}(B)$ is surjective.

Proposition 9.6. *Let X be a paracompact Hausdorff space and $\mathcal{F}$ be a sheaf on X. Suppose that there exists a family $(Z_j)_{j \in J}$ of subspaces such that the interior Z_j° cover X and such that for all j and every closed subspace A of X with $A \subseteq Z_j$ the restriction $\mathcal{F}(Z_j) \to \mathcal{F}(A)$ is surjective. Then $\mathcal{F}$ is soft.*

Proof. We first remark that one can replace $(Z_j)_j$ by a refinement. Let $(Y_i)_{i \in I}$ be a refinement of $(Z_j)_j$ such that $(Y_i^\circ)_i$ is an open covering. For all $i \in I$ let $j \in J$ with $Y_i \subseteq Z_j$. If A is a closed subspace of X with $A \subseteq Y_i$, then the surjectivity of $\mathcal{F}(Z_j) \to \mathcal{F}(Y_i) \to \mathcal{F}(A)$ implies the surjectivity of $\mathcal{F}(Y_i) \to \mathcal{F}(A)$.

Hence we may first replace $(Z_j)_j$ by the open covering $(Z_j^\circ)_j$ and then by a locally finite open covering $(U_i)_i$ because X is paracompact. By the shrinking lemma (Corollary 1.21) we find an open covering $(V_i)_{i \in I}$ with $\overline{V_i} \subseteq U_i$ for all i. As a subset of $\overline{V_i}$ is closed in X if and only if it is closed in $\overline{V_i}$, the restriction $\mathcal{F}_{|\overline{V_i}}$ is soft for all i by Remark 9.5.

Let $A \subseteq X$ be a closed subspace, $s \in \mathcal{F}(A)$. We want to extend s to X. For $J \subseteq I$ let $C_J := A \cup \bigcup_{i \in J} \overline{V_i}$. This is a closed subset of X because $(\overline{V_i})_i$ is locally finite (Appendix

Corollary 12.33). We have $C_I = X$. Define

$$\mathcal{E} := \big\{ (J, t_J) \, ; \, J \subseteq I, t_J \in \mathcal{F}(C_J) \text{ with } t_{J|A} = s \big\}.$$

Then $(\emptyset, s) \in \mathcal{E}$. Define a partial order on $\mathcal{E}$ by $(J, t) \le (J', t')$ if $J \subseteq J'$ and $t'_{|C_J} = t$.
If $F \subseteq \mathcal{E}$ is a totally ordered subset, then F has an upper bound: Set

$$\tilde{J} := \bigcup_{(J, t_J) \in F} J$$

and let $\tilde{t} \in \mathcal{F}(C_{\tilde{J}})$ be the unique section such that $\tilde{t}_{|F_J} = t_J$ (exists by Proposition 3.59).
Therefore we may apply Zorn's lemma. Let (J, t_J) be a maximal element of $\mathcal{E}$. It suffices
to show that $J = I$.

Assume that there exists $i \in I \setminus J$. As $\mathcal{F}_{|\overline{V_i}}$ is soft, there exists $t_i \in \mathcal{F}(\overline{V_i})$ such that
$t_{i|\overline{V_i} \cap C_J} = t_{J|\overline{V_i} \cap C_J}$. Let $\tilde{t} \in \mathcal{F}(\overline{V_i} \cup C_J)$ be the element that restricts to t_i and to t_J. Then
$(J \cup \{i\}, \tilde{t}) \in \mathcal{E}$ with $(J, t_J) < (J \cup \{i\}, \tilde{t})$. This is a contradiction. $\square$

Corollary 9.7. *Let X be a paracompact Hausdorff space and $\mathcal{F}$ be a sheaf on X. Then
$\mathcal{F}$ is soft if and only if every point $x \in X$ has a closed neighborhood N such that $\mathcal{F}_{|N}$ is
soft.*

Proof. The condition is necessary by Remark 9.5 and sufficient by Proposition 9.6. $\square$

Proposition 9.8. *Let X be a paracompact Hausdorff space and let $\mathcal{O}_X$ be a sheaf of (not
necessarily commutative) rings on X. Suppose that $\mathcal{O}_X$ is soft. Then every left and every
right $\mathcal{O}_X$-module is soft.*

Proof. It suffices to show the assertion for left modules. Let $A \subseteq X$ be closed and
$s \in \mathcal{F}(A)$. By Proposition 9.1 there exists $U \subseteq X$ open with $A \subseteq U$ and $\tilde{s} \in \mathcal{F}(U)$
extending s. As X is normal (Proposition 1.18), we find by Theorem 1.15 (iii) a closed
neighborhood C of A contained in U. Then A and ∂C are disjoint closed subsets. As $\mathcal{O}_X$
is soft, $\mathcal{O}(X) \to \mathcal{O}_X(A \cup \partial C)$ is surjective and we find $u \in \mathcal{O}_X(X)$ with $u_x = 1$ for all
$x \in A$ and $u_x = 0$ for all $x \in \partial C$. Then $\tilde{s}_{|C} u_{|C}$ extends s and can be extended to X by
zero. $\square$

Proposition 9.9 (Soft sheaves and partitions of sections). *Let X be a paracompact
Hausdorff space, let $\mathcal{A}$ be a soft sheaf of abelian groups on X and let $s \in \mathcal{A}(X)$. Then for
every open covering $(U_i)_{i \in I}$ of X there exists a family $(s_i)_{i \in I}$ of $s_i \in \mathcal{A}(X)$ such that:*

(a) $\operatorname{supp}(s_i) \subseteq U_i$ *for all $i \in I$.*
(b) *For all $x \in X$ there exists $x \in U \subseteq X$ open such that $\big\{ i \in I \, ; \, s_{i|U} \ne 0 \big\}$ is finite.*
(c) *For all $x \in X$ one has $s_x = \sum_{i \in I}(s_i)_x$ (a finite sum by (b)).*

Then $(s_i)_i$ is called a *partition of s subordinate to* $(U_i)_i$ and we write $s = \sum_{i \in I} s_i$.

Proof. (i). Assume first that $(U_i)_i$ is locally finite. By the shrinking lemma there exist closed subsets S_i with $S_i \subseteq U_i$ and $X = \bigcup_i S_i$.

For $J \subseteq I$ set $S_J := \bigcup_{i \in J} S_i$. Then S_J is closed because $(S_i)_i$ is locally finite (Appendix Corollary 12.33). Let $\mathcal{E}$ be the set of pairs $(J, (s_i)_{i \in J})$ with $J \subseteq I$, $s_i \in \mathcal{A}(X)$ with $\mathrm{supp}(s_i) \subseteq U_i$ for all $i \in J$ and such that $(\sum_{i \in J} s_i)_{|S_J} = s_{|S_J}$. Then $(\emptyset, \emptyset) \in \mathcal{E}$, in particular $\mathcal{E} \neq \emptyset$.

The set $\mathcal{E}$ is partially ordered by $(J, (s_i)_{i \in J}) \leq (K, (t_i)_{i \in K})$ if $J \subseteq K$ and $s_i = t_i$ for all $i \in J$. Clearly, every totally ordered subset of $\mathcal{E}$ has an upper bound. Hence $\mathcal{E}$ has a maximal element $(J, (s_i)_{i \in J})$ by Zorn's lemma. It suffices to show that $J = I$.

Assume there exists $\alpha \in I \setminus J$. There exists a unique section $t_\alpha \in \mathcal{A}((X \setminus U_\alpha) \cup (S_J \cup S_\alpha))$ such that

$$ t_\alpha = 0 \quad \text{on } X \setminus U_\alpha \quad \text{and} \quad t_\alpha = s - \sum_{i \in J} s_i \quad \text{on } S_J \cup S_\alpha $$

because $s - \sum_{i \in J} s_i = 0$ on $(X \setminus U_\alpha) \cap (S_J \cup S_\alpha) \subseteq S_J$. As $\mathcal{A}$ is soft, there exists an extension $s_\alpha \in \mathcal{A}(X)$ of t_α. Hence $(J \cup \{\alpha\}, (s_i)_{i \in J \cup \{\alpha\}}) \in \mathcal{E}$ is strictly larger than $(J, (s_i)_{i \in J})$. This is a contradiction.

(ii). In general, let $(V_j)_{j \in J}$ be a locally finite refinement of $(U_i)_i$ such that $V_j \subseteq U_{\rho(j)}$ for a map $\rho: J \to I$. Let $(t_j)_j$ be a partition of s subordinate to (V_j). Define $s_i := \sum_{j \in \rho^{-1}(i)} t_j$, which makes sense because $(V_j)_{j \in J}$ is locally finite. Then $(s_i)_{i \in I}$ is a partition of s subordinate to $(U_i)_i$. $\square$

In fact, the existence of such partitions characterizes soft sheaves (Problem 9.5).

9.2   Softness of Sheaves of Differentiable Functions

The next goal is to show that the structure sheaf of a real C^α-manifold with $\alpha \leq \infty$ is soft. This is essentially equivalent to the fact that there exist arbitrary fine partitions of 1 (see Corollary 9.13 and Problem 9.5). Hence it is not surprising that we start with a lemma that is also a standard step in the construction of smooth partitions of unity.

Lemma 9.10. *Let $m \in \mathbb{N}_0$, $U \subseteq \mathbb{R}^m$ open and $K \subseteq U$ compact. Then there exists a C^∞-function $\varphi: \mathbb{R}^m \to \mathbb{R}$ with $0 \leq \varphi \leq 1$ and $\varphi_{|K} = 1$, $\mathrm{supp}(\varphi) \subseteq U$.*

Proof. For $p \in \mathbb{R}^m$, $r > 0$ set $C_r(p) := \{ x \in \mathbb{R}^m \; ; \; \|x - p\|_\infty < r \}$. Then $(C_r(rp))_{p \in \mathbb{Z}^m}$ is a locally finite open cover of $\mathbb{R}^m$.

Let

$$ \theta: \mathbb{R} \to \mathbb{R}, \qquad x \mapsto \begin{cases} \exp\left(-\frac{1}{1-x^2}\right), & x \in (-1, 1); \\ 0, & x \in \mathbb{R} \setminus (-1, 1). \end{cases} $$

Then θ is a C^∞-function with $\mathrm{supp}(\theta) = [-1, 1]$. Moreover, define for $\varepsilon > 0$ and $p \in \mathbb{R}^m$

$$\theta_p^\varepsilon \colon \mathbb{R}^m \to \mathbb{R}, \qquad x \mapsto \prod_{i=1}^m \theta\left(\frac{x_i}{\varepsilon} - p_i\right).$$

This is also a C^∞-function and we have $\mathrm{supp}(\theta_p^\varepsilon) = \overline{C_\varepsilon(\varepsilon p)}$ and $\theta_p^\varepsilon{}_{|C_\varepsilon(\varepsilon p)} > 0$. Hence, $\theta^\varepsilon \colon \mathbb{R}^m \to \mathbb{R}, x \mapsto \sum_{p \in \mathbb{Z}^m} \theta_p^\varepsilon(x)$ (a finite sum for every $x \in \mathbb{R}^m$) is a C^∞-function with $\theta^\varepsilon > 0$. Hence $\varphi_p^\varepsilon := \theta_p^\varepsilon/\theta^\varepsilon$ is a well-defined C^∞-function and $\sum_{p \in \mathbb{Z}^m} \varphi_p^\varepsilon(x) = 1$ for every $x \in \mathbb{R}^m$ and $\mathrm{supp}(\varphi_p^\varepsilon) = \mathrm{supp}(\theta_p^\varepsilon)$.

As $K \subseteq U$ is compact, there exists $\varepsilon > 0$ such that $C_{2\varepsilon}(p) \subseteq U$ for every $p \in K$. Set $P_\varepsilon := \left\{ p \in \mathbb{Z}^m \; ; \; \mathrm{supp}(\varphi_p^\varepsilon) \cap K \neq \emptyset \right\}$ and define $\varphi \colon \mathbb{R}^m \to \mathbb{R}, x \mapsto \sum_{p \in P_\varepsilon} \varphi_p^\varepsilon(x)$. Since $\sum_{p \in \mathbb{Z}^m} \varphi_p^\varepsilon(x) = 1$ for every $x \in \mathbb{R}^m$, the definition of P_ε yields $\varphi(x) = \sum_{p \in P_\varepsilon} \varphi_p^\varepsilon(x) = 1$ for every $x \in K$. Also, we have $\mathrm{supp}(\varphi) \subseteq \bigcup_{p \in P_\varepsilon} \mathrm{supp}(\varphi_p^\varepsilon) \subseteq U$ because $\mathrm{supp}(\varphi_p^\varepsilon) = \overline{C_\varepsilon(p)} \subseteq C_{2\varepsilon}(p) \subseteq U$ for every $p \in K$. $\square$

Theorem 9.11. *Let M be a real paracompact Hausdorff C^α-premanifold with $\alpha \in \mathbb{N}_0 \cup \{\infty\}$. Then $\mathcal{C}_M^\alpha$ is a soft sheaf.*

Proof. Every point $p \in M$ has a chart (U, x) at p and a compact neighborhood $p \in C \subseteq U$. By Corollary 9.7 it suffices to show that $x_*(\mathcal{C}_{M|C}^\alpha)$ is a soft sheaf on the compact subset $x(C)$ of $\mathbb{R}^m$, where $m = \dim_p(M)$. Then every closed subset of $x(C)$ is compact. Hence it suffices to show that every C^α-function f defined on some open neighborhood U of a compact subset $K \subseteq \mathbb{R}^m$ can be extended to a C^α-function on $\mathbb{R}^m$ after possibly shrinking U. As K is compact, there exists $\varepsilon > 0$ such that $K' := \{ x \in \mathbb{R}^m \; ; \; d(x, K) \leq \varepsilon \} \subseteq U$ (d some metric on $\mathbb{R}^m$ defining the topology of $\mathbb{R}^m$).

Due to Lemma 9.10 there exists a C^α-function $g \colon \mathbb{R}^m \to \mathbb{R}$ such that $g_{|K'} = 1$ and such that $\mathrm{supp}(g) \subseteq U$ (we may even assume that $g \geq 0$). Let

$$h \colon \mathbb{R}^m \to \mathbb{R}, \qquad x \mapsto \begin{cases} (fg)(x), & x \in U; \\ 0, & x \in \mathbb{R}^n \setminus U. \end{cases}$$

Then h is the desired extension because $h = f$ on the open neighborhood $\{ x \in \mathbb{R}^m \; ; \; d(x, K) < \varepsilon \}$ of K. $\square$

Corollary 9.12. *Let M be a real paracompact Hausdorff C^∞-premanifold. Then the tangent sheaf $\mathcal{T}_M$ and the sheaves of differential forms Ω_M^r, $r \geq 0$, are soft sheaves.*

Proof. Theorem 9.11 and Proposition 9.8. $\square$

Finally, as we have for every soft sheaves and every global section arbitrary fine partitions, we obtain in particular the existence of partitions of unity.

Corollary 9.13 (Partition of unity). *Let M be a paracompact Hausdorff C^α-premanifold. For every open covering $(U_i)_i$ there exist $f_i \in \mathcal{C}^\alpha_M(M)$ such that:*

(a) $\operatorname{supp}(f_i) \subseteq U_i$ *and* $(\operatorname{supp} f_i)_{i \in I}$ *is locally finite.*

(b) $\sum_{i \in I} f_i = 1$.

Proof. Theorem 9.11 and Proposition 9.9. $\square$

9.3   Triviality of H^1 for Soft Sheaves and Applications

In the previous section we have seen that many interesting sheaves are soft. Now we show that the first cohomology vanishes for soft sheaves. This will allow us to prove several global results that a priori hold only locally.

Theorem 9.14. *Let X be a paracompact Hausdorff space and let $\mathcal{G}$ be a soft sheaf of groups on X. Then $H^1(X, \mathcal{G}) = 0$.*

Once we have defined higher cohomology groups in Chap. 10, we will show that $H^p(X, \mathcal{A}) = 0$ for all $p \geq 1$ and for every soft sheaf $\mathcal{A}$ of *abelian* groups (Proposition 10.17).

Proof. Let T be a $\mathcal{G}$-torsor. We have to show that $T(X) \neq \emptyset$ because then T is trivial by Proposition 7.4. Let $(U_i)_i$ be an open covering of X such that there exist $t_i \in T(U_i)$ for all i. As X is paracompact, we may assume that $(U_i)_i$ is locally finite. By the shrinking lemma we find $S_i \subseteq U_i$ such that S_i is closed in X and such that $\bigcup_i S_i = X$.

For $J \subseteq I$ set $S_J := \bigcup_{i \in J} S_i$ (note that S_J is closed in X because $(S_i)_i$ is locally finite) and define

$$\mathcal{E} := \{ (t, J) \; ; \; J \subseteq I, t \in T(S_J) \}.$$

Then $(*, \emptyset) \in \mathcal{E}$ and hence $\mathcal{E} \neq \emptyset$. It is an ordered set with partial order $(t, J) \leq (t', J')$ if $J \subseteq J'$ and $t = t'_{|S_J}$. As sections over the members of a locally finite closed covering can be glued (Proposition 3.59), every totally ordered subset of $\mathcal{E}$ has an upper bound. Hence there exists a maximal element (t, J) in $\mathcal{E}$ by Zorn's lemma. We claim that $J = I$ (then $T(X) \neq \emptyset$).

Assume that there exists $j \in I \setminus J$. Then there exists a unique $g \in \mathcal{G}(S_J \cap S_i)$ such that $t_{|S_J \cap S_i} = g t_{i|S_J \cap S_i}$. As $\mathcal{G}$ is soft, there exists $\tilde{g} \in \mathcal{G}(X)$, which extends g. Replacing t_i by $\tilde{g}_{|S_i} t_i$ we may assume that $t_{|S_J \cap S_i} = t_{i|S_J \cap S_i}$ and hence we may extend t to $S_i \cup S_J$; a contradiction to the maximality of (t, J). $\square$

We now apply Theorem 9.14 to real C^α-manifolds for $\alpha \leq \infty$ using the fact that their structure sheaf is soft. Note that all the results below apply slightly more generally to real paracompact Hausdorff C^α-premanifolds with $\alpha \leq \infty$.

Corollary 9.15. *Let M be a real C^α-manifold with $\alpha \leq \infty$ and let $\mathcal{E}$ be a $\mathcal{C}_M^\alpha$-module. Then $H^1(M, \mathcal{E}) = 0$.*

Proof. The sheaf $\mathcal{E}$ is soft by Theorem 9.11 and Proposition 9.8. Hence $H^1(M, \mathcal{E}) = 0$ by Theorem 9.14. $\qquad\square$

Corollary 9.16 (Smooth version of Tietze's extension theorem). *Let M be a real C^α-manifold with $\alpha \leq \infty$, $S \subseteq M$ a closed subspace. Then for every $f \in \mathcal{O}_S(S)$ (i.e., for every map $f: S \to \mathbb{R}$ such that for every point $s \in S$ there exist an open neighborhood U and $g \in \mathcal{C}_M^\alpha(U)$ with $g_{|U \cap S} = f_{|U \cap S}$) there exists $\tilde{f} \in \mathcal{C}_M^\alpha(M)$ such that $f = \tilde{f}_{|S}$.*

In particular, any C^α-function $S \to \mathbb{R}$ on a closed submanifold S of M can be extended to M.

Proof. Let $\mathcal{J}_S$ be the sheaf of C^α-functions vanishing on S (Remark 5.31) and consider the exact sequence (5.9) of $\mathcal{O}_M$-modules. The cohomology sequence (Proposition 7.22) yields an exact sequence $\mathcal{C}_M^\alpha(M) \to \mathcal{O}_S(S) \to H^1(M, \mathcal{J}_S)$. As $\mathcal{J}_S$ is an ideal in $\mathcal{C}_M^\alpha$, $H^1(M, \mathcal{J}_S) = 0$ by Corollary 9.15. Therefore $\mathcal{C}_M^\alpha(M) \to \mathcal{O}_S(S)$ is surjective. $\qquad\square$

More generally, a similar argument shows that we can always extend sections of vector bundles from closed subspaces to the whole manifold:

Corollary 9.17. *Let M be a real paracompact Hausdorff C^α-premanifold with $\alpha \in \mathbb{N}_0 \cup \{\infty\}$ and let (E, p) be any vector bundle over M. Then any section s of E over a closed subspace S of M can be extended to a section of E over M.*

Here by a section of E over S we mean a map $s: S \to E$ with $p \circ s = \mathrm{id}_S$ and such for all $p \in S$ there exists an open neighborhood U of p in M and a C^α-section t of E over U such that $t_{|U \cap S} = s_{|U \cap S}$.

Proof. Again let $\mathcal{J}_S$ be the sheaf of C^α-functions vanishing on S and let $\mathcal{E}$ be the sheaf of sections of E that is a locally free $\mathcal{C}_M^\alpha$-module. Then the global section s of the sheaf $\mathcal{E}/\mathcal{J}_S\mathcal{E}$ can be lifted to a global section of $\mathcal{E}$ because $H^1(M, \mathcal{J}_S\mathcal{E}) = 0$ by Corollary 9.15. $\qquad\square$

Corollary 9.18. *Let M be a real C^α-manifold with $\alpha \leq \infty$. Then composition with sign $\mathcal{O}_M^\times \to \{\pm 1\}_M \cong (\mathbb{Z}/2\mathbb{Z})_M$ induces on cohomology an isomorphism*

$$w: \mathrm{Pic}(M) \xrightarrow{\sim} H^1(M, (\mathbb{Z}/2\mathbb{Z})_M).$$

In particular $\mathrm{Pic}(M) = 1$ if M is simply connected.

For a line bundle L over M the cohomology class $w(L) \in H^1(M, (\mathbb{Z}/2\mathbb{Z})_M)$ is called the *(first) Stiefel-Whitney class* of L.

Proof. The first assertion follows from (8.16) and $H^1(M, \mathcal{C}_M^\alpha) = 0$. The second assertion follows from Proposition 7.5. $\square$

Example 9.19. Let $M := S^1$ considered as a real C^α-manifold with $\alpha \leq \infty$. Then $H^1(S^1, (\mathbb{Z}/2\mathbb{Z})_{S^1}) \cong \mathbb{Z}/2\mathbb{Z}$ (Example 7.10). Hence $\mathrm{Pic}(S^1) \cong \mathbb{Z}/2\mathbb{Z}$ has exactly two elements. As in Example 8.15 one sees that the two line bundles are the trivial line bundle and the *Möbius bundle* (which is the line bundle constructed as the Möbius band (Example 4.30) but with typical fiber $\mathbb{R}$ instead of $(-1, 1)$).

Corollary 9.20. *Let M be a real C^α-manifold with $\alpha \leq \infty$ and let E be a vector bundle over M. Then every subbundle of E is split.*

Proof. Remark 8.51 and Corollary 9.15. $\square$

Corollary 9.21. *Let M be a real C^α-manifold with $\alpha \leq \infty$, let E and F be vector bundles over M, and let $\varpi: E \to F$ be a surjective morphism of vector bundles. Then for all $U \subseteq M$ open the map on sections $E(U) \to F(U)$ is surjective.*

Proof. Let $w: \mathcal{E} \to \mathcal{F}$ be the surjective homomorphism of finite locally free $\mathcal{O}_M$-modules corresponding to ϖ (Proposition 8.49). Then $E(U) \to F(U)$ is identified with $w_U: \mathcal{E}(U) \to \mathcal{F}(U)$. The exact sequence of $\mathcal{C}_U^\alpha$-modules $0 \to \mathrm{Ker}(w)_{|U} \to \mathcal{E}_{|U} \to \mathcal{F}_{|U} \to 0$ yields a cohomology sequence $\mathcal{E}(U) \to \mathcal{F}(U) \to H^1(U, \mathrm{Ker}(w))$ and $H^1(U, \mathrm{Ker}(w)) = 0$ by Corollary 9.15. $\square$

9.4 Problems

Problem 9.1. Let X be a hereditarily paracompact Hausdorff space and let $A \subseteq X$ be locally closed. Show that for every soft sheaf $\mathcal{F}$ on X its restriction $\mathcal{F}_{|A}$ is a soft sheaf on A.

Problem 9.2. Let M be a real analytic or a complex premanifold of dimension > 0. Show that $\mathcal{O}_M$ is not soft.

Problem 9.3. Let X be a paracompact Hausdorff space and let $\mathcal{O}_X$ be a sheaf of (not necessarily commutative) rings on X. Show that $\mathcal{O}_X$ is soft if and only if every point of x has a neighborhood U such that for all disjoint closed subspaces A, B of U there exists a section $s \in \mathcal{O}_X(U)$ such that $s_{|A} = 1$ and $s_{|B} = 0$.

Problem 9.4. Let X be a paracompact Hausdorff space. Show that every colimit of soft abelian sheaves is again soft.

Problem 9.5. Let X be a paracompact Hausdorff space and let $\mathcal{A}$ be a sheaf of abelian groups on X. Show that the following assertions are equivalent:

(i) $\mathcal{A}$ is soft.
(ii) For every closed subspace A of X, every $s \in \mathcal{A}(A)$ and for every locally finite open covering $(U_i)_i$ of A in X there exist $s_i \in \mathcal{A}(X)$ for all i with $\mathrm{supp}(s_i) \subseteq U_i$ and $s_x = \sum_{i \in I} (s_i)_x$ for all $x \in A$.

Suppose that $\mathcal{A}$ is a sheaf of (not necessarily commutative) rings. Show that $\mathcal{A}$ is soft if and only if (ii) holds for $A = X$ and $s = 1$.

Problem 9.6. Let X be a paracompact Hausdorff space. A sheaf of abelian groups $\mathcal{A}$ on X is called *fine* if the sheaf $\mathcal{H}om_{\mathbb{Z}_X}(\mathcal{A}, \mathcal{A})$ is soft.

1. Show that a sheaf of abelian groups $\mathcal{A}$ on X is fine if and only if for every closed subset A of X and for every neighborhood U of A there exists an endomorphism of $\mathcal{A}$ that is 1 on A and 0 outside U.
2. Show that every fine sheaf is soft.
3. Let $\mathcal{O}_X$ be a soft sheaf of (not necessarily commutative) rings. Show that every $\mathcal{O}_X$-module is fine.
4. Let $\mathcal{A}$ be a fine sheaf of abelian groups on X. Show that for every sheaf $\mathcal{M}$ of abelian groups on X the tensor product (Problem 8.7) $\mathcal{A} \otimes_{\mathbb{Z}_X} \mathcal{M}$ is fine.

Problem 9.7. Let M be a real C^α-manifold with $\alpha \leq \infty$ and let $p \in M$. Show that forming the germ $\mathcal{O}_M(M) \to \mathcal{O}_{M,p}$ is surjective.

Problem 9.8. Let M be a real C^α-manifold with $\alpha \leq \infty$. Show that there exists a proper morphism $f \colon M \to \mathbb{R}$.

Problem 9.9. A commutative monoid M (written additively) is called *integral* if for all $m \in M$ the map $M \to M, n \mapsto m + n$ is injective. The integral monoids together with homomorphisms of monoids form a category and one has the notion of a sheaf of integral monoids.

1. Show that M is integral if and only if the canonical homomorphism $M \to M^{\mathrm{gp}}$ (Appendix Problem 13.22) is injective.
2. Generalize Proposition 9.9 to soft sheaves $\mathcal{A}$ of integral monoids.
3. Let M be a real paracompact Hausdorff C^α-premanifold with $\alpha \leq \infty$. Show that $U \mapsto \{ f \in \mathcal{C}_M^\alpha(U) \,;\, f(p) \geq 0 \ \forall \, p \in U \}$ is a soft sheaf of integral monoids. Deduce that in Corollary 9.13 one can assume that $f_i \geq 0$ for all i.

Problem 9.10. Let U be a tautological line bundle on $\mathbb{P}^1(\mathbb{R}) = \text{Grass}_1(\mathbb{R}^2)$ (Problem 8.28) and let $p: S^1 \to \mathbb{P}^1(\mathbb{R})$ be the canonical map. Show that the pullback $p^*(U)$ (Problem 8.17) is isomorphic to the Möbius bundle.

Problem 9.11. Let X be a paracompact Hausdorff space, let $\mathcal{F}$ be a presheaf of sets on X and $\tilde{\mathcal{F}}$ its sheafification. Suppose that for every locally finite open covering $(U_i)_i$ of X and for all $s_i \in \mathcal{F}(U_i)$ with $s_{i|U_i \cap U_j} = s_{j|U_i \cap U_j}$ for all i, j there exists $s \in \mathcal{F}(X)$ such that $s_{|U_i} = s_i$ for all i. Show that $\mathcal{F}(X) \to \tilde{\mathcal{F}}(X)$ is surjective.

Problem 9.12. Let Δ be a topological space. For every topological space X let $S_\Delta(X)$ be the set of continuous maps $\Delta \to X$. Let A be an abelian group and let $S^\Delta(X, A)$ be the abelian group of maps $S_\Delta(X) \to A$, called the group of *singular Δ-cochains of X with values in A*. If $U \subseteq V \subseteq X$ are open subspaces, one has an injection $S_\Delta(U) \hookrightarrow S_\Delta(V)$ and hence restriction maps $S^\Delta(V, A) \to S^\Delta(U, A)$ making $U \mapsto S^\Delta(U, A)$ into a presheaf of abelian groups. Let $U \mapsto \mathcal{S}^\Delta(U, A)$ be its sheafification.

1. Deduce from Problem 9.11 that $S^\Delta(X, A) \to \mathcal{S}^\Delta(X, A)$ is surjective.
2. Show that $\mathcal{S}^\Delta(X, A)$ is a fine sheaf (Problem 9.6).

Problem 9.13. Consider the smooth manifold $\mathbb{P}^n(\mathbb{R})$, $n \geq 1$. Show that $\text{Pic}(\mathbb{P}^n(\mathbb{R})) \cong \mathbb{Z}/2\mathbb{Z}$.

Hint: Problem 7.6.

Cohomology of Complexes of Sheaves

10

In this chapter we generalize for sheaves of abelian groups (or more generally, for $\mathcal{O}_X$-modules, where $(X, \mathcal{O}_X)$ is a ringed space) cohomology groups in degree 1 to higher degrees in such a way that we can extend the exact sequence of cohomology. The idea (motivated in Sect. 10.1) is to replace an $\mathcal{O}_X$-module by a quasi-isomorphic complex of such $\mathcal{O}_X$-modules whose cohomology we expect to be trivial (namely injective $\mathcal{O}_X$-modules), then to apply the functor of global sections to this complex to obtain a complex of $\Gamma(X, \mathcal{O}_X)$-modules, and finally take the cohomology of this complex. Of course, once one follows the strategy of first replacing an $\mathcal{O}_X$-module by a complex, it is only natural to start with an arbitrary complex of $\mathcal{O}_X$-modules[1]. This will yield the (hyper-)cohomology $H^p(X, \mathcal{F}^\bullet)$ of a complex $\mathcal{F}^\bullet$ defined in Sect. 10.2. Much more generally, one may view these constructions as a special case of the right derivation of a left exact functor between abelian categories, which is here applied to the functor of taking global sections of an $\mathcal{O}_X$-module.

In Sect. 10.3 we will study classes of acyclic $\mathcal{O}_X$-modules $\mathcal{F}$ (i.e., $H^p(X, \mathcal{F}) = 0$ for all $p \geq 1$). Injective $\mathcal{O}_X$-modules are acyclic, almost by definition. We will introduce flabby sheaves as a technical tool and we will also prove that soft sheaves are acyclic.

The cohomological techniques developed in the first three sections will allow us in Sect. 10.4 to define de Rham cohomology and to prove the de Rham isomorphism. We can also compute some cohomology groups of sheaves of holomorphic functions and get as a corollary the theorem of Mittag-Leffler from complex analysis in one variable.

After studying some functoriality properties of cohomology with respect to inverse and direct images in Sect. 10.5 we formulate and prove in the last section the proper base change theorem for bounded below complexes of abelian sheaves on arbitrary topological spaces.

---

[1] Even more natural would be to work systematically in a category of complexes, where quasi-isomorphisms are defined to be isomorphisms, i.e., in the derived category. But the required amount of homological algebra is beyond the scope of this book.

© Springer Fachmedien Wiesbaden 2016

T. Wedhorn, *Manifolds, Sheaves, and Cohomology*, Springer Studium Mathematik – Master, DOI 10.1007/978-3-658-10633-1_10

Notation: In this chapter $(X, \mathcal{O}_X)$ will denote a ringed space and we set $R := \mathcal{O}_X(X)$. If K is a commutative ring such that $(X, \mathcal{O}_X)$ is K-ringed, then R is a K-algebra.

Recall that the category of $\mathcal{O}_X$-modules and the category of R-modules are abelian (Appendix Definition 15.25).

10.1   Strategy for the Definition of Cohomology of Sheaves

Consider the additive functor (Appendix Definition 15.21)

$$\Gamma(X, \cdot) : (\mathcal{O}_X\text{-Mod}) \longrightarrow (R\text{-Mod}), \qquad \mathcal{F} \to \Gamma(X, \mathcal{F}) := \mathcal{F}(X).$$

It is left exact, i.e., it commutes with finite limits (Remark 8.37). Equivalently (by Appendix Remark 15.27), for every exact sequence

$$0 \to \mathcal{F}' \to \mathcal{F} \to \mathcal{F}''$$

of $\mathcal{O}_X$-modules the induced sequence

$$0 \to \Gamma(X, \mathcal{F}') \to \Gamma(X, \mathcal{F}) \to \Gamma(X, \mathcal{F}'')$$

is exact.

Remark and Definition 10.1 (Goal of the construction of cohomology). We would like to construct a so-called δ-*functor extending* $\Gamma(X, \cdot)$, i.e.,

(a) a family of additive functors $H^n(X, \cdot) : (\mathcal{O}_X\text{-Mod}) \longrightarrow (R\text{-Mod})$ for $n \geq 0$ such that $H^0(X, \cdot) = \Gamma(X, \cdot)$,

(b) for every exact sequence $0 \to \mathcal{F}' \to \mathcal{F} \to \mathcal{F}'' \to 0$ of $\mathcal{O}_X$-modules a family of homomorphisms of R-modules $\delta \colon H^n(X, \mathcal{F}'') \to H^{n+1}(X, \mathcal{F}')$, $n \geq 0$,

satisfying the following conditions:

1. For every exact sequence $0 \to \mathcal{F}' \to \mathcal{F} \to \mathcal{F}'' \to 0$ of $\mathcal{O}_X$-modules the sequence

$$\begin{aligned} 0 &\longrightarrow H^0(X, \mathcal{F}') \longrightarrow H^0(X, \mathcal{F}) \longrightarrow H^0(X, \mathcal{F}'') \\ &\xrightarrow{\delta} H^1(X, \mathcal{F}') \longrightarrow H^1(X, \mathcal{F}) \longrightarrow H^1(X, \mathcal{F}'') \\ &\xrightarrow{\delta} H^2(X, \mathcal{F}') \longrightarrow H^2(X, \mathcal{F}) \longrightarrow \ldots \end{aligned}$$

is exact.

2. For every morphism $(0 \to \mathcal{F}' \to \mathcal{F} \to \mathcal{F}'' \to 0) \longrightarrow (0 \to \mathcal{G}' \to \mathcal{G} \to \mathcal{G}'' \to 0)$ of exact sequences of $\mathcal{O}_X$-modules the diagram

$$
\begin{array}{ccc}
H^n(X, \mathcal{F}'') & \xrightarrow{\ \delta\ } & H^{n+1}(X, \mathcal{F}') \\
\downarrow & & \downarrow \\
H^n(X, \mathcal{G}'') & \xrightarrow{\ \delta\ } & H^{n+1}(X, \mathcal{G}')
\end{array}
$$

is commutative.

Moreover, to characterize the cohomology up to unique isomorphism, we would like this δ-functor $(H^n(X, \dots), \delta)_n$ to be *universal* in the following sense: If $(F^n, \delta)_{n \geq 0}$ is any δ-functor extending $\Gamma(X, \cdot)$ then there is a unique family of morphisms of functors $\varphi^n \colon H^n(X, \cdot) \to F^n$ compatible with δ such that $\varphi^0 = \mathrm{id}$.

Clearly, a universal δ-functor extending $\Gamma(X, \cdot)$ is unique up to unique isomorphism (once we have shown that it exists). It is then called the *right derived functor of $\Gamma(X, \cdot)$*.

Remark and Definition 10.2 (Strategy of the construction of cohomology). To get an idea of how to construct such a universal δ-functor extending $\Gamma(X, \cdot)$ we think backwards. Assume that we have already constructed the right derived functors $H^n(X, \cdot)$ of $\Gamma(X, \cdot)$.

An $\mathcal{O}_X$-module $\mathcal{I}$ is called Γ-*acyclic* if $H^n(X, \mathcal{I}) = 0$ for all $n > 0$.

Now let $\mathcal{F}$ be an arbitrary $\mathcal{O}_X$-module. Assume:

> There exists an exact sequence of $\mathcal{O}_X$-modules
> $$ 0 \to \mathcal{F} \to \mathcal{I}^0 \to \mathcal{I}^1 \to \dots \qquad (\heartsuit) $$
> such that $\mathcal{I}^n$ is Γ-acyclic for all $n \geq 0$.

In other words, $\mathcal{F}$ is quasi-isomorphic to the complex $\mathcal{I}^\bullet$ of Γ-acyclic $\mathcal{O}_X$-modules.

Claim: The cohomology modules can be calculated as follows:

$$ H^n(X, \mathcal{F}) = H^n\big(0 \to \Gamma(X, \mathcal{I}^0) \to \Gamma(X, \mathcal{I}^1) \to \Gamma(X, \mathcal{I}^2) \to \dots\big), \qquad (10.1) $$

where the right-hand side denotes the cohomology of a complex (Appendix Definition 15.2), i.e., the right-hand side is equal to

$$ R^n := \mathrm{Ker}(\Gamma(X, \mathcal{I}^n) \to \Gamma(X, \mathcal{I}^{n+1})) / \mathrm{Im}(\Gamma(X, \mathcal{I}^{n-1}) \to \Gamma(X, \mathcal{I}^n)). $$

Let us believe this claim for a moment. Then this suggests a way of *defining* right derived functors $H^n(X, \dots)$:

1. Find a class $\mathcal{I}$ of $\mathcal{O}_X$-modules that we expect to be always Γ-acyclic once we have defined $(H^n(X, \dots), \delta)_{n \geq 0}$.

2. Show that ($\heartsuit$) is possible for every $\mathcal{O}_X$-module $\mathcal{F}$ with $\mathcal{J}^n$ in $\mathcal{I}$.
3. Then define $H^n(X, \mathcal{F})$ via (10.1).

As exact sequences of the form $0 \to \mathcal{J} \to \mathcal{F} \to \mathcal{F}'' \to 0$ with $\mathcal{J}$ injective $\mathcal{O}_X$-module always split, it is plausible to take all injective $\mathcal{O}_X$-modules as this class. Then ($\heartsuit$) will be always possible by Lemma 10.3 below.

Let us show the claim. Denote the right-hand side of (10.1) by R^n. Let $n = 0$. Applying the left exact functor $\Gamma(X, \cdot)$ to the exact sequence $0 \to \mathcal{F} \to \mathcal{J}^0 \to \mathcal{J}^1$ we get the exact sequence

$$0 \to \Gamma(X, \mathcal{F}) \to \Gamma(X, \mathcal{J}^0) \to \Gamma(X, \mathcal{J}^1)$$

and hence

$$H^0(X, \mathcal{F}) = \Gamma(X, \mathcal{F}) = \mathrm{Ker}(\Gamma(X, \mathcal{J}^0) \to \Gamma(X, \mathcal{J}^1)) = R^0.$$

Assume now that $n \geq 1$. Set $\mathcal{J}^{-1} := \mathcal{K}^0 := \mathcal{F}$ and define for $p \geq 1$

$$\mathcal{K}^p := \mathrm{Coker}(\mathcal{J}^{p-2} \to \mathcal{J}^{p-1}) \cong \mathrm{Im}(\mathcal{J}^{p-1} \to \mathcal{J}^p) = \mathrm{Ker}(\mathcal{J}^p \to \mathcal{J}^{p+1}),$$

where the isomorphism is induced by $\mathcal{J}^{p-1} \to \mathcal{J}^p$. From the exact sequence $0 \to \mathcal{K}^p \to \mathcal{J}^p \to \mathcal{J}^{p+1}$ we get

$$H^0(X, \mathcal{K}^p) = \mathrm{Ker}(H^0(X, \mathcal{J}^p) \to H^0(X, \mathcal{J}^{p+1})). \qquad (*)$$

The short exact sequence

$$0 \to \mathcal{K}^p \to \mathcal{J}^p \to \mathcal{K}^{p+1} \to 0$$

yields a long exact sequence

$$0 \longrightarrow H^0(X, \mathcal{K}^p) \longrightarrow H^0(X, \mathcal{J}^p) \longrightarrow H^0(X, \mathcal{K}^{p+1})$$
$$\longrightarrow \cdots$$
$$\longrightarrow H^n(X, \mathcal{K}^p) \to \underbrace{H^n(X, \mathcal{J}^p)}_{=0} \to H^n(X, \mathcal{K}^{p+1})$$
$$\longrightarrow H^{n+1}(X, \mathcal{K}^p) \longrightarrow \underbrace{H^{n+1}(X, \mathcal{J}^p)}_{=0} \longrightarrow \cdots$$

and hence isomorphisms $H^n(X, \mathcal{K}^{p+1}) \xrightarrow{\sim} H^{n+1}(X, \mathcal{K}^p)$ for $n \geq 1$. As $\mathcal{F} = \mathcal{K}^0$ we see by induction that

$$H^n(X, \mathcal{F}) = H^n(X, \mathcal{K}^0) \cong \cdots \cong H^1(X, \mathcal{K}^{n-1})$$
$$\cong H^0(X, \mathcal{K}^n) / \mathrm{Im}(H^0(X, \mathcal{J}^{n-1}) \to H^0(X, \mathcal{K}^n)) \overset{(*)}{=} R^n.$$

10.2 Definition of Cohomology

The strategy outlined above suggests how to define cohomology of an $\mathcal{O}_X$-module $\mathcal{F}$ as follows:

1. Choose a quasi-isomorphism $\mathcal{F} \xrightarrow{qis} \mathcal{J}^\bullet$, where $\mathcal{J}^\bullet$ is a complex consisting of injective $\mathcal{O}_X$-module $\mathcal{J}^p$ for all p with $\mathcal{J}^p = 0$ for $p < 0$.
2. Apply the functor $\Gamma(X, \cdot)$ to $\mathcal{J}^\bullet$ and take the cohomology of complexes of $\cdots \rightarrow \Gamma(X, \mathcal{J}^p) \rightarrow \Gamma(X, \mathcal{J}^{p+1}) \rightarrow \ldots$.

But if we replace a single $\mathcal{O}_X$-module $\mathcal{F}$ by a complex, why not start with a whole complex of $\mathcal{O}_X$-modules $\mathcal{F}^\bullet$? Then it will turn out that the crucial property of $\mathcal{J}^\bullet$ is its K-injectivity (Appendix Definition 15.18). Hence the starting point of the construction of cohomology of arbitrary complexes of $\mathcal{O}_X$-modules is the following result, which ensures that step (1) is always possible.

Lemma 10.3. *The following abelian categories have injective and K-injective resolutions (Appendix Definition 15.28):*

1. *The abelian category of R-modules for a ring R.*
2. *The abelian category of $\mathcal{O}_X$-modules for a ringed space $(X, \mathcal{O}_X)$.*

We will not prove this lemma here but refer to [Spa] Theorem 4.5, [Stacks] Tag 079P, or [AJS] Theorem 5.4.

Lemma 10.3 in particular shows that if $\mathcal{F}^\bullet$ is bounded below, then we find a quasi-isomorphism $\mathcal{F}^\bullet \xrightarrow{qis} \mathcal{J}^\bullet$, where $\mathcal{J}^\bullet$ is bounded below and consists of injective $\mathcal{O}_X$-modules (in particular $\mathcal{J}^\bullet$ is K-injective by Appendix Proposition 15.19). This is the only case that we will apply in the rest of this book (and we refer to Appendix Problem 15.14 and Problem 10.1 for some hints on the existence of such injective resolutions). Hence in the sequel we usually give proofs only in this case or even only in the case where $\mathcal{F}^\bullet$ is concentrated in degree 0 and refer to the literature for the general arguments from homological algebra.

Now we define the sheaf cohomology of a complex of $\mathcal{O}_X$-modules as follows.

Definition and Theorem 10.4. *Let $(X, \mathcal{O}_X)$ be a ringed space and let $R := \Gamma(X, \mathcal{O}_X)$. Let $\mathcal{F}^\bullet$ be a complex of $\mathcal{O}_X$-modules. Choose a quasi-isomorphism $\mathcal{F}^\bullet \xrightarrow{qis} \mathcal{J}^\bullet$, where $\mathcal{J}^\bullet$ is a K-injective complex of $\mathcal{O}_X$-modules and define for $n \in \mathbb{Z}$ the n-th cohomology of $\mathcal{F}^\bullet$ by*

$$H^n(X, \mathcal{F}^\bullet) := H^n(\cdots \rightarrow \Gamma(X, \mathcal{J}^{p-1}) \rightarrow \Gamma(X, \mathcal{J}^p) \rightarrow \Gamma(X, \mathcal{J}^{p+1}) \rightarrow \ldots).$$

1. *This defines for all $n \in \mathbb{Z}$ additive functors*

$$H^n(X, \cdot) \colon K(\mathcal{O}_X) \rightarrow (R\text{-Mod})$$

such that for an $\mathcal{O}_X$-module $\mathcal{F}$ (considered as a complex concentrated in degree 0) one has a functorial isomorphism

$$H^0(X, \mathcal{F}) \xrightarrow{\sim} \Gamma(X, \mathcal{F}).$$

If the complex $\mathcal{F}^\bullet$ is bounded below, say there exists $a \in \mathbb{Z}$ such that $\mathcal{F}^p = 0$ for all $p < a$, then $H^n(X, \mathcal{F}^\bullet) = 0$ for all $n < a$.

2. *For every short exact sequence of complexes of $\mathcal{O}_X$-modules*

$$0 \to \mathcal{F}^\bullet \xrightarrow{u} \mathcal{G}^\bullet \xrightarrow{v} \mathcal{H}^\bullet \to 0 \tag{10.2}$$

(i.e., $0 \to \mathcal{F}^p \xrightarrow{u^p} \mathcal{G}^p \xrightarrow{v^p} \mathcal{H}^p \to 0$ is an exact sequence of $\mathcal{O}_X$-modules for all $p \in \mathbb{Z}$) there are connecting homomorphisms of R-modules $\delta \colon H^n(X, \mathcal{H}^\bullet) \to H^{n+1}(X, \mathcal{F}^\bullet)$ making the sequence

$$
\begin{aligned}
&\cdots \\
\xrightarrow{\delta}\ &H^n(X, \mathcal{F}^\bullet) \xrightarrow{H^n(X,u)} H^n(X, \mathcal{G}^\bullet) \xrightarrow{H^n(X,v)} H^n(X, \mathcal{H}^\bullet) \\
\xrightarrow{\delta}\ &H^{n+1}(X, \mathcal{F}^\bullet) \xrightarrow{H^{n+1}(X,u)} H^{n+1}(X, \mathcal{G}^\bullet) \xrightarrow{H^{n+1}(X,v)} H^{n+1}(X, \mathcal{H}^\bullet) \\
\xrightarrow{\delta}\ &\cdots
\end{aligned}
\tag{10.3}
$$

exact and this long exact cohomology sequence is functorial for morphisms of short exact sequences of complexes.

If $Z \subseteq X$ is any subspace, we simply write $H^n(Z, \mathcal{F}^\bullet)$ instead of $H^n(Z, \mathcal{F}^\bullet_{|Z})$.

One should not confuse cohomology as a complex $H^p(\mathcal{F}^\bullet)$ defined as the $\mathcal{O}_X$-module $\mathrm{Ker}(\mathcal{F}^p \to \mathcal{F}^{p+1})/\mathrm{Im}(\mathcal{F}^{p-1} \to \mathcal{F}^p)$ and the cohomology $H^n(X, \mathcal{F}^\bullet)$ just defined, which is an R-module. To distinguish these two notions, $H^n(X, \mathcal{F}^\bullet)$ is sometimes also called the *hypercohomology of the complex*, but we will not use this terminology.

For a sheaf of abelian groups $\mathcal{F}$, which is the same as a $\mathbb{Z}_X$-module, we now have for the moment two definitions of $H^1(X, \mathcal{F})$: the definition given here and the definition via torsors (or Čech cocycles) given in Chap. 7. We will see in Proposition 10.16 that these definitions coincide.

Before proving Theorem 10.4 we make the following remark.

Remark 10.5. Let $u \colon \mathcal{F}^\bullet \to \mathcal{G}^\bullet$ be a morphism of complexes of $\mathcal{O}_X$-modules. Let $\Gamma(X, \mathcal{F}^\bullet)$ be the complex of R-modules

$$\cdots \longrightarrow \Gamma(X, \mathcal{F}^p) \longrightarrow \Gamma(X, \mathcal{F}^{p+1}) \longrightarrow \cdots$$

and let $\Gamma(X, u) \colon \Gamma(X, \mathcal{F}^\bullet) \to \Gamma(X, \mathcal{G}^\bullet)$ be the morphism of complexes of R-modules given by $\Gamma(X, u)^p := \Gamma(X, u^p) \colon \Gamma(X, \mathcal{F}^p) \to \Gamma(X, \mathcal{G}^p)$. This defines a functor $(\mathrm{Com}(\mathcal{O}_X)) \to (\mathrm{Com}(R))$. Applying to a homotopy $u \simeq v$ of homomorphisms of complexes (Appendix Definition 15.4 and Remark 15.26) the functor $\Gamma(X, \cdot)$ yields a homotopy $\Gamma(X, u) \simeq \Gamma(X, v)$. Hence $\Gamma(X, \cdot)$ induces a functor of homotopy categories (Appendix Definition 15.6 and Remark 15.26)

$$\Gamma(X, \cdot) \colon K(\mathcal{O}_X) \to K(R).$$

In particular, $H^n(\Gamma(X, u)) = H^n(\Gamma(X, v))$ for homotopic morphisms u and v of complexes of $\mathcal{O}_X$-modules for all $n \in \mathbb{Z}$.

Proof (Proof of Theorem 10.4). (i). We first show that $H^n(X, \mathcal{F}^\bullet)$ depends up to unique isomorphism not on the choice of the K-injective resolution. Let $s \colon \mathcal{F}^\bullet \xrightarrow{qis} \mathcal{I}^\bullet$ and $t \colon \mathcal{F}^\bullet \xrightarrow{qis} \mathcal{J}^\bullet$ be two K-injective resolutions. Applying Appendix Lemma 15.20 to $u = \mathrm{id} \colon \mathcal{F}^\bullet \to \mathcal{F}^\bullet$ we obtain a unique isomorphism $w \colon \mathcal{I}^\bullet \xrightarrow{\sim} \mathcal{J}^\bullet$ in $K(\mathcal{O}_X)$ such that $w \circ s = t$ and hence a unique isomorphism

$$H^n(\Gamma(w)) \colon H^n(\Gamma(X, \mathcal{I}^\bullet)) \xrightarrow{\sim} H^n(\Gamma(X, \mathcal{J}^\bullet))$$

by Remark 10.5.

(ii). $H^n(X, \cdot)$ defines an additive functor $H^n(X, \cdot) \colon K(\mathcal{O}_X) \to (R\text{-Mod})$, again by Appendix Lemma 15.20 and Remark 10.5.

(iii). Assume that $\mathcal{F}^p = 0$ for all $p < a$. Then we may choose a K-injective resolution $\mathcal{F}^\bullet \xrightarrow{qis} \mathcal{I}^\bullet$, where $\mathcal{I}^p = 0$ for all $p < a$ (Lemma 10.3). Hence $H^n(X, \mathcal{F}^\bullet) = H^n(\Gamma(X, \mathcal{I}^\bullet)) = 0$ for all $n < a$.

(iv). Let $\mathcal{F}$ be an $\mathcal{O}_X$-module. Then a K-injective resolution $\mathcal{F} \xrightarrow{qis} \mathcal{I}^\bullet$ with $\mathcal{I}^p = 0$ for $p < 0$ is an exact sequence

$$\dots \longrightarrow 0 \longrightarrow \mathcal{F} \longrightarrow \mathcal{I}^0 \longrightarrow \mathcal{I}^1 \longrightarrow \dots.$$

As $\Gamma(X, \cdot)$ is left exact, we obtain

$$\Gamma(X, \mathcal{F}) = \mathrm{Ker}(\Gamma(X, \mathcal{I}^0) \to \Gamma(X, \mathcal{I}^1)) = H^0(\Gamma(X, \mathcal{I}^\bullet)).$$

(v). Proof of (2): If $\mathcal{F}^\bullet$, $\mathcal{G}^\bullet$, and $\mathcal{H}^\bullet$ are bounded below: We claim that there exists a commutative diagram in $(\mathrm{Com}(\mathcal{O}_X))$

$$
\begin{array}{ccccccccc}
0 & \longrightarrow & \mathcal{F}^\bullet & \longrightarrow & \mathcal{G}^\bullet & \longrightarrow & \mathcal{H}^\bullet & \longrightarrow & 0 \\
 & & \downarrow & & \downarrow & & \downarrow & & \\
0 & \longrightarrow & \mathcal{I}^\bullet & \longrightarrow & \mathcal{J}^\bullet & \longrightarrow & \mathcal{K}^\bullet & \longrightarrow & 0,
\end{array}
\qquad (*)
$$

such that:

1. The complexes in the lower row are bounded below and consist of injective $\mathcal{O}_X$-modules and the vertical arrows are quasi-isomorphisms.

2. The lower row is an exact sequence of complexes.

Let us briefly indicate the argument: By Lemma 10.3 we can find $\mathcal{F}^\bullet \xrightarrow{qis} \mathcal{I}^\bullet$ as desired. Forming the componentwise pushout in the abelian category $(\mathcal{O}_X\text{-Mod})$, we obtain a diagram as desired, but where $\mathcal{J}^\bullet$ and $\mathcal{K}^\bullet$ are arbitrary bounded below complexes. As $\mathcal{I}^\bullet$ consists componentwise of injective objects, the sequence $0 \to \mathcal{I}^\bullet \to \mathcal{J}^\bullet \to \mathcal{K}^\bullet \to 0$ is componentwise split. Hence we can assume that (10.2) is componentwise split and that the components of $\mathcal{A}^\bullet$ are injective (see also Appendix Problem 15.13). Then choose an injective resolution $\mathcal{H}^\bullet \xrightarrow{qis} \mathcal{K}^\bullet$ such that $\mathcal{H}^p \to \mathcal{K}^p$ is injective for all p. Set $\mathcal{J}^p := \mathcal{A}^p \oplus \mathcal{K}^p$. Then it is easy to see that there are homomorphisms $\delta^p : \mathcal{K}^p \to \mathcal{A}^{p+1}$ such that

$$d^p_{\mathcal{J}^\bullet} := \begin{pmatrix} d^p_{\mathcal{A}^\bullet} & \delta^p \\ 0 & d^p_{\mathcal{K}^\bullet} \end{pmatrix}$$

makes $\mathcal{J}^\bullet$ into a complex and such that if we define $\mathcal{G}^p \to \mathcal{J}^p$ as the sum of id_{A^p} and $\mathcal{H}^p \to \mathcal{K}^p$, then (*) commutes.

As $\mathcal{I}^p$ is injective, the exact sequences $0 \to \mathcal{I}^p \to \mathcal{J}^p \to \mathcal{K}^p \to 0$ even split. Hence this sequence stays exact after applying $\Gamma(X, \cdot)$. We obtain an exact sequence of complexes of R-modules

$$0 \to \Gamma(\mathcal{I}^\bullet) \longrightarrow \Gamma(\mathcal{J}^\bullet) \longrightarrow \Gamma(\mathcal{K}^\bullet) \longrightarrow 0.$$

Now we can apply Appendix Lemma 15.14 to obtain (10.3).

(vi). For the proof of (2) in general we refer to [Stacks] Tag 0152.   □

Remark 10.6. Let $\mathcal{F}^\bullet$ and $\mathcal{G}^\bullet$ be complexes of $\mathcal{O}_X$-modules. Then every quasi-isomorphism $u : \mathcal{F}^\bullet \to \mathcal{G}^\bullet$ yields an isomorphism

$$H^n(X, \mathcal{F}^\bullet) \xrightarrow{\sim} H^n(X, \mathcal{G}^\bullet).$$

Indeed, if $i : \mathcal{G}^\bullet \to \mathcal{I}^\bullet$ is a K-injective resolution of $\mathcal{G}^\bullet$, then $i \circ u$ is a K-injective resolution of $\mathcal{F}^\bullet$.

To achieve all goals formulated in Remark 10.1 we still have to show that if we restrict our construction of cohomology to a single $\mathcal{O}_X$-module, then we obtain a universal δ-functor. To make sense of this we will always consider an $\mathcal{O}_X$-module $\mathcal{F}$ as the complex of $\mathcal{O}_X$-modules

$$\cdots \to 0 \to 0 \to \mathcal{F} \to 0 \to 0 \to \ldots,$$

where $\mathcal{F}$ sits in degree 0 of the complex. Then we write $H^n(X, \mathcal{F})$ for its cohomology.

Corollary 10.7. *Restricting the functors $H^n(X, \ldots)$ to the category of $\mathcal{O}_X$-modules considered as complexes concentrated in degree 0 the family $(H^n(X, \cdot), \delta)_{n \geq 0}$ is a universal δ-functor extending $\Gamma(X, \cdot)$. Moreover, every injective $\mathcal{O}_X$-module $\mathcal{I}$ is Γ-acyclic (i.e., $H^n(X, \mathcal{I}) = 0$ for all $n > 0$).*

Proof. Theorem 10.4 immediately implies that $(H^n(X, \cdot), \delta)_{n \geq 0}$ is a δ-functor that extends $\Gamma(X, \cdot)$.

Let $\mathfrak{I}$ be an injective $\mathcal{O}_X$-module. If we consider $\mathfrak{I}$ as a complex concentrated in degree 0, then id: $\mathfrak{I} \to \mathfrak{I}$ is a K-injective resolution. Hence the definition of $H^n(X, \mathfrak{I})$ shows that $H^n(X, \mathfrak{I}) = 0$ for all $n > 0$.

It remains to show the universality of the δ-functor. Let $(T^n, \delta')_{n \geq 0}$ be a δ-functor extending $\Gamma(X, \cdot)$. We have to show that for all $n \geq 0$ and for all $\mathcal{O}_X$-modules $\mathcal{F}$ there exist unique homomorphisms $\varphi_{\mathcal{F}}^n \colon H^n(X, \mathcal{F}) \to T^n(\mathcal{F})$ compatible with δ and δ' and functorial in $\mathcal{F}$ such that $\varphi_{\mathcal{F}}^0 = \mathrm{id}_{\Gamma(X, \mathcal{F})}$. We proceed by induction on n.

The assertion is clear for $n = 0$. Let $n > 0$. Choose an injective homomorphism $i \colon \mathcal{F} \to \mathfrak{I}$ with $\mathfrak{I}$ an injective $\mathcal{O}_X$-module and set $\mathfrak{Q} := \mathrm{Coker}(i)$. We obtain a short exact sequence of $\mathcal{O}_X$-modules

$$0 \to \mathcal{F} \to \mathfrak{I} \to \mathfrak{Q} \to 0$$

and hence by induction hypothesis a commutative diagram

$$\begin{array}{ccccccccc}
\cdots \longrightarrow & H^{n-1}(X, \mathfrak{I}) & \longrightarrow & H^{n-1}(X, \mathfrak{Q}) & \stackrel{\delta}{\longrightarrow} & H^n(X, \mathcal{F}) & \longrightarrow & H^n(X, \mathfrak{I}) = 0 \\
 & \Big\downarrow \varphi_{\mathfrak{I}}^{n-1} & & \Big\downarrow \varphi_{\mathfrak{Q}}^{n-1} & & & & \\
\cdots \longrightarrow & T^{n-1}(\mathfrak{I}) & \longrightarrow & T^{n-1}(\mathfrak{Q}) & \stackrel{\delta'}{\longrightarrow} & T^n(\mathcal{F}) & &
\end{array}$$

with exact rows. Hence δ is surjective and $\mathrm{Ker}(\delta) \subseteq \mathrm{Ker}(\delta' \circ \varphi_{\mathfrak{Q}}^{n-1})$. Therefore there exists a unique $\varphi_{\mathcal{F}}^n \colon H^n(X, \mathcal{F}) \to T^n(\mathcal{F})$ making the diagram commutative. It is straight forward – albeit a bit cumbersome – to check that $\varphi_{\mathcal{F}}^n$ is functorial in $\mathcal{F}$. $\qquad\square$

For concrete calculations working with injective resolutions is often not advisable because injective modules tend to be quite large (e.g., non-zero injective abelian groups are never finitely generated $\mathbb{Z}$-modules by Appendix Problem 15.14). But in fact the strategy of Remark 10.2 required only to choose a resolution by a complex of Γ-acyclic modules:

Proposition 10.8. *Let $\mathcal{F}^\bullet$ be a bounded below complex of $\mathcal{O}_X$-modules and let $\mathcal{F}^\bullet \xrightarrow{qis} \mathcal{A}^\bullet$ be a quasi-isomorphism, where $\mathcal{A}^\bullet$ is a bounded below complex of $\mathcal{O}_X$-modules such that $\mathcal{A}^n$ is Γ-acyclic for all $n \in \mathbb{Z}$. Then*

$$H^n(X, \mathcal{F}^\bullet) = H^n(\cdots \to \Gamma(X, \mathcal{A}^p) \to \Gamma(X, \mathcal{A}^{p+1}) \to \dots).$$

In the next section we will see several examples of Γ-acyclic $\mathcal{O}_X$-modules.

Proof. If $\mathcal{F}^\bullet$ is concentrated in degree 0 and $\mathcal{A}^n = 0$ for all $n < 0$, we have seen the result in (10.1). For the general case see [Stacks] Tag 05TA. $\qquad\square$

All of the results above can be vastly generalized with (almost) verbatim the same proofs. Instead of the functor $\Gamma(X, \cdot)\colon (\mathcal{O}_X\text{-Mod}) \to (\Gamma(X, \mathcal{O}_X)\text{-Mod})$ one considers an arbitrary left exact functor between abelian categories

$$F\colon \mathcal{A} \to \mathcal{B}.$$

Then all of the above notions, statements, and proofs generalize – if(!) $\mathcal{A}$ has injective and K-injective resolutions. One obtains a universal δ-functor from $\mathcal{A}$ to $\mathcal{B}$ extending F, which is called the *right derived functor of F* and which is denoted by $R^n F$ (in this notation we would have $H^n(X, \cdot) = R^n \Gamma(X, \cdot)$). More generally one obtains functors $R^n F\colon K(\mathcal{A}) \to \mathcal{B}$ with all the properties above. And one has $R^n F(I) = 0$ for all injective objects I in $\mathcal{A}$ and all $n > 0$.

Examples are the following.

1. Let $f\colon (X, \mathcal{O}_X) \to (Y, \mathcal{O}_Y)$ be a morphism of ringed spaces. The left exactness of the functors $\Gamma(f^{-1}(V), \cdot)$, $V \subseteq Y$ open, shows that the functor $f_*\colon (\mathcal{O}_X\text{-Mod}) \to (\mathcal{O}_Y\text{-Mod})$ is left exact. We obtain right derived functors $R^n f_*$.

    This is in fact a generalization of the cohomology functors $H^n(X, \cdot)$ above. Taking for Y the one point space with $\mathcal{O}_Y(Y) = \Gamma(X, \mathcal{O}_X)$ one has $R^n f_* = H^n(X, \cdot)$ by identifying the category of $\mathcal{O}_Y$-modules and the category of R-modules.

2. For any abelian category $\mathcal{A}$ and for every object Z in $\mathcal{A}$ the functor

$$\mathcal{A} \to (\text{Ab}), \qquad X \mapsto \text{Hom}_{\mathcal{A}}(Z, X)$$

    is left exact. Hence if $\mathcal{A}$ has injective and K-injective resolutions, then there exist the right derived functors $X \mapsto \text{Ext}_{\mathcal{A}}^n(Z, X) := R^n \text{Hom}_{\mathcal{A}}(Z, X)$.

Finally, there exists a dual theory for right exact functors F by replacing all arrows in all diagrams by an arrow in the opposite direction: The dual notion of an injective object and a K-injective complex is called a *projective* object (see Appendix Problem 14.4) and a *K-projective* complex, respectively. One obtains left derived functors denoted by $L^n F$. Note however that if an abelian category has injective and K-injective (right) resolutions, then the dual assertion, namely that there always exist projective and K-projective (left) resolutions, might not be true. For instance, the category of R-modules (for any not necessarily commutative ring R) also has projective and K-projective resolutions ([Spa] Theorem C) but this is usually not the case for the category of $\mathcal{O}_X$-modules for a ringed space $(X, \mathcal{O}_X)$. Often it is possible to circumvent this problem by working solely with acyclic resolutions.

10.3   Acyclic Sheaves

We continue to denote by $(X, \mathcal{O}_X)$ a ringed space. Proposition 10.8 shows that we can define the right derived functor of a left exact functor F of a bounded below complex also via a resolution with F-acyclic objects. The following lemma will be the main tool to see whether a certain class of objects is F-acyclic.

Lemma 10.9. *Let* $F: \mathcal{A} \to \mathcal{B}$ *be a left exact functor between abelian categories and suppose that* $\mathcal{A}$ *has injective and K-injective resolutions. Let* $\mathcal{I} \subseteq \mathrm{Ob}(\mathcal{A})$ *such that the following conditions are satisfied:*

(a) *For every object X of $\mathcal{A}$ there exists an injective morphism $X \hookrightarrow M$ with $M \in \mathcal{I}$.*
(b) *If I is an injective object of $\mathcal{A}$, then $I \in \mathcal{I}$.*
(c) *Let*
$$0 \to X' \longrightarrow X \longrightarrow X'' \to 0$$
be an exact sequence in $\mathcal{A}$ with $X', X \in \mathcal{I}$. Then one has $X'' \in \mathcal{I}$ and $F(X) \to F(X'')$ is surjective.

Then every object in $\mathcal{I}$ is F-acyclic, i.e., $R^n F(I) = 0$ for all $I \in \mathcal{I}$ and $n \geq 1$.

Proof. Let $M \in \mathcal{I}$ and choose an exact sequence
$$0 \to M \longrightarrow I^0 \xrightarrow{d^0} I^1 \xrightarrow{d^1} I^2 \xrightarrow{d^2} \dots$$

with I^p injective for all p. As $I^0 \in \mathcal{I}$ by (b), we can apply (c) to the exact sequence $0 \to M \to I^0 \to \mathrm{Im}(d^0) \to 0$. This shows that $\mathrm{Im}(d^0) \in \mathcal{I}$ and that $0 \to F(M) \to F(I^0) \to F(\mathrm{Im}(d^0)) \to 0$ is exact. Then induction and the exact sequences $0 \to \mathrm{Im}(d^{p-1}) \to I^p \to \mathrm{Im}(d^p) \to 0$ show that $\mathrm{Im}(d^p) \in \mathcal{I}$ for all p and that $0 \to F(\mathrm{Im}(d^{p-1})) \to F(I^p) \to F(\mathrm{Im}(d^p)) \to 0$ is exact. For $n \geq 1$ we find therefore

$$\mathrm{Im}(F(I^{n-1}) \to F(I^n)) = F(\mathrm{Im}(d^{n-1})) = \mathrm{Ker}(F(I^n) \to F(I^{n+1}))$$

and hence $R^n F(M) = 0$. $\qquad\square$

The following class of sheaves will yield Γ-acyclic objects.

Definition 10.10. A sheaf $\mathcal{F}$ on a topological space X is called *flabby* or *flasque* if the restriction maps $\mathcal{F}(X) \to \mathcal{F}(U)$ are surjective for all open subsets $U \subseteq X$.

Flabby sheaves will allow construction of Γ-acyclic resolutions in an easy functorial way as follows.

Remark 10.11 (Godement resolution). Let $\mathcal{F}$ be a sheaf on a topological space. Define a sheaf $\mathcal{F}^{[0]}$ on X by $\mathcal{F}^{[0]}(U) := \prod_{x \in U} \mathcal{F}_x$ for $U \subseteq X$ open, where the restriction maps are given by the projections. The sheaf $\mathcal{F}^{[0]}$ is flabby and there is an injective morphism of sheaves

$$\iota_{\mathcal{F}} \colon \mathcal{F} \hookrightarrow \mathcal{F}^{[0]}, \qquad \mathcal{F}(U) \ni s \mapsto (s_x)_{x \in U} \in \mathcal{F}^{[0]}(U), \quad U \subseteq X \text{ open.}$$

Every morphism $\varphi \colon \mathcal{F} \to \mathcal{G}$ of sheaves induces a morphism of flabby sheaves $\varphi^{[0]} \colon \mathcal{F}^{[0]} \to \mathcal{G}^{[0]}$ with $\varphi_U^{[0]} = \prod_{x \in U} \varphi_x$. We obtain a functor $(\ )^{[0]}$ from the category of sheaves to the full subcategory of flabby sheaves. The morphism $\iota_{\mathcal{F}}$ is functorial in $\mathcal{F}$.

Now suppose that $\mathcal{F}$ is an $\mathcal{O}_X$-module. Then $\mathcal{F}^{[0]}$ is an $\mathcal{O}_X$-module: addition is given by the addition within the stalks and scalar multiplication of $a \in \mathcal{O}_X(U)$ on $(s_x)_{x \in U} \in \mathcal{F}^{[0]}(U)$ by $(a_x s_x)_{x \in U}$. We obtain a functor $(\ )^{[0]}$ from the category of $\mathcal{O}_X$-modules to the full subcategory of flabby $\mathcal{O}_X$-modules. The morphism $\iota_{\mathcal{F}}$ is a functorial homomorphism of $\mathcal{O}_X$-modules.

In particular we find for every $\mathcal{O}_X$-module a functorial exact sequence of $\mathcal{O}_X$-modules

$$0 \longrightarrow \mathcal{F} \overset{\iota}{\longrightarrow} \mathcal{F}^{[0]} \longrightarrow \operatorname{Coker}(\iota) \longrightarrow 0 \tag{10.4}$$

with $\mathcal{F}^{[0]}$ flabby. Applying the same argument to $\operatorname{Coker}(\iota)$ we obtain inductively a functorial exact sequence of $\mathcal{O}_X$-modules

$$0 \longrightarrow \mathcal{F} \overset{\iota}{\longrightarrow} \mathcal{F}^{[0]} \longrightarrow \mathcal{F}^{[1]} \longrightarrow \ldots, \tag{10.5}$$

where $\mathcal{F}^{[p]}$ is a flabby $\mathcal{O}_X$-module for all $p \geq 0$. This resolution of $\mathcal{F}$ is called *Godement resolution*.

Finally, the variant of Appendix Lemma 15.11 for the abelian category of $\mathcal{O}_X$-modules shows that given $a \in \mathbb{Z}$ there exists for every complex $\mathcal{F}^{\bullet}$ of $\mathcal{O}_X$-modules with $\mathcal{F}^p = 0$ for all $p < a$ a quasi-isomorphism $\mathcal{F}^{\bullet} \overset{qis}{\longrightarrow} \mathcal{A}^{\bullet}$, where $\mathcal{A}^p$ is a flabby $\mathcal{O}_X$-module for all p and $\mathcal{A}^p = 0$ for all $p < a$.

Proposition 10.12.

1. *Let $(X, \mathcal{O}_X)$ be a ringed space. Then every injective $\mathcal{O}_X$-module is flabby.*
2. *Let X be a paracompact Hausdorff space. Then every flabby sheaf $\mathcal{F}$ on X is soft.*

Proof. (1). Let $\mathcal{J}$ be an injective $\mathcal{O}_X$-module. By Remark 10.11 there exists an injective homomorphism of $\mathcal{O}_X$-module $i \colon \mathcal{J} \to \mathcal{G}$ where $\mathcal{G}$ is flabby. As $\mathcal{J}$ is injective, this makes $\mathcal{J}$ into a direct summand of $\mathcal{G}$. Hence $\mathcal{J}$ is flabby.

(2). Let $A \subseteq X$ be closed, $s \in \mathcal{F}(A)$. By Proposition 9.1 there exists an open neighborhood U of A and $\tilde{s} \in \mathcal{F}(U)$ extending s. As $\mathcal{F}$ is flabby, one can extend $\tilde{s}$ to X. $\qquad \square$

Proposition 10.13. *Let X be a topological space and let $\mathcal{F}$ be a flabby sheaf on X.*

1. *For every open subset U of X the restriction $\mathcal{F}_{|U}$ is again flabby.*
2. *Suppose that X is hereditarily paracompact and Hausdorff. Then for any subspace Z of X the restriction $\mathcal{F}_{|Z}$ is flabby again.*

Recall Proposition 1.13, which shows that any metrizable space satisfies the hypotheses of 2.

Proof. The first assertion is clear. To show 2 let $W \subseteq Z$ be open and let $s \in \mathcal{F}(W)$. By Proposition 9.1 2 we can extend s to a section s' over an open neighborhood U of W in X. As $\mathcal{F}$ is flabby, we can extend s' to a section $\tilde{s} \in \mathcal{F}(X)$. Then $\tilde{s}_{|Z} \in \mathcal{F}(Z)$ extends s. $\qquad\square$

Lemma 10.14. *Let X be a topological space.*

1. *Let $\mathcal{G}$ be a flabby sheaf of groups. Then every $\mathcal{G}$-torsor is trivial.*
2. *Let $1 \to \mathcal{G}' \to \mathcal{G} \to \mathcal{G}'' \to 1$ be an exact sequence of sheaves of groups and let $\mathcal{G}'$ and $\mathcal{G}$ be flabby. Then $\mathcal{G}''$ is flabby and $\mathcal{G}(U) \to \mathcal{G}''(U)$ is surjective for all $U \subseteq X$ open.*

Proof. 1. The argument is very similar to the proof that torsors for soft sheaves are trivial (Theorem 9.14) but as we can work with open sets we do not run into any topological difficulties requiring such tools as the shrinking lemma. Indeed, let T be a $\mathcal{G}$-torsor, let $(U_i)_{i \in I}$ be an open covering such that there exist $t_i \in T(U_i)$ for all i. For $J \subseteq I$ set $U_J := \bigcup_{i \in J} U_i$. Then $U_I = X$. Define $\mathcal{E} := \{(t, J) \,;\, J \subseteq I, t \in T(U_J)\}$. Then $\mathcal{E} \neq \emptyset$ because $(*, \emptyset) \in \mathcal{E}$. It is partially ordered by $(t, J) \leq (t', J')$ if $J \subseteq J'$ and $t'_{|U_J} = t$. As T is a sheaf, every totally ordered subset of $\mathcal{E}$ has an upper bound in $\mathcal{E}$. Hence there exists a maximal element (t, J) in $\mathcal{E}$ by Zorn's lemma. It suffices to show that $J = I$.

Assume there exists $i \in I \setminus J$ and let $g \in \mathcal{G}(U_J \cap U_i)$ with $t_{|U_J \cap U_i} = g t_{i|U_J \cap U_i}$. As $\mathcal{G}$ is flabby, we can extend g to $\tilde{g} \in \mathcal{G}(X)$. Replacing t_i by $\tilde{g}_{|U_i} t_i$ we may assume $t_{|U_J \cap U_i} = t_{i|U_J \cap U_i}$ and hence we can glue t and t_i to a section over $U_{J \cup \{i\}}$. This contradicts the maximality of (t, J).

2. As $\mathcal{G}'_{|U}$ is flabby for every open subset $U \subseteq X$, $H^1(U, \mathcal{G}'_{|U}) = 1$ by 1 (here we use the cohomology defined in Definition 7.3) and the cohomology sequence attached to the exact sequence $1 \to \mathcal{G}'_{|U} \to \mathcal{G}_{|U} \to \mathcal{G}''_{|U} \to 1$ shows that $\mathcal{G}(U) \to \mathcal{G}''(U)$ is surjective. Hence we obtain a commutative diagram

$$
\begin{array}{ccc}
\mathcal{G}(X) & \longrightarrow & \mathcal{G}''(X) \\
\downarrow & & \downarrow \\
\mathcal{G}(U) & \longrightarrow & \mathcal{G}''(U)
\end{array}
$$

with $\mathcal{G}(X) \to \mathcal{G}(U) \to \mathcal{G}''(U)$ a composition of surjective maps. Therefore $\mathcal{G}''(X) \to \mathcal{G}''(U)$ is surjective. This shows that $\mathcal{G}''$ is flabby. $\qquad\square$

Proposition 10.15. *A flabby $\mathcal{O}_X$-module is Γ-acyclic.*

Proof. This follows from Lemma 10.9 using Remark 10.11 and Lemma 10.14. □

Proposition 10.16. *Let X be a topological space and let $\mathcal{A}$ be a sheaf of abelian groups, which we may consider a $\mathbb{Z}_X$-module. Then $H^1(X, \mathcal{A})$ as defined in Definition 10.4 and $H^1(X, \mathcal{A})$ defined via torsors in Definition 7.3 coincide.*

The proof will show that there is a functorial isomorphism between these cohomology groups.

Proof. Denote for the moment the cohomology group defined via torsors in Definition 7.3 by $\tilde{H}^1(X, \cdot)$. By Remark 10.11 there exists a functorial exact sequence of sheaves of abelian groups $0 \to \mathcal{A} \to \mathcal{A}^0 \to \mathcal{C} \to 0$ with $\mathcal{A}^0$ flabby. Then $H^1(X, \mathcal{A}^0) = 0$ by Proposition 10.15 and $\tilde{H}^1(X, \mathcal{A}^0) = 0$ by Lemma 10.14 1. Hence the long exact cohomology sequence (10.3) and the long exact sequence for $\tilde{H}^1$ (7.11) imply that $H^1(X, \mathcal{A}) = \tilde{H}^1(X, \mathcal{A}) = \operatorname{Coker}(\mathcal{A}^0(X) \to \mathcal{C}(X))$. □

Proposition 10.17. *Let $(X, \mathcal{O}_X)$ be a ringed space such that X is paracompact and Hausdorff. Then every soft $\mathcal{O}_X$-module is Γ-acyclic.*

Proof. We check the conditions of Lemma 10.9 for the class $\mathcal{I}$ of soft $\mathcal{O}_X$-modules. By Remark 10.11 we can embed every $\mathcal{O}_X$-module into a flabby $\mathcal{O}_X$-module $\mathcal{A}$, which is soft by Proposition 10.12. The same proposition also shows that every injective $\mathcal{O}_X$-module is soft.

It remains to show Condition (c) of Lemma 10.9. Let $0 \to \mathcal{F}' \to \mathcal{F} \to \mathcal{F}'' \to 0$ be an exact sequence of $\mathcal{O}_X$-modules with $\mathcal{F}$ and $\mathcal{F}'$ soft. Then $\mathcal{F}(X) \to \mathcal{F}''(X)$ is surjective because $H^1(X, \mathcal{F}') = 0$ by Theorem 9.14. Let $A \subseteq X$ be closed. Then $\mathcal{F}'_{|A}$ is soft by Remark 9.5 and therefore the same argument proves that $\mathcal{F}(A) \to \mathcal{F}''(A)$ is surjective. Then the commutative diagram

$$
\begin{array}{ccc}
\mathcal{F}(X) & \longrightarrow & \mathcal{F}''(X) \\
\downarrow & & \downarrow \\
\mathcal{F}(A) & \longrightarrow & \mathcal{F}''(A)
\end{array}
$$

shows that $\mathcal{F}''(X) \to \mathcal{F}''(A)$ is surjective. Therefore $\mathcal{F}''$ is soft. □

Every complex of $\mathcal{O}_X$-modules $\mathcal{F}^\bullet$ is also a complex of sheaves of abelian groups (i.e., a complex of $\mathbb{Z}_X$-modules), which we call $^a\mathcal{F}^\bullet$ for the moment. Similarly, if M is a module over a ring R, we denote by ^aM the underlying additive group.

Proposition 10.18. *Let* $\mathcal{F}^\bullet$ *be a bounded below complex of* $\mathcal{O}_X$*-modules. Then we have*

$$H^n(X, {}^a\mathcal{F}^\bullet) = {}^a H^n(X, \mathcal{F}^\bullet).$$

Proof. By Remark 10.11 there exists a quasi-isomorphism $\mathcal{F}^\bullet \xrightarrow{qis} \mathcal{A}^\bullet$ of $\mathcal{O}_X$-module, where $\mathcal{A}^p$ is flabby for all p. By Proposition 10.8, $\Gamma(X, \mathcal{A}^\bullet)$ computes the cohomology of $\mathcal{F}^\bullet$. By Lemma 10.15 each $\mathcal{A}^p$ is also Γ-acyclic if considered as a $\mathbb{Z}_X$-module. Hence it also computes the cohomology of ${}^a\mathcal{F}^\bullet$. $\qquad\square$

10.4 Applications: Theorems of De Rham and of Mittag-Leffler

Definition and Remark 10.19 (De Rham Cohomology). Let M be a real C^α-premanifold with $\alpha \geq \infty$ or a complex premanifold. Then

$$\Omega_M^\bullet: \qquad \cdots \to 0 \to \Omega_M^0 \xrightarrow{d} \Omega_M^1 \xrightarrow{d} \Omega_M^2 \xrightarrow{d} \cdots$$

is a bounded below complex of $\mathbb{K}_M$-modules that is exact except in degree 0 (Corollary 8.69). For $n \geq 0$ we call

$$H_{\mathrm{DR}}^n(M) := H^n(M, \Omega_M^\bullet)$$

the *n-th de Rham cohomology of* M. The quasi-isomorphism (8.36)

$$\mathbb{K}_M \xrightarrow{qis} \Omega_M^\bullet$$

yields by Remark 10.6 for all $n \geq 0$ an isomorphism

$$H^n(M, \mathbb{K}_M) \xrightarrow{\sim} H_{\mathrm{DR}}^n(M). \tag{10.6}$$

If Ω_M^p is Γ-acyclic for all $p \geq 0$, it calculates its own cohomology by Proposition 10.8 and we obtain

$$H_{\mathrm{DR}}^n(M) = \frac{\{\text{closed } n\text{-forms on } M\}}{\{\text{exact } n\text{-forms on } M\}}. \tag{10.7}$$

If M is of dimension $m \in \mathbb{N}_0$, this implies in particular $H_{\mathrm{DR}}^n(M) = 0$ for $n > m$.

If M is a real C^∞-manifold, then Ω_M^p is soft for all $p \geq 0$ (Corollary 9.12) and hence Γ-acyclic (Proposition 10.17). Hence we obtain:

Theorem 10.20 (De Rham). *Let M be a real C^∞-manifold. Then for all $n \geq 0$*

$$H^n(M, \mathbb{R}_M) \cong H^n_{\mathrm{DR}}(M) \cong \frac{\{\text{closed } n\text{-forms on } M\}}{\{\text{exact } n\text{-forms on } M\}}. \tag{10.8}$$

Note that this result in particular implies the surprising fact that the question of whether every closed n-form is exact on M depends only on the underlying topological space of M. We will even see in Corollary 11.19 below that $H^n(M, \mathbb{R}_M) = H^n(M', \mathbb{R}_{M'})$ if M and M' are homotopy equivalent[2].

Remark 10.21. One can show that Ω^p_M is also Γ-acyclic for all $p \geq 0$ if M is a real analytic manifold (combine [Car] Théoreme 3 together with the fact that for every real analytic manifold there exists a closed embedding into some $\mathbb{R}^N$ by [Gra]). Hence one has the isomorphisms (10.8) also for real analytic manifolds M.

For complex manifolds M one still has that $H^n(M, \mathbb{C}_M) = H^n_{\mathrm{DR}}(M)$ by the Poincaré lemma (Remark 10.19) but (10.7) does *not* hold in general (see Problem 10.6 for $M = \mathbb{P}^1(\mathbb{C})$). Still, there are interesting special cases where one still has an analogue of (10.8). One of these cases relies on the following result.

Theorem 10.22. *Let $U \subseteq \mathbb{C}$ be open and $\mathcal{O}_U$ be the sheaf of holomorphic functions. Then $H^i(U, \mathcal{O}_U) = 0$ for all $i > 0$.*

We use the following standard fact from complex analysis (the "Dolbeault lemma"): For every $\mathbb{C}$-valued C^∞-function g on U there exists a $\mathbb{C}$-valued C^∞-function f on U such that $\frac{\partial f}{\partial \bar{z}} = g$. For a proof see [Hoe] Theorem 1.4.4 in the case considered here or [GuRo] Chap. VI, C, Lemma 1 for a version in an arbitrary complex dimension).

Proof. Let $C^\infty_{U;\mathbb{C}}$ be the sheaf of $\mathbb{C}$-valued C^∞-functions on U. Applying the Dolbeault lemma to open subsets of U we see that $\frac{\partial}{\partial \bar{z}} : C^\infty_{U;\mathbb{C}} \to C^\infty_{U;\mathbb{C}}$ is surjective. By the Cauchy–Riemann differential equations we obtain an exact sequence

$$0 \to \mathcal{O}_U \longrightarrow C^\infty_{U;\mathbb{C}} \xrightarrow{\frac{\partial}{\partial \bar{z}}} C^\infty_{U;\mathbb{C}} \to 0.$$

Hence $\mathcal{O}_U$ has a resolution by the complex

$$\cdots \to 0 \to \underbrace{C^\infty_{U;\mathbb{C}}}_{\text{degree } 0} \xrightarrow{\frac{\partial}{\partial \bar{z}}} \underbrace{C^\infty_{U;\mathbb{C}}}_{\text{degree } 1} \to 0 \to \cdots,$$

---

[2] This is formulated in a somewhat too dramatic way: The Poincaré lemma (Theorem 8.68), for which we did not give a proof, is in the C^∞-case often shown (for instance in the given reference) by proving that smooth maps between premanifolds that are homotopic via a smooth homotopy induce the same map on the right-hand side of (10.8). One then concludes that on a smoothly contractible C^∞-manifold every closed form is exact. Nevertheless it remains surprising that continuity of the homotopy is already sufficient.

which consists of soft (and hence Γ-acyclic) sheaves. Hence the cohomology is calculated by this complex. Clearly this implies $H^n(U, \mathcal{O}_U) = 0$ for $n \geq 2$. Moreover, the Dolbeault lemma (now applied to U) shows that $H^1(U, \mathcal{O}_U) = 0$. $\qquad\square$

Corollary 10.23. *Let $U \subseteq \mathbb{C}$ be open. Then for all $n \geq 0$*

$$H^n(U, \mathbb{C}_U) \cong H^n_{\mathrm{DR}}(U) \cong \frac{\{\text{closed holomorphic } n\text{-forms on } U\}}{\{\text{exact holomorphic } n\text{-forms on } U\}}. \tag{10.9}$$

As U has dimension 1 as complex manifold, $\Omega^n_U = 0$ for $n > 1$. In particular, all terms in (10.9) are 0 for $n > 1$.

Proof. By Theorem 10.22, $\mathcal{O}_U = \Omega^0_U$ is Γ-acyclic. Denote by z the coordinate on U. Then $f \mapsto f\,dz$ defines an isomorphism $\mathcal{O}_U \xrightarrow{\sim} \Omega^1_U$ of $\mathcal{O}_U$-modules, hence Ω^1_U is Γ-acyclic as well. We conclude by Remark 10.19. $\qquad\square$

Much more generally, one has for every Stein manifold (Theorem 5.45) X that $H^i(X, \mathcal{F}) = 0$ for all $i > 0$ and for all finite locally free $\mathcal{O}_X$-modules $\mathcal{F}$ (a special case of "Cartan's Theorem B", e.g., [GuRo], Chap. VIII, A, Theorem 14). Hence Corollary 10.23 holds if U is an arbitrary Stein manifold.

Corollary 10.24 (Theorem of Mittag-Leffler). *Let $U \subseteq \mathbb{C}$ be open and let $S \subseteq U$ be discrete and closed. For each $z_0 \in S$ let there be given a finite principal part of a Laurent series*

$$\sum_{i=1}^{r} a_{-i}(z - z_0)^{-i} \tag{*}$$

with $a_{-i} \in \mathbb{C}$ and r depending on z_0.

Then there exists a meromorphic function f on U such that f is holomorphic on $U \setminus S$ and such that for all $z_0 \in S$ the principal part of the Laurent series expansion at z_0 is given by (*).

Proof. Let $\mathcal{P}_U$ be the sheaf on U of functions $s \colon V \to \mathbb{C}^{(-\mathbb{N})}$, $V \subseteq U$ open, such that $\{z_0 \in V \; ; \; s(z_0) \neq 0\}$ is discrete and closed in V. We think of such a function s as a map that attaches to z_0 the principal part $\sum_{i \in -\mathbb{N}} a_i(z - z_0)^i$ if $s(z_0) = (a_{-1}, a_{-2}, a_{-3}, \dots)$. Let $\mathcal{M}_U$ be the sheaf of meromorphic functions on U. We obtain an exact sequence of sheaves of $\mathbb{C}$-vector spaces

$$0 \to \mathcal{O}_U \longrightarrow \mathcal{M}_U \xrightarrow{\;\text{principal part}\;} \mathcal{P}_U \to 0.$$

As $H^1(U, \mathcal{O}_U) = 0$ by Theorem 10.22, the map $\mathcal{M}_U(U) \to \mathcal{P}_U(U)$ is surjective. $\qquad\square$

10.5　Cohomology and Inverse and Direct Image

Remark 10.25. Recall that we defined a functor $f^{-1}:(\mathrm{Sh}(Y)) \to (\mathrm{Sh}(X))$ such that

$$(f^{-1}\mathcal{G})_x = \mathcal{G}_{f(x)} \tag{10.10}$$

for every sheaf $\mathcal{G}$ on Y and $x \in X$. More precisely, $f^{-1}\mathcal{G}$ was defined as the sheafification of the presheaf on X

$$(f^+\mathcal{G})(U) := \operatorname*{colim}_{V \supseteq f(U)} \mathcal{G}(V).$$

If $\mathcal{G}_1$ and $\mathcal{G}_2$ are sheaves on Y, then $f^+(\mathcal{G}_1 \times \mathcal{G}_2) = f^+(\mathcal{G}_1) \times f^+(\mathcal{G}_2)$ because filtered colimits commute with finite limits (Appendix Proposition 13.39). As sheafification commutes with finite limits, the inverse image functor f^{-1} commutes with finite limits. In other words, it is left exact.

　　In particular, if $\mathcal{G}$ is a sheaf of abelian groups on Y, then the group law $a:\mathcal{G} \times \mathcal{G} \to \mathcal{G}$ defines via functoriality a group law

$$f^{-1}(\mathcal{G}) \times f^{-1}(\mathcal{G}) = f^{-1}(\mathcal{G} \times \mathcal{G}) \xrightarrow{f^{-1}(a)} f^{-1}(\mathcal{G}).$$

Hence f^{-1} yields a functor

$$f^{-1}:(\mathbb{Z}_Y\text{-Mod}) \to (\mathbb{Z}_X\text{-Mod}).$$

Here we identify $(\mathbb{Z}_X\text{-Mod})$ with the category of sheaves of abelian groups on X. We obtain also functors

$$f^{-1}:(\mathrm{Com}(\mathbb{Z}_Y)) \to (\mathrm{Com}(\mathbb{Z}_X)), \qquad \text{and} \qquad f^{-1}:K(\mathbb{Z}_Y) \to K(\mathbb{Z}_X)$$

by defining $f^{-1}(\mathcal{G}^\bullet) := (\cdots \to f^{-1}(\mathcal{G}^p) \to f^{-1}(\mathcal{G}^{p+1}) \to \ldots)$.

Proposition 10.26. *The functor* $f^{-1}:(\mathbb{Z}_Y\text{-Mod}) \to (\mathbb{Z}_X\text{-Mod})$ *is exact. In particular, if* $u:\mathcal{G}^\bullet \to \mathcal{H}^\bullet$ *is a quasi-isomorphism in* $(\mathrm{Com}(\mathbb{Z}_Y))$, *then* $f^{-1}(u):f^{-1}\mathcal{G}^\bullet \to f^{-1}\mathcal{H}^\bullet$ *is a quasi-isomorphism.*

Proof. It is left exact by Remark 10.25 and right exact because it is left adjoint to the functor f_* (Proposition 3.49 and Remark 3.53) and in particular commutes with arbitrary colimits (Appendix Proposition 13.47). □

　　One can also use (10.10) to prove Proposition 10.26 with less abstract nonsense.

Remark and Definition 10.27. Let $\mathcal{G}^\bullet$ be a complex of sheaves of abelian groups on Y. Choose a K-injective resolution $\mathcal{G}^\bullet \xrightarrow{qis} \mathcal{I}^\bullet$. By Proposition 10.26 we obtain a quasi-isomorphism $f^{-1}\mathcal{G}^\bullet \xrightarrow{qis} f^{-1}\mathcal{I}^\bullet$. Now let $s\colon f^{-1}\mathcal{I}^\bullet \xrightarrow{qis} \mathcal{J}^\bullet$ be a K-injective resolution. We obtain for all $n \in \mathbb{Z}$ homomorphisms of abelian groups

$$
\begin{aligned}
f^{-1}\colon H^n(Y,\mathcal{G}^\bullet) &= H^n(\cdots \to \Gamma(Y,\mathcal{I}^p) \to \Gamma(Y,\mathcal{I}^{p+1}) \to \dots) \\
&\longrightarrow H^n(\cdots \to \Gamma(X,f^{-1}\mathcal{I}^p) \to \Gamma(X,f^{-1}\mathcal{I}^{p+1}) \to \dots) \\
&\xrightarrow{s} H^n(\cdots \to \Gamma(X,\mathcal{J}^p) \to \Gamma(X,\mathcal{J}^{p+1}) \to \dots) \\
&= H^n(X, f^{-1}\mathcal{G}^\bullet),
\end{aligned}
$$

where the first arrow is given by pullback of sections (Definition 3.57) in every degree. If $\mathcal{G}^\bullet$ is concentrated in degree 0, then we can assume that $\mathcal{I}^p = 0$ and $\mathcal{J}^p = 0$ for all $p < 0$. In particular we see that f^{-1} is the usual pullback of sections for $n = 0$ in this case.

Lemma 10.28. *Let $\mathcal{F}$ be an injective $\mathbb{Z}_X$-module. Then $f_*(\mathcal{F})$ is an injective $\mathbb{Z}_Y$-module.*

Proof. Let $\mathcal{G} \to \mathcal{H}$ be an injective homomorphism of $\mathbb{Z}_Y$-modules. Then $f^{-1}\mathcal{G} \to f^{-1}\mathcal{H}$ is injective. Therefore the map

$$
\operatorname{Hom}(\mathcal{H}, f_*\mathcal{I}) = \operatorname{Hom}(f^{-1}\mathcal{H}, \mathcal{I}) \to \operatorname{Hom}(f^{-1}\mathcal{G}, \mathcal{I}) = \operatorname{Hom}(\mathcal{G}, f_*\mathcal{I})
$$

is surjective. $\square$

The proof generalizes immediately to the case where f_* is an arbitrary functor $F\colon \mathcal{A} \to \mathcal{B}$ between abelian categories $\mathcal{A}$ and $\mathcal{B}$ that has an exact left adjoint functor.

Proposition 10.29. *Let $f\colon X \to Y$ be a continuous map, let $\mathcal{F}$ be a sheaf of abelian groups on X and assume that $R^q f_*(\mathcal{F}) = 0$ for all $q > 0$. Then we get an isomorphism $H^p(Y, f_*\mathcal{F}) \xrightarrow{\sim} H^p(X, \mathcal{F})$ for all $p \geq 0$.*

Proof. Let $0 \to \mathcal{F} \to \mathcal{I}^0 \to \mathcal{I}^1 \to \dots$ be an injective resolution. Then $R^q f_*(\mathcal{F})$ is by definition the complex cohomology of $f_*(\mathcal{I}^\bullet)$. Hence $R^q f_*(\mathcal{F}) = 0$ for all $q > 0$ implies that

$$
0 \to f_*(\mathcal{F}) \longrightarrow f_*(\mathcal{I}^0) \longrightarrow f_*(\mathcal{I}^1) \longrightarrow \dots
$$

is exact. By Lemma 10.28 the abelian sheaves $f_*(\mathcal{I}^p)$ are injective. Hence

$$
\begin{aligned}
H^p(Y, f_*(\mathcal{F})) &= H^p(\cdots \to 0 \to \Gamma(Y, f_*(\mathcal{I}^0)) \longrightarrow \Gamma(Y, f_*(\mathcal{I}^1)) \longrightarrow \dots) \\
&= H^p(\cdots \to 0 \to \Gamma(X, \mathcal{I}^0) \longrightarrow \Gamma(X, \mathcal{I}^1) \longrightarrow \dots) \\
&= H^p(X, \mathcal{F}).
\end{aligned}
$$
                                                                                        $\square$

Corollary 10.30. *Let X be a topological space, let $i: A \hookrightarrow X$ be the inclusion of a closed subspace, and let $\mathcal{F}$ be an abelian sheaf on A. Then $H^p(X, i_*\mathcal{F}) = H^p(A, \mathcal{F})$ for all $p \geq 0$.*

Proof. Then $i_*: (\mathbb{Z}_A\text{-Mod}) \to (\mathbb{Z}_X\text{-Mod})$ is exact by (3.10) and therefore $R^q i_* = 0$ for all $q > 0$. Hence we can apply Proposition 10.29.   □

Proposition 10.31 (Mayer–Vietoris sequence). *Let A and B closed subspaces of a topological space X. Then for every abelian sheaf $\mathcal{F}$ on X there is a long exact sequence*

$$\ldots \to H^p(A \cup B, \mathcal{F}) \to H^p(A, \mathcal{F}) \oplus H^p(B, \mathcal{F}) \to H^p(A \cap B, \mathcal{F})$$
$$\to H^{p+1}(A \cup B, \mathcal{F}) \to \ldots \tag{10.11}$$

Here we write $H^p(A, \mathcal{F})$ instead of $H^p(A, \mathcal{F}_{|A})$. Similarly for the other terms.

Proof. Let $i: A \to X$, $j: B \to X$, $k: A \cap B \to X$, and $l: A \cup B \to X$ be the inclusions. Let $\mathcal{G}$ be an abelian sheaf with $\mathcal{G}_x = 0$ for all $x \in X \setminus A$. Then $\mathcal{G} \to i_*(i^{-1}\mathcal{G})$ is an isomorphism by (3.10). Hence every morphism of sheaves $\mathcal{F} \to \mathcal{G}$ factors $\mathcal{F} \to i_*(i^{-1}\mathcal{F}) \to i_*(i^{-1}\mathcal{G}) \xrightarrow{\sim} \mathcal{G}$. As i factors through l, this shows that $\mathcal{F} \to i_*(i^{-1}\mathcal{F})$ factors through $l_*(l^{-1}\mathcal{F})$, similarly for j. These factorizations yield a morphism $l_*(l^{-1}\mathcal{F}) \to i_*(i^{-1}\mathcal{F}) \oplus j_*(j^{-1}\mathcal{F})$. The same argument shows that the composition of $\mathcal{F} \oplus \mathcal{F} \to \mathcal{F}$, $(s,t) \mapsto s - t$, and of $\mathcal{F} \to k_*(k^{-1}\mathcal{F})$ yields a morphism $i_*(i^{-1}\mathcal{F}) \oplus j_*(j^{-1}\mathcal{F}) \to k_*(k^{-1}\mathcal{F})$. Moreover, the sequence

$$0 \to l_*(l^{-1}\mathcal{F}) \to i_*(i^{-1}\mathcal{F}) \oplus j_*(j^{-1}\mathcal{F}) \to k_*(k^{-1}\mathcal{F}) \to 0$$

is exact on stalks by (3.10), hence it is exact. It induces the long exact sequence (10.11) by Corollary 10.30.   □

10.6 Proper Base Change

In this section we will prove the proper base change theorem. We will consider a commutative diagram of topological spaces

$$\begin{array}{ccc} W & \xrightarrow{\ p\ } & X \\ {\scriptstyle q}\downarrow & & \downarrow{\scriptstyle f} \\ Z & \xrightarrow{\ g\ } & Y. \end{array} \tag{10.12}$$

Let $\mathcal{F}$ be a sheaf of abelian groups on X. Then there is a functorial morphism of abelian sheaves

$$g^{-1}(f_*\mathcal{F}) \longrightarrow q_*(p^{-1}\mathcal{F}) \tag{10.13}$$

defined as follows. As the functor g^{-1} is left adjoint to g_* it suffices to define a functorial morphism $f_*\mathcal{F} \to g_*(q_*(p^{-1}\mathcal{F})) = f_*(p_*(p^{-1}\mathcal{F}))$. As p_* and p^{-1} are adjoint, $\mathrm{id}_{p^{-1}\mathcal{F}}$ corresponds to a morphism $\mathcal{F} \to p_*(p^{-1}\mathcal{F})$. Now we apply the functor f_* to this morphism.

The universal property of a fiber product yields a continuous map $W \to X \times_Y Z$, $w \mapsto (p(w), q(w))$ and (10.12) is called *cartesian* if this map is a homeomorphism.

Assume that this is the case and that g is the inclusion of a point $Z = \{y\}$ to Y. Then p is the inclusion $f^{-1}(y) \to X$ and q_* is given by the functor $\mathcal{G} \mapsto \Gamma(f^{-1}(y), \mathcal{G})$. Hence (10.13) is the morphism

$$(f_*\mathcal{F})_y \longrightarrow \Gamma\left(f^{-1}(y), \mathcal{F}_{|f^{-1}(y)}\right) \tag{10.14}$$

defined as the composition

$$f_*(\mathcal{F})_y = \underset{\substack{V \subseteq Y \text{ open} \\ V \ni y}}{\operatorname{colim}} \mathcal{F}(f^{-1}(V))$$

$$\overset{(A)}{\longrightarrow} \underset{\substack{U \subseteq X \text{ open} \\ U \supseteq f^{-1}(y)}}{\operatorname{colim}} \mathcal{F}(U) \tag{10.15}$$

$$\overset{(B)}{\longrightarrow} \mathcal{F}(f^{-1}(y)) = \Gamma\left(f^{-1}(y), \mathcal{F}_{|f^{-1}(y)}\right).$$

Remark 10.32. Suppose that (10.12) is cartesian. If f is proper (respectively separated), then q is proper (respectively separated).

Indeed, for "proper" this follows from Theorem 1.30 (iv): If $Z' \to Z$ is a continuous map, then $Z' \times_Z W = Z' \times_Y X$ and hence the projection $Z' \times_Z W \to Z'$ is closed. For "separated" we use Proposition 1.25 (ii). Let $z \in Z$ and let $w, w' \in q^{-1}(z)$ with $w \neq w'$. As (10.12) is cartesian, p induces a homeomorphism $q^{-1}(z) \overset{\sim}{\to} f^{-1}(g(z))$. In particular $p(w) \neq p(w')$. Hence there exist open disjoint neighborhoods U of $p(w)$ and U' of $p(w')$ in X. Then $p^{-1}(U)$ and $p^{-1}(U')$ are open disjoint neighborhoods of w and w' respectively in W. Hence $q^{-1}(z)$ is relatively Hausdorff in W.

Theorem 10.33 (Proper base change). *Let a cartesian diagram* (10.12) *be given and let $\mathcal{F}^\bullet$ be a bounded below complex of abelian sheaves on X. Suppose that* one *of the following hypotheses is satisfied:*

1. *The map f is separated and proper.*
2. *The spaces X and W are Hausdorff hereditarily paracompact (e.g., if X and W are manifolds or – more generally – are metrizable) and the maps f and q are closed.*

Then one has a functorial isomorphism

$$g^{-1}(R^n f_* \mathcal{F}^\bullet) \overset{\sim}{\longrightarrow} R^n q_* \left(p^{-1}\mathcal{F}^\bullet\right) \tag{10.16}$$

for all $n \geq 0$, which is given by (10.13) for $n = 0$.

For $Z = \{y\}$ with $y \in Y$ we obtain the following corollary.

Corollary 10.34. *Let $f: X \to Y$ be a continuous map of topological spaces. Let $\mathcal{F}^\bullet$ be a bounded below complex of sheaves of abelian groups on X and $y \in Y$. Denote by $i_y: f^{-1}(y) \to X$ the inclusion. Suppose that one of the following hypotheses is satisfied:*

1. *The map f is proper and separated.*
2. *The space X is Hausdorff and hereditarily paracompact and f is closed.*

Then one has a functorial isomorphism

$$R^n f_*(\mathcal{F}^\bullet)_y \xrightarrow{\sim} H^n\left(f^{-1}(y), i_y^{-1}(\mathcal{F}^\bullet)\right) \tag{10.17}$$

for all $n \geq 0$, which is given by (10.15) for $n = 0$.

For the proof of Theorem 10.33 we will use the following result.

Lemma 10.35. *Suppose that the hypotheses of Corollary 10.34 are satisfied and let $\mathcal{I}$ be a flabby abelian sheaf on X. Then $\mathcal{I}_{|f^{-1}(y)}$ is Γ-acyclic for all $y \in Y$.*

Proof. If Hypothesis 2 is satisfied, then $\mathcal{I}_{|f^{-1}(y)}$ is flabby by Proposition 10.13 2 and hence Γ-acyclic by Proposition 10.15. Now suppose that Hypothesis 1 is satisfied. Then $f^{-1}(y)$ is compact and relatively Hausdorff in X. We show that $\mathcal{I}_{|f^{-1}(y)}$ is soft and hence Γ-acyclic (Proposition 10.17). Indeed, if $A \subseteq f^{-1}(y)$ is closed then A is compact and relatively Hausdorff in X. Hence by Proposition 9.1 3 we can extend every section $s \in \mathcal{I}(A)$ to a section s' over an open neighborhood U of A in X. As $\mathcal{I}$ is flabby, we can extend s' to a section $\tilde{s}$ over X. Its restriction to $f^{-1}(y)$ extends s. $\square$

Proof (Proof of Theorem 10.33). *(i).* We first show the claim for $n = 0$ and for $\mathcal{F}^\bullet = \mathcal{F}$ concentrated in degree 0. We have to show that (10.13) is an isomorphism. This we can do on stalks. Therefore we may assume that $Z = \{y\}$ for some $y \in Y$. Hence it suffices to prove Corollary 10.34 for $n = 0$ and $\mathcal{F}^\bullet = \mathcal{F}$. We show that the morphisms (A) and (B) in (10.15) are isomorphisms.

Morphism (A) is an isomorphism because of Appendix Proposition 12.11 as both hypotheses imply that f is closed. To see that Morphism (B) is an isomorphism we can apply Proposition 9.1. If Hypothesis 2 is satisfied, then every neighborhood of $f^{-1}(y)$ is Hausdorff paracompact. If Hypothesis 1 is satisfied, then $f^{-1}(y)$ is compact because f is proper and a relatively Hausdorff subspace because f is separated.

(ii). Next we show that for every flabby abelian sheaf $\mathcal{I}$ on X the pullback $p^{-1}\mathcal{I}$ is acyclic for the functor q_*.

For every $z \in Z$ we have $(p^{-1}\mathcal{I})_{|q^{-1}(z)} = \mathcal{I}_{|f^{-1}(g(z))}$. Hence Lemma 10.35 shows that the restrictions of $\mathcal{G} := p^{-1}\mathcal{I}$ to the fibers of q are Γ-acyclic.

Let $0 \to \mathcal{G} \to \mathcal{J}^0 \to \mathcal{J}^1 \to \ldots$ be an injective resolution. We have to show that

$$0 \to q_* \mathcal{G} \to q_* \mathcal{J}^0 \to q_* \mathcal{J}^1 \to \ldots$$

is exact. We check this on stalks at all $z \in Z$. Now the hypotheses of Theorem 10.33 imply that the hypotheses of Corollary 10.34 are satisfied for the morphism q (for Hypothesis 1 use Remark 10.32). Hence we can apply Corollary 10.34 for the map q and $n = 0$ and for complexes concentrated in degree 0, which we already proved in (i). Therefore it suffices to see that

$$\begin{aligned} 0 \to &\Gamma\left(q^{-1}(z), \mathcal{G}_{|q^{-1}(z)}\right) \\ \to &\Gamma\left(q^{-1}(z), \mathcal{J}^0{}_{|q^{-1}(z)}\right) \to \Gamma\left(q^{-1}(z), \mathcal{J}^1{}_{|q^{-1}(z)}\right) \longrightarrow \ldots \end{aligned} \tag{$*$}$$

is exact. As $\mathcal{J}^p$ is flabby (Proposition 10.12), Lemma 10.35 shows that $\mathcal{J}^p{}_{|q^{-1}(z)}$ is Γ-acyclic for all p and all $z \in Z$. Hence $\mathcal{G}_{|q^{-1}(z)} \to \mathcal{J}^\bullet{}_{|q^{-1}(z)}$ computes the cohomology of $\mathcal{G}_{|q^{-1}(z)}$ by Proposition 10.8. As we have already seen that $\mathcal{G}_{|q^{-1}(z)}$ is Γ-acyclic, $(*)$ is exact.

(iii). We now conclude the proof. Choose an injective resolution $\mathcal{F}^\bullet \xrightarrow{qis} \mathcal{J}^\bullet$. In particular $\mathcal{J}^\bullet$ is bounded below and consists of flabby abelian sheaves (Proposition 10.12). By definition of $R^n f_*$ we have

$$R^n f_*(\mathcal{F}^\bullet) = H^n\left(\cdots \to f_* \mathcal{J}^p \to f_* \mathcal{J}^{p+1} \to \ldots\right).$$

As taking inverse images is exact (Proposition 10.26), we deduce

$$\begin{aligned} g^{-1}(R^n f_*(\mathcal{F}^\bullet)) &= H^n\left(\cdots \to g^{-1}(f_* \mathcal{J}^p) \longrightarrow g^{-1}(f_* \mathcal{J}^{p+1}) \to \ldots\right) \\ &\overset{(i)}{=} H^n\left(\cdots \to q_*(p^{-1} \mathcal{J}^p) \longrightarrow q_*(p^{-1} \mathcal{J}^{p+1}) \to \ldots\right) \end{aligned} \tag{$**$}$$

As p^{-1} is an exact functor, we see that $p^{-1} \mathcal{J}^\bullet$ is a resolution of $p^{-1} \mathcal{F}^\bullet$. It consists of q_*-acyclic sheaves by (ii). Hence $(**)$ computes $R^n q_*(p^{-1} \mathcal{F}^\bullet)$ by the analogue of Proposition 10.8 for the functor q_*. $\qquad\square$

Corollary 10.36. *Let X, Y, f be as in Corollary 10.34 and let $\mathcal{F}$ be an abelian sheaf on X. Assume that*

$$H^n\left(f^{-1}(y), i_y^{-1} \mathcal{F}\right) = 0 \tag{$*$}$$

for all $n \geq 1$ and $y \in Y$. Then

$$H^n(Y, f_* \mathcal{F}) = H^n(X, \mathcal{F}).$$

Proof. By Corollary 10.34, $(*)$ implies $R^n f_*(\mathcal{F}) = 0$ for all $n \geq 1$. Hence we can apply Proposition 10.29. $\qquad\square$

10.7   Problems

Problem 10.1. Let $(X, \mathcal{O}_X)$ be a ringed space. Show that every bounded below complex of $\mathcal{O}_X$-modules has an injective resolution.
Hint: Let $\mathcal{F}$ be an $\mathcal{O}_X$-module and for every $x \in X$ let $\mathcal{F}_x \hookrightarrow I_x$ be an injective $\mathcal{O}_{X,x}$-linear map to an injective $\mathcal{O}_{X,x}$-module I_x (Appendix Problem 15.14). Show that $U \mapsto \mathcal{I}(U) := \prod_{x \in U} I_x$ is an injective $\mathcal{O}_X$-module and that the canonical homomorphism $\mathcal{F} \to \mathcal{I}$ of $\mathcal{O}_X$-modules is injective. Conclude by Appendix Problem 15.11.

Problem 10.2. Let $\mathcal{A}$ and $\mathcal{A}'$ be abelian categories, let $F \colon \mathcal{A} \to \mathcal{A}'$ be a left exact functor and let $(F^n, \delta)_{n \geq 0}$ be a δ-functor extending F. Show that $(F^n, \delta)_{n \geq 0}$ is universal if for all $n \geq 1$ and for every object X in $\mathcal{A}$ there exists a monomorphism $u \colon X \to Y$ in $\mathcal{A}$ such that $F^n(u) = 0$.

Problem 10.3. Let X be a topological space, $x \in X$, A be an abelian group, and let $\mathcal{A}$ be the skyscraper sheaf in x with values A (Problem 3.6). Show that $H^p(X, \mathcal{A}) = 0$ for all $p > 0$.

Problem 10.4. Let $(X, \mathcal{O}_X)$ be a ringed space, let $\mathcal{F}^\bullet$ be a bounded below complex of $\mathcal{O}_X$-modules, and let $U, V \subseteq X$ be open. Show that there exists a *Mayer–Vietoris sequence*

$$\ldots \to H^p(U \cup V, \mathcal{F}^\bullet) \to H^p(U, \mathcal{F}^\bullet) \oplus H^p(V, \mathcal{F}^\bullet) \to H^p(U \cap V, \mathcal{F}^\bullet)$$
$$\to H^{p+1}(U \cup V, \mathcal{F}^\bullet) \to \ldots$$

Hint: Show that for every flabby sheaf $\mathcal{A}$ of abelian groups on X the sequence

$$0 \to \mathcal{A}(U \cup V) \xrightarrow{s \mapsto (s_{|U}, s_{|V})} \mathcal{A}(U) \oplus \mathcal{A}(V) \xrightarrow{(s,t) \mapsto s-t} \mathcal{A}(U \cap V) \to 0$$

is exact.

Problem 10.5. Let A be an abelian group, $n \geq 1$, $p \geq 0$ and write $H^p(X, A)$ instead of $H^p(X, A_X)$. Let $A_2 := \{a \in A \; ; \; 2a = 0\}$. Show that

$$H^p(\mathbb{P}^n(\mathbb{R}), A) \cong \begin{cases} A, & \text{if } p = 0 \text{ or if } p = n \text{ and } n \text{ odd}; \\ A/2A, & \text{if } p \text{ is even and } 0 < p \leq n; \\ A_2, & \text{if } p \text{ is odd and } 0 < p < n; \\ 0, & \text{otherwise} \end{cases}$$

and

$$H^p(\mathbb{P}^n(\mathbb{C}), A) \cong \begin{cases} A, & \text{if } p \text{ is even and } 0 \leq p \leq 2n; \\ 0, & \text{otherwise}. \end{cases}$$

Problem 10.6. Let $X := \mathbb{P}^1(\mathbb{C})$.

1. Show that $H^0_{\mathrm{DR}}(X) \cong H^2_{\mathrm{DR}}(X) \cong \mathbb{C}$ and $H^1_{\mathrm{DR}}(X) = 0$. Deduce that $H^2_{\mathrm{DR}}(X)$ is not isomorphic to the quotient of the space of closed holomorphic 2-forms by the space of exact holomorphic 2-forms.
2. Show that $H^0(X, \mathcal{O}_X) \cong H^1(X, \Omega^1_X) \cong \mathbb{C}$ and $H^q(X, \Omega^p_X) = 0$ for all $(p, q) \notin \{(0,0), (1,1)\}$.

Hint: Problem 10.4.

Problem 10.7. Generalize the results of Problem 10.6 (2) and show that

$$H^q\left(\mathbb{P}^n(\mathbb{C}), \Omega^p_{\mathbb{P}^n(\mathbb{C})}\right) \cong \begin{cases} \mathbb{C}, & \text{if } 0 \leq p = q \leq n; \\ 0, & \text{otherwise.} \end{cases}$$

Problem 10.8. Let M be a real C^α-manifold with $\alpha \leq \infty$ and let $\mathrm{Pic}_\mathbb{C}(M)$ be the group of complex C^α-line bundles on M. Show that the complex exponential sequence induces an isomorphism of abelian groups $\mathrm{Pic}_\mathbb{C}(M) \xrightarrow{\sim} H^2(M, \mathbb{Z}_M)$.

Problem 10.9. Let $f: X \to Y$ be a continuous map and let $\mathcal{F}^\bullet$ be a bounded below complex of abelian sheaves on X. Show that $R^n f_*(\mathcal{F}^\bullet)$ is the sheaf associated with the presheaf $V \mapsto H^n(f^{-1}(V), \mathcal{F}^\bullet)$ for all $n \in \mathbb{Z}$.

Problem 10.10. Let M be a real simply connected manifold. Show that every closed 1-form is exact.

Problem 10.11. Show that on the real C^∞-manifold S^1 there exists a non-exact 1-form ω that is unique up to multiplication with $\lambda \in \mathbb{R}^\times$. What is ω?

Problem 10.12. Let $f: X \to Y$. Show that the proper direct image functor $f_!$ (Problem 3.16) from the category of abelian sheaves on X to the category of abelian sheaves on Y is a left exact functor. Its right derived functor by $R^n f_!$ is called *higher direct image with compact support*. If f is the unique map to the space Y consisting of a single point, then we consider $R^n f_!(\mathcal{F}^\bullet)$ as an abelian group for every complex $\mathcal{F}^\bullet$ of abelian sheaves. This abelian group is called the *n-th cohomology with compact support* and it is denoted by $H^n_c(X, \mathcal{F}^\bullet)$. We write $\Gamma_c(X, \cdot)$ instead of $H^0_c(X, \cdot)$.

Prove for locally compact Hausdorff spaces an analogue of the Mayer–Vietoris sequence for closed subspaces (Proposition 10.31) for the cohomology with compact support.

Problem 10.13. Let X be a locally compact Hausdorff space (respectively a paracompact Hausdorff space), $U \subseteq X$ open, $Z := X \setminus U$. Show that the exact sequence in Problem 3.18 yields for every abelian sheaf $\mathcal{F}$ a long exact sequence of cohomology with compact support (Problem 10.12)

$$\cdots \to H_c^n(U, \mathcal{F}_{|U}) \to H_c^n(X, \mathcal{F}) \to H_c^n(Z, \mathcal{F}_{|Z}) \to H_c^{n+1}(U, \mathcal{F}_{|U}) \to \cdots$$

(respectivey a long exact sequence

$$\cdots \to H^n(U, \mathcal{F}_{|U}) \to H^n(X, \mathcal{F}) \to H^n(Z, \mathcal{F}_{|Z}) \to H^{n+1}(U, \mathcal{F}_{|U}) \to \cdots)$$

Problem 10.14. Let X be a paracompact locally compact Hausdorff space and let $\mathcal{F}$ be an abelian sheaf. Show that the following assertions are equivalent:

(i) $\mathcal{F}$ is soft.
(ii) $H^n(X, j_!(\mathcal{F}_{|U})) = 0$ for all $n \geq 1$ and for every open topological embedding $j: U \to X$.
(iii) $H^1(X, j_!(\mathcal{F}_{|U})) = 0$ for every open topological embedding $j: U \to X$.

Here $j_!$ is the proper direct image (Problem 3.16).
Hint: Problem 10.13.

Problem 10.15. Let X be a locally compact Hausdorff space. A sheaf $\mathcal{F}$ on X is called *c-soft* if the restriction map $\mathcal{F}(X) \to \mathcal{F}(K)$ is surjective for every compact subspace K of X.

1. Show that for an abelian sheaf $\mathcal{F}$ the following assertions are equivalent:
    (i) $\mathcal{F}$ is c-soft.
    (ii) $H_c^n(X, j_!(\mathcal{F}_{|U})) = 0$ (Problem 10.12) for all $n \geq 1$ and for every open topological embedding $j: U \to X$.
    (iii) $H_c^1(X, j_!(\mathcal{F}_{|U})) = 0$ for every open topological embedding $j: U \to X$.
    Hint: Problem 10.13.
2. Show that every colimit of c-soft abelian sheaves on X is a c-soft abelian sheaf.
3. Show that the functors $H_c^n(X, \cdot)$ commute with colimits.

Problem 10.16. Let M be a real C^α-manifold with $\alpha \leq \infty$. Show that any soft sheaf on M is c-soft (Problem 10.15) and deduce that $H_c^n(M, \mathcal{E}) = 0$ for all $n \geq 1$ and every $\mathcal{C}_M^\alpha$-module $\mathcal{E}$.

Problem 10.17. Let $U \subseteq \mathbb{R}$ be open and connected.

1. Show that
$$0 \longrightarrow \mathbb{R}_U \longrightarrow \mathcal{C}_U^1 \xrightarrow{f \mapsto f'} \mathcal{C}_U^0 \longrightarrow 0$$
is an exact sequence of $\mathbb{R}$-vector spaces.

2. Show that $H_c^n(U, \mathbb{R}_U) = \mathbb{R}$ (Problem 10.12) for $n = 1$ and $H_c^n(U, \mathbb{R}_U) = 0$ for $n \neq 1$.

    Hint: The sequence $0 \to \Gamma_c(U, \mathcal{C}_U^1) \xrightarrow{f \mapsto f'} \Gamma_c(U, \mathcal{C}_U^0) \xrightarrow{\int} \mathbb{R} \to 0$ is exact, where $\int f := \int_{-\infty}^{\infty} f(x)\, dx$.

Problem 10.18. Let X be a topological space and $A \subseteq X$ a closed subspace. For every sheaf of abelian groups $\mathcal{F}$ define

$$\Gamma_A(X, \mathcal{F}) := \{\, s \in \Gamma(X, \mathcal{F}) \,;\, \mathrm{supp}(s) \subseteq A \,\}.$$

Show that $\Gamma_A(X, \cdot)$ is a left exact functor from the category of abelian sheaves to the category of abelian groups. For every complex $\mathcal{A}^\bullet$ of abelian sheaves $H_A^p(X, \mathcal{A}^\bullet) := R^p \Gamma_A(X, \mathcal{A}^\bullet)$ is called the *local cohomology group with support in A*.

Problem 10.19. Let X be a topological space, $A \subseteq X$ closed and $U := X \setminus A$. Show that for every bounded below complex of abelian sheaves $\mathcal{A}^\bullet$ one has an exact sequence

$$\cdots \to H_A^p(X, \mathcal{A}^\bullet) \to H^p(X, \mathcal{A}^\bullet) \to H^p(U, \mathcal{A}^\bullet) \to H_A^{p+1}(X, \mathcal{A}^\bullet) \to \ldots,$$

where $H_A^p(X, \cdot)$ denotes the local cohomology with support in A (Problem 10.18).

Problem 10.20. Let $\mathcal{A}$ be an abelian category with injective and K-injective resolutions. Let $M^\bullet$ and $N^\bullet$ be complexes of objects in $\mathcal{A}$ and choose a K-injective resolution $N^\bullet \xrightarrow{qis} I^\bullet$. For $p \in \mathbb{Z}$ the abelian group

$$\mathrm{Ext}_{\mathcal{A}}^p(M^\bullet, N^\bullet) := \mathrm{Hom}_{K(\mathcal{A})}(M^\bullet, I^\bullet[p]),$$

is called the *Ext group of $M^\bullet$ and $N^\bullet$*. Here the shift $[p]$ is defined in Appendix Problem 15.1.

1. Show that if $M^\bullet \xrightarrow{qis} J^\bullet$ is a quasi-isomorphism, then

$$\mathrm{Ext}_{\mathcal{A}}^p(M^\bullet, N^\bullet) = \mathrm{Hom}_{K(\mathcal{A})}(J^\bullet, I^\bullet[p]).$$

2. Show that the group $\mathrm{Ext}_{\mathcal{A}}^p(M^\bullet, N^\bullet)$ does not depend on the choice of the K-injective resolution of $N^\bullet$ up to unique isomorphism and that $M^\bullet \mapsto \mathrm{Ext}_{\mathcal{A}}^p(M^\bullet, N^\bullet)$ and $N^\bullet \mapsto \mathrm{Ext}_{\mathcal{A}}^p(M^\bullet, N^\bullet)$ define functors $K(\mathcal{A})^{\mathrm{opp}} \to (\mathrm{Ab})$ and $K(\mathcal{A}) \to (\mathrm{Ab})$ that send quasi-isomorphisms of complexes to isomorphism of abelian groups.

3. Show that for all objects M and N of $\mathcal{A}$ (considered as complexes concentrated in degree 0) one has $\mathrm{Ext}_{\mathcal{A}}^0(M, N) = \mathrm{Hom}_{\mathcal{A}}(M, N)$.

4. Show that for $p, q \in \mathbb{Z}$ and for all complexes $M^\bullet, N^\bullet, P^\bullet$ of objects in $\mathcal{A}$ and all K-injective resolutions $M^\bullet \xrightarrow{qis} J^\bullet, N^\bullet \xrightarrow{qis} I^\bullet, P^\bullet \xrightarrow{qis} K^\bullet$ the composition

$$\mathrm{Hom}_{K(\mathcal{A})}(J^\bullet, I^\bullet[p]) \times \mathrm{Hom}_{K(\mathcal{A})}(K^\bullet, J^\bullet[q]) \to \mathrm{Hom}_{K(\mathcal{A})}(K^\bullet, I^\bullet[p+q]),$$
$$(v, u) \mapsto v[q] \circ u$$

yields a well-defined $\mathbb{Z}$-bilinear map

$$\cup : \mathrm{Ext}^p_{\mathcal{A}}(M^\bullet, N^\bullet) \times \mathrm{Ext}^q_{\mathcal{A}}(P^\bullet, M^\bullet) \to \mathrm{Ext}^{p+q}_{\mathcal{A}}(P^\bullet, N^\bullet), \qquad (10.18)$$

which is called the *Yoneda product*. Show that the Yoneda product is associative and makes $\bigoplus_p \mathrm{Ext}^p_{\mathcal{A}}(M^\bullet, M^\bullet)$ into a ring.

5. Show that for every object M of $\mathcal{A}$ the p-th right derived functor of the left exact functor $\mathcal{A} \to (\mathrm{Ab})$, $X \mapsto \mathrm{Hom}_{\mathcal{A}}(M, X)$, is the functor $X \mapsto \mathrm{Ext}^p_{\mathcal{A}}(M, X)$.

Problem 10.21. Let $(X, \mathcal{O}_X)$ be a ringed space. Show that for all $\mathcal{O}_X$-modules $\mathcal{F}$ and all $p \in \mathbb{Z}$ there is a functorial isomorphism $H^p(X, \mathcal{F}) \xrightarrow{\sim} \mathrm{Ext}^p_{\mathcal{O}_X}(\mathcal{O}_X, \mathcal{F})$ (Problem 10.20). Deduce that the Yoneda product (10.18) endows $\bigoplus_p H^p(X, \mathcal{O}_X)$ with the structure of a ring and $\bigoplus_p H^p(X, \mathcal{F})$ with the structure of a right module over the ring $\bigoplus_p H^p(X, \mathcal{O}_X)$. The scalar multiplication

$$H^p(X, \mathcal{F}) \times H^q(X, \mathcal{O}_X) \to H^{p+q}(X, \mathcal{F}), \qquad (\xi, \alpha) \mapsto \xi \cup \alpha$$

is called the *cup product*.

Problem 10.22. Let M be a real manifold. Show that via the isomorphism $\bigoplus_p H^p(M, \mathbb{K}_M) \xrightarrow{\sim} \bigoplus_p H^p_{\mathrm{DR}}(M)$ the cup product (Problem 10.21) on the left-hand side corresponds to the exterior product $(\omega, \eta) \mapsto \omega \wedge \eta$ of differential forms on the right-hand side.

Problem 10.23. Let X be a topological space and let $1 \to \mathcal{A} \to \mathcal{G} \to \mathcal{H} \to 1$ (*) be an exact sequence of sheaves of groups on X such that $\mathcal{A}$ is in the center of $\mathcal{G}$. Show that there exists a functorial morphism of pointed sets $\partial : H^1(X, \mathcal{H}) \to H^2(X, \mathcal{A})$ such that $H^1(X, \mathcal{G}) \to H^1(X, \mathcal{H}) \xrightarrow{\partial} H^2(X, \mathcal{A})$ is an exact sequence of pointed sets and such that ∂ is the usual connecting morphism if (*) is a sequence of sheaves of abelian groups.
Hint: Apply $( )^{[0]}$ to (*).

Problem 10.24. Let M be a real manifold. Show that every principal $\mathrm{PGL}_n(\mathbb{R})$-bundle (respectively every principal $\mathrm{PGL}_n(\mathbb{C})$-bundle, where we consider $\mathrm{PGL}_n(\mathbb{C})$ as a real Lie group) over M comes from a principal $\mathrm{GL}_n(\mathbb{R})$-bundle (respectively a principal $\mathrm{GL}_n(\mathbb{C})$-bundle) by the functoriality of bundles (Remark 8.23) if and only if $H^2(M, (\mathbb{Z}/2\mathbb{Z})_M) = 0$ (respectively $H^3(M, \mathbb{Z}_M) = 0$).
Hint: Use Problem 10.23 and the exponential sequence.
Remark: In Corollary 11.14 we will see $H^2(M, (\mathbb{Z}/2\mathbb{Z})_M) = H^3(M, \mathbb{Z}_M) = 0$ if M is a contractible manifold.

Cohomology of Constant Sheaves

11

In this chapter we focus on cohomology of constant sheaves. We start by defining singular (co)homology and then show that for a locally contractible space X singular cohomology with values in a ring R and the cohomology of the constant sheaf R_X are equal. We deduce that R_X is acyclic if X is contractible and locally contractible.

Combining this result for $X = [0, 1]$ and the proper base change theorem we deduce homotopy invariance of cohomology of constant sheaves for continuous maps between arbitrary topological spaces in Sect. 11.3. We conclude the chapter with some easy applications.

Notation: Let R always be a commutative ring and let X be a topological space.

11.1 Singular Cohomology

In this section we define singular cohomology and show that it vanishes on contractible spaces.

Definition 11.1. Let $n \in \mathbb{N}_0$ and set $[n] := \{0, \dots, n\}$. Let $e_i \in \mathbb{R}^{[n]} = \mathbb{R}^{n+1}$ be the i-th standard unit vector for $i = 0, \dots, n$.

1. The *n-dimensional standard simplex* is

$$\Delta^n := \Delta[n] := \left\{ (t_0, \dots, t_n) \in \mathbb{R}^{n+1} \; ; \; \sum_{i=0}^{n} t_i = 1, t_i \geq 0 \right\}.$$

Then $\Delta^0 = \{1\} \subseteq \mathbb{R}$. We identify

$$\Delta^1 = \{ (t_0, t_1) \; ; \; 0 \leq t_0 = 1 - t_1 \leq 1 \} \xrightarrow{\sim} [0, 1], \qquad (t_0, t_1) \mapsto t_1.$$

© Springer Fachmedien Wiesbaden 2016
T. Wedhorn, *Manifolds, Sheaves, and Cohomology*, Springer Studium Mathematik – Master,
DOI 10.1007/978-3-658-10633-1_11

Δ^2 is an equilateral triangle, Δ^3 is a tetrahedron.

2. Every weakly increasing map $\alpha: [m] \to [n]$ induces an affine map

$$\Delta(\alpha): \Delta[m] \to \Delta[n], \qquad \sum_{i=0}^{m} t_i e_i \mapsto \sum_{i=0}^{m} t_i e_{\alpha(i)}.$$

Then $\Delta(\alpha \circ \beta) = \Delta(\alpha) \circ \Delta(\beta)$ and $\Delta(\mathrm{id}) = \mathrm{id}$.

3. For $0 \le i \le n$ let $\delta_i^n: [n-1] \to [n]$ be the weakly increasing injective map that misses the value i. Then $\delta_j^{n+1} \circ \delta_i^n = \delta_i^{n+1} \circ \delta_{j-1}^n$ for $i < j$ (both compositions miss i and j). Set

$$d_i^n := \Delta(\delta_i^n): \Delta[n-1] \to \Delta[n].$$

This maps $\Delta[n-1]$ injectively on the "i-th face of $\Delta[n]$". For instance

$$d_0^1(1) = (0, 1) \in \Delta^1 \qquad d_1^1(1) = (1, 0) \in \Delta^1.$$

By functoriality one has

$$d_j^{n+1} \circ d_i^n = d_i^{n+1} \circ d_{j-1}^n \qquad \text{for } i < j. \tag{11.1}$$

Definition 11.2. Let X be a topological space, $n \in \mathbb{N}_0$.

1. A continuous map $\sigma: \Delta^n \to X$ is called a *(singular) n-simplex in X*. The set of n-simplices in X is denoted by $\Sigma_n(X)$. The *i-th face of* σ is $\sigma \circ d_i^n$ (an $(n-1)$-simplex).
2. Denote by $S_n(X)$ the free abelian group with basis the set of singular n-simplices in X. Set $S_n(X) := 0$ for $n < 0$. An element of $x \in S_n(X)$ is called a *singular n-chain*. We think of x as a formal finite linear combination $x = \sum_\sigma n_\sigma \sigma, n_\sigma \in \mathbb{Z}$.
3. Let G be an abelian group. A *singular n-cochain with values in G* is a map $\Sigma_n(X) \to G$. Equivalently, it is a group homomorphism $S_n(X) \to G$. We denote the group of singular n-cochains with values in G by $S^n(X, G) = \mathrm{Hom}_{\mathbb{Z}}(S_n(X), G)$.
4. Define boundary operators for all $q \in \mathbb{Z}$:

$$\partial_q: S_q(X) \to S_{q-1}(X), \qquad \sigma \mapsto \sum_{i=0}^{q} (-1)^i \left(\sigma \circ d_i^q\right),$$

$$\partial^q: S^q(X, G) \to S^{q+1}(X, G), \qquad \partial^q(\varphi) = (-1)^{q+1}(\varphi \circ \partial_{q+1}).$$

Example 11.3.

1. One can identify $\Sigma_0(X) = X$ (as a set), $S_0(X) = \mathbb{Z}^{(X)}$ and

$$S^0(X, G) = \{\varphi: X \to G \text{ map}\}.$$

2. Associating to $\sigma \in \Sigma_1(X)$ the path $[0, 1] \to X, t \mapsto \sigma(t, 1-t)$ yields an identification of $\Sigma_1(X)$ and the set of paths in X. Hence $\varphi \in S^1(X, G)$ is a map that attaches to each path in X an element in G. For $\sigma \in \Sigma_1(X)$, considered as a path $[0, 1] \to X$, one has

$$\partial_1(\sigma) = \sigma(1) - \sigma(0) \in S_0(X).$$

For $\varphi \in S^0(X, G)$ considered as a map $X \to G$ one has

$$\partial^0(\varphi)(\sigma) = -\varphi(\partial_1(\sigma)) = \varphi(\sigma(0)) - \varphi(\sigma(1))$$

for a path σ in X. In particular

$$\mathrm{Ker}(\partial^0) = \{\varphi \colon X \to G \mid \varphi(x) = \varphi(y) \text{ if } x \text{ and } y \text{ are in the same} \atop \text{path-connected component}\}. \tag{11.2}$$

Lemma 11.4. *For all $q \in \mathbb{Z}$ one has $\partial_q \circ \partial_{q-1} = 0$ and $\partial^q \circ \partial^{q-1} = 0$. In particular we obtain complexes*

$$\cdots \xrightarrow{\partial_3} S_2(X) \xrightarrow{\partial_2} S_1(X) \xrightarrow{\partial_1} S_0(X) \longrightarrow 0 \longrightarrow \cdots, \tag{11.3}$$

$$\cdots \longrightarrow 0 \longrightarrow S^0(X, G) \xrightarrow{\partial^0} S^1(X, G) \xrightarrow{\partial^1} S^2(X, G) \xrightarrow{\partial^2} \cdots \tag{11.4}$$

Proof. It suffices to show $\partial_q \circ \partial_{q-1} = 0$. We decompose

$$\partial_q \partial_{q-1} \sigma = \sum_{j=0}^{q} \sum_{i=0}^{q-1} (-1)^{i+j} \left(\sigma \circ d_j^q \circ d_i^{q-1} \right)$$

into the parts $\sum_{i<j}$ and $\sum_{i \geq j}$. When we rewrite the first sum using (11.1), the result is the negative of the second sum. $\qquad\square$

Definition 11.5. We define for all $n \in \mathbb{Z}$ the *singular homology*

$$H_n^{\mathrm{sing}}(X) := H_n^{\mathrm{sing}}(X, \mathbb{Z}) := \frac{\mathrm{Ker}(\partial_n \colon S_n(X) \to S_{n-1}(X))}{\mathrm{Im}(\partial_{n+1} \colon S_{n+1}(X) \to S_n(X))}$$

and the *singular cohomology* or *Betti cohomology with coefficients in G*

$$H_{\mathrm{sing}}^n(X, G) := \frac{\mathrm{Ker}(\partial^n \colon S^n(X, G) \to S^{n+1}(X, G))}{\mathrm{Im}(\partial^{n-1} \colon S^{n-1}(X, G) \to S^n(X, G))}.$$

Hence both the fundamental group and $H_1^{\mathrm{sing}}(X, \mathbb{Z})$ consist of equivalence classes of paths. These groups are related as follows (see Problem 11.5 or [tDi] (9.2.1)).

Remark 11.6. Let X be a path connected space, $x_0 \in X$. Attaching to the homotopy class $[\gamma] \in \pi_1(X, x_0)$ its class in $H_1^{\mathrm{sing}}(X, \mathbb{Z})$ yields a well-defined isomorphism of groups

$$\pi_1(X, x_0)^{\mathrm{ab}} \xrightarrow{\sim} H_1^{\mathrm{sing}}(X, \mathbb{Z}).$$

Here for any group G we denote by G^{ab} its maximal abelian quotient, i.e., $G^{\mathrm{ab}} = G/[G, G]$, where the commutator subgroup $[G, G]$ is the subgroup of G generated by all commutators $ghg^{-1}h^{-1}$, which is a normal subgroup of G.

Definition and Remark 11.7. Let R be a commutative ring.

1. We also define $H_n^{\mathrm{sing}}(X, R)$ by replacing $S_n(X)$ by the free R-modules $S_n(X, R) := R^{(\Sigma_n(X))}$ with basis the set of singular n-simplices in X. Then

$$\cdots \xrightarrow{\partial_3} S_2(X, R) \xrightarrow{\partial_2} S_1(X, R) \xrightarrow{\partial_1} S_0(X, R) \longrightarrow 0 \longrightarrow \cdots \qquad (11.5)$$

   is a complex of R-modules and we obtain the *singular homology with values in R*, denoted by $H_n^{\mathrm{sing}}(X, R)$. Then $S_n(X) = S_n(X, \mathbb{Z})$ and $S_n(X, R) = S_n(X) \otimes_{\mathbb{Z}} R$.

2. If $G = R$, then (11.4) is a complex of R-modules with $S^n(X, R) = S_n(X, R)^{\vee}$ (dual as R-module) and ∂^q is up to the sign $(-1)^{q+1}$ the dual map of ∂_{q+1}. In particular, $H_{\mathrm{sing}}^n(X, R)$ is an R-module.

Example 11.8. Let R be a commutative ring. Then $H_0^{\mathrm{sing}}(X, R) \cong R^{(\pi_0(X))}$, where $\pi_0(X)$ is the set of path-connected components of X.

As forming the dual of an R-module is in general only left exact (and not an exact functor), the relation between the dual of the homology of a complex of modules and the homology of the complex of the dual modules is in general quite complicated and best expressed via spectral sequences. We use only the following general fact for complexes of free R-modules (e.g., [BouA3] Chap. X, §5.6, Cor. 3 and Cor. 4 de Théoreme 3). Variants of the version given here are sometimes grouped under the *universal coefficient theorem for singular cohomology*.

Proposition 11.9. *Let $n \in \mathbb{Z}$ and let R be a commutative ring. Assume* one *of the following two conditions:*

(a) *R is a principal ideal domain and that $H_{n-1}^{\mathrm{sing}}(X, R)$ is a free R-module.*
(b) *$H_m^{\mathrm{sing}}(X, R)$ is a free R-module for all $m \le n - 1$.*

Then

$$H_{\mathrm{sing}}^n(X, R) \cong H_n^{\mathrm{sing}}(X, R)^{\vee} = \mathrm{Hom}_R\left(H_{\mathrm{sing}}^n(X, R), R\right). \qquad (11.6)$$

Corollary 11.10.

1. *Let K be a field. For all $n \in \mathbb{Z}$ one has $H^n_{\mathrm{sing}}(X, K) \cong H^{\mathrm{sing}}_n(X, K)^\vee$.*
2. *Let R be a commutative ring. If $n \leq 1$ then one has $H^n_{\mathrm{sing}}(X, R) \cong H^{\mathrm{sing}}_n(X, R)^\vee$ because $H^{\mathrm{sing}}_m(X, R)$ is a free R-module for all $m < 1$ by Example 11.8.*

We encourage the reader to prove 1 directly using the fact that every subspace U of a K-vector space V is a direct summand and hence any linear map on U can be extended to V.

Definition and Remark 11.11. Let Y be a topological space. Define the *cone over Y* by

$$C(Y) := (Y \times [0, 1])/\sim,$$

where $(y, t) \sim (y', t')$ if $y = y'$ and $t = t' < 1$ or if $t = t' = 1$. Endow $C(Y)$ with the quotient topology. Then $C(\Delta^n) = \Delta^{n+1}$.

Proposition 11.12. *Let X be a contractible space and let R be a commutative ring. Then one has for all $n > 0$*

$$H^{\mathrm{sing}}_n(X, R) = H^n_{\mathrm{sing}}(X, R) = 0.$$

More generally, one can show that if $f, g: X \to Y$ are homotopic continuous maps of topological spaces, then they induce homotopic morphisms of complexes $S_\bullet(X, R) \to S_\bullet(Y, R)$. In particular, they induce the same map on the singular homology. This gives a connection between the topological notion of homotopy and the algebraic notion of homotopy between morphism of complexes.

Proof. Set $I := [0, 1]$. By Proposition 11.9 it suffices to show that $H^{\mathrm{sing}}_n(X, R) = 0$ for all $n > 0$. Let $p \in X$ and let $F: X \times I \to X$ be a homotopy of id_X and the constant map $X \to X, x \mapsto p$.

Define a morphism of complexes $\varepsilon: S_\bullet(X, R) \to S_\bullet(X, R)$ by $\varepsilon_n := 0$ for $n \neq 0$ and $\varepsilon_0(\sum_{x \in X} n_x x) := (\sum_x n_x)p$ (where we identify 0-simplices in X and points of X). It suffices to show that there exists a homotopy between the morphism of complexes ε and $\mathrm{id}_{S_\bullet(X,R)}$ (then they induce the same maps on $H^{\mathrm{sing}}_n(X, R)$ and hence $\mathrm{id}_{H^{\mathrm{sing}}_n(X,R)} = 0$ for all $n > 0$).

Let $\sigma: \Delta^n \to X$ be an n-simplex in X. Then $F \circ (\sigma \times \mathrm{id}_I): \Delta^n \times I \to X$ factors through $C(\Delta^n) = \Delta^{n+1}$ and hence induces an $(n + 1)$-simplex $h_n(\sigma): \Delta^n \to X$. This defines homomorphisms $h_n: S_n(X, R) \to S_{n+1}(X, R)$. One easily checks $(h_n)_n$ is the desired homotopy (note that the numbering for homotopies is reversed because the differential in the complexes here lowers the degree). $\qquad\square$

11.2   Cohomology and Singular Cohomology

We now compare singular cohomology with values in a commutative ring R and sheaf cohomology of the sheaf of locally constant R-valued functions. In the sequel we will write $H^n(X, R)$ instead of $H^n(X, R_X)$ for all $n \geq 0$.

Theorem 11.13. *Let X be a locally contractible space (e.g., a premanifold) and let R be a commutative ring. Then there is for all $n \in \mathbb{Z}$ an isomorphism $H^n(X, R) \cong H^n_{\mathrm{sing}}(X, R)$ of R-modules.*

Proof. (i). For $V \subseteq U \subseteq X$ open we have $\Sigma_n(V) \subseteq \Sigma_n(U)$ and we define a surjective map

$$S^n(U, R) \longrightarrow S^n(V, R), \qquad \varphi \mapsto \varphi_{|\Sigma_n(V)}.$$

Let $\mathcal{S}^n_R$ be the sheafification of the presheaf of abelian groups $U \mapsto S^n(U, R)$. Then one easily checks that for $U \subseteq X$ open one has

$$\mathcal{S}^n_R(U) = S^n(U, R)/S^n(U, R)_0, \tag{*}$$

where

$$S^n(U, R)_0 = \left\{ \varphi \in S^n(U, R) \; ; \; \exists \text{ covering } (V_i)_i \text{ of } U \text{ with } \varphi_{|\Sigma_n(V_i)} = 0 \text{ for all } i \right\}.$$

As $S^n(U, R) \longrightarrow S^n(V, R)$ is surjective for $V \subseteq U$, (*) shows in particular that the sheaves $\mathcal{S}^n_R$ are flabby.

(ii). The map ∂^n yield morphisms of sheaves of abelian groups $\partial^n \colon \mathcal{S}^n_R \longrightarrow \mathcal{S}^{n+1}_R$ and we obtain a complex

$$\cdots \longrightarrow 0 \longrightarrow \mathcal{S}^0_R \xrightarrow{\partial^0} \mathcal{S}^1_R \xrightarrow{\partial^1} \mathcal{S}^2_R \xrightarrow{\partial^2} \cdots \tag{**}$$

of $\mathbb{Z}_X$-modules. As X is locally path connected, a map $X \to R$ is locally constant if and only if it is constant on path-connected components. Hence (11.2) shows that $\mathrm{Ker}(\partial^0) = R_X$.

Moreover, the complex (**) is exact in all degrees > 0. Let $x \in X$. By hypothesis on X we can compute stalks of presheaves in a point x by taking colimits only over open contractible neighborhoods of x. Now Proposition 11.12 shows that for all open contractible neighborhoods U of x the complex of cochains $S^\bullet(U, R)$ is exact in degree > 0. This implies that $\mathcal{S}^\bullet_R$ is exact on stalks and hence exact in degree > 0. Therefore we have seen that

$$R_X \xrightarrow{qis} \mathcal{S}^\bullet_R$$

is a flabby resolution of R_X and hence

$$H^n(X, R) = H^n \left(\cdots \to \mathcal{S}^p_R(X) \to \mathcal{S}^{p+1}_R(X) \to \cdots \right).$$

(iii). Let $p: S^\bullet(X, R) \to S_R^\bullet(X)$ be the morphism of complexes induced by (*). It remains to show that p is an isomorphism in the homotopy category $K(R)$. This follows from the "theorem of small chains" ([tDi] Theorem 9.4.5). □

Corollary 11.14. *Let X be a contractible locally contractible space. Then $H^n(X, R) = 0$ for all $n > 0$ and for every commutative ring R.*

Proof. Proposition 11.12 and Theorem 11.13. □

Corollary 11.15. [1] *Let M be a contractible real C^∞-manifold, $n \geq 1$. Then every closed n-form on M is exact.*

Proof. We have

$$\frac{\{\text{closed } n\text{-forms on } M\}}{\{\text{exact } n\text{-forms on } M\}} \overset{\text{Theorem 10.20}}{=} H^n(M, \mathbb{R}) \overset{\text{Corollary 11.14}}{=} 0.$$

□

By Remark 10.21, Corollary 11.15 also holds for real analytic manifolds.

Let us give a proof of the Weierstraß theorem from complex analysis using Corollary 11.14.

Example 11.16. Let X be a connected 1-dimensional complex manifold (for instance an open subset of $\mathbb{C}$). Let $U \subseteq X$ be open. A *divisor on U* is an element $D = (n_z)_{z \in U} \in \mathbb{Z}^U$ such that $\mathrm{supp}(D) := \{z \in U \; ; \; n_z \neq 0\}$ is a closed and discrete subspace of U. Denote the set of divisors on U by $\mathcal{D}_X(U)$. We obtain an abelian sheaf $\mathcal{D}_X$ of $\mathbb{Z}$-valued functions on X with stalks $\mathcal{D}_{X,x} = \mathbb{Z}$ for all $x \in X$.

Let $\mathcal{M}_X^\times$ be the sheaf of those meromorphic functions f on open subspaces U of X such that $1/f$ exists as meromorphic function (i.e., a meromorphic function f on U is in $\mathcal{M}_X^\times(U)$ if $f_{|W} \neq 0$ for every connected component W of U). Let $z_0 \in X$. To describe the stalk $\mathcal{M}_{X,z_0}^\times$ we choose a chart of X at z_0 and hence may assume that $X \subseteq \mathbb{C}$ is open. Then $\mathcal{M}_{X,z_0}^\times$ is the abelian group of non-vanishing Laurent series with finite principal parts $\sum_{n \geq m} a_n (z - z_0)^n$, $a_m \neq 0$, that converge in $U \setminus \{z_0\}$ for some open neighborhood U of z_0 in $\mathbb{C}$.

For every $f \in \mathcal{M}_X^\times(U)$ let $\mathrm{div}_U(f) := (\mathrm{ord}_z(f))_{z \in U} \in \mathcal{D}(U)$, where $\mathrm{ord}_z(f)$ denotes the order of vanishing of f at z (equal to $-k \in \mathbb{Z}_{<0}$ if z is a pole of order k for f). We obtain a morphism of abelian sheaves $\mathrm{div}: \mathcal{M}_X^\times \to \mathcal{D}_X$, which is given on stalks by sending

---

[1] Note that this corollary is something of a circular statement because it uses the Poincaré lemma for smooth manifolds, which is often proved by showing this corollary first.

$\sum_{n \geq m} a_n (z - z_0)^n$ with $a_m \neq 0$ to $m \in \mathbb{Z}$. In particular, it is surjective. Therefore we get an exact sequence of abelian sheaves

$$1 \longrightarrow \mathcal{O}_X^\times \longrightarrow \mathcal{M}_X^\times \xrightarrow{\text{div}} \mathcal{D}_X \longrightarrow 0.$$

The long cohomology sequence shows that for every divisor D on X there exists a meromorphic function f on X with $\text{div}(f) = D$ if and only if $\text{Pic}(X) = H^1(X, \mathcal{O}_X^\times) \to H^1(X, \mathcal{M}_X^\times)$ is injective.

To analyze $\text{Pic}(X)$ we use the complex exponential sequence (Example 8.11). It yields an exact sequence

$$\cdots \to H^1(X, \mathcal{O}_X) \to \text{Pic}(X) \to H^2(X, \mathbb{Z}) \to H^2(X, \mathcal{O}_X) \to \cdots$$

If $X \subseteq \mathbb{C}$ is open, then $H^i(X, \mathcal{O}_X) = 0$ for all $i \geq 1$ (Theorem 10.22) and we obtain an isomorphism $\text{Pic}(X) \xrightarrow{\sim} H^2(X, \mathbb{Z})$. In particular it follows from Corollary 11.14 that $\text{Pic}(X) = 1$ if X is contractible[2]. Hence in this case every divisor on X comes from a meromorphic function.

11.3 Homotopy Invariance of Cohomology

Lemma 11.17. *Let X be a topological space and let $p : X \times [0, 1] \to X$ be the projection. Then the pullback maps (Definition 10.27)*

$$p^{-1} : H^n(X, R) \xrightarrow{\sim} H^n(X \times [0, 1], R)$$

are isomorphism of R-modules for all $n \geq 0$ and for every commutative ring R.

Proof. As $[0, 1]$ is Hausdorff and compact, p is proper and separated (Proposition 10.32). Moreover we have $p_*(R_{X \times [0,1]}) = R_X$ (Example 3.44).

By Corollary 11.14 we have $H^n(p^{-1}(x), R) = H^n([0, 1], R) = 0$ for all $n > 0$ and $x \in X$. Therefore the result follows from Corollary 10.36. $\square$

Theorem 11.18 (Homotopy Invariance of Cohomology). *Let X and Y be topological spaces, and let $f, g : X \to Y$ be homotopic continuous maps. Then*

$$f^{-1} = g^{-1} : H^n(Y, R) \to H^n(X, R)$$

for all $n \geq 0$ and for every commutative ring R.

---

[2] In fact it can be shown that $H^2(X, \mathbb{Z}) = 1$ for every open subspace X of $\mathbb{C}$ (see for instance [Ive] VI, 6.8., considering X as an open subset of $\mathbb{P}^1(\mathbb{C})$).

Proof. For $t = 0, 1$ let $i_t : X \to X \times [0, 1]$, $i_t(x) = (x, t)$. Let $p : X \times [0, 1] \to X$ be the first projection. Then $p \circ i_t = \mathrm{id}_X$. Hence the composition

$$H^n(X, R) \xrightarrow{p^{-1}} H^n(X \times [0, 1], R) \xrightarrow{i_t^{-1}} H^n(X, R)$$

is the identity. As p^{-1} is an isomorphism by Lemma 11.17, we see that $i_0^{-1} = i_1^{-1}$ and that both maps are isomorphisms.

Let F be a homotopy between f and g. Then $F \circ i_0 = f$ and $F \circ i_1 = g$ and hence $f^{-1} = i_0^{-1} \circ F^{-1} = i_1^{-1} \circ F^{-1} = g^{-1}$. $\qquad\square$

Corollary 11.19. *Let X and Y be topological spaces and let $f : X \to Y$ and $g : Y \to X$ be continuous maps such that $g \circ f$ is homotopic to id_X and such that $g \circ f$ is homotopic to id_Y. Then f and g induce mutually inverse isomorphisms*

$$H^n(X, R) \cong H^n(Y, R)$$

for all $n \geq 0$ and for every commutative ring R.

11.4 Applications

Example 11.20. Let $n \in \mathbb{N}$ and let R be a commutative ring. Let

$$S^n = \left\{ x \in \mathbb{R}^{n+1} \; ; \; x_0^2 + \cdots + x_n^2 = 1 \right\}.$$

Define $S_{\pm}^n = \{ x \in S^n \; ; \; \pm x_n \geq 0 \}$. Each of these half spheres is homeomorphic to the closed unit ball in $\mathbb{R}^n$ and in particular contractible. Moreover, $S_+^n \cap S_-^n = S^{n-1}$. Hence the Mayer–Vietoris sequence (Proposition 10.31) has the form

$$0 \to H^0(S^n, R) \to H^0\left(S_+^n, R\right) \oplus H^0\left(S_-^n, R\right) \to H^0(S^{n-1}, R) \to H^1(S^n, R)$$
$$\to 0 \to \cdots \to 0 \to H^{p-1}(S^{n-1}, R) \xrightarrow{\sim} H^p(S^n, R) \to 0 \to \ldots \qquad (11.7)$$

Hence we see

$$H^p(S^1, R) = \begin{cases} R, & \text{for } p = 0, 1; \\ 0, & \text{otherwise.} \end{cases}$$

Moreover we have $H^0(S^n, R) = R$ because S^n is connected. Hence (11.7) shows by induction that for all $n \geq 1$ one has

$$H^p(S^n, R) = \begin{cases} R, & \text{for } p = 0, n; \\ 0, & \text{otherwise.} \end{cases} \qquad (11.8)$$

Corollary 11.21. *Let $n \geq 2$. Let $x_0 \in \mathbb{R}^n$ and $X := \mathbb{R}^n \setminus \{x_0\}$. Then*

$$H^p(X, R) = \begin{cases} R, & \text{for } p = 0, n-1; \\ 0, & \text{otherwise,} \end{cases} \tag{11.9}$$

for every commutative ring R. In particular we see that X is not contractible.

Proof. We may assume that $x_0 = 0$. The inclusion $i \colon S^{n-1} \hookrightarrow X := \mathbb{R}^n \setminus \{0\}$ is a homotopy equivalence (Example 2.4). Hence the claim follows from Example 11.20 via homotopy invariance (Corollary 11.19). $\qquad\square$

Corollary 11.22. *Let $f \colon \mathbb{R}^n \to \mathbb{R}^m$ be a homeomorphism. Then $n = m$.*

Proof. The homeomorphism f induces a homeomorphism $\mathbb{R}^n \setminus \{0\} \xrightarrow{\sim} \mathbb{R}^m \setminus \{f(0)\}$ and hence an isomorphism $H^p(\mathbb{R}^n \setminus \{0\}, \mathbb{Z}) \xrightarrow{\sim} H^p(\mathbb{R}^m \setminus \{f(0)\}, \mathbb{Z})$ for all p. Now the claim follows from Corollary 11.21 $\qquad\square$

Corollary 11.23 (Brouwer's fixed point theorem). *For $n \in \mathbb{N}$ let $D^n := \{x \in \mathbb{R}^n \; ; \; \|x\|_2 \leq 1\}$ be the closed unit ball. Then every continuous map $f \colon D^n \to D^n$ has a fixed point (i.e., there exists $x \in D^n$ with $f(x) = x$).*

Proof. Assume that there exists a continuous map $f \colon D^n \to D^n$ without fixed points. Then we can construct a continuous map $r \colon D^n \to S^{n-1}$ as follows. Let $r(x)$ denote the point, where the half line starting in $f(x)$ towards x intersects S^{n-1}. Let $i \colon S^{n-1} \to D^n$ be the inclusion. As $r \circ i = \text{id}$, the map

$$i^{-1} \colon H^{n-1}(D^n, \mathbb{Z}) \to H^{n-1}(S^{n-1}, \mathbb{Z})$$

is surjective. But for $n \geq 2$ one has $H^{n-1}(D^n, \mathbb{Z}) = 0$ because D^n is contractible and $H^{n-1}(S^{n-1}, \mathbb{Z}) = \mathbb{Z}$ by (11.8). For $n = 1$ one has $H^0(D^1, \mathbb{Z}) = \mathbb{Z}$ and $H^0(S^0, \mathbb{Z}) \cong \mathbb{Z}^2$. In both cases we obtain a contradiction! $\qquad\square$

For $n = 1$, D^1 is a closed interval and Brouwer's fixed point theorem is also a standard application of the intermediate value theorem.

11.5 Problems

In the problem section, R always denotes a commutative ring.

Problem 11.1. What is the chain complex $S_\bullet(X)$ for X consisting of a single point? What are $H_n^{\text{sing}}(X)$ and $H_{\text{sing}}^n(X, R)$?

Problem 11.2. Let X be a topological space and let $(X_i)_i$ be its family of path compo-nents. Show that $H_n^{\text{sing}}(X) = \bigoplus_i H_n^{\text{sing}}(X_i)$ for all $n \geq 0$.

Problem 11.3. Suppose that $R \neq 0$. Show that $H_{\text{sing}}^0(X^0, R) \neq H^0(X^0, R)$ with X^0 defined in Problem 2.7.

Problem 11.4. Let R be a commutative ring without $\mathbb{Z}$-torsion (e.g., if R is a $\mathbb{Q}$-algebra). Show that $H_n^{\text{sing}}(X, R) \cong H_n^{\text{sing}}(X) \otimes_{\mathbb{Z}} R$ (*).

Give an example such that (*) does not hold for $R = \mathbb{Z}/2\mathbb{Z}$.

Hint: R is a flat $\mathbb{Z}$-module (Appendix Problem 14.9). For the example see Problem 7.6.

Problem 11.5. Let X be a topological space, let $x_0 \in X$. Show that $\pi_1(X, x_0)^{\text{ab}} \cong H_1^{\text{sing}}(X, \mathbb{Z})$. The following steps might be useful:

1. Let $\sigma, \tau : [0, 1] \to X$ paths with $\sigma(1) = \tau(0)$. Let $\omega \in \Sigma_2(X)$, $(t_0, t_1, t_2) \mapsto (\sigma \cdot \tau)(t_1/2 + t_2)$. Show that $\partial_2(\omega) = \sigma + \tau - \sigma \cdot \tau$.
2. Let $H : \Delta^1 \times [0, 1] \to X$ be a homotopy $\sigma \simeq \tau$ of paths $\sigma, \tau : \Delta^1 \to X$ relative to $\{0, 1\}$. Show that H factors through the quotient map $\Delta^1 \times [0, 1] \to \Delta^2$, $(t_0, t_1, t) \mapsto (t_0, t_1(1-t), t_1 t)$ and hence yields an element $\eta \in \Sigma_2(X)$. Show that $\partial_2(\eta) = c + \sigma - \tau$, where c is a constant map.
3. Deduce from (1) and (2) that attaching to a path σ in X with start and end point x_0 its class in $H_1^{\text{sing}}(X, \mathbb{Z})$ yields a well-defined group homomorphism $h_{X, x_0} : \pi_1(X, x_0)^{\text{ab}} \to H_1^{\text{sing}}(X, \mathbb{Z})$, where $(\ )^{\text{ab}}$ denotes the maximal abelian quotient.
4. Suppose that X is path connected. Choose for every $x \in X$ a path γ_x from x_0 to x. Let $\sigma \in \Sigma_1(X)$ and let $k(\sigma)$ be the closed path $(\gamma_{\sigma(0)} \cdot \sigma) \cdot \gamma_{\sigma(1)}^-$. Show that the linear extension $\tilde{k}$ of k to $\text{Ker}(\partial_1)$ yields an inverse to h_{X, x_0}.
   Hint: To show that $\tilde{k}$ maps $\text{Im}(\partial_2)$ to 0 use that Δ^2 is contractible and deduce that for $\omega \in \Sigma_2(X)$ one has $\omega_2 \cdot \omega_0 = \omega_1$ if $\omega_i := \omega \circ d_i^2$ is the i-th face of ω.

Problem 11.6. Show that $H_c^n(\mathbb{R}^n, R) = R$ (Problem 10.12) and $H_c^p(\mathbb{R}^n, R) = 0$ for $p \neq n$.

Hint: Problem 10.13.

Problem 11.7. What is $\dim_{\mathbb{R}}(H_{\text{DR}}^p(\mathbb{R}^n \setminus \{0\}))$ for $n \geq 1$, $p \geq 0$?

Problem 11.8. Let X be a topological space. Then

$$b_k(X) := \dim_{\mathbb{Q}} H_k^{\text{sing}}(X, \mathbb{Q}) \in \mathbb{N}_0 \cup \{\infty\}, \qquad k \geq 0$$

is called the k-th *(rational) Betti number* of X. Suppose that $b_k(X) < \infty$ for all k and that $b_k(X) \neq 0$ for only finitely many k. Then

$$\chi(X) := \sum_{k \geq 0} (-1)^k b_k(X)$$

is called the *Euler characteristic of* X. Determine the Betti numbers of S^n, of $\mathbb{P}^n(\mathbb{R})$ and of $\mathbb{P}^n(\mathbb{C})$ and deduce that

$$\chi(S^n) = \begin{cases} 2, & \text{if } n \text{ is even;} \\ 0, & \text{if } n \text{ is odd;} \end{cases} \qquad \chi(\mathbb{P}^n(\mathbb{R})) = \begin{cases} 1, & \text{if } n \text{ is even;} \\ 0, & \text{if } n \text{ is odd;} \end{cases}$$

and $\chi(\mathbb{P}^n(\mathbb{C})) = n + 1$. Show that there exist no integers $n, m \geq 1$ such that $\mathbb{P}^n(\mathbb{R})$ and $\mathbb{P}^m(\mathbb{C})$ are homotopy equivalent. Show that S^k is not homotopically equivalent to any projective space for $k \geq 3$. What is with $k = 1, 2$?
Hint: Problem 10.5.

Problem 11.9. Let $n \geq 1$. Show that $\mathrm{End}_{\mathbb{Z}}(H^n(S^n, \mathbb{Z})) = \mathbb{Z}$. For every continuous map $f: S^n \to S^n$ let $\deg(f) := f^{-1} \in \mathrm{End}_{\mathbb{Z}}(H^n(S^n, \mathbb{Z})) = \mathbb{Z}$.

1. Show that $\deg(f)$ depends only on the homotopy class of f and that $\deg(f \circ g) = \deg(f)\deg(g)$ for all continuous maps $f, g: S^n \to S^n$.
2. Identify $S^1 = \{z \in \mathbb{C} ; |z| = 1\}$. Show that $\deg(z \mapsto z^k) = k$ for all $k \in \mathbb{Z}$.
3. Let $f: S^n \to S^n$ be induced by an orthogonal matrix $S \in O_{n+1}(\mathbb{R})$. Show that $\deg(f) = \det(S)$.
4. Let $f: S^n \to S^n$ be without fixed points. Show that f is homotopic to $-\mathrm{id}_{S^n}$ and deduce $\deg(f) = (-1)^{n+1}$.
5. Show that a continuous map $f: S^n \to S^n$ is homotopic to a constant map if and only if it is not surjective. Show that in this case there exist $x, y \in S^n$ with $f(x) = x$ and $f(y) = -y$.
6. Let G be a non-trivial group that acts freely on S^n. Show that $G \cong \mathbb{Z}/2\mathbb{Z}$ if n is even.

Problem 11.10. Show the *hairy ball theorem*. Let $X = S^n$ with n even. Show that every continuous section s of the tangent bundle $T_X \to X$ has a zero.
Hint: Consider s as a continuous map $s: S^n \to \mathbb{R}^{n+1}$. If s has no zero then one can assume that $s(x) \in S^n$ for all $x \in S^n$. Then use Problem 11.9.

Appendix A: Basic Topology

12

In this appendix some basic notions in point set topology are recalled. We assume that the reader has already some familiarity with the concepts in this chapter and in particular with the notion of a metric space. For the convenience of the reader we included some proofs such that this chapter might also serve as a rather brisk introduction to point set topology. For a detailed introduction to this topic we refer to [BouGT1] and [BouGT2].

12.1 Topological Spaces

Definition 12.1 (Topological space). Let X be a set. A *topology on X* is a set $\mathcal{T}$ of subsets of X such that $\mathcal{T}$ is stable under arbitrary unions and under finite intersections.

The pair $(X, \mathcal{T})$, usually denoted simply by X, is called a *topological space* and the subsets in $\mathcal{T}$ are called *open*. A subset of X is called *closed* if its complement in X is open.

If $\mathcal{T}$ and $\mathcal{T}'$ are topologies on X, then we say that $\mathcal{T}$ is *coarser* than $\mathcal{T}'$ or, equivalently, that $\mathcal{T}'$ is *finer* than $\mathcal{T}$ if $\mathcal{T} \subseteq \mathcal{T}'$.

Note that the properties of a topology $\mathcal{T}$ on a set X in particular imply that $\mathcal{T}$ contains the union and the intersection of the empty family of sets in $\mathcal{T}$. Hence every topology contains the empty set $\emptyset$ and the set X.

If S is any set of subsets of a set X, then the intersection of all topologies on X containing S is a topology. This is then the smallest topology containing S, called the *topology generated by S*.

Example 12.2 (Discrete topology). A simple example of topologies on a set X is the set of all subsets of X. This topology is called the *discrete topology* on X. As every subset is a union of subsets consisting of a single element, a topological space X carries the discrete

© Springer Fachmedien Wiesbaden 2016
T. Wedhorn, *Manifolds, Sheaves, and Cohomology*, Springer Studium Mathematik – Master,
DOI 10.1007/978-3-658-10633-1_12

topology if and only if the sets $\{x\}$ are open for all $x \in X$. In other words, the discrete topology is the topology generated by $\{\, \{x\} \, ; \, x \in X \,\}$.

Example 12.3 (Metric spaces and normed spaces as topological spaces). Let (X, d) be a metric space. For $x_0 \in X$ and $r \in \mathbb{R}^{>0}$ set

$$B_r(x_0) := \{\, x \in X \, ; \, d(x, x_0) < r \,\}, \qquad B_{\leq r} := \{\, x \in X \, ; \, d(x, x_0) \leq r \,\}.$$

Recall that a subset U of X is called *open* if for all $x \in U$ there exists $r \in \mathbb{R}^{>0}$ such that $B_r(x) \subseteq U$. Then the open subsets form a topology on X, called the *topology induced by the metric d*. In the sequel we endow every metric space with the induced topology.

A topological space $X = (X, \mathcal{T})$ is called *metrizable* if there exists a metric d on X such that $\mathcal{T}$ is the topology induced by d.

Every norm on a real or complex vector space V yields a metric and hence induces a topology on V. Equivalent norms induce the same topology.

If V is finite-dimensional, then all norms are equivalent. Hence there exists in this case a unique topology on V induced by some norm. We will usually endow V with this topology.

Definition 12.4 (Neighborhoods and basis of a topology). Let X be a topological space.

1. Let $Y \subseteq X$ be a subset. A subset V of X is called a *neighborhood of Y* if there exists an open subset U of X such that $Y \subseteq U \subseteq V$. A *fundamental system of neighborhoods of Y* is a set S of neighborhoods of Y such that for every neighborhood V of Y there exists $W \in S$ with $W \subseteq V$. For $x \in X$, a neighborhood of $\{x\}$ (respectively a fundamental system of neighborhoods of $\{x\}$) is also called a *neighborhood of x* (respectively a *fundamental system of neighborhoods of x*).
2. A set $\mathcal{B}$ of open subsets of X is called a *basis of the topology* if every open subset of X is a union of sets in $\mathcal{B}$.

A subset $\mathcal{B}$ of the topology on X is a basis if and only if $\mathcal{B}$ contains a fundamental system of open neighborhoods of every $x \in X$.

Remark 12.5. Let X be a set. Let S be a set of subsets of X and let $\mathcal{T}$ be the topology generated by S. Then $\mathcal{T}$ consists of arbitrary unions of finite intersections of sets in S. In particular the set of intersections of all finite families of sets in S is a basis of the topology $\mathcal{T}$.

In a metric space the open balls $B_r(x_0)$, for $r \in \mathbb{R}^{>0}$, $x_0 \in X$, form a basis of the topology.

12.2 Continuous, Open, and Closed Maps

Definition 12.6 (Continuous maps, homeomorphisms). Let X and Y be topological spaces and let $f\colon X \to Y$ be a map. Let $x_0 \in X$. Then f is called *continuous in x_0* if for every neighborhood V of $f(x_0)$ its inverse image $f^{-1}(V)$ is a neighborhood of x_0.

The map f is called *continuous* if the following equivalent conditions hold:

(i) The map f is continuous in all $x_0 \in X$.
(ii) For every open subset $V \subseteq Y$ its inverse image $f^{-1}(V)$ is open in X.
(iii) For every closed subset $B \subseteq Y$ its inverse image $f^{-1}(B)$ is closed in X.

A continuous map $f\colon X \to Y$ is called a *homeomorphism* if there exists a continuous map $g\colon Y \to X$ such that $g \circ f = \mathrm{id}_X$ and $f \circ g = \mathrm{id}_Y$.

When we write that $f\colon X \to Y$ is a continuous map, it will be always tacitly understood that X and Y are topological spaces.

The composition of continuous maps is continuous. Let S be a set of subsets of Y generating the topology on Y. Then a map $f\colon X \to Y$ between topological spaces is continuous if and only if $f^{-1}(V)$ is open in X for all $V \in S$.

A bijective continuous map $f\colon X \to Y$ is not necessarily a homeomorphisms because its inverse map $f^{-1}\colon Y \to X$ might not be continuous (see Problem 12.10 for examples).

Example 12.7 (Continuity for metric spaces). For maps between metric spaces, condition (i) for continuity can be expressed by the Weierstraß' ε-δ condition: A map $f\colon (X, d) \to (X', d')$ between metric spaces is continuous if and only if for all $\varepsilon > 0$ and for all $x_0 \in X$ there exists $\delta > 0$ such that $d(x_0, x) < \delta$ implies $d'(f(x_0), f(x)) < \varepsilon$ for all $x \in X$.

Example 12.8 (Continuity of linear maps). Let $\mathbb{K}$ be either $\mathbb{R}$ or $\mathbb{C}$ and let V and W be finite-dimensional vector spaces over $\mathbb{K}$ endowed with the unique topology given by a norm. Then it is a standard result from basic analysis that every linear map $u\colon V \to W$ is continuous. In particular, every bijective linear map is a homeomorphism.

More generally and more precisely, let $V_1, \ldots, V_r$ and W be finite-dimensional vector spaces over $\mathbb{K}$ and let $\alpha\colon V_1 \times \cdots \times V_r \to W$ be an r-multilinear map (i.e., α is linear in each component V_i). Then there exist norms on V_i and W such that

$$\|\alpha(v_1, \ldots, v_r)\| \leq \|v_1\| \cdots \|v_r\|$$

for all $v_i \in V_i$, $i = 1, \ldots, r$. This implies that α is continuous by Example 12.7.

Example 12.9 (Extended real line). Let $\overline{\mathbb{R}} := \mathbb{R} \cup \{\pm\infty\}$ be the extended real line. Define a map

$$\varphi \colon \overline{\mathbb{R}} \to [-1, 1], \qquad \varphi(x) := \begin{cases} -1, & x = -\infty; \\ \frac{x}{1+|x|}, & x \in \mathbb{R}; \\ 1, & x = \infty. \end{cases}$$

It is easy to check that φ is bijective. We endow $\overline{\mathbb{R}}$ with a metric via transport of structure, i.e., we define

$$d \colon \overline{\mathbb{R}} \times \overline{\mathbb{R}} \to \mathbb{R}, \qquad d(x, y) := |\varphi(x) - \varphi(y)|.$$

Then d is a metric on $\overline{\mathbb{R}}$ and we obtain an induced topology on $\overline{\mathbb{R}}$. As φ preserves the metrics (by definition), it is a homeomorphism of topological spaces.

It is easy to check, that $\mathbb{R} \subseteq \overline{\mathbb{R}}$ is an open subset with respect to this topology and that the topology induced on this subspace is the usual topology on $\mathbb{R}$ (but note that the restriction of d is not the standard metric on $\mathbb{R}$).

Definition 12.10 (Open and closed maps). Let X and Y be topological spaces and let $f \colon X \to Y$ be a map. Then f is called *open* (respectively *closed*) if for every open (respectively closed) subset Z of X its image $f(Z)$ is open (respectively closed) in Y.

The composition of open (respectively closed) maps is again open (respectively closed). Every homeomorphism is open and closed. Conversely, a bijective continuous map is a homeomorphism if and only if it is open (or, equivalently, closed).

Let $\mathcal{B}$ be a basis of the topology on X. As the image of a union of subsets is the union of the images of these subsets, a map $f \colon X \to Y$ is open if and only if $f(U)$ is open in Y for all $U \in \mathcal{B}$.

Proposition 12.11. *A continuous map $f \colon X \to Y$ is closed if and only if for all $y \in Y$ and all open neighborhoods U of $f^{-1}(y)$ in X there exists an open neighborhood V of y in Y such that $f^{-1}(V) \subseteq U$.*

Proof. Let f be closed. Let $y \in Y$, U an open neighborhood of $f^{-1}(y)$. Then $B := f(X \setminus U)$ is closed in Y and $y \notin B$. Take $V := Y \setminus B$. Then $f^{-1}(V) = X \setminus f^{-1}(B) \subseteq U$.

Conversely, the condition is also sufficient. Let $A \subseteq X$ be closed. We have to show that $W := Y \setminus f(A)$ is open in Y. Choose $y \in W$. Then $U := X \setminus A$ is an open neighborhood of $f^{-1}(y)$. Hence there exists $y \in V \subseteq Y$ open such that $f^{-1}(V) \subseteq U$ and hence $V \subseteq W$. □

12.3 Closure and Interior

Definition 12.12 (Closure, interior, and boundary). Let X be a topological space, $Y \subseteq X$ a subset. The set

$$\overline{Y} := \bigcap_{\substack{A \text{ closed} \\ Y \subseteq A}} A$$

is called the *closure of* Y (in X). It is the smallest closed subset of X that contains Y. The set

$$Y^\circ := \bigcup_{\substack{U \text{ open} \\ U \subseteq Y}} U$$

is called the *interior of* Y (in X). It is the greatest open subset of X that is contained in Y. The set

$$\partial Y := \overline{Y} \setminus Y^\circ$$

is called the *boundary of* Y (in X).

Remark 12.13. Let X be a topological space and let Y be a subset of X. Then

1. $(X \setminus Y)^\circ = X \setminus \overline{Y}$.
2. $\overline{Y} = \{ x \in X \ ; \ \text{for all neighborhoods } U \text{ of } x \text{ one has } U \cap Y \neq \emptyset \}$. In the definition of the right-hand set it suffices if U runs through a fundamental system of neighborhoods of x.

Example 12.14. Let $(V, \| \cdot \|)$ be a normed real or complex vector space. Then for all $r \in \mathbb{R}^{>0}$ and $x \in X$ one has

$$\overline{B_r(x)} = B_{\leq r}(x), \qquad \partial B_r(x_0) = \{ x \in V \ ; \ \|x - x_0\| = r \}.$$

Analogous equalities do not hold in a general metric space (Exercise 12.5).

Proposition 12.15. *Let X and Y be topological spaces and let $f : X \to Y$ be a map. If f is continuous, then for every subset A of X one has $f(\overline{A}) \subseteq \overline{f(A)}$.*

The converse also holds (Problem 12.9).

Proof. If B is a closed subset of Y containing $f(A)$, then $f^{-1}(B)$ is closed in X and contains A. Hence $f^{-1}(B)$ contains $\overline{A}$, in other words $f(\overline{A}) \subseteq B$. Therefore $f(\overline{A})$ is contained in the intersection of all such B. $\qquad\square$

Definition and Remark 12.16 (Dense subsets). Let X be a topological space. A subset Y of X is called *dense in* X, if the following conditions (which are equivalent by Remark 12.13 2) are satisfied:

(i) The closure of Y in X is equal to X.
(ii) For every non-empty open subset U of X one has $U \cap Y \neq \emptyset$.

12.4   Construction of Topological Spaces

Definition 12.17 (Inverse and direct image topology). Let $(X_i)_{i \in I}$ be a family of topological spaces.

1. Let Y be a set and let $f_i : Y \to X_i$ be maps for $i \in I$. Then the topology on Y generated by $\bigcup_{i \in I} \{ f_i^{-1}(U_i) \; ; \; U_i \subseteq X_i \text{ open} \}$ is called the *inverse image of the topology of the* $(X_i)_i$ *under* $(f_i)_i$.
   It is the coarsest topology on Y such that all maps f_i are continuous. In other words, it satisfies the following universal property: If $h : W \to Y$ is a map from a topological space W, then h is continuous if and only if $f_i \circ h$ is continuous for all $i \in I$.
2. Let Z be a set and $g_i : X_i \to Z$ be maps for $i \in I$. Then $\bigcap_{i \in I} \{ W \subseteq Z \; ; \; g_i^{-1}(W)$ is open in $X_i \}$ is a topology on Z, called the *direct image of the topology of the* $(X_i)_i$ *under* $(g_i)_i$.
   It is the finest topology on Z such that all maps g_i are continuous. In other words, it satisfies the following universal property: If $h : Z \to W$ is a map to a topological space W, then h is continuous if and only if $h \circ g_i$ is continuous for all $i \in I$.

A particularly important example is the subspace topology.

Example 12.18 (Subspaces). If Y is a subset of topological space X and $f : Y \to X$ is the inclusion, then the inverse topology of X under f is simply called the *induced topology on* Y and the topological space Y is called a *subspace of* X.

In this case, the open (respectively closed) subsets of Y are those of the form $A \cap Y$ for $A \subseteq X$ open (respectively closed).

Let Y be a subset of a finite-dimensional vector space V over $\mathbb{R}$ or over $\mathbb{C}$. Then we always endow Y with the topology induced by the unique topology on V that is given by some norm on V.

Definition 12.19 (Topological embedding). A map $j : Y \to X$ between topological spaces is called a *topological embedding*, if j yields a homeomorphism $Y \to j(Y)$, where $j(Y)$ is endowed with the topology induced by the topology on X. An embedding $j : Y \to X$ is called *open* (respectively *closed*) if j is an open (respectively a closed) map.

Remark 12.20.

1. A topological embedding is injective and continuous. The converse does not hold in general (Problem 12.10).
2. An injective continuous map $j: Y \to X$ is a topological embedding if and only if for every open subset V of Y there exists an open subset U of X such that $j^{-1}(U) = V$.
3. The composition of two topological embeddings is again a topological embedding (by 2).
4. Let $j: Y \to X$ be a continuous map such that there exists a continuous map $p: Y \to X$ with $p \circ j = \mathrm{id}_Y$ (i.e., j is a section of p), then for every open $V \subseteq Y$ one has $j^{-1}(p^{-1}(V)) = V$. Hence j is a topological embedding by 2.
5. A topological embedding $j: Y \to X$ is open (respectively closed) if and only if $j(Y)$ is an open (respectively a closed) subset of X.

A very important example of the direct image topology are quotient spaces.

Example 12.21 (Quotient spaces). Let X be a topological space and let $g: X \to Z$ be a surjective map. Then the direct image topology of g is also called the *quotient topology* and the topological space Z is called a *quotient space of X*.

A subset A of Z is open (respectively closed) if and only if $g^{-1}(A)$ is open (respectively closed) in X. The universal property of the quotient topology is the following. Let Y be a topological space. Then a map $f: Z \to Y$ is continuous if and only if $f \circ g$ is continuous.

An other important special case of the inverse image topology is the product topology.

Definition and Remark 12.22 (Product space). Let I be a set and let $(X_i)_{i \in I}$ be a family of topological spaces. Let $X := \prod_{i \in I} X_i$ be the cartesian product and let $p_i: X \to X_i$ be the i-th projection. The inverse image on X of the topology of the $(X_i)_i$ under $(p_i)_i$ is called the *product topology* and the set X endowed with this topology is called the *product of the topological spaces $(X_i)_i$*.

Hence the product topology is the coarsest topology such that the projections p_i are continuous for all i. A basis of the product topology is given by open subsets of the form $\prod_{i \in I} U_i$, where $U_i \subseteq X_i$ is open and $U_i = X_i$ for all but finitely many $i \in I$.

As openness of maps can be checked on a basis of the topology, one sees that the projections p_i are open for all i.

Proposition 12.23 (Product and closure). *Let $(X_i)_{i \in I}$ be a family of topological spaces and let $A_i \subseteq X_i$ be subsets for all i. Then the closure of $\prod_{i \in I} A_i$ in the product space $\prod_{i \in I} X_i$ is $\prod_{i \in I} \overline{A}_i$.*

Proof. Let $x = (x_i)_i$ be in the closure of $\prod_{i \in I} A_i$. Then for all $i \in I$ its projection $p_i(x) = x_i$ lies in $\overline{A}_i$ (Proposition 12.15) and hence $x \in \prod_i \overline{A}_i$.

Conversely, let $y = (y_i)_i \in \prod_i \overline{A_i}$. The sets of the form $\prod_i U_i$ containing y and with $U_i \subseteq X_i$ open for all $i \in i$ and $U_i = X_i$ for all but finitely many $i \in I$ form a fundamental system of neighborhoods of y. For every $i \in I$ the open neighborhood U_i of y_i contains a point $a_i \in A_i$. Hence $\prod_i U_i$ contains $(a_i)_i \in \prod_{i \in I} A_i$. This shows that y is in the closure of $\prod_{i \in I} A_i$. □

Corollary 12.24. *Let $(X_i)_{i \in I}$ be a family of topological spaces and let $A_i \subseteq X_i$ be non-empty subsets for all i. Then $\prod_{i \in I} A_i$ is closed (respectively dense) in $\prod_{i \in I} X_i$ if and only if A_i is closed (respectively dense) in X_i for all $i \in I$.*

Remark 12.25. Let (X_1, d_1) and (X_2, d_2) be metric spaces. Then

$$d((x_1, x_2), (y_1, y_2)) := \max\{d_1(x_1, y_1), d_2(x_2, y_2)\}$$

defines a metric on $X_1 \times X_2$. It is easy to check that d induces the product topology. In particular, the product of two metrizable spaces is again metrizable.

More generally, one can show that the product of countably many metrizable spaces is again metrizable (Problem 12.11).

Definition and Remark 12.26 (Sum of topological spaces). Let $(X_i)_{i \in I}$ be a family of topological spaces. Let X be the disjoint union of all the X_i and let $\iota_i : X_i \to X$ be the inclusion. Endow the set X with the direct image topology of the X_i under ι_i. We obtain a topological space X such that $X_i \subseteq X$ is open and closed for all $i \in I$. The space X is called the *sum* or the *coproduct* of $(X_i)_{i \in I}$ and it is denoted by $\coprod_{i \in I} X_i$.

The sum of topological spaces is a special case of the following more general construction of gluing topological spaces. Let $(U_i)_{i \in I}$ be a family of topological spaces, for all $i, j \in I$ fix a subset $U_{ij} \subseteq U_i$, and for all $i, j \in I$ a continuous map $\varphi_{ji} : U_{ij} \to U_{ji}$ such that:

(a) $U_{ii} = U_i$ and $\varphi_{ii} = \mathrm{id}_{U_i}$ for all $i \in I$,
(b) the *cocycle condition* holds: $\varphi_{kj} \circ \varphi_{ji} = \varphi_{ki}$ on $U_{ij} \cap U_{ik}$ for all triples (i, j, k) of I.

In the cocycle condition we implicitly assume that $\varphi_{ji}(U_{ij} \cap U_{ik}) \subseteq U_{jk}$, such that the composition is meaningful.

Proposition 12.27 (Gluing of topological spaces). *Suppose that U_{ij} is open (respectively closed) in U_i for all $i, j \in I$. Then there exists a topological space X together with morphisms $\psi_i : U_i \to X$, such that:*

1. *The map ψ_i is an open (respectively a closed) embedding for all $i \in I$.*
2. *$\psi_j \circ \varphi_{ji} = \psi_i$ on U_{ij} for all i, j.*

3. $X = \bigcup_i \psi_i(U_i)$.
4. $\psi_i(U_i) \cap \psi_j(U_j) = \psi_i(U_{ij}) = \psi_j(U_{ji})$ for all $i, j \in I$.

Furthermore, $(X, (\psi_i)_i)$ has the following universal property: If Z is a topological space, and for all $i \in I$, $\xi_i : U_i \to Z$ is a continuous map such that $\xi_j \circ \varphi_{ji} = \xi_i$ on U_{ij} for all $i, j \in I$, then there exists a unique continuous map $\xi : X \to Z$ with $\xi \circ \psi_i = \xi_i$ for all $i \in I$.

In particular, $(X, (\psi_i)_i)$ is uniquely determined up to unique isomorphism.

Proof. First note that the cocycle conditions for the triples (i, j, i) and (j, i, j) show that φ_{ij} and φ_{ji} are mutually inverse homeomorphisms. To define the underlying topological space of X, we start with the sum $\coprod_{i \in I} U_i$ of the U_i and define an equivalence relation $\sim$ on it as follows. Points $x_i \in U_i$, $x_j \in U_j$, $i, j \in I$, are equivalent, if and only if $x_i \in U_{ij}$, $x_j \in U_{ji}$ and $x_j = \varphi_{ji}(x_i)$. The conditions (a) and (b) imply that $\sim$ is in fact an equivalence relation. As a set, define X to be the set of equivalence classes,

$$X := \coprod_{i \in I} U_i / \sim .$$

The natural maps $\psi_i : U_i \to X$ are injective, we have $\psi_i(U_{ij}) = \psi_i(U_i) \cap \psi_j(U_j)$ for all $i, j \in I$, and properties (2) and (3) hold. We endow X with the direct image topology of the U_i under the ψ_i. Hence a subset V of X is open (respectively closed) if and only if $\psi_i^{-1}(V)$ is open (respectively closed) in U_i for all $i \in I$. The universal property then follows from the universal property of the direct image topology.

To show that ψ_i is an open (respectively a closed) embedding we have to show that for all $i \in I$ a subset A_i of U_i is open (respectively closed) in U_i if and only if for all $j \in J$ the subset $\psi_j^{-1}(\psi_i(A_i))$ is open (respectively closed) in U_j. By choosing $j = i$ one sees that the condition is clearly necessary. Conversely, if A_i is open (respectively closed) in U_i, then we have $\psi_j^{-1}(\psi_i(A_i)) = \varphi_{ji}(A_i \cap U_{ij})$ by construction. This set is open (respectively closed) in U_j because φ_{ji} is a homeomorphism and U_{ji} is open (respectively closed) in U_j. $\square$

The following construction of fiber products of topological spaces plays an important role in several places in the book.

Definition 12.28 (Fiber products). Let $f : X \to S$ and $g : Y \to S$ be continuous maps of topological spaces. Define

$$X \times_S Y := \{(x, y) \in X \times Y ; f(x) = g(y)\}$$

endowed with the inverse topology of X and Y under the projections $p : X \times_S Y \to X$ and $q : X \times_S Y \to Y$. This is also the topology induced by the product topology on $X \times Y$. The topological space $X \times_S Y$ is called the *fiber product* of X and Y with respect to f and g.

Example 12.29. Let $f: X \to S$ and $g: Y \to S$ be continuous maps of topological spaces. Let $p: X \times_S Y \to X$ and $q: X \times_S Y \to Y$ be the projections.

1. Suppose that S consists of a single point. Then $X \times_S Y = X \times Y$.
2. Suppose that Y consists of a single point y and set $s := g(y)$. Then p induces a homeomorphism $X \times_S Y \xrightarrow{\sim} f^{-1}(s)$, where the fiber $f^{-1}(s)$ is endowed with the subspace topology of X.
3. More generally, for arbitrary f and g the projection p induces for all $y \in Y$ a homeomorphism $q^{-1}(y) \xrightarrow{\sim} f^{-1}(g(y))$.
4. Let $h: X \to Y$ be a continuous map such that $g \circ h = f$. Then its *graph* $\Gamma_h: X \to X \times_S Y$, $x \mapsto (x, h(x))$ is a topological embedding by Remark 12.20 4 because it is a section of the projection $X \times_S Y \to X$.
5. Applying 4 to the special case $X = Y$, $h = \mathrm{id}_X$, the graph of id_X

$$\Delta_f: X \to X \times_S X = \{ (x, x') \; ; \; f(x) = f(x') \}, \qquad x \mapsto (x, x)$$

is called the *diagonal of* f. The map f is injective if and only if Δ_f is a homeomorphism.

12.5 Local Properties

Definition 12.30 (Coverings). Let X be a topological space and let Y be a subset of X. A family $(A_i)_{i \in I}$ of subsets of X is called a *covering of Y in X* if $Y \subseteq \bigcup_{i \in I} A_i$.

A covering $(A_i)_{i \in I}$ of Y in X is called an *open covering* (respectively a *closed covering*) if all A_i are open (respectively closed) in X.

In the sequel we will see many examples of properties that are local in the sense that they can be checked after "restriction to an open covering". Of course one has to explain what this means precisely. We make this precise in some cases. Let **P** be a property of maps of topological spaces.

We say that **P** is *local on the target* (respectively *local on the source*), if for every map $f: X \to Y$ of topological spaces and for every open covering $(V_i)_{i \in I}$ of Y (respectively for every open covering $(U_i)_{i \in I}$ of X) the map f has property **P** if and only if its restriction $f^{-1}(V_i) \to V_i$ (respectively its restriction $U_i \to Y$) has property **P** for all i. Let us give some easy examples.

1. The following properties are local on the target and local on the source: "continuous", "open".
2. The following properties are local on the target: "injective", "surjective", "bijective", "homeomorphism", "closed", "open topological embedding", "closed topological embedding".

As the properties "open topological embedding" and "closed topological embedding" are both local on the target, one sees by considering inclusions of subspaces that for every topological space X and every open covering $(U_i)_i$ of X a subset A of X is open (respectively closed) in X if and only if $A \cap U_i$ is open (respectively closed) in U_i for all $i \in I$.

In fact there is a stronger result. To formulate it we need another example of a local notion, which plays an important role when the notion of paracompact spaces is defined.

Definition 12.31 (Locally finite covering). Let X be a topological space. A family of subsets $(A_\lambda)_{\lambda \in \Lambda}$ is called *locally finite* if there exists an open covering $(U_i)_{i \in I}$ of X such that each U_i meets only finitely many of the A_λ.

Proposition 12.32. *Let $(A_i)_{i \in I}$ be a family of subsets of a topological space X satisfying one of the following properties:*

(i) *The interiors A_i° of A_i cover X.*
(ii) *$(A_i)_{i \in I}$ is a locally finite closed covering of X.*

Then a subset B of X is open (respectively closed) in X if and only if $B \cap A_i$ is open (respectively closed) in A_i for all i.

Proof. We have to show that the condition is sufficient. As $(X \setminus B) \cap A_i = A_i \setminus (B \cap A_i)$, it suffices by passing to the complement to consider either the case that for all i the subset $B \cap A_i$ of A_i is open or that for all i it is closed.

Suppose that (i) is satisfied and suppose that $B \cap A_i$ is open in A_i for all i. Then $B \cap A_i^\circ$ is open in A_i° and hence open in X. By hypothesis, $B = \bigcup_i (B \cap A_i^\circ)$ and hence B is open.

Suppose that (ii) is satisfied and suppose that $B \cap A_i$ is closed in A_i for all i. As A_i is closed in X, $B \cap A_i$ is closed in X. To see that B is closed we may work locally on X. The hypothesis implies that $(B \cap A_i)_i$ is locally finite and working locally on X we may assume that all but finitely many of the $B \cap A_i$ are empty. But then $B = \bigcup_i (B \cap A_i)$ is closed. $\square$

Corollary 12.33. *Let $(A_i)_{i \in I}$ be a locally finite family of closed subsets of a topological space. Then $\bigcup_{i \in I} A_i$ is closed in X.*

Corollary 12.34. *Let $(A_i)_i$ be a covering of a topological space X such that the interiors A_i° of A_i cover X or which is a locally finite closed covering. Let Y be a topological space and let $f : X \to Y$ be a map such that $f_{|A_i} : A_i \to Y$ is continuous for all i. Then f is continuous.*

We introduce two local notions of properties already defined.

Definition and Remark 12.35 (Locally closed subsets). Let X be a topological space and let $Z \subseteq X$ be a subset. Then Z is called *locally closed* if the following equivalent conditions are satisfied:

(i) There exists a family $(U_i)_{i \in I}$ of open subsets $U_i \subseteq X$ such that $Z \subseteq \bigcup_{i \in I} U_i$ and such that $Z \cap U_i$ is closed in U_i.
(ii) Z is an open subset of its closure $\bar{Z}$ in X.
(iii) Z is the intersection of an open and a closed subset of X.

Proof. "(ii) $\Rightarrow$ (iii) $\Rightarrow$ (i)" is clear.

"(i) $\Rightarrow$ (ii)". Let $(U_i)_i$ as in (i). As $Z \cap U_i$ is closed in U_i, we have $Z \cap U_i = \bar{Z} \cap U_i$, which is open in $\bar{Z}$. Hence $Z = \bigcup_i (Z \cap U_i)$ is open in $\bar{Z}$. $\square$

Definition 12.36 (Local homeomorphisms). A continuous map $f : X \to Y$ is called a *local homeomorphism* if there exists an open covering $(U_i)_i$ of X such that $f_{|U_i} : U_i \to Y$ is an open topological embedding.

In other words, $f : X \to Y$ is a local homeomorphism if and only if there exist open coverings $(U_i)_i$ of X and $(V_i)_i$ of $f(X)$ in Y such that f induces for all i a homeomorphism $U_i \xrightarrow{\sim} V_i$.

Remark 12.37. Let $f : X \to Y$ be a continuous map.

1. Every local homeomorphism is an open map. In particular, every injective local homeomorphism is an open embedding.
2. If f is a local homeomorphism, then the diagonal $\Delta_f : X \to X \times_Y X$ is an open embedding. Indeed, let $(U_i)_i$ be an open covering of X such that $f_{|U_i}$ is an open embedding for all i. Then $\Delta_{f|U_i}$ is an open embedding (with image $U_i \times_Y U_i$) for all i. As Δ_f is injective, this shows that Δ_f is an open embedding.
3. Let $g : Y \to Z$ be a local homeomorphism and suppose that $g \circ f$ is an open embedding. Then f is injective and a local homeomorphism. Hence f is an open embedding.

12.6   Hausdorff Spaces

Definition and Proposition 12.38 (Hausdorff spaces). A topological space X is called *Hausdorff* or a T_2-*space* if it satisfies the following equivalent properties:

(i) For all $x, y \in X$ with $x \neq y$ there exist neighborhoods U of x and V of y such that $U \cap V = \emptyset$.
(ii) The diagonal $\Delta_X := \{ (x, x) \; ; \; x \in X \}$ is closed in $X \times X$.
(iii) For every topological space Y and for all continuous maps $f_1, f_2 : Y \to X$ the set $\{ y \in Y \; ; \; f_1(y) = f_2(y) \}$ is closed in Y.

Proof. "(i) $\Rightarrow$ (ii)". For $(x, y) \in (X \times X) \setminus \Delta_X$ clearly one has $x \neq y$ so that there exist neighborhoods U of x and V of y in X such that $U \cap V = \emptyset$. Therefore $U \times V$ is a neighborhood of (x, y) in $X \times X$ that is contained in $(X \times X) \setminus \Delta_X$. Hence $(X \times X) \setminus \Delta_X$ is open in $X \times X$.

"(ii) $\Rightarrow$ (i)". For $x, y \in X$ with $x \neq y$ we have $(x, y) \in (X \times X) \setminus \Delta_X$. As $(X \times X) \setminus \Delta_X$ is open, there exists a neighborhood of (x, y) in $(X \times X) \setminus \Delta_X$ and by definition of the product topology we can chose it to be of the form $U \times V$, where U and V are open neighborhoods of x and y respectively. Then $U \cap V = \emptyset$.

"(ii) $\Rightarrow$ (iii)". Let $f: Y \to X \times X$ be the continuous map $y \mapsto (f_1(x), f_2(x))$. Since $\Delta_X \subseteq X \times X$ is closed and f is continuous, $f^{-1}(\Delta_X) = \{ y \in Y \; ; \; f_1(y) = f_2(y) \}$ is closed in Y.

"(iii) $\Rightarrow$ (ii)". Applying (iii) to the projections $p_1, p_2: X \times X \to X$ yields that $\{ (x, y) \in X \times X \; ; \; p_1((x, y)) = p_2((x, y)) \} = \Delta_X$ is closed in $X \times X$. $\square$

Every metrizable topological space is Hausdorff. Every subspace of a Hausdorff topological space is Hausdorff. If $(X_i)_i$ is a family of Hausdorff topological spaces, then $\prod_{i \in I} X_i$ is Hausdorff.

Property (iii) in Proposition 12.38 for Hausdorff spaces implies the following result.

Proposition 12.39. *Let X and Y be topological spaces, let $f, g: Y \to X$. Suppose that X is Hausdorff and that there exists a dense subset D of Y such that $f_{|D} = g_{|D}$. Then $f = g$.*

12.7   Connected Spaces and Locally Constant Maps

Definition 12.40. A non-empty topological space X is called *connected* if X and $\emptyset$ are the only subsets of X that are both closed and open.

Note that we explicitly excluded the empty space to be connected.

Proposition 12.41 (Images and products of connected spaces).

1. *Let $f: X \to Y$ be a continuous surjective map. If X is connected, then Y is connected.*
2. *Let $(X_i)_{i \in I}$ be a family of topological spaces. Then the product space $\prod_{i \in I} X_i$ is connected if and only if X_i is connected for all $i \in I$.*

Proof. 1. Let $A \subseteq Y$ be open and closed. Then $f^{-1}(A) \subseteq X$ is also open and closed because f is continuous and we have $f^{-1}(A) = \emptyset$ or $f^{-1}(A) = X$ because X is connected. Since f is surjective, we have $A = f(f^{-1}(A))$ and therefore $A = \emptyset$ or $A = Y$. Thus Y is connected.

2. Assume that $\prod_{i \in I} X_i$ is connected. For all $j \in I$ the j-th projection $p_j :$
$\prod_{i \in I} X_i \to X_j$ is continuous and surjective and hence X_j is connected by 1.

Conversely, assume that X_i is connected for every $i \in I$. Let $\emptyset \neq U \subseteq \prod_{i \in I} X_i$
be open and closed. Let $x \in X$ and $y \in U$ such that x and y differ in at most a single
component, i.e., there exists $i_0 \in I$ with $x_i = y_i$ for every $i \in I \setminus \{i_0\}$. Then

$$
j : X_{i_0} \to \prod_{i \in I} X_i, \qquad z \mapsto (z_i)_i \quad z_i := \begin{cases} x_i, & i \in I \setminus \{i_0\}; \\ z, & i = i_0; \end{cases}
$$

is a section of p_{i_0} and therefore a topological embedding. By assumption X_{i_0} is connected
so that $j(X_{i_0})$ is connected as well. Hence, $y \in j(X_{i_0}) \cap U \subseteq j(X_{i_0})$ is open and closed so
that $j(X_{i_0}) \cap U = j(X_{i_0})$. Since $x \in j(X_{i_0})$ we have $x \in U$. Using induction, this result
can be extend to the case where x and y differ in finitely many components. Now there
exists a non-empty open subset $W \subseteq U$ of the form $W = \prod_{i \in I} W_i$ such that $W_i = X_i$
for all but finitely many $i \in I$. This implies that for every point $x \in X$ there exists a point
$y \in W$ such that x and y differ only in finitely many components. Thus $W = X$ by the
above argument and hence $U = X$. This shows that X is connected.   □

Remark 12.42. Let A be a connected subspace of a topological space X. Then every
subspace B of X with $A \subseteq B \subseteq \bar{A}$ is connected. Indeed, A is dense in B and hence
every non-empty open and closed subset Z of B meets A. As A is connected we find
$A \cap Z = A$ and hence $Z = B$, again because A is dense in B.

Recall that an *interval in* $\mathbb{R}$ is a subset I of $\mathbb{R}$ such that for all $x, y \in I$ and $z \in \mathbb{R}$ with
$x < z < y$ one has $z \in I$.

Proposition 12.43 (Connected subspaces of $\mathbb{R}$). *A subspace of $\mathbb{R}$ is connected if and
only if it is an interval.*

Proof. Let $A \subseteq \mathbb{R}$ be a connected subset and assume that A is not an interval. Then there
exist $a, b \in A$ and $c \in \mathbb{R} \setminus A$ with $a < c < b$. Then $A \cap (-\infty, c)$ and $A \cap (c, \infty)$ are
disjoint non-empty open subsets of A such that their union is A. Thus they are both open
and closed in A. Contradiction.

Let $A \subseteq \mathbb{R}$ be an interval and assume that there exists $U \subseteq A$ open and closed with
$\emptyset \neq U \neq A$. Then $A \setminus U$ has the same properties and after possibly replacing U by $A \setminus U$
we find $a \in U, b \in A \setminus U$ with $a < b$. Because A is an interval we have $[a, b] \subseteq A$ so that
$V := U \cap [a, b] \subsetneq [a, b]$ is open, closed and non-empty. Then V is bounded and closed
in $\mathbb{R}$ so that $\beta := \sup(V)$ has to be an element of V. In particular this implies $\beta \neq b$
and it is clear that we must have $\beta \neq a$. Since $V \subseteq [a, b]$ is open and $a < \beta < b$, there
exists $\epsilon > 0$ with $(\beta - \epsilon, \beta + \epsilon) \subseteq V$. This contradicts $\beta = \sup(V)$, hence A has to be
connected.   □

Proposition 12.44. *Let X be a topological space, let $(A_i)_{i \in I}$ be a family of connected subspaces such that $\bigcap_i A_i \neq \emptyset$. Then $A := \bigcup_{i \in I} A_i$ is a connected subspace of X.*

Proof. Let $x \in \bigcap_i A_i$. Let $U \subseteq A$ be open and closed. After possibly passing to $A \setminus U$ we may assume that $x \in U$. Then $U \cap A_i$ is open and closed in A_i and non-empty. As A_i is connected, $U \cap A_i = A_i$ for all i and hence $U = A$. $\qquad\square$

Definition 12.45 (Connected components). Let X be a topological space and let $x \in X$. Then the union of all connected subspaces containing x is again connected by Proposition 12.44. It is therefore the largest connected subspace containing x. It is called the *connected component of x*.

Remark 12.46. Let X be a topological space.

1. For $x, y \in X$ write $x \sim y$ if y belongs to the component of x. This relation is clearly reflexive and symmetric. Moreover, it is transitive by Proposition 12.44 and hence an equivalence relation. The equivalence classes are the connected components of X.
2. As the closure of a connected set is again connected (Remark 12.42), the connected components of X are closed.

In general, connected components are not open (Problem 12.20), see however Proposition 2.13 (or Problem 12.21 and Problem 12.22) for a positive result.

Definition and Remark 12.47 (Locally constant maps). Let X be a topological space, M a set, and $t: X \to M$ a map.

1. The map t is called *constant* if $t(x) = t(y)$ for all $x, y \in X$.
2. The map t is called *locally constant* if the following two equivalent conditions are satisfied:
   (i) There exists an open covering $(U_i)_i$ of X such that $t_{|U_i}$ is constant for all $i \in I$.
   (ii) The map t is continuous if we endow M with the discrete topology.

Proposition 12.48. *A topological space X is connected if and only if every locally constant map f on X is constant.*

Proof. Let X be connected and let $f: X \to S$ be locally constant. Then $f(X)$ is connected in the discrete space S (Proposition 12.41 1) and hence consists of at most a single point.

Conversely, assume that X is not connected. Hence there exist non-empty disjoint open subsets $A, B \subseteq X$ whose union is X. Then the map f from X to a discrete space with two elements a and b with $f(A) = \{a\}$ and $f(B) = \{b\}$ is continuous. $\qquad\square$

12.8 Compact Spaces

Definition 12.49 ((Locally) compact spaces). Let X be a topological space.

1. X is called *compact* if every open covering of X has a finite subcovering.
2. X is called *locally compact* if every point has a fundamental system of compact neighborhoods.

The definitions of "compact" and "locally compact" differ within the literature:

- Often compact spaces as defined above are called quasi-compact. Then compact spaces are defined as quasi-compact spaces that are also Hausdorff.
- Locally compact spaces are very often defined as spaces such that each point has a compact neighborhood – a condition that is strictly weaker than our definition of "locally compact" (Problem 12.32). For Hausdorff spaces both of these definitions of "locally compact" are equivalent (Proposition 12.54).

Example 12.50. Let V be a normed vector space over $\mathbb{R}$ or over $\mathbb{C}$. Then a basic result of analysis shows that V is locally compact if and only if V is finite-dimensional (see also Problem 12.28). In this case the Heine–Borel theorem shows that a subset of V is compact if and only if it is closed in V and bounded (see also Problem 12.27).

Proposition 12.51. *Let X be a compact topological space.*

1. *Every closed subspace of X is compact.*
2. *Let $f\colon X \to Y$ be a continuous surjective map of topological spaces. Then Y is compact.*

Proof. 1. Let X be a compact topological space, $Y \subseteq X$ a closed subspace and $(U_i)_i$ an open covering of Y. By definition of the induced topology there exist $V_i \subseteq X$ open with $V_i \cap Y = U_i$ so that we obtain an open covering of X by adding the open subset $X \setminus Y$ to $(V_i)_i$. Since X is compact, we find finitely many $i_1, \ldots, i_n$ such that $X = (X \setminus Y) \cup V_{i_1} \cup \cdots \cup V_{i_n}$ and therefore $Y = X \cap Y = U_{i_1} \cup \cdots \cup U_{i_n}$ is a finite subcovering of $(U_i)_i$.

2. Let $(V_j)_{j \in J}$ be an open covering of Y. Then $(f^{-1}(V_j))_{j \in J}$ is an open covering of X. As X is compact, there exist $j_1, \ldots, j_n \in J$ such that $X = \bigcup_{i=1}^{n} f^{-1}(V_{j_i})$. Hence $Y \subseteq \bigcup_{i=1}^{n} V_{j_i}$. Therefore Y is compact. $\square$

Proposition 12.52. *Let $f\colon X \to Y$ be a continuous map. Suppose that X is compact and that Y is Hausdorff. Then the subspace $f(X)$ of Y is closed in Y and compact.*

Proof. As Y is Hausdorff, the subspace $f(X)$ is Hausdorff. By Proposition 12.51 2, $f(X)$ is also compact.

It remains to show that $Y \setminus f(X)$ is open in Y. Let $y \in Y \setminus f(X)$. For all $z \in f(X)$ there exist open neighborhoods U_z of z and V_z of y with $U_z \cap V_z = \emptyset$ (because Y is Hausdorff). Then $(U_z)_z$ is an open covering of $f(X)$. As $f(X)$ is compact we find $z_1, \ldots, z_n \in f(X)$ with $f(X) \subseteq \bigcup_{1 \leq i \leq n} U_{z_i}$. Then $\bigcap_{1 \leq i \leq n} V_{z_i} \subseteq Y \setminus f(X)$ is a neighborhood of y. Hence $Y \setminus f(X)$ is open in Y. $\qquad\square$

Corollary 12.53. *Let X be compact and let Y be Hausdorff. Then every injective continuous map $f \colon X \to Y$ is a closed embedding. In particular, every bijective continuous map $X \to Y$ is a homeomorphism.*

Proof. It suffices to show that f is a closed map. Let $A \subseteq X$ be a closed subspace. As X is compact, A is compact (Proposition 12.51). Hence $f(A)$ is closed in Y by Proposition 12.52. $\qquad\square$

Proposition 12.54. *A Hausdorff topological space X is locally compact if and only if every point of X has a compact neighborhood.*

Proof. We have to show that the condition is sufficient. Let $x \in X$ and let C be a compact neighborhood of x. As X is Hausdorff, C is closed in X by Corollary 12.53. Let V be any neighborhood of x. We have to show that V contains a compact neighborhood of x. Replacing V by the interior of $V \cap C$ we may assume that $V \subseteq C$ and that V is open in X. Then $C \setminus V$ is closed in C and hence compact. As X and hence C is Hausdorff, for each $z \in C \setminus V$ there exist in C open neighborhoods W_z of z and U_z of x such that $W_z \cap U_z = \emptyset$. As $C \setminus V$ is compact, there exist finitely many $z_1, \ldots, z_n \in C \setminus V$ such that $C \setminus V \subseteq W := \bigcup_{1 \leq i \leq n} W_{z_i}$. Then $U := \bigcap_{1 \leq i \leq n} U_{z_i}$ is an open neighborhood of x in C not meeting W. Therefore its closure $\bar{U}$ is contained in $C \setminus W$ and hence in V. Moreover, $\bar{U}$ is compact (being closed in C) and a neighborhood because U is open in V and V is open in X. $\qquad\square$

Proposition 12.55. *Let X be a topological space and let S be a set of open subsets of X that generates the topology on X. Then X is compact if and only if every covering of X by elements in S has a finite subcovering.*

Proof. The condition is clearly necessary. To show that it is sufficient assume that X is not compact. Denote by $\mathcal{T}$ the topology on X. As X is not compact, the set

$$\mathbb{U} := \left\{ \mathcal{A} \subseteq \mathcal{T} \;;\; X = \bigcup_{U \in \mathcal{A}} U \text{ and } \bigcup_{U \in \mathcal{A}_0} U \neq X \text{ for every finite subset } \mathcal{A}_0 \text{ of } \mathcal{A} \right\}$$

is non-empty. The set is partially ordered by inclusion and if $\mathbb{U}_0$ is a totally ordered subset of $\mathbb{U}$, then $\bigcup_{\mathcal{A} \in \mathbb{U}_0} \mathcal{A} \in \mathbb{U}$. Hence we can apply Zorn's lemma (see Proposition 13.28) and see that there exists a maximal element $\mathcal{A}$ in $\mathbb{U}$. Set $\mathcal{B} := \mathcal{A} \cap S$.

To obtain a contradiction it remains to show that $\mathcal{B}$ is a covering of X (then $\mathcal{B}$ is a covering of X by elements in S that has no finite subcovering, otherwise $\mathcal{A}$ also would have a finite subcovering). Assume that there exists $x \in X \setminus \bigcup_{U \in \mathcal{B}} U$. Choose $U \in \mathcal{A}$ with $x \in U$. As S generates the topology, we find $V_1, \dots, V_r \in S$ such that $x \in \bigcap_{1 \le i \le r} V_i \subseteq U$. As V_j contains x we have $V_j \notin \mathcal{A}$. Now $\mathcal{A} \cup \{V_j\}$ has a finite subcovering by the maximality of $\mathcal{A}$, i.e., for all j we find a finite union W_j of elements in $\mathcal{A}$ such that $W_j \cup V_j = X$. Hence $(\bigcap_{1 \le j \le r} V_j) \cup (\bigcup_{1 \le j \le r} W_j) = X$ and hence $X = U \cup (\bigcup_{1 \le j \le r} W_j)$. Hence $\mathcal{A}$ has a finite subcovering. This contradiction proves that $\mathcal{B}$ is a covering. $\square$

Theorem 12.56 (Theorem of Tychonoff). *Let $(X_i)_{i \in I}$ be a family of compact spaces. Then the product space $\prod_{i \in I} X_i$ is compact.*

Proof. Let $p_i \colon X := \prod_{i \in I} X_i \to X_i$ be the projection. The topology on $\prod_{i \in I} X_i$ is generated by open subsets of the form $p_i^{-1}(U_i)$ with $U_i \subseteq X_i$ open. Let $\mathcal{A}$ be a covering consisting of such sets. By Proposition 12.55 it suffices to show that $\mathcal{A}$ has a finite subcovering.

For all $i \in I$ set $\mathcal{A}_i := \{U \subseteq X_i \text{ open} \,;\, p_i^{-1}(U) \in \mathcal{A}\}$. As $\mathcal{A}$ is a covering consisting of sets of the form $p_i^{-1}(U_i)$, there exists $j \in I$ such that $\mathcal{A}_j$ is a covering of X_j. As X_j is compact, there exist $U_1, \dots, U_r \in \mathcal{A}_j$ such that $X_j = \bigcup_{1 \le k \le r} U_k$. Then $\{p_j^{-1}(U_k) \,;\, k = 1, \dots, r\}$ is an open covering of X, which is a subcovering of $\mathcal{A}$ by definition of $\mathcal{A}_j$. $\square$

12.9 Topological Groups

Definition 12.57 (Topological group). A *topological group* is a set G endowed with the structure of a topological space and of a group such that the maps

$$G \times G \to G, \quad (g, g') \mapsto gg', \qquad G \to G, \quad g \mapsto g^{-1}$$

are continuous (where we endow $G \times G$ with the product topology).

A group G endowed with a topology is a topological group if and only if the map $G \times G \to G, (g, g') \mapsto gg'^{-1}$, is continuous.

Let G be a topological group, $a \in G$. Then left translation $g \mapsto ag$ and right translation $g \mapsto ga$ are homeomorphisms $G \to G$. In particular the topology of G is uniquely determined by a fundamental system of neighborhoods of one element of G.

Let G be a topological group and let $H \subseteq G$ be a subgroup. We endow the set of cosets G/H with the quotient topology (Example 12.21).

Proposition 12.58. *Let G be a topological group, $H \subset G$ a subgroup.*

1. *The canonical homomorphism $G \to G/H$ is open.*
2. *The closure $\overline{H}$ is a subgroup of G.*
3. *If H is locally closed in G, then H is closed in G.*
4. *If H contains a non-empty open subset U of G, then H is open and closed in G.*
5. *H is open in G (respectively closed in G) if and only if G/H is discrete (respectively Hausdorff).*
6. *Let G_1 and G_2 be topological groups and let $H_i \subseteq G_i$ be subgroups. Then the canonical bijective map $\alpha\colon (G_1 \times G_2)/(H_1 \times H_2) \longrightarrow G_1/H_1 \times G_2/H_2$ is a homeomorphism.*
7. *If H is a normal subgroup, G/H is a topological group.*

Applying 5 to $H = \{e\}$ one sees that G is discrete (respectively Hausdorff) if and only if $\{e\}$ is open (respectively closed) in G.

Proof. 1. Let $U \subseteq G$ be open. Then $p^{-1}(p(U)) = \bigcup_{h \in H} hU$ is open in G. Hence $p(U)$ is open by definition of the quotient topology.

2. Let $a\colon G \times G \to G$ be the continuous map $(g, h) \mapsto gh^{-1}$. Then

$$a(\overline{H} \times \overline{H}) = a(\overline{H \times H}) \subseteq \overline{a(H \times H)} = \overline{H}.$$

This shows 2.

3. We may assume that H is open and dense in G (by replacing G by the subgroup $\overline{H}$). Then for all $g \in G$ the two cosets gH and H have non-empty intersection hence they are equal, i.e., $g \in H$.

4. We have $H = \bigcup_{h \in H} hU$, hence H is open G. Therefore it is also closed by 3.

5. The quotient G/H is discrete if and only if gH is open in G for all $g \in G$ if and only if H is open in G because left translation is a homeomorphism.

If G/H is Hausdorff, then $eH \in G/H$ is a closed point and its inverse image H in G is closed. Conversely, if H is closed, then $H = HeH$ is a closed point in the quotient space $H \backslash G/H$. Hence its inverse image under the continuous map $G/H \times G/H \to H \backslash G/H$, $(g_1 H, g_2 H) \mapsto H g_2^{-1} g_1 H$ is closed. But this is the diagonal of $G/H \times G/H$.

6. The map α is continuous by the universal property of the quotient topology (Example 12.21). Moreover, $G_1 \times G_2 \to G_1/H_1 \times G_2/H_2$ is open by 1. Hence α is also open.

7. The composition of $G \times G \to G$, $(g, g') \mapsto gg'^{-1}$ and of $G \to G/H$ induces a continuous map $G/H \times G/H \cong (G \times G)/(H \times H) \to G/H$ by the universal property of the quotient topology (Example 12.21). $\qquad\square$

Corollary 12.59. *Let G be a connected topological group and let U be a non-empty open subset of G. Then U generates the group G.*

Proof. By Proposition 12.58 4, the subgroup H generated by U is open and closed in G. Hence $H = G$ because G is connected. $\qquad\square$

Proposition 12.60. *Let G be a topological group with neutral element e. Let G^0 be the connected component of e. Then G^0 is a closed normal subgroup of G.*

The subgroup G^0 is called the *identity component of G*.

Proof. For all $g \in G^0$ the subset $g^{-1}G^0$ is connected and contains e. Therefore $g^{-1}h \in G^0$ for all $g, h \in G^0$, which shows that G^0 is a subgroup. This subgroup is stable under all homeomorphisms $G \to G$ sending e to e, in particular G^0 is stable under inner automorphisms. Thus G^0 is normal. Finally G^0 is closed because all connected components are closed (Remark 12.46 2). $\qquad\square$

12.10 Problems

Problem 12.1. Classify (up to homeomorphism) all topological spaces with ≤ 2 points.

Problem 12.2. Let X be a set and let $\mathcal{B}$ be a set of subsets of X. Show that the following assertions are equivalent:

(i) $\mathcal{B}$ is a basis of a topology on X.
(ii) One has $X = \bigcup_{U \in \mathcal{B}} U$ and for all $U, V \in \mathcal{B}$ and $x \in U \cap V$ there exists $W \in \mathcal{B}$ with $x \in W \subseteq U \cap V$.

Problem 12.3. Let X be a set. Show that $d(x, y) := 0$ for $x = y$ and $d(x, y) := 1$ for $x \neq y$ defines a metric on X whose induced topology is the discrete topology. This metric is also called the *trivial metric*.

Problem 12.4. Let p be a prime number. Fix a real number $0 < \rho < 1$. Define the *p-adic absolute value* on $\mathbb{Q}$ by $|0|_p := 0$ and $|p^n \frac{a}{b}|_p := \rho^n$ for $n \in \mathbb{Z}$, $a, b \in \mathbb{Z} \setminus \{0\}$ such that p does not divide ab.

1. Show that the topology on $\mathbb{Q}$ (the *p-adic topology*) induced by the *p-adic metric* $d_p(x, y) := |x - y|_p$ does not depend on the choice of ρ.
2. Show that for different prime numbers p and ℓ the p-adic and the ℓ-adic topology on $\mathbb{Q}$ are different. Show that they are also different from the topology on $\mathbb{Q}$ induced by usual absolute value.

Problem 12.5. Show that the trivial metric (Problem 12.3) and the p-adic metric on $\mathbb{Q}$ (Problem 12.4) both give examples of metric spaces, where the closure of $B_r(x)$ is not $B_{\leq r}(x)$ in general.

Problem 12.6. Let (X, d) be a metric space. Show that $d' \colon X \times X \to \mathbb{R}$, $d'(x, y) :=$ $\min\{d(x, y), 1\}$ is a metric on X and that d and d' induce the same topology on X.

Problem 12.7. A topological space is called a T_0-*space* or a *Kolmogorov space* if for every pair of distinct points one of them has a neighborhood not containing the other. Let X be a set, $p \in X$. Define a subset U of X to be open if $U = \emptyset$ or if $p \in U$. Show that this defines a topology on X (called the *particular point topology on X*). Show that X is a T_0-space and that the closure of $\{p\}$ is X (in particular $\{p\}$ is not closed if X has more than a single element).

Problem 12.8. Let X be a topological space. Write $x \sim y$ if an open subset of X contains x if and only if it contains y. Show that $\sim$ is an equivalence relation and that $X/\sim$, endowed with the quotient topology, is a T_0-space.

Problem 12.9. Let X and Y be topological spaces. Let $f \colon X \to Y$ be a map. Show that the following assertions are equivalent:

(i) f is continuous.
(ii) For every subset A of X one has $f(\overline{A}) \subseteq \overline{f(A)}$.

Problem 12.10. Show that the following maps are bijective and continuous but not a homeomorphism:

1. The map $\mathrm{id}_{\mathbb{R}} \colon \mathbb{R}_d \to \mathbb{R}$, where $\mathbb{R}_d$ is the set of real numbers endowed with the discrete topology.
2. The map $[0, 1) \to S^1 := \{z \in \mathbb{C} \; ; \; |z| = 1\}$, $t \mapsto e^{2\pi i t}$.

Problem 12.11. Let $(X_n, d_n)_n$ be a countable family of metric spaces each also considered as a topological space. Let $X = \prod_n X_n$ be the product space. Show that

$$d \colon X \times X \to \mathbb{R}, \qquad (x, y) \mapsto \sum_n \frac{1}{2^n} \frac{d_n(x_n, y_n)}{1 + d_n(x_n, y_n)}$$

defines a metric on X that induces the product topology.

Problem 12.12. Let X be a topological space.

1. Let $Z \subseteq Y \subseteq X$ be subspaces. Show that Z is dense in X if and only if Z is dense in Y and Y is dense in X.
2. Let Y be a subspace and let $(U_i)_i$ be an open covering of X. Show that Y is dense in X if and only if $Y \cap U_i$ is dense in U_i for all i.

Problem 12.13. Let $f : X \rightarrow Y$ be a continuous map. Let $(B_i)_i$ be a family of subsets of Y and suppose (a) that the interiors cover Y or (b) that $(B_i)_i$ is a locally finite closed covering. Show that f is open (respectively closed) if and only if its restriction $f^{-1}(B_i) \rightarrow B_i$ is open (respectively closed) for all i.

Problem 12.14. Let $(X_i)_{i \in I}$ be a family of non-empty topological spaces. Show that the product space $\prod_{i \in I} X_i$ is Hausdorff if and only if X_i is Hausdorff for all $i \in I$.

Problem 12.15. Let X be a topological space and let $(U_i)_{i \in I}$ be an open covering.

1. Give an example where U_i is Hausdorff for all $i \in I$ but X is not Hausdorff.
2. Show that X is Hausdorff if and only if for all $i, j \in I$ the set $\{ (x, x) \, ; \, x \in U_i \cap U_j \}$ is closed in $U_i \times U_j$.

Problem 12.16. Let X be a set. Define a subset U of X to be open if $U = \emptyset$ or if $X \setminus U$ is a finite set.

1. Show that this defines a topology on X. It is called the *cofinite topology on X*.
2. Let X be endowed with the cofinite topology. Show that the following assertions are equivalent:
   (i) X is discrete.
   (ii) X is finite
   (iii) X is Hausdorff.
3. Show that X is compact.
4. Let X be infinite. Show that every non-empty open subspace of X is connected.

Problem 12.17. Let $f : X \rightarrow Y$ be a surjective map of topological spaces. Suppose that f is open *or* closed. Show that Y has the quotient topology induced by X.

Problem 12.18. Let $f : X \rightarrow Y$ be a surjective map of topological spaces and suppose that Y has the quotient topology induced by X. Show that if Y is connected and all fibers of f are connected, then X is connected.

Problem 12.19. A topological space X is called *totally disconnected* if the connected component of each point consists of the point alone.

1. Show that the subspace $\mathbb{Q}$ of $\mathbb{R}$ is totally disconnected.
2. Show that $\mathbb{Q}$ endowed with the p-adic topology (Problem 12.4) is totally disconnected.
3. Let X be a topological space, let $\bar{X}$ be the set of connected components endowed with the quotient topology with respect to the natural map $X \rightarrow \bar{X}$. Show that $\bar{X}$ is totally disconnected.

Problem 12.20.

1. Let X be a topological space. Show that the connected component of $x \in X$ is contained in the intersection of all open and closed subsets of X containing x.
2. Consider the following subspace of $\mathbb{R}^2$:

$$X := \{ (1/n, y) ; \, n \in \mathbb{N}, y \in [-1, 1] \} \cup \{ (0, y) ; \, y \in [-1, 1] \setminus \{0\} \}.$$

Show that there exists $x \in X$ whose connected component is different from the intersection of the open and closed subsets of X containing x.

Problem 12.21. Let X be a topological space. Show that the following three conditions are equivalent:

(i) The connected components of X are open.
(ii) For every $x \in X$ there exists a neighborhood of x that is contained in every subset that is both open and closed and that contains x.
(iii) For every $x \in X$ the intersection of all open and closed subsets containing x is an open subset of X.

Problem 12.22. A topological space X is called *locally connected* if every point of X has a fundamental system of connected neighborhoods.

1. Show that a topological space X is locally connected if and only if every connected component of every open subspace is open.
   Hint: Problem 12.21.
2. Show that every quotient space of a locally connected space is again locally connected.

Problem 12.23. Let $(X_i)_{i \in I}$ be a family of non-empty topological spaces and let $X = \prod_i X_i$ be the product space. Show that X is locally connected (Problem 12.22) if and only if each X_i is locally connected and X_i is connected for all but finitely many i.

Problem 12.24. Let $(X, \leq)$ be a totally ordered set (Definition 13.20) and endow X with the topology generated by sets of the form $X_{<a} := \{ x \in X ; \, x < a \}$ and $X_{>a} := \{ x \in X ; \, x > a \}$ with $a \in X$. This topology is called the *order topology on X*.

1. Show that a basis of the topology is given by the open intervals, i.e., sets of the form $X_{<a}$, $X_{>a}$ and $(a, b) := \{ x \in X ; \, a < x < b \}$ for $a, b \in X$.
2. Show that X is a Hausdorff space.
3. Show that X is totally disconnected (Problem 12.19) if X is well ordered (Definition 13.24).

Problem 12.25. Let (X, d) be a metric space. Show that if $C \subseteq X$ is compact, then C is closed and bounded in X. Show that the converse does not hold in general.

Problem 12.26. Let $a, b \in \mathbb{R}$ with $a < b$. Show that $[a, b]$ is compact (without using the Heine–Borel Theorem).

Problem 12.27. Let V be a finite-dimensional $\mathbb{K}$-vector space and let $\| \ \|$ be a norm on V. Show the Heine–Borel Theorem: A subspace of V is compact if and only if it is closed in V and bounded.
Hint: Problem 12.26.

Problem 12.28. Let $(V, \| \ \|)$ be a normed $\mathbb{K}$-vector space. Show that the following assertions are equivalent:

 (i) V is locally compact.
(ii) $B_{\leq 1}(0)$ is compact.
(iii) V is finite-dimensional.

Hint: Show that if $B_{\leq 1}(0) \subseteq \bigcup_{i=1}^{n} B_{\frac{1}{2}}(v_i)$, then $\{v_1, \ldots, v_n\}$ is a generating system of V to see that (ii) implies (iii).

Problem 12.29 (One-point compactification or Alexandroff compactification). Let X be a topological space and set $X^* := X \cup \{\infty\}$, with $\infty \notin X$. Define a subset U of X^* to be open if U is an open subset of X or if U contains ∞ and $X^* \setminus U$ is a closed and compact subspace of X.

1. Show that this defines a topology on X^*, that the inclusion $X \rightarrow X^*$ is an open embedding, and that X^* is compact. The space X^* is called the *one-point compactification* or *Alexandroff compactification* of X.
2. Show that X^* is Hausdorff if and only if X is Hausdorff and locally compact.

Problem 12.30. Show that no point of $\mathbb{Q}$ (considered as a subspace of $\mathbb{R}$) has a compact neighborhood. In particular $\mathbb{Q}$ is not locally compact.

Problem 12.31. Let X be a topological space and let A be a subset of X. Then A is called *relatively compact in X* if A is contained in a compact subspace of X.

Now suppose that X is Hausdorff. Show that a A is relatively compact in X if and only if $\bar{A}$ is compact.

Problem 12.32. Consider the following conditions on a topological space X:

(i) Every point of X has a compact neighborhood.
(ii) Every point of X has a closed compact neighborhood.
(iii) X is locally compact.

1. Show that (ii) and (iii) each imply (i).
2. Show that the one-point compactification of $\mathbb{Q}$ (Problem 12.29) satisfies (ii) but not (iii).
3. Show that the particular point topology on an infinite set (Problem 12.7) satisfies (iii) but not (ii).
4. Suppose that X is Hausdorff. Show that (i), (ii), and (iii) are equivalent.

Problem 12.33. Let X be a Hausdorff space.

1. Show that every locally compact subspace of X is locally closed.
2. Suppose that X is locally compact. Show that every locally closed subspace of X is locally compact.

Problem 12.34. Let $(X_i)_i$ be a family of non-empty topological spaces. Show that the product space $\prod_i X_i$ is locally compact and Hausdorff if and only if each space X_i is locally compact and Hausdorff and X_i is compact for all but finitely many i.

Problem 12.35. Let X be a topological space. A subset A of X is called *nowhere dense* if $(\bar{A})^\circ = \emptyset$. Show that the union of a locally finite family of nowhere dense subsets is again nowhere dense. Show that the boundary of a closed or of an open subset is nowhere dense.

Problem 12.36. Let G be a topological group and let H be a normal subgroup. Show that its closure $\bar{H}$ is a normal subgroup of G.

Problem 12.37. Let G be a topological group and let G^0 be its identity component. Show that G/G^0 is totally disconnected (Problem 12.19).

Problem 12.38. Let G be a Hausdorff locally compact topological group and let H be a closed subgroup.

1. Show that G/H is Hausdorff and locally compact.
2. Show that the identity component G^0 of G is the intersection of all open subgroups of G.
3. Show that the connected components of G/H are the closures of the images of the connected components of G.

Appendix B: The Language of Categories

13

The goal of this appendix is to make the reader familiar with the language of categories. It is by no means a rigorous introduction to category theory. We will explain the most important notions and give many examples. There will be almost no proofs in this chapter. Many of the claims are easy to check and are left to the reader as "exercise". A complete exposition can for instance be found in the book [KS2] by M. Kashiwara and P. Schapira. In particular it contains proofs of the statements that we leave as "exercises". For some of the more difficult results explicit references are given.

We start by introducing categories and functors. Important concepts are the notion of an equivalence of categories and, more generally, the notion of adjoint functors. We then give a short reminder on several notions of ordered sets and prove the principles of transfinite induction and transfinite recursion. In the last section we introduce limits and colimits in categories. This allows us to unify several important constructions such as kernels, products, fiber products, and final objects (which are all special cases of limits) or cokernels, coproducts, pushouts, and initial objects (which are all special cases of colimits).

13.1 Categories

Definition 13.1. A *category* C consists of:[1]

---

[1] Here we ignore all set-theoretical issues. To avoid set-theoretical difficulties one should work with a fixed universe in the sense of [KS2] Definition 1.1.1 and assume that the sets of morphisms between two objects always lies in the given universe. Then a category C is called *small* if $\mathrm{Ob}(C)$ is in that universe. Moreover it then will be sometimes necessary to pass to a bigger universe, for instance when considering the category of functors between two categories (see Definition 13.13). Finally one adds also the axiom that every set is an element of some universe to the axioms of Zermelo–Fraenkel set theory. We refer to [SGA4] Exp. I for details. Alternatively one can also work with classes as explained in [Sch].

© Springer Fachmedien Wiesbaden 2016

T. Wedhorn, *Manifolds, Sheaves, and Cohomology*, Springer Studium Mathematik – Master,
DOI 10.1007/978-3-658-10633-1_13

(a) a set $\text{Ob}(C)$ of *objects*,

(b) for any two objects X and Y a set $\text{Hom}_C(X, Y) = \text{Hom}(X, Y)$ of *morphisms* from X to Y,

(c) for any three objects X, Y, Z a *composition map*

$$\text{Hom}(X, Y) \times \text{Hom}(Y, Z) \to \text{Hom}(X, Z), \qquad (f, g) \mapsto g \circ f$$

such that all morphisms sets are disjoint (hence every morphism $f \in \text{Hom}(X, Y)$ has unique *source* X and a unique *target* Y) and such that:

1. the composition of morphisms is associative (i.e., for all objects X, Y, Z, W and for all $f \in \text{Hom}(X, Y), g \in \text{Hom}(Y, Z), h \in \text{Hom}(Z, W)$ one has $h \circ (g \circ f) = (h \circ g) \circ f$,

2. for all objects X there exists a morphism $\text{id}_X \in \text{Hom}(X, X)$, called the *identity of X* such that for all objects Y and for all morphisms $f \in \text{Hom}(X, Y)$ and $g \in \text{Hom}(Y, X)$ one has $f \circ \text{id}_X = f$ and $\text{id}_X \circ g = g$.

A category is called *finite* if it has only finitely many objects and morphisms.

Remark 13.2. For every object X of a category C the morphism id_X is uniquely determined by Condition 2 for if id_X and $\tilde{\text{id}}_X$ are identities of X, then $\tilde{\text{id}}_X = \tilde{\text{id}}_X \circ \text{id}_X = \text{id}_X$.

Definition and Remark 13.3. A morphism $f \colon X \to Y$ in a category is called an *isomorphism* if there exists a morphism $g \colon Y \to X$ such that $f \circ g = \text{id}_Y$ and $g \circ f = \text{id}_X$. We often write $f \colon X \xrightarrow{\sim} Y$ to indicate that f is an isomorphism. We also write $X \cong Y$ and say that X and Y are *isomorphic* if there exists an isomorphism $X \xrightarrow{\sim} Y$.

An morphism (respectively an isomorphism) with the same source and target X is called an *endomorphism* (respectively an *automorphism*) of X. Composition yields the structure of a monoid on the set $\text{End}_C(X)$ of endomorphisms of X and the structure of a group on the set $\text{Aut}_C(X)$ of automorphisms.

Definition 13.4. A *subcategory* of a category C is a category C' such that every object of C' is an object of C and such that $\text{Hom}_{C'}(X', Y') \subseteq \text{Hom}_C(X', Y')$ for any pair (X', Y') of objects of C', compatibly with composition of morphisms and identity elements. The subcategory C' is called *full* if $\text{Hom}_{C'}(X', Y') = \text{Hom}_C(X', Y')$ for all objects X' and Y' of C'.

Example 13.5.

1. (Sets) the category of sets: Objects are sets, for two sets X and Y a morphism is a map $X \to Y$, composition in the category is the usual composition of maps, and the identity of a set X is the usual identity map id_X. An isomorphism in (Sets) is simply a bijective map.

2. (Grp) the category of groups: Objects are groups, morphisms are group homomorphisms, composition is the usual composition of group homomorphisms, and the identity is the usual identity. An isomorphism in (Grp) is a group isomorphism.

3. (Ab) is the full subcategory of (Grp) of abelian groups: Objects are abelian groups,
morphisms are group homomorphisms, composition is the usual composition of group
homomorphisms, and the identity is the usual identity.

4. Fix a group G. Then (G-Sets) denotes the category of G-sets: Objects are sets with
a left action by the group G and morphisms are G-equivariant maps. By (Sets-G) we
denote the category of sets with a right action by G.

5. (Ring) the category of rings: Objects are rings (always assumed to have a unit) and
morphisms are ring homomorphisms (preserving the unit).

6. (Top) the category of topological spaces: Objects are topological spaces and morphisms are continuous maps. An isomorphism is a homeomorphism.

7. Let R be a ring. Then (R-Mod) denotes the category of left R-modules: Objects are
left R-modules, morphisms are R-linear maps. An isomorphism is a bijective R-linear
map because its inverse is automatically R-linear again.

   If $R = k$ is a field, we obtain the category of k-vector spaces, usually denoted by
(k-Vec) instead of (k-Mod).

Definition 13.6. Let C be a category. A morphism $f: X \to Y$ in C is called a *monomorphism* (respectively *epimorphism*) if for all objects Z in C and all morphisms $g, h: Z \to X$
(respectively $g, h: Y \to Z$) one has

$$f \circ g = f \circ h \Rightarrow g = h \quad (\text{respectively} \quad g \circ f = h \circ f \Rightarrow g = h).$$

Example 13.7. In the category of sets (respectively of groups), monomorphisms are injective maps (respectively group homomorphisms) and epimorphisms are surjective maps
(respectively group homomorphisms).

In the category of rings the inclusion $\mathbb{Z} \to \mathbb{Q}$ is a monomorphism and an epimorphism. This gives in particular an example of a morphism that is a monomorphism and an
epimorphism but not an isomorphism.

Definition 13.8. For every category C the *opposite category*, denoted by C^{opp}, is the
category with the same objects as C and where for two objects X and Y of C^{opp} we set
$\text{Hom}_{C^{\text{opp}}}(X, Y) := \text{Hom}_C(Y, X)$ with the obvious composition law.

Definition 13.9. Let C and $\mathcal{D}$ be categories. We define the *product category* $C \times \mathcal{D}$ by
$\text{Ob}(C \times \mathcal{D}) := \text{Ob}(C) \times \text{Ob}(\mathcal{D})$ and $\text{Hom}_{C \times \mathcal{D}}((X_1, Y_1), (X_2, Y_2)) := \text{Hom}_C(X_1, X_2) \times$
$\text{Hom}_C(Y_1, Y_2)$. Composition of morphisms is defined componentwise.

13.2   Functors

Definition 13.10. Given categories C and $\mathcal{D}$, a *(covariant) functor* $F: C \to \mathcal{D}$ is given by attaching to each object C of C an object $F(C)$ of $\mathcal{D}$, and to each morphism $f: C \to C'$ in C a morphism $F(f): F(C) \to F(C')$, compatible with composition of morphisms, i.e., $F(g \circ f) = F(g) \circ F(f)$ whenever the composition is defined, and preserving identity elements, i.e., $F(\mathrm{id}_C) = \mathrm{id}_{F(C)}$.

If $F: C \to \mathcal{D}$ and $G: \mathcal{D} \to \mathcal{E}$ are functors, we write $G \circ F: C \to \mathcal{E}$ for the composition. A covariant functor $C^{\mathrm{opp}} \to \mathcal{D}$ is also called a *contravariant functor from C to D*.

By a "functor" we will always mean a covariant functor. If we write $F: C \to \mathcal{D}$, then we always mean that F is covariant. A contravariant functor F from C to $\mathcal{D}$ will always be denoted by $F: C^{\mathrm{opp}} \to \mathcal{D}$. It attaches to each object C in C an object $F(C)$ in $\mathcal{D}$ and to each morphism $f: C \to C'$ in C a morphism $F(f): F(C') \to F(C)$ such that $F(g \circ f) = F(f) \circ F(g)$ and $F(\mathrm{id}_C) = \mathrm{id}_{F(C)}$.

Remark 13.11. Let $F: C \to \mathcal{D}$ be a functor. Let C and C' be objects in C and let $f: C \to C'$ be an isomorphism in C. Then $F(f)$ is an isomorphism.

Indeed, as f is an isomorphism, there exists a morphism $g: C' \to C$ such that $g \circ f = \mathrm{id}_C$ and $f \circ g = \mathrm{id}_{C'}$. Hence $F(g) \circ F(f) = F(g \circ f) = F(\mathrm{id}_C) = \mathrm{id}_{F(C)}$ and similarly $F(f) \circ F(g) = \mathrm{id}_{F(C')}$. Therefore $F(f)$ is an isomorphism.

Example 13.12. A simple example is the functor that "forgets" some structure. An example is the functor $F: (\mathrm{Grp}) \to (\mathrm{Sets})$ that attaches to every group the underlying set and that sends every group homomorphism to itself (but now considered as a map of sets).

Definition and Remark 13.13. For two functors $F, G: C \to \mathcal{D}$ we call a family of morphisms $\alpha(S): F(S) \to G(S)$ for every object S of C *functorial in S* if for every morphism $f: T \to S$ in C the diagram

$$
\begin{array}{ccc}
F(T) & \xrightarrow{\ \alpha(T)\ } & G(T) \\
{\scriptstyle F(f)}\downarrow & & \downarrow{\scriptstyle G(f)} \\
F(S) & \xrightarrow{\ \alpha(S)\ } & G(S)
\end{array}
$$

commutes. We also say that α is a *morphism of functors* $F \to G$.

If $\beta: G \to H$ is a second morphism of functors, we define the composition $\beta \circ \alpha$ by $(\beta \circ \alpha)(S) = \beta(S) \circ \alpha(S)$. With this notion of morphism we obtain the category of all functors from C to $\mathcal{D}$, denoted by $\mathrm{Func}(C, \mathcal{D})$ or $\mathcal{D}^C$. The identity id_F is given by $\mathrm{id}_{F(S)}$, where S runs through all objects of C. In particular we obtain the notion of an isomorphism $F \xrightarrow{\sim} G$ of functors.

Example 13.14. Let k be a field.

1. For every k-vector space V let $V^\vee := \mathrm{Hom}_k(V,k)$ be its dual space and for every k-linear map $u: V \to W$ let $u^\vee: W^\vee \to V^\vee$, $\lambda \mapsto \lambda \circ u$ be its dual homomorphism. We obtain a functor $(\ )^\vee: (k\text{-Vec})^{\mathrm{opp}} \to (k\text{-Vec})$.

2. For every k-vector space V one has the biduality homomorphism of k-vector space

$$\iota_V: V \to (V^\vee)^\vee, \qquad v \mapsto (\lambda \mapsto \lambda(v)), \qquad v \in V, \lambda \in V^\vee.$$

Moreover, if $u: V \to W$ is k-linear, then the diagram

$$
\begin{array}{ccc}
V & \xrightarrow{\ \iota_V\ } & (V^\vee)^\vee \\
{\scriptstyle u}\downarrow & & \downarrow{\scriptstyle (u^\vee)^\vee} \\
W & \xrightarrow{\ \iota_W\ } & (W^\vee)^\vee
\end{array}
$$

is commutative. In other words, the ι_V for V a k-vector space form a morphism of functors $\mathrm{id}_{(k\text{-Vec})} \to (^\vee \circ {}^\vee)$.

3. If V is a finite-dimensional vector space, then $V^\vee$ if finite-dimensional and ι_V is an isomorphisms. Hence $(\ )^\vee$ restricts to a functor $(k\text{-vec}) \to (k\text{-vec})$, where $(k\text{-vec})$ denotes the full subcategory of $(k\text{-Vec})$ of finite-dimensional k-vector spaces, and $V \mapsto \iota_V$ is an isomorphism of functors $\mathrm{id}_{(k\text{-vec})} \xrightarrow{\sim} (^\vee \circ {}^\vee)$.

Definition 13.15. Let $F: C \to \mathcal{D}$ be a functor.

1. F is called *faithful* (respectively *fully faithful*) if for all objects X and Y of C the map $\mathrm{Hom}_C(X,Y) \to \mathrm{Hom}_\mathcal{D}(F(X), F(Y))$, $f \mapsto F(f)$ is injective (respectively bijective).
2. F is called *essentially surjective* if for every object Y of $\mathcal{D}$ there exists an object X of C and an isomorphism $F(X) \cong Y$.
3. F is called an *equivalence of categories* if it is fully faithful and essentially surjective.

There are analogous notions for contravariant functors. A contravariant functor that is an equivalence of categories is sometimes also called an *anti-equivalence of categories*.

Theorem and Definition 13.16. *A functor $F: C \to \mathcal{D}$ is an equivalence of categories if and only if there exists a quasi-inverse functor G, i.e., a functor $G: \mathcal{D} \to C$ such that $G \circ F \cong \mathrm{id}_C$ and $F \circ G \cong \mathrm{id}_\mathcal{D}$.*

Proof. [KS2] Proposition 1.3.13. □

Definition 13.17. Let C and $\mathcal{D}$ be categories and let $F\colon C \to \mathcal{D}$ and $G\colon \mathcal{D} \to C$ be functors. Then G is said to be *right adjoint* to F and F is said to be *left adjoint* to G if for all objects X in C and Y in $\mathcal{D}$ there exists a bijection

$$\mathrm{Hom}_C(X, G(Y)) \cong \mathrm{Hom}_\mathcal{D}(F(X), Y),$$

which is functorial in X and in Y.

Example 13.18. If $F\colon C \to \mathcal{D}$ is an equivalence of categories, then a quasi-inverse functor $G\colon \mathcal{D} \to C$ is right adjoint and left adjoint to F. Indeed for all objects X in C and Y in $\mathcal{D}$ we have bijections, functorial in X and in Y

$$\mathrm{Hom}_C(X, G(Y)) \xrightarrow{\sim} \mathrm{Hom}_C(G(F(X)), G(Y)) \xrightarrow{\sim} \mathrm{Hom}_\mathcal{D}(F(X), Y).$$

This shows that F is left adjoint to G. A similar argument shows that F is also right adjoint to G.

Example 13.19. Let k be a field and let $F\colon (k\text{-Vec}) \to (\text{Sets})$ be the forgetful functor. Then F is right adjoint to the functor G that sends a set I to the k-vector space

$$k^{(I)} = \{ (x_i)_{i \in I} \;;\; x_i \in k, \, x_i = 0 \text{ for all but finitely many } i \}$$

and which sends a map $a\colon I \to J$ to the unique k-linear map $G(a)\colon k^{(I)} \to k^{(J)}$ such that $G(a)(e_i) = e_{a(i)}$ where $(e_i)_{i \in I}$ (respectively $(e_j)_{j \in J}$) is the standard basis of $k^{(I)}$ (respectively $k^{(J)}$).

Indeed for every set I and every k-vector space V we have a bijection, functorial in I and in V,

$$\mathrm{Hom}_{(k\text{-Vec})}(k^{(I)}, V) \xrightarrow{\sim} \mathrm{Hom}_{(\text{Sets})}(I, V), \qquad f \mapsto (i \mapsto f(e_i)).$$

13.3 Digression: Ordered Sets, Transfinite Induction, and Zorn's Lemma

Definition 13.20 (Ordered sets). Let I be a set.

1. A relation $\leq$ on I is called a *partial preorder* or simply a *preorder*, if $i \leq i$ for all $i \in I$ and $i \leq j$, $j \leq k$ imply $i \leq k$ for all $i, j, k \in I$.
2. A partial preorder $\leq$ on I is called a *partial order* or simply an *order* if $i \leq j$ and $j \leq i$ imply $i = j$ for all $i, j \in I$.
3. A partial preorder $\leq$ is called *filtered* if $I \neq \emptyset$ and if for all $i, j \in I$ there exists a $k \in I$ with $i \leq k$ and $j \leq k$.

4. A partial order $\leq$ is called a *total order* if for all $i, j \in I$ one has $i \leq j$ or $j \leq i$.

If $\leq$ is a partial preorder on a set I and $i, j \in I$, we write $i < j$ if $i \leq j$ and $i \neq j$; we write $i \geq j$ if $j \leq i$, and $i > j$ if $j < i$. We also set $I_{\leq i} := \{ j \in I ; j \leq i \}$. Similarly we define $I_{<i}$, $I_{\geq i}$, and $I_{>i}$.

If $(I, \leq)$ is a preordered set, then every subset J endowed with the induced relation (i.e., with the restriction of $\leq$ to $J \times J$) is again preordered. If I is partially (respectively totally) ordered, so is J. Every total order is filtered.

Example 13.21.

1. Let I be any set. The *discrete order* in I is defined by $i \leq j \Leftrightarrow i = j$. This is a partial order. It is neither filtered nor a total order if I consists of more than one element.
2. The *chaotic order* on a set I is the preorder such that $i \leq j$ for all $i, j \in I$.
3. The real numbers together with the usual ordering form a totally ordered set.
4. For $n, m \in \mathbb{N}$ we write $n \mid m$ if n divides m (i.e., there exists $k \in \mathbb{N}$ such that $m = kn$). Then $\mid$ is a filtered partial order on $\mathbb{N}$, which is not a total order.
5. Let X be a topological space, let $x \in X$. Let $\mathcal{U}(x)$ be the set of open neighborhoods of x in X. Endow $\mathcal{U}(x)$ with the opposite of the inclusion relation, i.e., $U \leq V$ if $V \subseteq U$. Then $\leq$ is a filtered partial order on $\mathcal{U}(x)$.

Definition 13.22. Let $(I, \leq)$ be a partially ordered set.

1. An element $i_0 \in I$ is called *minimal* (respectively *maximal*) if there exists no element $i \in I$ with $i < i_0$ (respectively $i > i_0$).
2. An element $i_0 \in I$ is called the *smallest element* (respectively the *greatest element*) if $i \geq i_0$ (respectively $i \leq i_0$) for all $i \in I$.
3. Let $S \subseteq I$ be a subset. An element $i_0 \in I$ is called *upper bound of S* (respectively *lower bound of S*) if $i \leq i_0$ (respectively $i \geq i_0$) for all $i \in S$.

A smallest and a greatest element of a partially ordered set is unique (if it exists). If I is totally ordered, then every minimal (respectively maximal) element is the smallest (respectively greatest) element.

Example 13.23. Consider the sets $\mathbb{N}$ and $\mathbb{N}_{\geq 2}$ with the dividing order (Example 13.21 4). In $\mathbb{N}_{\geq 2}$ the prime numbers are the minimal elements but $\mathbb{N}_{\geq 2}$ has no smallest element. In $\mathbb{N}$, 1 is a smallest element with respect to $\mid$.

Definition 13.24 (Well orders). A partial order $\leq$ on a set I is called a *well order* if every non-empty subset of I has a smallest element.

A well order is always a total order: For $i, j \in I$, the set $\{i, j\}$ has a smallest element. Every subset of a well-ordered set is again well-ordered with the induced order. Every element i of a well-ordered set is either the greatest element or it has a *successor* $i+$, which is defined to be the smallest element of $\{ j \in I \; ; \; j > i \}$. We then have $I_{<i+} = I_{\leq i}$.

The natural numbers with the usual order are a well-ordered set. On the real numbers the usual order is not a well order (the open interval $(0, 1)$ has no smallest element). We have the following principle of *transfinite induction*[2].

Proposition 13.25 (Transfinite induction). *Let $(I, \leq)$ be a partially ordered set such that every non-empty subset of I has a minimal element (e.g., if $(I, \leq)$ is well ordered). Let $J \subseteq I$ be a subset such that for all $i \in I$ one has*

$$I_{<i} \subseteq J \Rightarrow i \in J.$$

Then $I = J$.

Very often transfinite induction is applied to prove that a property $\mathbf{P}(i)$, depending on $i \in I$, holds for all $i \in I$ by setting $J := \{i \in I \; ; \; \mathbf{P}(i) \text{ holds}\}$. The special case $I = \mathbb{N}$ with the usual order corresponds to the standard mathematical induction.

Proof. Assume $I \setminus J \neq \emptyset$. By hypothesis there exists a minimal element i_0 in $I \setminus J$. But then $I_{<i_0} \subseteq J$ and hence $i_0 \in J$ by our hypothesis. This is a contradiction. $\square$

A method related to transfinite induction is the method of *transfinite recursion*, which we state as follows.

Proposition 13.26 (Transfinite recursion). *Let $(I, \leq)$ be a well-ordered set, let X be a set, and let W be the set of pairs (i, f^i), where $i \in I$ and $f^i \colon I_{<i} \to X$ is a map. Let $S \subseteq W$ be a subset and let $G \colon S \to X$ be a map satisfying the following two conditions:*

(a) *If $(i, f^i) \in S$, then $(j, f^i{}_{|I_{<j}}) \in S$ for all $j \in I_{<i}$.*
(b) *If $(i, f^i) \in S$ and i is not the greatest element of I, then $(i+, (f^i)_+) \in S$, where $(f^i)_+ \colon I_{<i+} \to X$ with $(f^i)_{+|I_{<i}} := f^i$ and $(f^i)_+(i) := G(i, f^i)$.*

Then there exists a unique map $f \colon I \to X$ such that $(i, f_{|I_{<i}}) \in S$ and $f(i) = G(i, f_{|I_{<i}})$ for all $i \in I$.

---

[2] Partial orders satisfying the hypothesis of Proposition 13.25 are also called *noetherian*. Hence Proposition 13.25 is also called the principle of *noetherian induction*.

For instance, to construct a family of subsets $(A_i)_{i \in I}$ of some set Z indexed by a well-ordered set, it suffices to construct A_i from $(A_j)_{j < i}$: Let X be the power set of Z, consider $(A_j)_{j < i}$ as a map $A^i : I_{<i} \to X$, and let G be the map that attaches to (i, A^i) the set A_i. Often the construction of A_i from $(A_j)_{j < i}$ will not be possible for an arbitrary family $(A_j)_{j < i}$ of subsets and one has to restrict to a certain subset S of such families.

Proof. Uniqueness. Let $f \neq \tilde{f}$ be two such maps and let $i \in I$ be the smallest element such that $f(i) \neq \tilde{f}(i)$. Then $f(i) = G(i, f_{|I_{<i}}) = G(i, \tilde{f}_{|I_{<i}}) = \tilde{f}(i)$. This is a contradiction.

Existence. Let us call a subset J of I a *segment* if $j \in J$, $i \in I$, and $i \leq j$ imply $i \in J$. Clearly the union of segments is again a segment. If $J \neq I$ let $i \in I$ be the smallest element in $I \setminus J$. Then $J = I_{<i}$. In other words, the segments $\neq I$ are the subsets $J = I_{<i}$ for some $i \in I$. The set of segments of I is well ordered by inclusion.

Let S be the set of segments J of I such that there exists a map $f_J : J \to X$ such that $(j, f_{|I_{<j}}) \in S$ and $f(j) = G(j, f_{|I_{<j}})$ for all $j \in J$. As J is well ordered, the above uniqueness argument shows that f_J is unique if it exists. If $J \subseteq K$ are segments in S, then $f_{K|J} = f_J$. Hence the union of segments in S is again a segment in S. Moreover suppose $I_{<i} \in S$ for some $i \in I$. Then $I_{<i+} \in S$ with $f_{I_{<i+}} := (f_{I_{<i}})_+$.

We conclude by showing that S contains every segment of I. Assume that there exists a segment J of I with $J \notin S$. As the set of segments is well ordered we may assume that J is the smallest segment with $J \notin S$. If J has no greatest element, then J is the union of the smaller segments $J_{<j}$ for $j \in J$. This is not possible because S is stable under union. But if J has greatest element j, then $J = J_{<j+} \in S$ because $J_{<j} \in S$. This is a contradiction. $\qquad\square$

Transfinite induction and recursion is particularly useful because of the following result.

Proposition 13.27 (Well-ordering theorem). *Every set can be endowed with a well order.*

The well-ordering theorem is equivalent to the axiom of choice and also to Zorn's Lemma:

Proposition 13.28 (Zorn's Lemma). *Let $(I, \leq)$ be a ordered set such that every well-ordered subset S has an upper bound in I. Then there exist maximal elements in I.*

Very often Zorn's Lemma is formulated with the stronger hypothesis that every totally ordered subset has an upper bound.

We do not prove here that the axiom of choice implies the well-ordering theorem and Zorn's Lemma but refer to [BouTS] III §2.3, Theorem 1 and [BouTS] III §2.4, Proposition 4.

13.4   Limits and Colimits

In this section $\mathcal{I}$ always denotes a small category.

Definition 13.29. Let C be a category.

1. An $\mathcal{I}$-*diagram in* C is a functor $X : \mathcal{I} \to C$. A *morphism of* $\mathcal{I}$-*diagrams* is a morphism of functors. We obtain the category of $\mathcal{I}$-diagrams in C, denoted by $C^{\mathcal{I}}$. We often write X_i instead of $X(i)$ for an object i in $\mathcal{I}$.

2. A *colimit* or *inductive limit* of an $\mathcal{I}$-diagram $X : \mathcal{I} \to C$ is an object

$$\operatorname*{colim}_{\mathcal{I}} X = \operatorname*{colim}_{i \in \mathcal{I}} X_i = \varinjlim_{i \in \mathcal{I}} X_i$$

in C together with morphisms $s_i : X_i \to \operatorname{colim}_{\mathcal{I}} X$ in C for all objects i in $\mathcal{I}$ such that:

(a) for every morphism $\varphi : i \to j$ in $\mathcal{I}$ one has $s_i = s_j \circ X(\varphi)$,

(b) for every object Z in C and for all morphisms $t_i : X_i \to Z$ such that for all morphisms $\varphi : i \to j$ in $\mathcal{I}$ one has $t_i = t_j \circ X(\varphi)$ there exists a unique morphism $q : \operatorname{colim}_{\mathcal{I}} X \to Z$ such that $t_i = t \circ s_i$.

3. A *limit* or *projective limit* of a diagram $X : \mathcal{I} \to C$ is an object

$$\lim_{\mathcal{I}} X = \lim_{i \in \mathcal{I}} X_i = \varprojlim_{i \in \mathcal{I}} X_i$$

in C together with morphisms $p_i : \lim_{\mathcal{I}} X \to X_i$ in C for all objects $i \in \mathcal{I}$ such that:

(a) for every morphism $\varphi : j \to i$ in $\mathcal{I}$, one has $p_i = X(\varphi) \circ p_j$,

(b) for every object Z in C and for all morphism $q_i : Z \to X_i$ such that $q_i = X(\varphi) \circ q_j$ for all morphism $\varphi : j \to i$ in $\mathcal{I}$, there exists a unique morphism $q : Z \to \lim_{\mathcal{I}} X$ such that $q_i = p_i \circ q$ for all $i \in \operatorname{Ob}(\mathcal{I})$.

Remark 13.30. Limits and colimits are (if they exist) unique up to unique isomorphism by the uniqueness requirement in the definition.

Remark 13.31. Suppose that in a category C all colimits of $\mathcal{I}$-diagrams exist. Every morphism of functors $X : \mathcal{I} \to C$ to $Y : \mathcal{I} \to C$ induces a morphism $\operatorname{colim}_{\mathcal{I}} X \to \operatorname{colim}_{\mathcal{I}} Y$ and we obtain a functor

$$\operatorname*{colim}_{\mathcal{I}} : C^{\mathcal{I}} \longrightarrow C.$$

Similarly we have a functor $\lim_{\mathcal{I}} : C^{\mathcal{I}} \to C$ if in C all limits of $\mathcal{I}$-diagrams exist.

Remark 13.32. Every preordered set I can be made into a category, again denoted by I. The objects are the elements of I and for two elements $i, j \in I$ the set of morphisms $\operatorname{Hom}_I(i, j)$ consists of one element if $i \leq j$ and is empty otherwise. There is a unique way to define a composition law making I into a category.

An I-diagram in a category C is then also called an *inductive system in C indexed by I*. It is a family $((X_i)_{i \in I}, (\varphi_{ji})_{i \leq j})$, where X_i is an object in C for all $i \in I$ and $\varphi_{ji} \colon X_i \to X_j$ is a morphism in C for all $i, j \in I$ with $i \leq j$ such that $\varphi_{ii} = \operatorname{id}_{X_i}$ for all i and $\varphi_{kj} \circ \varphi_{ji} = \varphi_{ki}$ for all $i \leq j \leq k$.

An I^{opp}-diagram in C is also called a *projective system in C indexed by I*. Of course, one can switch between the notions of inductive and projective systems by taking the opposite order. Usually one considers colimits of inductive systems and limits of projective systems.

In a given category C only some limits or colimits may exist. A category in which arbitrary limits (respectively colimits) exist is called *complete* (respectively *cocomplete*). A category in which limits (respectively colimits) of all $\mathcal{I}$-diagrams exist for arbitrary finite categories $\mathcal{I}$ is called *finitely complete* (respectively *finitely cocomplete*). The category of sets is complete and cocomplete as the following example shows.

Example 13.33. Let $\mathcal{I}$ be a small category, $I := Ob(\mathcal{I})$, and let $X \colon \mathcal{I} \to$ (Sets) be an $\mathcal{I}$-diagram in the category of sets.

1. The limit $\lim_{\mathcal{I}} X$ exists in (Sets) and can be described by

$$\lim_{\mathcal{I}} X = \left\{ (x_i)_{i \in I} \in \prod_{i \in I} X_i \ ; \ \forall\, \varphi \colon i \to j \text{ in } \mathcal{I} \colon X(\varphi)(x_i) = x_j \right\}. \tag{13.1}$$

For $j \in I$ the map $p_j \colon \lim_{\mathcal{I}} X \to X_j$ is given by the projection $(x_i)_{i \in I} \mapsto x_j$.
2. The colimit $\operatorname{colim}_{\mathcal{I}} X$ exists in (Sets) and can be described by

$$\operatorname{colim}_{\mathcal{I}} X = \left(\coprod_{i \in I} X_i \right) / \sim, \tag{13.2}$$

where $\left(\coprod_{i \in I} X_i \right)$ is the disjoint union of the sets X_i and where $\sim$ is the equivalence relation generated by the relation $x_i \sim x_j$ if $x_i \in X_i$, $x_j \in X_j$ and $X(\varphi)(x_i) = x_j$ for some $\varphi \colon i \to j$. For $j \in I$ the map $s_j \colon X_j \to \operatorname{colim}_{\mathcal{I}} X$ is given by attaching to $x_j \in X_j$ the equivalence class of $x_j \in \coprod_{i \in I} X_i$.

The definition of the equivalence relation $\sim$ in the description of the colimit as an equivalence relation generated by a relation makes it somewhat cumbersome to work with $\sim$. We will use the colimit usually in the case that the index category $\mathcal{I}$ is filtered in the sense of Definition 13.37. In that case the description of $\sim$ simplifies (see Example 13.38).

Remark 13.34. Let $X: \mathcal{I} \to C$, $i \mapsto X_i$ be a diagram in a category C. Then the universal property of $\lim X_i$ and $\operatorname{colim} X_i$ can also by definition be described as follows.

An object $\lim_{\mathcal{I}} X$ in C together with morphisms $p_i: \lim_{\mathcal{I}} X \to X_i$ for all objects i of $\mathcal{I}$ is a limit of X in C if and only if for all objects Y in C the map

$$\operatorname{Hom}_C(Y, \lim_{\mathcal{I}} X_i) \xrightarrow{u \mapsto (p_i \circ u)_i} \lim_{\mathcal{I}} \operatorname{Hom}_C(Y, X_i)$$

is bijective, where the right-hand side denotes the limit in the category of sets.

Similarly, an object $\operatorname{colim}_{\mathcal{I}} X$ in C together with morphisms $s_i: X_i \to \operatorname{colim}_{\mathcal{I}} X$ for all objects i of $\mathcal{I}$ is a colimit of X in C if and only if for all objects Y in C the map

$$\operatorname{Hom}_C(\operatorname{colim}_{\mathcal{I}} X_i, Y) \xrightarrow{u \mapsto (u \circ s_i)_i} \lim_{\mathcal{I}} \operatorname{Hom}_C(X_i, Y)$$

is bijective.

The category of topological spaces is also complete and cocomplete.

Example 13.35. If $X: \mathcal{I} \to$ (Top) is a diagram in the category of topological spaces, then $\lim_{\mathcal{I}} X$ exists in (Top). The underlying set of $\lim_{\mathcal{I}} X$ is the limit of sets described in (13.1) and its topology is the subspace topology of the product topology.

Similarly $\operatorname{colim}_{\mathcal{I}} X$ exists in (Top). Its underlying set is given by (13.2) and its topology is the quotient topology of the natural topology on $\coprod_i X_i$.

One can also show that the category of groups is complete and cocomplete (Problem 13.11 and Problem 13.21). In Sect. 14.1 we will see that the category of left R-modules (R a fixed ring) is complete and cocomplete.

Remark 13.36. Let $\mathcal{I}$ and $\mathcal{J}$ be (small) categories and let C be a category such that limits (respectively colimits) of all $\mathcal{I}$-diagrams and all $\mathcal{J}$-diagrams in C exist. Let $X: \mathcal{I} \times \mathcal{J} \to C$, $(i, j) \mapsto X_{ij}$ be a diagram in C.

Then because of the definition of limits (respectively colimits) via a universal property one obtains that $\lim_{\mathcal{I} \times \mathcal{J}} X$ (respectively $\operatorname{colim}_{\mathcal{I} \times \mathcal{J}} X$) exists and one has isomorphisms

$$\lim_{i,j} X_{ij} \cong \lim_i \lim_j X_{ij} \cong \lim_j \lim_i X_{ij}$$

$$\left(\text{resp.} \quad \operatorname{colim}_{i,j} X_{ij} \cong \operatorname{colim}_i \operatorname{colim}_j X_{ij} \cong \operatorname{colim}_j \operatorname{colim}_i X_{ij} \right).$$

Definition 13.37. A category $\mathcal{I}$ is called *filtered* if $\operatorname{Ob}(\mathcal{I})$ is non-empty and if the following two conditions are satisfied:

(a) For all objects i and j in $\mathcal{I}$ there exists an object k and morphisms $i \to k$ and $j \to k$.
(b) For all objects i and j and all morphisms $f, g \colon i \to j$ there exists a morphism $h \colon j \to k$ such that $h \circ f = h \circ g$.

For instance, a partially ordered set I is filtered if and only if the attached category is a filtered category.

Example 13.38. Let $\mathcal{I}$ be a filtered category and let $X \colon \mathcal{I} \to$ (Sets) be an $\mathcal{I}$-diagram in the category of sets. In this case one has $\mathrm{colim}_\mathcal{I} X = (\coprod_{i \in I} X_i)/\sim$ where for $x_i \in X_i$ and $x_j \in X_j$ one defines $x_i \sim x_j$ if there exist morphisms $\varphi \colon i \to k$ and $\psi \colon j \to k$ such that $X(\varphi)(x_i) = X(\psi)(x_j)$ (the properties of a filtered category imply that $\sim$ is an equivalence relation).

Filtered colimits in the category of sets commute with finite limits ([KS2] Theorem 3.1.6).

Proposition 13.39. *Let $\mathcal{I}$ be a filtered category, let $\mathcal{J}$ be a finite category and let $X \colon \mathcal{I} \times \mathcal{J} \to$ (Sets) be a diagram in the category of sets. Then the universal properties of limit and colimit yield an isomorphism in the category of sets*

$$\mathrm{colim}_\mathcal{I} \lim_\mathcal{J} X \xrightarrow{\sim} \lim_\mathcal{J} \mathrm{colim}_\mathcal{I} X.$$

There is also a converse to this proposition (Problem 13.18).
There are several important special cases of limits and colimits.

Definition 13.40 (Products and coproducts). Let I be a set with the discrete order (i.e., $i \le j \Leftrightarrow i = j$). Consider I as a category (equivalent to I^{opp}) and let $X \colon I \to C$ be a diagram in a category C.

1. Then $\prod_{i \in I} X_i := \lim_I X$ is called the *product of X* (if it exists). It is equipped with morphisms $p_j \colon \prod_{i \in I} X_i \to X_j$ for all $j \in I$, called *projections*. It is characterized by the following universal property: For every object Z in C and every family $(f_i \colon Z \to X_i)_{i \in I}$ of morphisms there exists a unique morphism $f \colon Z \to \prod_i X_i$ such that $f_i = p_i \circ f$.
2. Dually, $\coprod_{i \in I} X_i := \mathrm{colim}_I X$ is called the *coproduct of X* or the *direct sum of X* (if it exists).

For $I = \{1, \dots, n\}$ we also write $X_1 \times \cdots \times X_n$ for the product and $X_1 \amalg \cdots \amalg X_n$ for the coproduct.

Example 13.41 (Final and initial objects). Take $I = \emptyset$ in Definition 13.40. Then there is a unique I-diagram in every category C. Its limit (respectively its colimit) is an object P in C such that for every object X in C there exists a unique morphism $X \to P$ (respectively $P \to X$). The object P is called a *final object* (respectively an *initial object*). It is unique up to unique isomorphism (if it exists).

Example 13.42. In the category of sets (respectively the category of topological spaces) products and coproducts have the usual description:

1. The product of a family $(X_i)_{i \in I}$ of sets (respectively topological spaces) is the usual cartesian product X of sets (respectively of topological spaces) together with its projections $p_i \colon X \to X_i$. For any set (respectively topological space) Z and any family of maps (respectively continuous maps) $(f_i \colon Z \to X_i)_{i \in I}$ there is a unique map (respectively a unique continuous map) $f \colon Z \to X$ such that $p_i \circ f = f_i$, namely the map (respectively the continuous map) $z \mapsto (f_i(z))_{i \in I}$. The final object is the *singleton*, i.e., the set (respectively topological space) consisting of a single element.
2. The coproduct of the family $(X_i)_{i \in I}$ is the disjoint union of the sets X_i (respectively the sum of the topological spaces X_i in the sense of Definition 12.26).
   The initial object is $\emptyset$.

Remark and Definition 13.43 (Fiber products and pushouts). Let $\mathcal{I}$ be the category with three objects j, i_1, and i_2 and whose only morphisms except the identities are two morphisms $i_1 \to j$ and $i_2 \to j$. We represent $\mathcal{I}$ schematically by

$$i_1 \longrightarrow j \longleftarrow i_2$$

Then an $\mathcal{I}$-diagram X in a category C is a diagram of morphisms in C of the form

$$X_1 \xrightarrow{f_1} Y \xleftarrow{f_2} X_2.$$

The limit of X, if it exists, is called the *fiber product of X_1 and X_2 over Y*. It consists of an object denoted by $X_1 \times_Y X_2$ and morphisms $p_i \colon X_1 \times_Y X_2 \to X_i$ for $i = 1, 2$, called *projections*. It is characterized by the following universal property: For every object Z in C and all morphisms $g_1 \colon Z \to X_1$ and $g_2 \colon Z \to X_2$ such that $f_1 \circ g_1 = f_2 \circ g_2$ there exists a unique morphism $g \colon Z \to X_1 \times_Y X_2$ such that $p_i \circ g = g_i$ for $i = 1, 2$. We indicate this universal property by the following diagram:

$$(13.3)$$

The morphism g is also denoted by $(g_1, g_2)_Y$.

Dually, there is the notion of a *pushout* in a category C, which is the colimit of an inductive system $\mathcal{J} \to C$, where $\mathcal{J}$ is represented schematically by

$$\bullet \longleftarrow \bullet \longrightarrow \bullet.$$

The pushout of a $\mathcal{J}$-diagram $X_1 \longleftarrow Y \longrightarrow X_2$ in C is denoted by $X_1 \amalg_Y X_2$.

Example 13.44. Let $f: X \to Y$ be a morphism in a category and assume that the fiber product of

$$X \xrightarrow{\ f\ } Y \xleftarrow{\ f\ } X.$$

exists. Then $\Delta_f := (\mathrm{id}_X, \mathrm{id}_X)_Y: X \to X \times_Y X$ is called the *diagonal of f*.

Remark and Definition 13.45. Let C be a category and $f: X \to Y$ a morphism in C. Then f is a monomorphism if and only if the fiber product $X \times_Y X$ exists and the diagonal $\Delta_f: X \to X \times_Y X$ is an isomorphism.

One has a similar characterization of epimorphisms via pushouts.

Example 13.46. Consider the category of topological spaces. Then fiber products always exist in this category and are given by fiber products in the sense of Definition 12.28. The diagonal of a continuous map $f: X \to Y$ is the diagonal $\Delta_f: X \to X \times_Y X, x \mapsto (x, x)$.

Let $F: C \to \mathcal{D}$ be a functor between categories. Let $X: \mathcal{I} \to C$ be a diagram in C such that the limits $\lim_{\mathcal{I}} X$ and $\lim_{\mathcal{I}}(F \circ X)$ exist in C and $\mathcal{D}$, respectively. For every object i in $\mathcal{I}$ the morphism $\lim_{\mathcal{I}} X \to X_i$ induces by application of F a morphism $F(\lim_{\mathcal{I}} X) \to F(X_i)$. The family of these morphisms corresponds by the universal property of $\lim_{\mathcal{I}}(F \circ X)$ to a morphism

$$F(\lim_{\mathcal{I}} X) \to \lim_{\mathcal{I}}(F \circ X). \tag{13.4}$$

We say that F *commutes with limits* if for every diagram $X: \mathcal{I} \to C$ such that its limit $\lim_{\mathcal{I}} X$ exists in C, the limit of $F \circ X$ exists in $\mathcal{D}$ and the morphism (13.4) is an isomorphism.

Dually, there is the notion a functor that *commutes with colimits*.

Proposition 13.47. *Let $F: C \to \mathcal{D}$ be a functor.*

1. *Suppose that F is right adjoint to some functor $G: \mathcal{D} \to C$. Then F commutes with limits.*
2. *Dually, suppose that F is left adjoint to some functor. Then F commutes with colimits.*

Proof. [KS2] Proposition 2.1.10. $\qquad\qquad\qquad\qquad\qquad\qquad\qquad\qquad\quad\square$

Definition 13.48. Let C be a category that is finitely cocomplete, i.e., for all diagrams $X: \mathcal{I} \to C$, where $\mathcal{I}$ is a finite category, the colimit $\operatorname{colim}_{\mathcal{I}} X$ exists in C. Then a functor $F: C \to C'$ is called *right exact* if it commutes with finite colimits.

Similarly we define for a finitely complete category C a functor $F: C \to C'$ to be *left exact* if it commutes with finite limits.

Remark 13.49. Let C be a finitely complete category and let $F: C \to \mathcal{D}$ be a left exact functor. As finite products, final objects, and fiber products are all special cases of finite limits, F commutes with all these constructions. Moreover, the characterization of monomorphisms in Remark 13.45 implies that F also maps monomorphisms in C to monomorphisms in $\mathcal{D}$.

Example 13.50. Let $\mathcal{I}$ be a filtered category. Then Proposition 13.39 shows that the category of $(\mathrm{Sets})^{\mathcal{I}}$ of $\mathcal{I}$-diagrams of sets is finitely complete and that $\operatorname{colim}_{\mathcal{I}}: (\mathrm{Sets})^{\mathcal{I}} \to$ (Sets) (Remark 13.31) is left exact.

In particular $\operatorname{colim}_{\mathcal{I}}$ sends monomorphisms in $(\mathrm{Sets})^{\mathcal{I}}$ to monomorphism in (Sets) (i.e., to an injective map).

13.5   Problems

Problem 13.1. Let $(I, \leq)$ and $(J, \leq)$ be ordered sets. A map $f: I \to J$ is called *increasing* if $i \leq i'$ implies $f(i) \leq f(i')$ for all $i, i' \in I$. Show that ordered sets (as objects) and increasing functions (as morphisms) define a category.

Problem 13.2. Let C be a category and $\mathrm{id}_C: C \to C$ the identity functor. Show that the monoid $\mathrm{End}_{\mathrm{Func}(C,C)}(\mathrm{id}_C)$ is commutative.

Problem 13.3. Let $(I, \leq)$ be an ordered set. Let $\mathcal{T}_r$ be the set of subsets $U \subseteq I$ satisfying the following condition: For all $i \in U$ and $j \in I$ with $i \leq j$ one has $j \in U$.

1. Show that $\mathcal{T}_r$ is a topology on I, the so-called *right order topology on I*.
2. Show that a map $f: I \to J$ between partially ordered sets $(I, \leq)$ and $(J, \leq)$ is increasing if and only if it is continuous for the right order topologies on I and J. In particular we obtain a fully faithful functor T from the category of ordered sets (where morphisms are increasing maps) to the category of topological spaces.
3. Show that T yields an equivalence of the category of ordered sets with the category of topological spaces X that are T_0-spaces (Problem 12.7) such that the intersection of an arbitrary family of open subsets is again open (considered as a full subcategory of the category of topological spaces).

Similarly, one can also define the *left order topology on I* in which the open subsets U are those such that for all $i \in U$ and $j \in I$ with $i \geq j$ one has $j \in U$.

Problem 13.4. Show that the construction in Problem 12.8 yields a functor from the category of topological spaces to the full subcategory of T_0-spaces that is left adjoint to the inclusion functor. Deduce that any limit of T_0-spaces in the category of topological spaces is again a T_0-space.

Problem 13.5. Let C be a (small) category and let id: $C \to C$ be the identity functor. Show that if colim_C id exists in C, then it is a final object of C.

Problem 13.6. Let G be a group. Convince yourself that one can consider G as a category with a single object whose set of endomorphisms is G and whose composition law is given by multiplication in G.

1. Show that a G-diagram in the category of sets is the same as a G-set X (i.e., X is a set endowed with a left G-action). Show that a morphisms of G-diagrams is the same as a G-equivariant map between G-sets.
2. Let X be a G-set considered as a G-diagram. Show that $\lim_G X = X^G := \{x \in X ; gx = x \ \forall g \in G\}$ and that $\mathrm{colim}_G X = G\backslash X$ (the set of G-orbits).

Problem 13.7. A (small) category in which every morphism is an isomorphism is called a *groupoid*. A category C is called *connected* if it is non-empty and if for all pairs of objects X and Y in C there exists a finite sequence $X = X_0, X_1, \ldots, X_n = Y$ of objects X_i in C such that for all $i = 1, \ldots, n$ one of the sets $\mathrm{Hom}_C(X_{i-1}, X_i)$ or $\mathrm{Hom}_C(X_i, X_{i-1})$ is non-empty.

Now let C be a non-empty category. Show that C is a connected groupoid if and only if C is equivalent to the category defined by a group (Problem 13.6).

Problem 13.8. Let 1 and C be categories. Let X be an object in C and let $\varepsilon_X : 1 \to C$ be the *constant functor with value* X (i.e., $\varepsilon_X(i) = X$ for every object i in 1 and $\varepsilon_X(\varphi) = \mathrm{id}_X$ for every morphism φ in 1). Suppose that 1 is connected (Problem 13.7). Show that $X \xrightarrow{\sim} \lim_1 \varepsilon_X$ and $\mathrm{colim}_1 \varepsilon_X \xrightarrow{\sim} X$.

Problem 13.9. Let 1 be the category with two objects 0 and 1 whose only morphisms other than the identities are two morphisms $0 \to 1$. Hence an 1-diagram X in a category C is simply a pair of two morphisms $f, g : X_0 \to X_1$ in C. We call $\mathrm{Ker}(f, g) := \lim_1 X$ (respectively $\mathrm{Coker}(f, g) := \mathrm{colim}_1 X$) the *equalizer* (respectively *coequalizer*) of f and g (if it exists).

1. Let $g, h : X_0 \to X_1$ be a pair of morphisms in C such that $\mathrm{Ker}(f, g)$ (respectively $\mathrm{Coker}(f, g)$) exists. Let $i : \mathrm{Ker}(f, g) \to X_0$ (respectively $p : X_1 \to \mathrm{Coker}(f, g)$) be the natural morphism. Show that i is a monomorphism (respectively that p is an epimorphism).
2. Let C be the category of sets and let $f, g : X_0 \to X_1$ be maps of sets. Show that $\mathrm{Ker}(f, g) = \{x \in X_0 ; f(x) = g(x)\}$. Describe $\mathrm{Coker}(f, g)$.

Problem 13.10. Let C be a category.

1. Show that C is complete if and only if all products and all equalizers (Problem 13.9) exist in C.
2. Show that the following assertions are equivalent:
   (i) C is finitely complete.
   (ii) C admits all finite products and all equalizers.
   (iii) C admits all fiber products and has a final object.

Dualize the above statements to obtain criteria for C to be (finitely) cocomplete.

Problem 13.11. Let C be either the category of all groups, of all rings, or of all left R-modules (R a fixed ring). Show that all these categories are complete and that the forgetful functor from C to the category of sets commutes with limits.

Problem 13.12.

1. Show that in the category (Top) of topological spaces a continuous map $f : X \to Y$ is a monomorphism (respectively an epimorphism) if and only if f is injective (respectively surjective).
2. Let C be the category of all Hausdorff spaces (considered as a full subcategory of (Top)). Show that again the monomorphisms in C are the injective continuous maps but that every continuous map with dense image is an epimorphism. Are there other epimorphisms?

Problem 13.13. Let $\mathcal{I}$ be a category and let $X : \mathcal{I} \to$ (Top) be an $\mathcal{I}$-diagram in the category of topological spaces. Show that if X_i is Hausdorff for all i in $\mathcal{I}$, then $\lim_{\mathcal{I}} X$ is Hausdorff.

Problem 13.14. Let $\mathcal{I}$ be a small category and let $i \mapsto X_i$ be an $\mathcal{I}$-diagram of topological spaces.

1. Suppose that X_i is Hausdorff and compact for all objects i of $\mathcal{I}$. Show that $\lim_i X_i$ (limit in the category of topological spaces) is Hausdorff and compact.
2. Let $\mathcal{I} = \mathbb{N}^{\mathrm{opp}}$ be the category attached to the set $\mathbb{N}$ endowed with the opposite of the usual order. For $i \in \mathbb{N}$ let $X_i = \mathbb{N}$ be endowed with the unique topology such that the sets $\{1\}, \dots, \{i\}$ are open and closed in X_i and such that the only open subsets of $\mathbb{N}_{>i}$ are $\emptyset$ and $\mathbb{N}_{>i}$ itself. Let the transition maps be the identity of $\mathbb{N}$.
   Show that X_i is compact but that $\lim_i X_i = \mathbb{N}$ with the discrete topology. In particular $\lim_i X_i$ is not compact.

Problem 13.15. Let G be an abelian group, let $\mathcal{F}(G)$ be the set of finitely generated subgroups of G and let $C(G)$ be the set of cyclic subgroups of G. Endow both sets with the partial order given by $\subseteq$ and consider them as categories. Let $\nu\colon \mathcal{F}(G) \to$ (Grp) and $\mu\colon C(G) \to$ (Grp) be the inclusion functors.

1. Show that the universal property of the colimit yields maps

$$f\colon \operatorname*{colim}_{\mathcal{F}(G)} \nu \to G, \qquad g\colon \operatorname*{colim}_{C(G)} \mu \to G.$$

2. Show that $\mathcal{F}(G)$ is filtered and that f is an isomorphism.
3. Show that g is always surjective and that it is an isomorphism if the inclusion order on $C(G)$ is filtered.
4. Show that g is not an isomorphism for $G = \mathbb{Z}/2\mathbb{Z} \times \mathbb{Z}/2\mathbb{Z}$.

Problem 13.16. Let $\mathcal{I}$ be a category. Show that $\mathcal{I}$ is filtered if and only if for every finite category $\mathcal{J}$ and any functor $F\colon \mathcal{J} \to \mathcal{I}$ there exists an object i of $\mathcal{I}$ such that $\lim_{j \in \mathcal{J}} \operatorname{Hom}_{\mathcal{I}}(F(j), i) \neq \emptyset$.

Problem 13.17. Show that the categories (Sets) and (Sets)^{opp} are not equivalent.

Problem 13.18. Let $\mathcal{I}$ be a category such that for every finite category $\mathcal{J}$ and every diagram $X\colon \mathcal{I} \times \mathcal{J} \to$ (Sets) in the category of sets the map

$$\operatorname*{colim}_{\mathcal{I}} \lim_{\mathcal{J}} X \longrightarrow \lim_{\mathcal{J}} \operatorname*{colim}_{\mathcal{I}} X.$$

is bijective. Show that $\mathcal{I}$ is filtered.

Problem 13.19. Describe the gluing of topological spaces as a colimit.

Problem 13.20. Let R be a commutative ring. Show that each of the following forgetful functors has a left adjoint functor:

1. The functor (Top) $\to$ (Sets) sending a topological space to its underlying set.
2. The functor from the category $(R\text{-Alg})$ of commutative R-algebras to (Sets) sending an R-algebra to its underlying set.
3. The functor from $(R\text{-Alg})$ to the category of commutative monoids sending an R-algebra to $(R, \cdot)$.
4. The functor $(R\text{-Alg}) \to$ (Ab) sending an R-algebra to its underlying additive group $(R, +)$.

Problem 13.21. The category of groups is cocomplete:

1. Show that the forgetful functor from the category of groups to the category of sets has a left adjoint F: (Sets) $\to$ (Grp). Show that the canonical map of sets $S \to F(S)$ is injective and its image generates the group $F(S)$. For a set S the group $F(S)$ is called the *free group generated by* S.

2. Show that every group G is isomorphic to a quotient of the form $F(X)/N$ for some set X and some normal subgroup N of $F(X)$. If $R \subseteq F(X)$ is a subset such that N is the smallest normal subgroup containing R, then one says that G *is generated by* X *with relators* R and we write $G = \langle X; R \rangle$.

3. Let $(G_i)_{i \in I}$ be a family of groups. Show that their coproduct exists in the category of groups. It is often called the *free product of the* G_i.
   Hint: If $G_i = \langle X_i; R_i \rangle$, show that $\langle \coprod_i X_i; \coprod_i R_i \rangle$ is a coproduct of the G_i.

4. Show that the coequalizer of every pair of group homomorphisms $f, g \colon G_0 \to G_1$ exists in the category of groups. Deduce that the category of groups is cocomplete (use Problem 13.10).

Problem 13.22. Let (Mon) be the category of monoids and monoid homomorphisms.

1. For M a monoid let $M^\times := \{m \in M ; \exists n \in M \colon mn = nm = e\}$ be the set of invertible elements. Show that $M^\times$ is a group and that every homomorphism of monoids $M \to N$ induces a group homomorphism $M^\times \to N^\times$. We obtain a functor $(\ )^\times \colon (\text{Mon}) \to (\text{Grp})$. Show that this functor is right adjoint to the inclusion functor (Grp) $\to$ (Mon).

2. For M a commutative monoid (written additively) let M^{gp} be its *Grothendieck group*, i.e., $M^{\mathrm{gp}} = (M \times M)/\!\sim$ with $(m, m') \sim (n, n') :\Leftrightarrow \exists s \in M \colon s + m + n' = s + m' + n$. Show that $\sim$ is indeed an equivalence relation, that the componentwise addition on $M \times M$ induces the structure of an abelian group on M^{gp}, and that $M \to M^{\mathrm{gp}}$, $m \mapsto [m, 0]$ is a homomorphism of monoids, called *canonical*. Show that every homomorphism of commutative monoids $M \to N$ induces a homomorphism of abelian groups $M^{\mathrm{gp}} \to N^{\mathrm{gp}}$. We obtain a functor $(\ )^{\mathrm{gp}}$ from the category of commutative monoids to the category of abelian groups. Show that this functor is left adjoint to the inclusion functor.

Appendix C: Basic Algebra

14

14.1 The Category of Modules over a Ring

Modules

Let R be a ring (with unit, but not necessarily commutative). We recall very briefly some notions about R-modules: A *left R-module* is a "vector space over R", i.e., it is an abelian group $(M, +)$ together with a *scalar multiplication*

$$R \times M \to M, \qquad (a, m) \mapsto am$$

such that for all $a, b \in R$, $m, m' \in M$ one has

$$(a+b)m = am + bm, \qquad a(m+m') = am + am', \qquad (ab)m = a(bm), \qquad 1m = m.$$

In the sequel we will usually speak simply of *R-modules*. For R-modules M and N, a *homomorphism of R-modules* or an *R-linear map* from M to N is a map $u \colon M \to N$ such that for all $a \in R$, $m, m' \in M$ one has

$$u(am + m') = au(m) + u(m').$$

We obtain the categories of R-modules denoted by $(R\text{-Mod})$, where composition is defined as the usual composition of maps. There is a unique way to define a scalar multiplication on the trivial group $\{0\}$. This R-module is called the *zero module* and it is denoted by 0. Every bijective R-linear map is an isomorphism because the inverse of a bijective R-linear map is automatically R-linear again.

The set of homomorphisms of R-modules $M \to N$ is denoted by $\mathrm{Hom}_R(M, N)$. For $u, w \in \mathrm{Hom}_R(M, N)$ we define $u + w \in \mathrm{Hom}_R(M, N)$ by $(u + w)(m) := u(m) + w(m)$. This defines the structure of an abelian group on $\mathrm{Hom}_R(M, N)$. Let

$$C := \mathrm{Cent}(R) := \{ a \in R \; ; \; ab = ba \text{ for all } b \in R \}.$$

© Springer Fachmedien Wiesbaden 2016
T. Wedhorn, *Manifolds, Sheaves, and Cohomology*, Springer Studium Mathematik – Master,
DOI 10.1007/978-3-658-10633-1_14

be the *center* of the ring R. This is a commutative subring of R. For $a \in C$ the map $au: M \to N$, $au(m) := a(u(m))$ is again R-linear and we obtain the structure of a C-module on the abelian group $\mathrm{Hom}_R(M, N)$.

Example 14.1.

1. If $R = k$ is a field, an R-module is simply a k-vector space.
2. If A is an abelian group, it is a $\mathbb{Z}$-module by

$$na := \underbrace{a + \cdots + a}_{n \text{ times}} \quad \text{for } n \geq 0, \qquad na := (-n)a \quad \text{for } n < 0.$$

Conversely, forgetting the scalar multiplication makes every $\mathbb{Z}$-module into an abelian group. Therefore we see that it is the same to give an abelian group or to give a $\mathbb{Z}$-module.

Let M be an R-module. A subset N of M is called an R-*submodule* if N is a subgroup of $(M, +)$ and if $an \in N$ for all $a \in R, n \in N$. Then addition and scalar multiplication of M induce on N the structure of an R-module.

If $(N_i)_{i \in I}$ is a family of R-submodules of M, then $\bigcap_i N_i$ is an R-submodule. In particular there exists for every subset S a smallest R-submodule of M, containing S. It is called the R-*submodule generated by* S and it is denoted by $\langle S \rangle$ or $\langle S \rangle_R$.

A subset $S \subseteq M$ such that $\langle S \rangle = M$ is called a *generating set of* M. The R-module M is called a *finitely generated* if there exists a finite generating set of M.

Let N be an R-submodule of M. Then the scalar multiplication of M induces a scalar multiplication of the quotient group M/N making M/N into an R-module, called the *quotient module of M by N*.

Categorical Constructions of Modules
Let $u: M \to N$ be a homomorphism of R-modules. Let $N' \subseteq N$ be an R-submodule. Then $u^{-1}(N')$ is an R-submodule of M. In particular

$$\mathrm{Ker}(u) := \{ m \in M \; ; \; u(m) = 0 \}$$

is an R-submodule of M, called the *kernel of* u. As for groups one sees that u is injective if and only if $\mathrm{Ker}(u) = 0$.

Let $M' \subseteq M$ be an R-submodule. Then $u(M')$ is an R-submodule of N. In particular $\mathrm{Im}(u) := u(M)$ is an R-submodule of N, called the *image of* u.

We call $\mathrm{Coker}(u) := N/\mathrm{Im}(u)$ and $\mathrm{Coim}(u) := M/\mathrm{Ker}(u)$ the *cokernel* and the *coimage* of u, respectively. Then u is surjective if and only $\mathrm{Coker}(u) = 0$.

As for groups one sees that every R-linear map $u: M \to N$ induces an isomorphism of R-modules

$$\bar{u}: \mathrm{Coim}(u) \xrightarrow{\sim} \mathrm{Im}(u).$$

We now construct limits and colimits in the category of R-modules. In particular we will see that the category of R-modules is complete and cocomplete.

Let $(M_i)_{i \in I}$ be a family of modules. The cartesian product $\prod_i M_i$ with componentwise addition and scalar multiplication is an R-module. Together with the projections $\mathrm{pr}_j \colon \prod_i M_i \to M_j$ it is a product in the category of R-modules in the sense of Definition 13.40. In other words, for every R-module N the map

$$\mathrm{Hom}_R\left(N, \prod_{i \in I} M_i\right) \xrightarrow{\; u \mapsto (\mathrm{pr}_i \circ u)_i \;} \prod_{i \in I} \mathrm{Hom}_{(R\text{-Mod})}(N, M_i) \qquad (14.1)$$

is bijective. Note that (14.1) is C-linear, where C is the center of R.

More generally, let $\mathcal{I}$ be a small category and let $M \colon \mathcal{I} \to (R\text{-Mod})$, $i \mapsto M_i$, be an $\mathcal{I}$-diagram of R-modules. Then

$$\lim_{\mathcal{I}} M := \left\{ (m_i)_{i \in I} \in \prod_{i \in I} M_i \; ; \; \forall \, \varphi \colon i \to j \text{ in } \mathcal{I} \colon M(\varphi)(m_i) = m_j \right\}$$

is an R-submodule of $\prod_{i \in I} M_i$ and the projections $\lim_{\mathcal{I}} M \to M_i$ yield a bijection

$$\mathrm{Hom}_R\left(N, \lim_{\mathcal{I}} M\right) \xrightarrow{\;\sim\;} \lim_{\mathcal{I}} \mathrm{Hom}_R(N, M_i), \qquad (14.2)$$

which shows that $\lim_{\mathcal{I}} M$ is indeed the limit of the $\mathcal{I}$-diagram M (Remark 13.34). Again, (14.2) is a C-linear map.

To construct colimits in the category of R-modules, we proceed as in the category of sets (Remark 13.33): We define them as suitable quotients of coproducts. Hence let us construct coproducts (called direct sums for modules) for a family of modules $(M_i)_{i \in I}$ first. The submodule $\bigoplus_{i \in I} M_i$ of $\prod_i M_i$ consisting of tuples $(m_i)_i$ with $m_i = 0$ for all but finitely many $i \in I$ is called the *direct sum of* $(M_i)_i$. Together with the maps $s_j \colon M_j \to \bigoplus_{i \in I} M_i$, $m_j \mapsto (m_i)_i$, where $m_i = 0$ for all $i \neq j$, it is a coproduct in the category of R-modules: For every R-module N and every family $(u_i \colon M_i \to N)_i$ of R-linear maps there exists a unique R-linear map $u \colon \bigoplus_{i \in I} M_i \to N$ such that $u \circ s_i = u_i$ for all $i \in I$. In other words, the map

$$\mathrm{Hom}_{(R\text{-Mod})}\left(\bigoplus_{i \in I} M_i, N\right) \longrightarrow \prod_{i \in I} \mathrm{Hom}_{(R\text{-Mod})}(M_i, N), \qquad (14.3)$$

$$u \mapsto (u \circ s_i)_i$$

is bijective. An inverse map is given by $(u_i)_i \mapsto ((m_i)_i \mapsto \sum_{i \in I} u_i(m_i))$ (a finite sum because all but finitely many m_i are zero).

Now let $M \colon \mathcal{I} \to (R\text{-Mod})$, $i \mapsto M_i$, be an $\mathcal{I}$-diagram of R-modules. For all objects j of $\mathcal{I}$ we consider M_j as a submodule of $\bigoplus_{i \in \mathrm{Ob}(\mathcal{I})} M_i$ via $s_j \colon M_j \to \bigoplus_{i \in \mathrm{Ob}(\mathcal{I})} M_i$. Let N

be the submodule of $\bigoplus_{i\in\mathrm{Ob}(\mathcal{I})} M_i$ generated by all $m_i - m_j$ with $m_i \in M_i$, $m_j \in M_j$ and $M(\varphi)(m_i) = m_j$ for some morphism $\varphi\colon i \to j$ in $\mathcal{I}$. Let

$$\operatorname*{colim}_{\mathcal{I}} M := \left(\bigoplus_{i\in\mathrm{Ob}(\mathcal{I})} M_i \right) / N$$

and let $t_i\colon M_i \to \operatorname*{colim}_{\mathcal{I}} M$ be the composition of s_i with the projection. Then composition with all t_i yields for all R-modules a bijective map

$$\mathrm{Hom}_R\left(\operatorname*{colim}_{\mathcal{I}} M, N \right) \xrightarrow{\sim} \lim_{\mathcal{I}} \mathrm{Hom}_R(M_i, N) \qquad\qquad (14.4)$$

and hence $\operatorname*{colim}_{\mathcal{I}} M$ is indeed the colimit in the category of R-modules. Again, (14.4) is C-linear, where C is the center of R.

If I is a finite set, then $\bigoplus_{i\in I} M_i = \prod_{i\in I} M_i$. For $I = \emptyset$ we obtain the zero module, which is an initial and a final object in the category of R-modules.

If $M_i = M$ for all i, then we write M^I (respectively $M^{(I)}$) instead of $\prod_{i\in I} M_i$ (respectively $\bigoplus_{i\in I} M_i$). For every integer $n \geq 0$ we set $M^n := M^{\{1,\dots,n\}} = M^{(\{1,\dots,n\})}$. In particular we have for $M = R$ the R-modules R^I, $R^{(I)}$ and R^n. We also usually write $e_j := s_j(1) \in R^{(I)}$.

Example 14.2. Consider the following diagram of R-modules:

$$M \underset{0}{\overset{u}{\rightrightarrows}} N.$$

Then the limit of this diagram consists of pairs $(m, n) \in M \times N$ such that $u(m) = n$ and $n = 0$. Hence it is identified with $\mathrm{Ker}(u)$.

The colimit of this diagram is the quotient of $M \oplus N$ by the submodule E generated of all elements of the form $(m, 0) - (0, n)$ with $u(m) = n$ and of all elements of the form $(m, 0)$. Then E is the kernel of the projection $M \oplus N \to N/\mathrm{Im}(u)$. Hence the colimit is identified with $\mathrm{Coker}(u)$.

Note that the forgetful functor $(R\text{-Mod}) \to (\text{Sets})$ commutes with limits. This we can also deduce from Proposition 13.47 because the forgetful functor has a right adjoint as we see in the next subsection. The forgetful functor does not commute with arbitrary colimits as the example of coproducts shows. It does commute with filtered colimits (Problem 14.3).

Free Modules and Matrices

As with vector spaces (Example 13.19) one sees that the functor $I \mapsto R^{(I)}$ is left adjoint to the forgetful functor $(R\text{-Mod}) \to (\text{Sets})$. In other words, we have for every R-module a bijection, functorial in I and in M,

$$\mathrm{Hom}_{(R\text{-Mod})}(R^{(I)}, M) \xrightarrow{\sim} \mathrm{Hom}_{(\mathrm{Sets})}(I, M) = M^I,$$
$$u \mapsto (u(e_i))_{i \in I},$$

(14.5)

where $e_i \in R^{(I)}$ is the tuple whose j-th entry is δ_{ij} for all $j \in I$. Hence to give a family of elements $\underline{m} = (m_i)_i$ with $m_i \in M$ is equivalent to giving an R-linear map $u_{\underline{m}} \colon R^{(I)} \to M$. One can also view (14.5) as a special case of (14.3).

We call $\underline{m}$ *linearly independent* (respectively a *generating system*, respectively a *basis*) if $u_{\underline{m}}$ is injective (respectively surjective, respectively an isomorphism).

An R-module is called *free* if it admits a basis, i.e., if it is isomorphic to an R-module of the form $R^{(I)}$ for some set I.

In contrast to vector spaces, R-modules are usually not free (Problem 14.6).

Remark 14.3. Let $(m_i)_{i \in I}$ be a generating system (respectively a linearly independent system) of an R-module M and let $u \colon M \to N$ be a surjective (respectively injective) homomorphism of A-modules. Then $(u(m_i))_{i \in I}$ is a generating system (respectively a linearly independent system) of N.

As for linear maps between vector spaces one can express linear maps between free modules by matrices after choosing a basis. For integers $m, n \geq 0$ we denote by $M_{n \times m}(R)$ the set of matrices with n rows and m columns and with coefficients in R. We also set $M_n(R) := M_{n \times n}(R)$. The usual addition and scalar multiplication of matrices makes $M_{n \times m}(R)$ into an R-module. It is a free R-module: A basis is given by the matrices E_{ij} with all entries equal to 0 except the (i, j)-th entry which is equal to 1.

One has the usual matrix multiplication ($p \geq 0$ some integer)

$$M_{p \times n}(R) \times M_{n \times m}(R) \longrightarrow M_{p \times m}(R),$$

$$((a_{ij})_{i,j}, (b_{jk})_{j,k}) \mapsto \left(\sum_{j=1}^{n} a_{ij} b_{jk} \right)_{i,k}.$$

Note that here the order of multiplication of summands usually matters if R is not commutative.

Let M and N be R-modules such that there exists a basis $(e_1, \ldots, e_m)$ of M and a basis $(f_1, \ldots, f_n)$ of N. Let $u \colon M \to N$ be an R-linear map. Then for every $j = 1, \ldots, m$ there exist unique a_{ij}, $i = 1, \ldots, m$ such that

$$u(e_j) = \sum_{i=1}^{n} a_{ij} f_i$$

and $(a_{ij})_{i,j} \in M_{n \times m}(R)$ is the *matrix of u with respect to the bases (e_i) and (f_j)*. As for vector spaces, composition of linear maps is expressed by multiplication of the corresponding matrices.

Exact Sequences

A sequence of homomorphisms of R-modules

$$\ldots \longrightarrow M_{i-1} \xrightarrow{u_{i-1}} M_i \xrightarrow{u_i} M_{i+1} \longrightarrow \ldots$$

is called a *complex* (respectively is called an *exact*) if $u_i \circ u_{i-1} = 0$ (respectively if $\text{Im}(u_{i-1}) = \text{Ker}(u_i)$) for all i. A *short exact sequence* is an exact sequence of the form

$$0 \longrightarrow M' \longrightarrow M \longrightarrow M'' \longrightarrow 0.$$

Remark 14.4. Let R be a ring and let M_i, $i = 1, 2, \ldots$ be R-modules.

1. The sequence $0 \to M_1 \xrightarrow{f} M_2$ is exact if and only if f is injective. The sequence $M_1 \xrightarrow{f} M_2 \to 0$ is exact if and only if f is surjective.
2. Let $M_1 \xrightarrow{f_1} M_2 \xrightarrow{f_2} M_3 \xrightarrow{f_3} M_4$ be an exact sequence of R-modules. Then f_2 induces an isomorphism of R-modules $\text{Coker}(f_1) \xrightarrow{\sim} \text{Ker}(f_3)$. In particular, f_1 is surjective if and only if f_3 is injective.

Definition and Remark 14.5. A short exact sequence $0 \to M' \xrightarrow{i} M \xrightarrow{p} M'' \to 0$ of R-modules is called *split* if the following equivalent conditions are satisfied:

(i) There exists a section $s: M'' \to M$ of p (i.e., a homomorphism of R-modules $s: M'' \to M$ such that $p \circ s = \text{id}_{M''}$).
(ii) There exists an R-submodule $\tilde{M}''$ of M such that $p_{|\tilde{M}''}: \tilde{M}'' \to M''$ is an isomorphism.
(iii) There exists a retraction $r: M \to M'$ of i (i.e., a homomorphism of R-modules $r: M \to M'$ such that $r \circ i = \text{id}_{M'}$).

In this case one has isomorphisms of R-modules

$$M' \oplus M'' \cong M,$$
$$(m', m'') \mapsto i(m') + s(m''),$$
$$(r(m), p(m)) \mapsfrom m.$$

Example 14.6. Every short exact sequence $0 \to M' \xrightarrow{i} M \xrightarrow{p} M'' \to 0$ of R-modules with M'' a free R-module is split (more generally, this holds whenever M'' is a projective R-module, see Problem 14.4). Let $(m_i'')_{i \in I}$ be a basis of M''. As p is surjective, there exist $m_i \in M$ with $p(m_i) = m_i''$ for all i. Let $s: M'' \to M$ be the unique R-linear map with $s(m_i'') = m_i$ for all i. Then s is a section of p.

14.2   Multilinear Algebra and Tensor Products

In this section we always denote by A a *commutative* ring and by M an A-module.

Modules of Multilinear Maps

Let $r \in \mathbb{N}_0$ and let $(N_i)_{1 \leq i \leq r}$ be a finite family of A-modules. A map

$$\alpha \colon N_1 \times \cdots \times N_r \to M$$

is called *r-multilinear* if it is linear in every component. The sum of two r-multilinear maps and the product of an r-multilinear map with a scalar in A is again an r-multilinear map (for the second assertion we need that A is commutative). Hence the set of all such r-multilinear maps obtains a structure of an A-module, denoted by $L_A(N_1, \ldots, N_r; M)$.

For $r = 1$, we obtain the A-module $\operatorname{Hom}_A(N, M) := L_A(N; M)$ consisting of all A-linear maps $N \to M$.

A 0-multilinear map is an arbitrary map $\varepsilon \colon \{0\} \to M$. Identifying ε with $\varepsilon(0)$ gives an identification of 0-linear M-valued maps and elements in M.

Finally, 2-multilinear maps are also called *bilinear*.

Example 14.7. For every A-module the map $\operatorname{Hom}_A(A, M) \to M$, $u \mapsto u(1)$ is an isomorphism of A-modules, which we will often use to identify $\operatorname{Hom}_A(A, M)$ and M.

The formation of $\operatorname{Hom}_A(M, N)$ is functorial in M and in N. Let $u \colon M \to M'$ be an A-linear map. Then

$$\operatorname{Hom}_A(u, N) \colon \operatorname{Hom}_A(M', N) \to \operatorname{Hom}_A(M, N), \qquad v \mapsto v \circ u$$

is A-linear. We obtain a contravariant functor $\operatorname{Hom}_A(\cdot, N)$ from the category of A-modules to itself.

Similarly, one has a (covariant) functor $\operatorname{Hom}_A(M, \cdot) \colon (A\text{-Mod}) \to (A\text{-Mod})$.

Remark 14.8. Let $(M_i)_{i \in I}$ be a family of A-modules and let N be an A-module. The bijection (14.3) is an isomorphism of A-modules. As multilinear maps are defined as componentwise linear, it follows that for all integers $1 \leq s \leq r$ and for all A-modules $P_1, \ldots, P_{s-1}, P_{s+1}, \ldots, P_r$ there is an isomorphism of A-modules, functorial if M_i, P_j and N,

$$L_A\left(P_1, \ldots, P_{s-1}, \bigoplus_{i \in I} M_i, P_{s+1}, \ldots, P_r; N\right)$$
$$\xrightarrow{\sim} \prod_{i \in I} L_A(P_1, \ldots, P_{s-1}, M_i, P_{s+1}, \ldots, P_r; N). \tag{14.6}$$

Definition and Remark 14.9 (Dual module). For every A-module M the A-module $M^\vee := \mathrm{Hom}_A(M, A)$ is called the *dual of M*. As usual there is a functorial map

$$\iota_M: M \to (M^\vee)^\vee, \qquad m \mapsto (\lambda \mapsto \lambda(m)). \tag{14.7}$$

We have $A^\vee = A$ by Example 14.7. Hence for every set I we have

$$(A^{(I)})^\vee = A^I$$

by (14.6).

In particular we see for that every A-module M with a basis of $n \in \mathbb{N}_0$ elements its dual $M^\vee$ has also a basis with n elements. Moreover the biduality map ι_M is an isomorphism of A-modules.

Algebras

An *A-algebra (with unit)* is an A-module C together with a bilinear map $C \times C \to C$, $(c, d) \mapsto c \cdot d$, such that $(C, +, \cdot)$ is a ring. Hence we always assume that the multiplication in an algebra is associative and has a unit. If the ring multiplication is also commutative, C is called a *commutative algebra*. As usual we will often write cd instead of $c \cdot d$.

Remark 14.10. An A-algebra C is the same as a ring C together with a homomorphism of rings $\iota: A \to C$ such that the image of ι is contained in the center $\mathrm{Cent}(C) := \{ c \in C \; ; \; cd = dc \text{ for all } d \in C \}$ of C. Such a ring C becomes an A-module via the scalar multiplication $(a, c) \mapsto \iota(a)c$ for $a \in A$, $c \in C$. Conversely, given an A-algebra structure on C, the ring homomorphism ι is defined by $\iota(a) := a1_C$ for $a \in A$.

As $\mathbb{Z}$-modules are nothing but abelian groups, $\mathbb{Z}$-algebras are nothing but rings.

A map $\varphi: C \to C'$ between A-algebras C and C' is called a *homomorphism of A-algebras* if

(a) φ is A-linear,
(b) $\varphi(cd) = \varphi(c)\varphi(d)$ for all $c, d \in C$ and $\varphi(1) = 1$.

We obtain the category of A-algebras, denoted by $(A\text{-Alg})$.

Our primary source of algebras will be algebras of $\mathbb{R}$-valued or $\mathbb{C}$-valued functions, for instance:

Example 14.11. Let X be a topological space and denote by $\mathcal{C}_{X;\mathbb{K}}(X)$ the set of continuous maps $X \to \mathbb{K}$ (we refer to Example 3.5 for an explanation for this cumbersome notation). Then $\mathcal{C}_{X;\mathbb{K}}(X)$ is a $\mathbb{K}$-algebra with the usual addition of functions, scalar multiplication of a function with a real number, and multiplication of functions.

If $F: X \to Y$ is a continuous map of topological spaces,

$$F^*: \mathcal{C}_{Y;\mathbb{K}}(Y) \to \mathcal{C}_{X;\mathbb{K}}(X), \qquad g \mapsto g \circ F$$

is a homomorphism of $\mathbb{K}$-algebras.

Example 14.12. Let M be an A-module. Then the composition of linear maps defines on the A-module $\mathrm{End}_A(M) = \mathrm{Hom}_A(M, M)$ the structure of an A-algebra. If M admits a finite basis $(e_1, \ldots, e_n)$, then attaching to a linear map its matrix with respect to $(e_i)_i$ yields an isomorphism of A-algebras $\mathrm{End}_A(M) \xrightarrow{\sim} M_n(A)$, where the multiplication on the A-module $M_n(A)$ is given by matrix multiplication.

Tensor Products

Let M and N be A-modules. Then there exists an A-module $M \otimes_A N$ and an A-bilinear map $\tau: M \times N \to M \otimes_A N$, denoted by $(m, n) \mapsto m \otimes n$, which is universal, i.e., for all A-modules P and for all A-bilinear maps $b: M \times N \to P$ there exists a unique A-linear map $\bar{b}: M \otimes_A N \to P$ such that $b(m, n) = \bar{b}(m \otimes n)$ for all $(m, n) \in M \times N$. The pair $(M \otimes_A N, \tau)$ is determined uniquely up to unique isomorphism by this definition. The A-module $M \otimes_A N$ is called the *tensor product of M and N over A*. We will also often write $M \otimes N$ if it is clear over which ring we form the tensor product.

The construction of $M \otimes_A N$ is rather straightforward: For every A-module P, the set $\mathrm{Hom}_A(A^{(M \times N)}, P)$ is the set of all maps $\alpha: M \times N \to P$ (14.5). Such a map α is bilinear if and only if

$$\alpha(am + m', n) - (a\alpha(m, n) + \alpha(m', n)) = 0$$
$$\text{and} \quad \alpha(m, an + n') - (a\alpha(m, n) + \alpha(m, n')) = 0$$

for all $a \in A$, $m, m' \in M$, $n, n' \in N$. Hence, if we denote by Q the A-submodule of $A^{(M \times N)}$ generated by all elements of the form

$$e_{(am+m', n)} - (a e_{(m,n)} + e_{(m', n)}) \quad \text{and} \quad e_{(m, an+n')} - (a e_{(m,n)} + e_{(m, n')}),$$

then an A-linear map $u: A^{(M \times N)} \to P$ corresponds to a bilinear map α if and only if $u(Q) = 0$. Hence we can simply set $M \otimes_A N := A^{(M \times N)}/Q$ and define $m \otimes n$ as the image of $e_{(m,n)}$ in $M \otimes_A N$.

Remark 14.13. As the $e_{(m,n)}$ for $m \in M$ and $n \in N$ generate $A^{(M \times N)}$, their images $m \otimes n$ in $M \otimes_A N$ form a generating system of $M \otimes_A N$ (Remark 14.3) (but usually not every element of $M \otimes_A N$ is of this form).

In particular, every A-linear map $u: M \otimes_A N \to P$ is already uniquely determined by the elements $u(m \otimes n)$ for $m \in M$ and $n \in N$. Conversely, a map given by specifying elements $u(m \otimes n)$ for all $m \in M$ and $n \in N$ in an A-module P extends to a well-defined A-linear map $M \otimes_A N \to P$ if and only if

$$u(am \otimes n) = u(m \otimes an) = au(m \otimes n),$$
$$u((m + m') \otimes n) = u(m \otimes n) + u(m' \otimes n), \qquad (14.8)$$
$$u(m \otimes (n + n')) = u(m \otimes n) + u(m \otimes n')$$

for all $a \in A, m, m' \in M$, and $n, n' \in N$.

The maps $m \otimes n \mapsto n \otimes m$ and $n \otimes m \mapsto m \otimes n$ define mutually inverse isomorphisms, functorial in M and N,

$$M \otimes_A N \cong N \otimes_A M, \qquad (14.9)$$

which we often use to identify these modules. Similarly, the map

$$M \otimes (N \otimes P) \to (M \otimes N) \otimes P, \qquad m \otimes (n \otimes p) \mapsto (m \otimes n) \otimes p \qquad (14.10)$$

defines an A-linear isomorphism, which we use to identify these modules. We simply write $M \otimes_A N \otimes_A P$ or $M \otimes N \otimes P$ and $m \otimes n \otimes p$ for $m \in M, n \in N, p \in P$. We use similar notations for the tensor product of finitely many modules.

Remark 14.14. For A-modules $M_1, \ldots, M_r$ the r-multilinear map

$$\tau \colon M_1 \times \cdots \times M_r \to M_1 \otimes \cdots \otimes M_r, \qquad (m_1, \ldots, m_r) \mapsto m_1 \otimes \cdots \otimes m_r$$

is universal, i.e., for every A-module N the A-linear map

$$\mathrm{Hom}_A(M_1 \otimes \cdots \otimes M_r, N) \longrightarrow L_A(M_1, \ldots, M_r; N), \qquad u \mapsto u \circ \tau \qquad (14.11)$$

is bijective. In other words, for every r-multilinear map $\alpha \colon M_1 \times \cdots \times M_r \to N$ there exists a unique linear map $u \colon M_1 \otimes \cdots \otimes M_r \to N$ such that $u \circ \tau = \alpha$. One simply checks that

$$u \colon m_1 \otimes \cdots \otimes m_r \mapsto \alpha(m_1, \ldots, m_r)$$

is well defined.

The tensor product is functorial. Let $u \colon M \to M'$ and $v \colon N \to N'$ be A-linear maps. Then $m \otimes n \mapsto u(m) \otimes v(n)$ defines a well-defined A-linear map

$$u \otimes v \colon M \otimes_A N \longrightarrow M' \otimes_A N'.$$

We obtain a functor

$$(\cdot) \otimes N \colon (A\text{-Mod}) \longrightarrow (A\text{-Mod}), \qquad M \mapsto M \otimes_A N, \qquad u \mapsto u \otimes \mathrm{id}_N . \qquad (14.12)$$

It is isomorphic to the analogously defined functor $N \otimes (\cdot)$ by (14.9).

Remark 14.15. Suppose that $u: M \to M'$ and $v: N \to N'$ are surjective. Then $u \otimes v$ is surjective because its image contains the generating set of all $m \otimes n$ for $m \in M$ and $n \in N$.

Even if $v = \mathrm{id}_N$, it is in general not true that $u \circ \mathrm{id}_N$ is injective if u is injective (Problem 14.8; see however Problem 14.9).

Remark 14.16. The maps $A \otimes_A M \to M$, $a \otimes m \mapsto am$, and $M \to A \otimes_A M$, $m \mapsto 1 \otimes m$ yield mutually inverse isomorphisms

$$A \otimes_A M \cong M, \tag{14.13}$$

which we will usually use to identify these two modules.

Remark 14.17. As we carefully recorded the universal property of all our constructions with modules, we can use this to show several useful compatibilities. The key idea is the fact that two A-modules M and N are isomorphic if and only if for every A-module P there is a bijection of sets

$$\mathrm{Hom}_A(M, P) \cong \mathrm{Hom}_A(N, P), \qquad \text{functorial in } P. \tag{14.14}$$

Indeed the condition is clearly necessary. Conversely, choosing $P = N$ in (14.14), id_N corresponds to a map $u: M \to N$ and choosing $P = M$ in (14.14), id_M corresponds to a map $v: N \to M$. The functoriality in P implies that $u \circ v = \mathrm{id}_N$ and $v \circ u = \mathrm{id}_M$.

As an example of this principle we show that tensor products commute with direct sums. Let $(M_i)_i$ be a family of A-modules and let N be an A-module. Then there is a functorial isomorphism

$$\left(\bigoplus_{i \in I} M_i \right) \otimes_A N \cong \bigoplus_{i \in I} (M_i \otimes_A N). \tag{14.15}$$

Indeed, let P be an A-module. Then there are isomorphisms of A-modules (in particular bijective maps), functorial in P,

$$
\begin{aligned}
\mathrm{Hom}_A\left(\left(\bigoplus_{i \in I} M_i \right) \otimes_A N, P \right) &\underset{(14.11)}{\cong} L_A\left(\bigoplus_{i \in I} M_i, N; P \right) \\
&\underset{(14.6)}{\cong} \prod_{i \in I} L_A(M_i, N; P) \\
&\underset{(14.11)}{\cong} \prod_{i \in I} \mathrm{Hom}_A(M_i \otimes_A N, P) \\
&\underset{(14.6)}{\cong} \mathrm{Hom}_A\left(\bigoplus_{i \in I} (M_i \otimes_A N), P \right).
\end{aligned}
\tag{14.16}
$$

As all the isomorphisms in (14.16) are also functorial in the M_i and in N, the isomorphism (14.15) is functorial in the M_i and in N.

Remark 14.18. Let M and N be free A-modules with basis $\underline{m} = (m_i)_{i \in I}$ and $\underline{n} = (n_j)_{j \in J}$, respectively. Hence the maps $u_{\underline{m}} \colon A^{(I)} \to M$, $(a_i)_{i \in I} \mapsto \sum_{i \in I} a_i m_i$ and $u_{\underline{n}}$ are isomorphisms of A-modules. Then we can combine several of the above isomorphisms to obtain an isomorphism

$$M \otimes_A N \cong A^{(I)} \otimes_A A^{(J)} \underset{(14.15)}{\cong} \left(A \otimes_A A \right)^{(I \times J)} \underset{(14.13)}{\cong} A^{(I \times J)}$$

sending $m_i \otimes n_j$ to $e_{(i,j)}$. In particular, $M \otimes_A N$ is again a free module with basis $(m_i \otimes n_j)_{(i,j) \in I \times J}$.

Another useful fact is that the tensor product and the Hom-functor are adjoint to each other. This is based on the easy observation that for all A-modules M, N, and P the map

$$L_A(M, N; P) \longrightarrow \operatorname{Hom}_A(M, \operatorname{Hom}_A(N, P)), \qquad \alpha \mapsto (m \mapsto (n \mapsto \alpha(m, n))$$

is an isomorphism of A-modules, functorial in M, N, and P: An inverse is given by $u \mapsto ((m, n) \mapsto u(m)(n))$. As $L_A(M, N; P) = \operatorname{Hom}_A(M \otimes N, P)$ we obtain a functorial isomorphism

$$\operatorname{Hom}_A(M \otimes N, P) \cong \operatorname{Hom}_A(M, \operatorname{Hom}_A(N, P)). \tag{14.17}$$

In particular, the covariant functor $(\cdot) \otimes N$ from $(A\text{-Mod})$ to itself is left adjoint to the functor $\operatorname{Hom}_A(N, \cdot)$. Hence Proposition 13.47 shows that $(\cdot) \otimes N$ commutes with colimits and is in particular right exact. This observation also gives a new argument, why tensor products commute with direct sums because the direct sum is, as a coproduct in the category of modules, a special case of a colimit.

There is also a functorial homomorphism of A-modules

$$\operatorname{Hom}_A(M, P) \otimes_A N \longrightarrow \operatorname{Hom}_A(M, P \otimes_A N),$$
$$u \otimes n \mapsto (m \mapsto u(m) \otimes n). \tag{14.18}$$

This is an isomorphism if M or N is free with a finite basis. Indeed, assume that $M \cong A^n$ for some integer $n \geq 0$. As (14.18) is functorial, we may assume that $M = A^n$. As the functors $\operatorname{Hom}_A(\cdot, P)$ and $\operatorname{Hom}_A(\cdot, P \otimes_A N)$ commute with finite direct sums (Remark 14.8), we may assume $M = A$. Now the claim follows from Remark 14.7. The proof in the case that $N \cong A^n$ is similar using that tensor products commute with direct sums (14.15) and that $Q \otimes_A A = Q$ for every A-module Q (14.13).

Taking $P = A$ in (14.18) we obtain a functorial homomorphism of A-modules

$$\delta \colon M^\vee \otimes_A N \longrightarrow \operatorname{Hom}_A(M, N), \tag{14.19}$$

which is an isomorphism if M or N are free with finite basis.

For every A-module Q we can compose the A-linear map $\mathrm{Hom}_A(Q, \delta)$ with (14.17) and obtain a functorial homomorphism of A-modules

$$\mathrm{Hom}_A(Q, M^\vee \otimes_A N) \longrightarrow \mathrm{Hom}_A(Q \otimes_A M, N), \qquad (14.20)$$

which is an isomorphism if M or N are free with a finite basis.

Example 14.19. Let M be a free module with a finite basis. Then (14.19) yields an isomorphism

$$\mathrm{End}_A(M) := \mathrm{Hom}_A(M, M) \overset{\sim}{\to} M^\vee \otimes_A M. \qquad (14.21)$$

Composition with the linear map $M^\vee \otimes M \to A$, $(\lambda, m) \mapsto \lambda(m)$, yields an A-linear map

$$\mathrm{tr} \colon \mathrm{End}_A(M) \longrightarrow A, \qquad (14.22)$$

called the *trace map*. Via the choice of a basis $(e_1, \ldots, e_n)$ of M we identify $\mathrm{End}_A(M)$ and the matrix algebra $M_n(A)$. If $(x^1, \ldots, x^n)$ is the dual basis of $M^\vee$, then $x^i \otimes e_j \in M^\vee$ corresponds via (14.21) to the matrix E_{ij}, which has everywhere 0 coefficients except in (i, j)-th place, where its coefficient is 1. Hence the above trace map coincides with the usual trace map $M_n(A) \to A$ because both maps are linear and agree on the basis $(E_{ij})_{i,j}$ of $M_n(A)$.

14.3   Tensor Algebra, Exterior Algebra, Symmetric Algebra

Tensor Algebra
For $r \geq 0$ let

$$T^r(M) := T_A^r(M) := M^{\otimes r} := \underbrace{M \otimes_A \cdots \otimes_A M}_{r \text{ times}}$$

be the *r-th tensor power of M*. Then $T^0(M) = A$ and $T^1(M) = M$. Define the *tensor algebra of M*

$$T(M) := T_A(M) := \bigoplus_{r \geq 0} T^r(M)$$

with A-algebra structure defined as follows. For $m := m_1 \otimes \cdots \otimes m_r \in T^r(M)$ and $n := n_1 \otimes \cdots \otimes n_s \in T^s(M)$ we define the product by

$$m \otimes n := m_1 \otimes \cdots \otimes m_r \otimes n_1 \otimes \cdots \otimes n_s \in T^{r+s}(M),$$

then extend by linearity. This is well defined and makes $T(M)$ into an associative A-algebra with 1 (but not necessarily commutative).

The A-algebra $T(M)$ is a *graded A-algebra*, i.e., for $w \in T^i(M)$, $v \in T^j(M)$ one has $w \otimes v \in T^{i+j}(M)$ for all $i, j \geq 0$.

If $u: M \to N$ is a homomorphism of A-modules, then $m_1 \otimes \cdots \otimes m_r \mapsto u(m_1) \otimes \cdots \otimes u(m_r)$ defines an A-linear map $T^r(u): T^r(M) \to T^r(N)$. The map

$$T(u) := \bigoplus T^r(u): T(M) \to T(N)$$

is a homomorphism of A-algebras.

The formation of $T^r(u)$ and of $T(u)$ is compatible with composition and sends identities to identities. Hence we obtain functors

$$T^r: (A\text{-Mod}) \to (A\text{-Mod}), \qquad T: (A\text{-Mod}) \to (A\text{-Alg}).$$

Remark 14.20. Denote by $\iota: M \overset{\sim}{\to} T^1(M) \subseteq T(M)$ the inclusion. Then $(T(M), \iota)$ has the following universal property: For every associative A-algebra B with 1 the map

$$\operatorname{Hom}_{(A\text{-Alg})}(T(M), B) \to \operatorname{Hom}_{(A\text{-Mod})}(M, B), \qquad \varphi \mapsto \varphi \circ \iota$$

is bijective. In other words, the functor $M \mapsto T(M)$ is left adjoint to the forgetful functor $(A\text{-Alg}) \to (A\text{-Mod})$.

Exterior Powers
Let M and N be A-modules. An r-multilinear map $\alpha: M^r \to N$ is called *alternating* if $\alpha(m_1, \ldots, m_r) = 0$ whenever there exists $i \neq j$ with $m_i = m_j$.

Let $Q_r \subseteq T^r(M)$ be the A-submodule generated by the set

$$E_r := \left\{ m_1 \otimes \cdots \otimes m_r \in T^r(M) \; ; \text{ there exist } 1 \leq i \neq j \leq r \text{ with } m_i = m_j \right\}.$$

Then an r-multilinear map $\alpha: M^r \to N$ is alternating if and only if the corresponding A-linear map $u: T^r(M) \to N$ sends Q_r to 0. We set

$$\Lambda^r(M) := \Lambda^r_A(M) := T^r(M)/Q_r$$

and call it the *r-th exterior power of M*. The image of $m_1 \otimes \cdots \otimes m_r$ in $\Lambda^r(M)$ is denoted by $m_1 \wedge \cdots \wedge m_r$. We have $Q_0 = Q_1 = 0$ and hence $\Lambda^0(M) = A$ and $\Lambda^1(M) = M$.

By construction, $\Lambda^r(M)$ has the following universal property. Let

$$\pi: M^r \to \Lambda^r(M), \qquad (m_1, \ldots, m_r) \mapsto m_1 \wedge \cdots \wedge m_r.$$

Then π is alternating and for every A-module N the map

$$\operatorname{Hom}_{(A\text{-Mod})}(\Lambda^r(M), N) \to \{\alpha: M^r \to N \; ; \; \alpha \text{ alternating}\}, \qquad u \mapsto u \circ \pi$$

is bijective.

The direct sum

$$\Lambda(M) := \Lambda_A(M) := \bigoplus_{r \geq 0} \Lambda^r(M) = T(M) / \bigoplus_{r \geq 0} Q_r$$

is called the *exterior algebra of* M. The multiplication on the A-algebra $T(M)$ induces a multiplication on $\Lambda(M)$ making the A-module $\Lambda(M)$ into an A-algebra. We denote the multiplication in $\Lambda(M)$ by $(\omega, \eta) \mapsto \omega \wedge \eta$.

As elements of the form $m_1 \otimes \cdots \otimes m_r$, $m_i \in M$, generate $T^r(M)$, their images $m_1 \wedge \cdots \wedge m_r$ generate $\Lambda^r(M)$. A map u given on these elements with values in an A-module N extends (necessarily uniquely) to an A-linear map $u \colon \Lambda^r(M) \to N$ if and only if u satisfies linearity conditions in each component (as in (14.8)) and if u sends every element in E_r to 0.

The A-algebra $\Lambda(M)$ is a graded A-algebra that is *graded commutative*, i.e., for $\omega \in \Lambda^i(M)$, $\eta \in \Lambda^j(M)$ one has

$$\omega \wedge \eta = (-1)^{ij} \eta \wedge \omega.$$

Let $u \colon M \to N$ be a homomorphism of A-modules. Then the homomorphism $T(u)$ of A-algebras and the A-linear maps $T^r(u)$ induce a homomorphism of A-algebras $\Lambda(u) \colon \Lambda(M) \to \Lambda(N)$ and A-linear maps $\Lambda^r(u) \colon \Lambda^r(M) \to \Lambda^r(N)$ given by $m_1 \wedge \cdots \wedge m_r \mapsto u(m_1) \wedge \cdots \wedge u(m_r)$.

We obtain a functor $M \mapsto \Lambda(M)$ from the category $(A\text{-Mod})$ to the category of associative graded commutative A-algebras and functors $\Lambda^r \colon (A\text{-Mod}) \to (A\text{-Mod})$.

Remark 14.21. Let M be an A-module. As $\Lambda(M)$ is graded commutative one has for $m_1, m_2 \in M$ that $m_1 \wedge m_2 = -m_2 \wedge m_1$ in $\Lambda^2(M)$. As the group of permutations S_r is generated by transpositions of i and $i + 1$ and as $\operatorname{sgn} \colon S_r \to \{\pm 1\}$ is a homomorphism of groups, we deduce that for all $m_1, \ldots, m_r \in M$ and for every $\sigma \in S_r$ we have

$$m_{\sigma(1)} \wedge \cdots \wedge m_{\sigma(r)} = \operatorname{sgn}(\sigma) m_1 \wedge \cdots \wedge w_r. \tag{14.23}$$

For finitely generated free modules, the exterior power can be described as follows ([BouA1] III, §7.8 and §7.9). Let $r, n \geq 0$ be integers. Define

$$L^r(n) := \{ (i_1, \ldots, i_r) \, ; \, 1 \leq i_1 < \cdots < i_r \leq n \}.$$

Let E be an A-module, $e_1, \ldots, e_n \in E$. For $J = (i_1, \ldots, i_r) \in L^r(n)$ define $e_J := e_{i_1} \wedge \cdots \wedge e_{i_r} \in \Lambda^r(E)$.

Proposition 14.22. *Let E be a free A-module of rank $n \in \mathbb{N}$.*

1. *Let $r \in \mathbb{N}_0$. The exterior power $\Lambda^r(E)$ is a free A-module of rank $\binom{n}{r}$ (with $\binom{n}{r} = 0$ if $r > n$). For $1 \le r \le n$ a family $(e_1, \ldots, e_n)$ of elements in E is a basis of E if and only if $(e_J)_{J \in L^r(n)}$ is a basis of $\Lambda^r(E)$.*
2. *Let F be a free A-module with basis $(f_1, \ldots, f_m)$ with $m \in \mathbb{N}_0$ and let $u: E \to F$ be an A-linear map given by $u(e_j) = \sum_{i=1}^{m} a_{ij} f_i$ for a matrix $(a_{ij}) \in M_{m \times n}(A)$. Then the matrix of $\Lambda^r(u)$ with respect to the bases $(e_J)_{J \in L^r(n)}$ of $\Lambda^r(E)$ and $(f_I)_{I \in L^r(m)}$ of $\bigwedge^r(F)$ is the matrix*

$$(\det(A_{I,J}))_{I \in L^r(m), J \in L^r(n)},$$

   where $I = \{i_1 < \cdots < i_r\}$, $J = \{j_1 < \cdots < j_r\}$ and

$$A_{I,J} = (a_{i_\lambda, j_\mu})_{\substack{1 \le \lambda \le r \\ 1 \le \mu \le r}}$$

   the submatrix of A consisting only of rows (respectively columns) numbered by elements in I (respectively in J).
3. *In particular, if $r = n = m$, then $\det(E) := \Lambda^n(E)$ and $\det(F) := \Lambda^n(F)$ are free A-modules of rank 1 and $\det(u) := \Lambda^n(u)$ is given by $\det(A)$ with respect to the bases $e_{(1,\ldots,n)}$ and $f_{(1,\ldots,n)}$.*

Let E be a free A-module of rank $n \in \mathbb{N}$ and let $u: E \to E$ be an A-linear endomorphism. Then $\det(u)$ is an endomorphism of the free A-module $\det(E)$ of rank 1 and hence is given by multiplication with an element in A that we call again $\det(u)$.

Proposition 14.23. *The endomorphism u is bijective if and only if $\det(u)$ is a unit in A.*

Proof. Let $(e_1, \ldots, e_n)$ be a basis of E and set $x_i := u(e_i)$. Then

$$x_1 \wedge \cdots \wedge x_n = \det(u)(e_1 \wedge \cdots \wedge e_n). \qquad (*)$$

The linear map u is bijective if and only if $(x_1, \ldots, x_n)$ is a basis of E. By Proposition 14.22 1 this is the case if and only if $(x_1 \wedge \cdots \wedge x_n)$ is a basis of the free A-module $\Lambda^n(E)$ of rank 1. As $(e_1 \wedge \cdots \wedge e_n)$ is a basis, $(*)$ shows that $(x_1 \wedge \cdots \wedge x_n)$ is a basis if and only if $\det(u)$ is a unit in A. $\qquad \square$

Remark 14.24. Let M be an A-module and let $\alpha_1, \ldots, \alpha_r: M \to A$ be A-linear maps. Then

$$M^r \longrightarrow A, \qquad (m_1, \ldots, m_r) \mapsto \det((\alpha_i(m_j))_{1 \le i,j \le r})$$

is an alternating r-multilinear map and hence corresponds to an A-linear map $\Lambda^r(M) \to A$, which we denote by $\alpha_1 \wedge \cdots \wedge \alpha_r$. The map

$$(M^\vee)^r \to \Lambda^r(M)^\vee, \qquad (\alpha_1, \ldots, \alpha_r) \mapsto \alpha_1 \wedge \cdots \wedge \alpha_r$$

is r-multilinear and alternating. Hence it corresponds to an A-linear map

$$\Lambda^r(M^\vee) \longrightarrow \Lambda^r(M)^\vee. \tag{14.24}$$

Now suppose that M is finitely generated and free. We claim that (14.24) is then an isomorphism. Indeed, let $(e_1, \ldots, e_n)$ be a basis of M, and let $(\varepsilon_1, \ldots, \varepsilon_n)$ be the dual basis of $M^\vee$. Then (14.24) sends for $J \in L^r(n)$ the J-th basis vector ε_J of $\Lambda^r(M^\vee)$ to the J-th basis vector in the dual basis of the basis $(e_J)_{J \in L^r(n)}$ of $\Lambda^r(M)$. In particular our claim follows.

Symmetric Products

A similar construction as for alternating multilinear maps also works for symmetric multilinear maps. Again let M and N be A-modules. An r-multilinear map $\alpha: M^r \to N$ is called *symmetric* if $\alpha(m_1, \ldots, m_r) = \alpha(m_{\sigma(1)}, \ldots, m_{\sigma(r)})$ for all permutations σ in the symmetric group S_r of $\{1, \ldots, r\}$.

Let $P_r \subseteq T^r(M)$ be the A-submodule generated by all elements of the form

$$(m_1 \otimes \cdots \otimes m_r) - (m_{\sigma(1)} \otimes \cdots \otimes m_{\sigma(r)})$$

with $m_i \in M$ and $\sigma \in S_r$. Again an r-multilinear map $\alpha: M^r \to N$ is symmetric if and only if the corresponding A-linear map $u: T^r(M) \to N$ sends P_r to 0. We set

$$\mathrm{Sym}^r(M) := \mathrm{Sym}_A^r(M) := T^r(M)/P_r$$

and call it the *r-th symmetric power of M*. The image of $m_1 \otimes \cdots \otimes m_r$ in $\mathrm{Sym}^r(M)$ is denoted by $m_1 m_2 \cdots m_r$. We have $P_0 = P_1 = 0$ and hence $\mathrm{Sym}^0(M) = A$ and $\mathrm{Sym}^1(M) = M$. The A-module $\mathrm{Sym}^r(M)$ has again a universal property, now with respect to symmetric r-multilinear maps. As above, the direct sum

$$\mathrm{Sym}(M) := \mathrm{Sym}_A(M) := \bigoplus_{r \geq 0} \mathrm{Sym}^r(M)$$

is a graded A-algebra, called the *symmetric algebra of M* It is commutative in the usual sense. Again all these constructions are functorial yielding functors Sym from (A-Mod) to the category of commutative graded algebras and Sym^r from (A-Mod) to itself. Finally, by [BouA1] III, §6.6, Theorem 1 we have the following result about the symmetric powers of free modules.

Proposition 14.25. *Let M be a free module with basis $(e_i)_{i \in I}$. Let $\mathbb{N}_0^{(I)}$ be the set of maps $\beta: I \to \mathbb{N}_0$ such that $\beta(i) = 0$ for all but finitely many i. Write*

$$e^\beta := \prod_{i \in I} e_i^{\beta(i)},$$

product in the commutative A-algebra $\text{Sym}(M)$. *Then* $(e^\beta)_{\beta \in \mathbb{N}_0^{(I)}}$ *is a basis of the A-module* $\text{Sym}(M)$. *For all* $r \geq 0$ *the elements* e^β *with* $\sum_{i \in I} \beta(i) = r$ *form a basis of* $\text{Sym}^r(M)$.

14.4 Modules over Commutative Rings

Now let A be a commutative ring. We denote by $A^\times := \{a \in A \, ; \, \exists b \in A : ab = 1\}$ its group of units.

Recall that an *ideal of A* is an A-submodule of A. In other words, an ideal $\mathfrak{a}$ of A is a subgroup $\mathfrak{a}$ of $(A, +)$ such that $ba \in \mathfrak{a}$ for all $b \in A$ and $a \in \mathfrak{a}$.

Trivial examples of ideals of A are $\{0\}$ and A itself. For an ideal $\mathfrak{a}$ of A one has $\mathfrak{a} = A$ if and only if $1 \in \mathfrak{a}$. The commutative ring A is a field if and only if $A \neq \{0\}$ and $\{0\}$ and A are the only ideals of A.

Remark 14.26. If $\mathfrak{a}$ is an ideal of A then the multiplication $A \times A \to A$ induces a map $A/\mathfrak{a} \times A/\mathfrak{a} \to A/\mathfrak{a}$ on the quotient group $A/\mathfrak{a}$ making $A/\mathfrak{a}$ again into a commutative ring. The projection $\pi \colon A \to A/\mathfrak{a}$ is then a ring homomorphism.

Sending an ideal $\bar{\mathfrak{b}}$ of $A/\mathfrak{a}$ to $\pi^{-1}(\bar{\mathfrak{b}})$ yields a bijection between the set of ideals in $A/\mathfrak{a}$ and the set of ideals in A containing $\mathfrak{a}$.

Definition 14.27. Let $\mathfrak{a} \subseteq A$ be an ideal and M an A-module. The *product of $\mathfrak{a}$ and M* is the A-submodule of M defined by

$$\mathfrak{a}M := \left\{ \sum_{i=1}^n a_i m_i \, ; \, n \in \mathbb{N}_0, a_i \in \mathfrak{a}, m_i \in M \right\}.$$

Proposition 14.28. *Let* $\mathfrak{a} \subseteq A$ *be an ideal and* M *an A-module. Then the map* $u \colon M \to M \otimes_A A/\mathfrak{a}$, $m \mapsto m \otimes 1$ *induces an isomorphism of A-modules* $M/\mathfrak{a}M \xrightarrow{\sim} M \otimes_A (A/\mathfrak{a})$.

Proof. For $a \in A$ or $m \in M$ let $\bar{a}$ (respectively $\bar{m}$) be its image in $A/\mathfrak{a}$ (respectively in $M/\mathfrak{a}M$). For $a \in \mathfrak{a}$ and $m \in M$ we have $(a \cdot m) \otimes 1 = m \otimes (a \cdot 1) = m \otimes \bar{a} = 0$ so that $\mathfrak{a}M$ is contained in the kernel of u. Denote by $i \colon M/\mathfrak{a}M \to M \otimes_A (A/\mathfrak{a})$ the induced homomorphism. Conversely, define a map $M \times (A/\mathfrak{a}) \to M/\mathfrak{a}M$ by $(m, \bar{a}) \mapsto \overline{am}$. This assignment is well defined and A-bilinear so that it induces an A-linear map $j \colon M \otimes_A (A/\mathfrak{a}) \to M/\mathfrak{a}M$.

Now $j \circ i = \text{id}_{M/\mathfrak{a}M}$ hence i must be injective. Moreover i is also surjective because $M \otimes_A (A/\mathfrak{a})$ is generated by elements of the form $m \otimes \bar{a} = (a \cdot m) \otimes \bar{1}$ for $m \in M$ and $a \in A$. Hence i and j are mutually inverse isomorphisms. $\square$

Definition 14.29. An ideal $\mathfrak{m} \subsetneq A$ is called *maximal* if there exists no proper ideal $\mathfrak{a} \subsetneq A$ with $\mathfrak{m} \subsetneq \mathfrak{a}$.

Remark 14.30. Fix an ideal $\mathfrak{a} \subsetneq A$. If $(\mathfrak{a}_i)_{i \in I}$ is a family of proper ideals of A, totally ordered by inclusion such that $\mathfrak{a} \subseteq \mathfrak{a}_i$ for all $i \in I$, then $\bigcup_i \mathfrak{a}_i$ is a proper ideal of A containing $\mathfrak{a}$. Hence we can employ Zorn's lemma 13.28 to see that every proper ideal $\mathfrak{a}$ of a commutative ring is contained in a maximal ideal. Applying this to $\mathfrak{a} = \{0\}$, we see in particular there always exist maximal ideals of A if $A \neq \{0\}$.

Remark 14.31. An ideal $\mathfrak{a}$ of A is maximal if and only if $A/\mathfrak{a}$ is a field (Remark 14.26).

Definition 14.32. Let $\varphi \colon A \to B$ be a homomorphism of commutative rings. For every A-module M the *base change* of M along φ is defined as $M_B := M \otimes_A B$. There is a unique A-linear map $M_B \times B \to M_B$ with $(m \otimes b, \tilde{b}) \mapsto m \otimes (b \cdot \tilde{b})$, making M_B into a B-module.

If $f \colon M \to N$ is an A-linear map, then $f_B := f \otimes \mathrm{id}_B \colon M_B \to N_B$ is B-linear.

These constructions define a functor $(\ )_B$ from the category of A-modules to the category of B-modules.

For commutative rings $A \neq \{0\}$ free A-modules have a well-defined rank:

Definition and Remark 14.33 (Rank of a free module). Let $A \neq 0$ and let M be a free A-module, hence $M \cong A^{(I)}$ for some set I. Then the cardinality of I is called the *rank of M*. It is denoted by $\mathrm{rk}_A(M)$.

To see that the rank is well defined it suffices to prove that the existence of an A-linear isomorphism $f \colon A^{(I)} \xrightarrow{\sim} A^{(J)}$ implies that I and J have the same cardinality. As $A \neq 0$, we can choose a maximal ideal $\mathfrak{m} \subset A$ and set $k := A/\mathfrak{m}$. One has $k^{(I)} \cong A^{(I)} \otimes_A k$ because tensor products commute with direct sums. Hence

$$f_k \colon k^{(I)} \cong A^{(I)} \otimes_A k \xrightarrow{\sim} A^{(J)} \otimes_A k \cong k^{(J)}$$

is also an isomorphism (of k-vector spaces). Since k is a field (Remark 14.31), it is a standard result of linear algebra that I and J have the same cardinality.

For free modules over non-commutative rings the rank is in general not well defined: There exist non-commutative rings B and an isomorphism of B-modules $B \xrightarrow{\sim} B^2$ (Problem 14.28).

Example 14.34. Let M and N be free A-modules of finite rank m and n respectively and let $r \geq 0$ be an integer. Above we have expressed bases of $M \otimes N$, $\Lambda^r(M)$, and $\mathrm{Sym}^r(A)$ in terms of bases of M and N. In particular we see that

$$\mathrm{rk}_A(M \oplus N) = m + n, \qquad \mathrm{rk}_A(M \otimes_A N) = mn,$$
$$\mathrm{rk}_A(\mathrm{Hom}_A(M, N)) = mn, \qquad \mathrm{rk}_A(M^\vee) = m,$$
$$\mathrm{rk}_A(\Lambda^r(M)) = \binom{m}{r}, \qquad \mathrm{rk}_A(\mathrm{Sym}^r(M)) = \binom{m + r - 1}{r}.$$

Remark 14.35. Let M be a free A-module and let $\varphi\colon A \to B$ be a homomorphism of commutative rings. Choose an isomorphism $f\colon M \xrightarrow{\sim} A^{(I)}$ for some set I. Its base change via φ is an isomorphism $f_B\colon M_B \xrightarrow{\sim} B^{(I)}$ because tensor products commute with direct sums. Hence M_B is a free B-module of the same rank as the free A-module M.

Definition 14.36 (Local ring). A commutative ring A is called *local* if it has a unique maximal ideal $\mathfrak{m} \subset A$. The field $A/\mathfrak{m}$ is called the *residue field of A*.

In particular a local ring is always non-zero.

Remark 14.37. A commutative ring A is local if and only if $A \setminus A^\times$ is an ideal. In this case $\mathfrak{m} := A \setminus A^\times$ is the unique maximal ideal of A. In particular for $a \in \mathfrak{m}$ we have $1 + a \in A^\times$.

Proposition 14.38 (Nakayama's lemma). *Let A be a local ring with residue field k, let M' and M be A-modules, and let $f\colon M' \to M$ be an A-linear map. Suppose that M is a finitely generated A-module. Then f is surjective if and only if the induced map $\bar{f}\colon M'/\mathfrak{m}M' \to M/\mathfrak{m}M$ is surjective.*

Proof. We have a commutative diagram

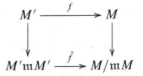

where vertical maps are the projections. In particular the condition is necessary. Now let $\bar{f}$ be surjective. Let $N := \mathrm{Coker}(f)$. Our aim is to show $N = 0$. As $\bar{f}$ is surjective, we have $N/\mathfrak{m}N = 0$ and hence $N = \mathfrak{m}N$. As a quotient of M, N is again finitely generated. Hence it remains to show that for a finitely generated A-module N the equality $N = \mathfrak{m}N$ implies $N = 0$.

Assume $N \neq 0$ and $N = \mathfrak{m}N$. Since N is finitely generated, there exists a minimal set of generators $n_1, \ldots, n_r \in N$ for $r \in \mathbb{N}$. As $n_r \in N = \mathfrak{m}N$, we find $a_1, \ldots, a_n \in \mathfrak{m}$ with $n_r = a_1 n_1 + \ldots + a_r n_r$. Since A is local, we find $1 - a_r \in A^\times$ so that $n_r = (1 - a_r)^{-1}(a_1 n_1 + \ldots + n_{r-1}n_{r-1})$ and therefore $n_1, \ldots, n_{r-1}$ already generate N. This contradicts the minimality of $r \in \mathbb{N}$. $\square$

Proposition 14.39. *Let A be a local ring with maximal ideal $\mathfrak{m}$, $k := A/\mathfrak{m}$. Let E be a free A-module of finite rank and let $u\colon E \to E$ be an A-linear endomorphism. Then u is an automorphism if and only if the k-linear map $\bar{u}\colon E/\mathfrak{m}E \to E/\mathfrak{m}E$ induced by u is an automorphism.*

Proof. The determinant $\det(\bar{u}) \in k$ is the image of $\det(u) \in A$. Hence $\det(\bar{u})$ is a unit in k (i.e., $\det(\bar{u}) \neq 0$) if and only if $\det(u)$ is a unit in A (i.e., $\det(u) \notin \mathfrak{m}$). We conclude by Proposition 14.23. $\qquad\square$

The conclusion of Proposition 14.39 also holds for arbitrary finitely generated A-modules E: If $\bar{u}$ is an automorphism, then u is surjective by Nakayama's lemma. Then conclude by Problem 14.30.

14.5 Problems

In the problem section let R be always a ring and A a commutative ring.

Problem 14.1. Let R be a ring. Show that a morphism in the category of R-modules is a monomorphism (respectively an epimorphism) if and only if it is injective (respectively surjective).

Problem 14.2. Let $u, v \colon M \to N$ be a homomorphisms of R-modules. Show that $\mathrm{Ker}(u, v) := \mathrm{Ker}(u - v)$ (respectively that $\mathrm{Coker}(u, v) := \mathrm{Coker}(u - v)$) is an equalizer (respectively a coequalizer) in the category of R-modules (Problem 13.9). Deduce that the category of R-modules is complete and cocomplete (use Problem 13.10).

Problem 14.3. Let I be a small filtered category and let $M \colon I \to (R\text{-Mod})$ be an I-diagram of R-modules. By composition with the forgetful functor $(R\text{-Mod}) \to (\text{Sets})$ we also obtain an I-diagram $\tilde{M}$ of sets. Define on $\mathrm{colim}_I \tilde{M} = (\coprod_{i \in \mathrm{Ob}(I)} \tilde{M}_i)/\sim$ an addition and a scalar multiplication as follows. For $a \in R$ and $m, m' \in \mathrm{colim}_I \tilde{M}$ choose representatives $\tilde{m}_i \in \tilde{M}_i = M_i$ and $\tilde{m}'_j \in \tilde{M}_j = M_j$ and choose morphisms $\varphi \colon i \to k$ and $\psi \colon j \to k$ in I for some object k in I. Define $m + m'$ as the equivalence class of $\tilde{M}(\varphi)(\tilde{m}_i) + \tilde{M}(\psi)(\tilde{m}'_j)$. Define am as the equivalence class of $a\tilde{m}_i$. Show that this defines the structure of a left R-module on $\mathrm{colim}_I \tilde{M}$, that the canonical maps $s_i \colon M = \tilde{M}_i \to \mathrm{colim}_I \tilde{M}$ are R-linear, and that $(\mathrm{colim}_I \tilde{M}, (s_i)_i)$ is a colimit in the category of R-modules.

Problem 14.4. Let P be an R-module. Show that the following conditions on P are equivalent:

(i) Every short exact sequence of R-modules of the form $0 \to M \to N \to P \to 0$ is split.
(ii) For every surjective homomorphism $p \colon \tilde{M} \to M$ of R-modules and for every R-linear map $u \colon P \to M$ there exists an R-linear map $\tilde{u} \colon P \to \tilde{M}$ such that $p \circ \tilde{u} = u$.
(iii) P is a direct summand of a free module (i.e., there exists an R-module P' such that $P \oplus P'$ is a free R-module).

If P satisfies these conditions, P is called *projective*.

Show that an R-module is projective and finitely generated if and only if it is a direct summand of a finitely generated free R-module.

Problem 14.5. Let P be an A-module. Show that P is projective and finitely generated (Problem 14.4) if and only if the homomorphism $P^\vee \otimes_A P \to \mathrm{End}_A(P)$ (14.19) is an isomorphism.

Problem 14.6. A non-zero commutative ring R is called an *integral domain* if for $a, b \in R$ the equality $ab = 0$ implies $a = 0$ or $b = 0$. Suppose now that R is an integral domain. An R-module M is called *torsion free* if for $a \in R$ and $m \in M$ the equality $am = 0$ implies $a = 0$ or $m = 0$.

1. Show that $\mathbb{Z}$ and $k[T]$ (k a field) are integral domains. Let $n \geq 1$ be an integer. Show that $\mathbb{Z}/n\mathbb{Z}$ is an integral domain if and only if n is a prime number.
2. Show that every free R-module is torsion free.
3. Let $a \in R$ with $a \neq 0$. Show that the R-module $R/(a)$ is not torsion free (and hence not free).
4. The *field of fractions of* R, denoted by $\mathrm{Frac}(R)$, is the set of pairs (a, s) with $a \in A$ and $s \in A \setminus \{0\}$ modulo the equivalence relation $(a, s) \sim (b, t) \Leftrightarrow at = bs$. The equivalence class of (a, s) is denoted by $\frac{a}{s}$. Define addition and multiplication as usual for fractions and show that this is well defined and endows $\mathrm{Frac}(R)$ with the structure of a field. Observe that $\mathrm{Frac}(\mathbb{Z}) = \mathbb{Q}$.
5. Show that the map $\iota \colon R \to \mathrm{Frac}(R)$, $a \mapsto \frac{a}{1}$ is an injective ring homomorphism satisfying the following universal property: If K is a field and $\varphi \colon R \to K$ is an injective ring homomorphism, then there exists a unique homomorphism $\varphi^0 \colon \mathrm{Frac}(R) \to K$ such that $\varphi^0 \circ \iota = \varphi$.
6. Show that $\mathrm{Frac}(R)$ is a torsion free R-module. Show that $\mathrm{Frac}(R)$ is not a free R-module if R is not a field.

Problem 14.7. Let $n, m \in \mathbb{Z}$ be non-zero and let d be a greatest common divisor of m and n. Show that the rings (considered as $\mathbb{Z}$-algebras) $\mathbb{Z}/m\mathbb{Z} \otimes_{\mathbb{Z}} \mathbb{Z}/n\mathbb{Z}$ and $\mathbb{Z}/d\mathbb{Z}$ are isomorphic (in particular $\mathbb{Z}/m\mathbb{Z} \otimes_{\mathbb{Z}} \mathbb{Z}/n\mathbb{Z} = 0$ if m and n are coprime).

Problem 14.8. For integers $r, n \geq 2$ consider the inclusion $u \colon r\mathbb{Z}/nr\mathbb{Z} \hookrightarrow \mathbb{Z}/nr\mathbb{Z}$ of $\mathbb{Z}/nr\mathbb{Z}$-modules. Show that $u \otimes \mathrm{id}_{\mathbb{Z}/r\mathbb{Z}}$ is not injective.

Problem 14.9. Let E be an A-module.

1. Show that the following assertions are equivalent:
    (i) For every injective A-linear map $u \colon M \to M'$ the map $u \otimes \mathrm{id}_E \colon M \otimes E \to M' \otimes E$ is injective.

(ii) For every short exact sequence $0 \to M' \xrightarrow{i} M \xrightarrow{p} M'' \to 0$ of A-modules the sequence $0 \to M' \otimes E \xrightarrow{i \otimes \mathrm{id}_E} M \otimes E \xrightarrow{p \otimes \mathrm{id}_E} M'' \otimes E \to 0$ is exact. In this case E is called *flat*.

2. Show that every free A-module is flat. Deduce that every projective A-module (Problem 14.4) is flat.
3. Suppose that A is an integral domain (Problem 14.6). Show that a flat A-module is torsion free.

Problem 14.10. Let M and N be A-modules. An r-multilinear map $\alpha: M^r \to N$ is called *skew-symmetric* if $\alpha(m_{\sigma(1)}, \ldots, m_{\sigma(r)}) = \mathrm{sgn}(\sigma)\alpha(m_1, \ldots, m_r)$ for every permutation $\sigma \in S_r$ and for all $m_1, \ldots, m_r \in M$.

1. Show that every alternating r-multilinear map $M^r \to N$ is skew-symmetric.
2. Show that an r-multilinear map $\alpha: M^r \to N$ is alternating if and only if $\alpha(m_1, \ldots, m_r) = 0$ whenever there exists $1 \leq i \leq r-1$ such that $m_i = m_{i+1}$.
3. Show that every skew-symmetric r-multilinear map $M^r \to N$ is alternating if the map $N \to N, n \mapsto 2n$ is injective.

The injectivity of $N \to N, n \mapsto 2n$, is satisfied if 2 is a unit in A (for instance if A is a commutative algebra over a field of characteristic $\neq 2$).

Problem 14.11. Let M be an A-module and let $r \geq 0$ be an integer. Show that

$$m_1 \wedge \cdots \wedge m_r \mapsto \sum_{\sigma \in S_r} \mathrm{sgn}(\sigma)(m_{\sigma(1)} \otimes \cdots \otimes m_{\sigma(r)})$$

defines an A-linear map $a_r: \Lambda^r(M) \to T^r(M)$. If $p_r: T^r(M) \to \Lambda^r(M)$ is the canonical projection, then $p_r \circ a_r = (r!) \, \mathrm{id}_{\Lambda^r(M)}$. Show that a_r is injective if M is a free A-module or if $r!$ is invertible in A. Give an example of a ring A and an A-module M such that a_2 is not injective.

Problem 14.12. Let M be an A-module, $r \geq 0$ an integer, $\omega \in \Lambda^r(M)$. Show that if r is odd, then $\omega \wedge \omega = 0$. Give an example of a module M and an element $\omega \in \Lambda^2(M)$ such that $\omega \wedge \omega \neq 0$.

Problem 14.13. Let M be an A-module generated by a single element. Show that $\mathrm{Sym}(M) = T(M)$ and that $\Lambda^r(M) = 0$ for all $r \geq 2$.

Problem 14.14. Let R be a commutative ring, let $n \geq 2$ be an integer, and let $A = R[T_1, \ldots, T_n]$ be the ring of polynomials over R in n indeterminates. Let $\mathfrak{a} := (T_1, \ldots, T_n)$ be the ideal in A generated by $T_1, \ldots, T_n$. Show that $\Lambda_A^n(\mathfrak{a}) = A/\mathfrak{a} = R$.

Problem 14.15. Let $u: M \to N$ be a homomorphism of A-modules.

1. Suppose that u is surjective. Show that $T^r(u)$, $\Lambda^r(u)$, and $\mathrm{Sym}^r(u)$ are surjective for all $r \geq 0$.
2. Show that an analogous assertion for "injective" does not hold.
   Hint: For T^r take the inclusion $\mathbb{Z}$-modules $u: 2\mathbb{Z}/4\mathbb{Z} \to \mathbb{Z}/4\mathbb{Z}$ and use Problem 14.13 for Sym^r. For Λ^r use Problem 14.14.
3. Suppose that u is injective and $u(M)$ is a direct summand of N (i.e., there exists a submodule N' of N such that $N = u(M) \oplus N'$). Show that $T^r(u)$, $\Lambda^r(u)$, and $\mathrm{Sym}^r(u)$ are injective for all $r \geq 0$ and their images are direct summands (see also Problem 14.17).

Problem 14.16. Let M be a projective A-module (Problem 14.4), $m_1, \ldots, m_r \in M$. Show that $(m_1, \ldots, m_r)$ is linearly independent if and only if there exists no $0 \neq a \in A$ such that $a m_1 \wedge \cdots \wedge m_r = 0$.
   Hint: Consider first the case that M is a free A-module.

Problem 14.17. Let $u: M \to N$ be an injective homomorphism of projective A-modules (Problem 14.4). Show that $\Lambda^r(u)$ is injective for all $r \geq 0$.

Problem 14.18. Let $m, n \in \mathbb{N}_0$ and $u: A^m \to A^n$ be an A-linear map. Show that if u is injective (respectively surjective), then $m \leq n$ (respectively $m \geq n$).
   Hint: One could use Problem 14.17 and Problem 14.15.

Problem 14.19. Let M be a free A-module or rank $d \in \mathbb{N}_0$ and let $u: M \to M$ be an A-linear endomorphism. Show that u is injective if and only if $\Lambda^d(u)$ is injective.

Problem 14.20. Let $d \in \mathbb{N}_0$ and let M be an A-module generated by elements $m_1, \ldots, m_d$. Suppose that there exists an A-linear injection $A^d \hookrightarrow M$. Show that $(m_1, \ldots, m_d)$ is a basis of M.
   Hint: Use Problem 14.19.

Problem 14.21. For every A-module M let $\iota_M: M \xrightarrow{\sim} T^1(M) \subseteq T(M)$ be the inclusion. Show that for every homomorphism $u: M \to N$ of A-modules $T(u)$ is the unique homomorphism of A-algebras such that

$$
\begin{array}{ccc}
M & \xrightarrow{\;u\;} & N \\
\downarrow{\scriptstyle \iota_M} & & \downarrow{\scriptstyle \iota_N} \\
T(M) & \xrightarrow{\;T(u)\;} & T(N)
\end{array}
$$

commutes. Formulate and prove an analogous statement for the exterior algebra and the symmetric algebra.

Problem 14.22. Let M be an A-module and let $\iota'_M : M \xrightarrow{\sim} \Lambda^1(M) \hookrightarrow \Lambda(M)$ be the inclusion. Show that for every A-algebra C and for every A-linear map $u : M \to C$ with $u(m)^2 = 0$ for all $m \in M$ there exists a unique homomorphism of A-algebras $\tilde{u} : \Lambda(M) \to C$ such that $u = \tilde{u} \circ \iota'_M$.

Problem 14.23. Let A be an integral domain and let K be its field of fractions (Problem 14.6). Show that $\Lambda^2_A(K) = 0$.

Problem 14.24. Let M be an A-module and $r \in \mathbb{N}_0$.

1. Suppose that $\Lambda^r(M) = 0$ (respectively that $\Lambda^r(M)$ is a finitely generated A-module). Show that $\Lambda^s(M)$ has the same property for all $s \geq r$.
2. Let $u : M \to N$ be a homomorphism of A-modules and suppose that $\Lambda^r(u) = 0$. Show that $\Lambda^s(u) = 0$ for all $s \geq r$.

Problem 14.25. Let M be an A-module, let $r \geq 0$ be an integer and assume that $\Lambda^r(M)$ is generated by $d \in \mathbb{N}_0$ elements. Show that $\Lambda^s(M) = 0$ for all $s > r + d$.

Problem 14.26. Let M be an A-module, let N be a free A-submodule of M of rank $d \in \mathbb{N}$, and let (e) be a basis of the free A-module $\Lambda^d(N)$ of rank 1. Suppose that N is a direct summand of M. Show that an element $m \in M$ is contained in N if and only if $e \wedge m = 0$.

Problem 14.27. Show that the functor $M \mapsto \mathrm{Sym}(M)$ is left adjoint to the forgetful functor from the category of commutative A-algebras to $(A\text{-Mod})$.

Problem 14.28. Let k be a field, let V be a k-vector space with basis $(e_i)_{i \in \mathbb{N}}$, and let $R = \mathrm{End}_k(V)$ be the ring of endomorphisms. Define $f_1, f_2 \in R$ by $f_1(e_i) = e_{2i-1}$ and $f_2(e_i) = e_{2i}$. Show that (f_1, f_2) is an R-basis of the R-left module R. Deduce that $R^n \cong R^m$ for all $n, m \in \mathbb{N}$.

Problem 14.29. Show the following version of Nakayama's lemma. Let M be a finitely generated module over a commutative ring A, let $\mathfrak{a} \subseteq A$ be an ideal. If $M = \mathfrak{a}M$, then there exists $a \in A$ such that $aM = 0$ and $a \equiv 1 \bmod \mathfrak{a}$.

Problem 14.30. Let M be a finitely generated module over a commutative ring A and let $u : M \to M$ be a surjective A-linear endomorphism. Show that u is bijective.
Hint: View M as an $A[T]$-module by setting $T \cdot m := u(m)$ and use Problem 14.29.

Note that injective A-linear endomorphisms u are not necessarily bijective (e.g., $M = A = \mathbb{Z}$ and $u : n \mapsto 2n$).

Problem 14.31. Show that every finitely generated direct summand of a finitely generated free module over a local ring A is again free. In other words (Problem 14.4), every finitely generated projective A-module is free.

Appendix D: Homological Algebra

15

In this appendix we recall some basic notions from homological algebra. We start by defining complexes of modules and the notion of homotopy between morphisms of complexes. This is in fact the cohomological shadow of the topological notion of homotopy (see the remark after Proposition 11.12). One obtains the homotopy category of modules, which is a fundamental notion in homological algebra. After a short interlude on diagram chases we introduce a central notion for the definition of cohomology: injective modules and K-injective complexes. Until then all notions were explained for modules over a ring, but in fact they make sense much more generally in arbitrary abelian categories. This is explained in the last section of this appendix.

Notation: If not otherwise specified, R always denotes a ring, not necessarily commutative. An R-module is always a left R-module.

15.1   Homotopy Category of Modules

Definition 15.1 (Complexes). A sequence

$$M^\bullet = (\cdots \to M^{i-1} \xrightarrow{d^{i-1}} M^i \xrightarrow{d^i} M^{i+1} \to \dots)$$

of homomorphisms of R-modules is called a *complex* if $d^i \circ d^{i-1} = 0$ for all $i \in \mathbb{Z}$ (equivalently $\mathrm{Im}(d^{i-1}) \subseteq \mathrm{Ker}(d^i)$). Sometimes we write d^i_M instead of d^i.

A *morphism of complexes of R-modules* $u \colon M^\bullet \to N^\bullet$ is a family $u^i \colon M^i \to N^i$ of homomorphisms of R-modules such that for all $i \in \mathbb{Z}$ the diagram

$$\begin{array}{ccc} M^i & \xrightarrow{d^i_M} & M^{i+1} \\ \left\downarrow{\scriptstyle u^i}\right. & & \left\downarrow{\scriptstyle u^{i+1}}\right. \\ N^i & \xrightarrow{d^i_N} & N^{i+1} \end{array} \tag{15.1}$$

© Springer Fachmedien Wiesbaden 2016
T. Wedhorn, *Manifolds, Sheaves, and Cohomology*, Springer Studium Mathematik – Master,
DOI 10.1007/978-3-658-10633-1_15

commutes. We obtain the category $(\mathrm{Com}(R))$ of complexes of R-modules.

A complex $M^\bullet$ is called *bounded below* if there exists $a \in \mathbb{Z}$ such that $M^i = 0$ for all $i < a$.

We consider every R-module M as a complex $M^\bullet$ with $M^0 = M$ and $M^i = 0$ for all $i \neq 0$.

Definition 15.2 (Cohomology and quasi-isomorphisms). Let $M^\bullet$ be a complex of R-modules. Then

$$H^i(M^\bullet) := \mathrm{Ker}(d^i)/\mathrm{Im}(d^{i-1})$$

is called the *cohomology of the complex* $M^\bullet$. For every morphism $u\colon M^\bullet \to N^\bullet$ of complexes of R-modules the commutativity of (15.1) shows $u^i(\mathrm{Ker}(d_M^i)) \subseteq \mathrm{Ker}(d_N^i)$ and $u^i(\mathrm{Im}(d_M^{i-1})) \subseteq \mathrm{Im}(d_N^{i-1})$. Hence u^i induces a homomorphism of R-modules

$$H^i(u)\colon H^i(M) \to H^i(N)$$

for all $i \in \mathbb{Z}$.

The morphism u is called a *quasi-isomorphism* or shorter *qis* if $H^i(u)$ is an isomorphism for all $i \in \mathbb{Z}$.

Example 15.3. Let $\cdots \to M^{i-1} \xrightarrow{d^{i-1}} M^i \xrightarrow{d^i} M^{i+1} \to \ldots$ be an exact sequence. Fix $j \in \mathbb{Z}$. Then the vertical arrows of

$$\begin{array}{ccccccc}
\cdots \longrightarrow & M^{j-1} & \xrightarrow{d^{j-1}} & M^j & \longrightarrow & 0 & \longrightarrow \cdots \\
& \downarrow{\scriptstyle u^{j-1}:=0} & & \downarrow{\scriptstyle u^j:=d^j} & & \downarrow{\scriptstyle u^{j+1}:=0} & \\
\cdots \longrightarrow & 0 & \longrightarrow & M^{j+1} & \xrightarrow{d^{j+1}} & M^{j+2} & \longrightarrow \cdots
\end{array}$$

yield a quasi-isomorphism of complexes: This is clear in degree $\neq j$. In degree j, d^j induces an isomorphism $M^j/\mathrm{Im}(d^{j-1}) = M^j/\mathrm{Ker}(d^j) \xrightarrow{\sim} \mathrm{Im}(d^j) = \mathrm{Ker}(d^{j+1})$.

Definition 15.4 (Homotopy of morphisms of complexes). Let $M^\bullet$ and $N^\bullet$ be complexes of R-modules and let $u, v\colon M^\bullet \to N^\bullet$ be two morphisms of complexes.

1. A *homotopy* h between u and v is a family of homomorphisms $h^i\colon M^i \to N^{i-1}, i \in \mathbb{Z}$, such that
   $$u^i - v^i = d_N^{i-1} \circ h^i + h^{i+1} \circ d_M^i.$$

2. The morphisms u and v are called *homotopic* if there exists a homotopy between u and v. We then write $u \simeq v$.

Remark 15.5. Let $M^\bullet$ and $N^\bullet$ be complexes of R-modules.

1. The relation $\simeq$ is an equivalence relation on $\mathrm{Hom}_{(\mathrm{Com}(R))}(M^\bullet, N^\bullet)$: $h = 0$ defines a homotopy between u and u. If h is a homotopy between u and v, then $-h$ is a homotopy between v and u. If h is a homotopy between u and v, and h' is a homotopy between v and w, then $h + h'$ is a homotopy between u and w.
2. If $u \simeq v$ then $u + w \simeq v + w$ for every $w \in \mathrm{Hom}_{(\mathrm{Com}(R))}(M^\bullet, N^\bullet)$.
3. Let $t: L^\bullet \to M^\bullet$, $u, v: M^\bullet \to N^\bullet$, $w: N^\bullet \to P^\bullet$ be morphisms of complexes of R-modules. Let h be a homotopy between u and v. Then $(w^{i-1} \circ h^i \circ t^i)_{i \in \mathbb{Z}}$ is a homotopy between $w \circ u \circ t$ and $w \circ v \circ t$.

Remark 15.5 allows us to define:

Definition 15.6 (Homotopy category of modules). The *homotopy category of complexes of R-modules* is the following category $K(R)$:

(a) Objects are complexes of R-modules.
(b) For two complexes $M^\bullet$ and $N^\bullet$ we set

$$\mathrm{Hom}_{K(R)}(M^\bullet, N^\bullet) := \mathrm{Hom}_{(\mathrm{Com}(R))}(M^\bullet, N^\bullet)/ \simeq .$$

Addition of morphism of complexes endows $\mathrm{Hom}_{K(R)}(M^\bullet, N^\bullet)$ with the structure of an abelian group.
(c) The identity is the identity morphism of complexes. Composition is given by composition of morphisms of complexes (this is well defined by Remark 15.5).

Proposition 15.7. *Let $u, v: M^\bullet \to N^\bullet$ be morphisms of complexes of R-modules that are homotopic. Then*

$$H^p(u) = H^p(v): H^p(M^\bullet) \to H^p(N^\bullet)$$

for all $p \in \mathbb{Z}$.

In particular we see that it makes sense to say that a morphism in $K(R)$ is a quasi-isomorphism.

Proof. By considering $u - v$ it suffices to show: If there exists a homotopy h of u and 0, then $H^p(u) = 0$ for all p. Let $i: \mathrm{Ker}(d_M^p) \to M^p$ be the inclusion. As $u^p = d_N^{p-1} \circ h^p + h^{p+1} \circ d_M^p$ we find that $u^p \circ i = d_N^{p-1} \circ h^p \circ i$ and hence factors through $\mathrm{Im}(d_N^{p-1})$. Hence $H^p(u) = 0$. $\qquad\square$

Remark 15.8. One also has the notion of a complex with decreasing numbering, i.e., of complexes $M_\bullet$ of the form $\cdots \to M_{i+1} \xrightarrow{d_{i+1}} M_i \xrightarrow{d_i} M_{i-1} \to \ldots$ with $d_i \circ d_{i+1} = 0$. If we want to distinguish between these notions, we call complexes with increasing numbering (i.e., complexes as defined in Definition 15.1) *cochain complexes* and complexes with decreasing numbering *chain complexes*. All of the above definitions and results can also be made for chain complexes.

For the proof of the next two less trivial results we refer to [GeMa] Chap. III, §4, Theorem 4.

Lemma 15.9. *Every diagram in $K(R)$ of the form*

$$
\begin{array}{ccc}
M^\bullet & \xrightarrow{\ \ s\ \ } & N^\bullet \\
{\scriptstyle u}\downarrow & {\scriptstyle qis} & \\
P^\bullet, & &
\end{array}
$$

where s is a quasi-isomorphism, can be completed to a diagram

$$
\begin{array}{ccc}
M^\bullet & \xrightarrow{\ \ s\ \ } & N^\bullet \\
{\scriptstyle u}\downarrow & {\scriptstyle qis} & \downarrow{\scriptstyle v} \\
P^\bullet & \xrightarrow[{\scriptstyle qis}]{\ \ t\ \ } & Q^\bullet,
\end{array}
\tag{15.2}
$$

that commutes in $K(R)$ and such that t is a quasi-isomorphism.

It is in general not possible to find the commutative diagram (15.2) in $(\mathrm{Com}(R))$.

Lemma 15.10. *Let $w\colon I^\bullet \to J^\bullet$ be a morphism of complexes. Then the following assertions are equivalent:*

(i) *There exists a quasi-isomorphism $s\colon M^\bullet \xrightarrow{qis} I^\bullet$ such that $w \circ s \simeq 0$.*
(ii) *There exists a quasi-isomorphism $t\colon J^\bullet \xrightarrow{qis} N^\bullet$ such that $t \circ w \simeq 0$.*

Finally, we will use the following result for which we refer to [Stacks] Tag 05T6.

Lemma 15.11. *Let $\mathcal{I}$ be a set of objects in $(R\text{-Mod})$ containing 0 such that for every R-module M there exists an injective homomorphism $M \to I$ of R-modules with $I \in \mathcal{I}$. Let $a \in \mathbb{Z}$. Then for every bounded below complex $M^\bullet$ of R-modules with $M^p = 0$ for $p < a$ there exists a quasi-isomorphism $M^\bullet \xrightarrow{qis} I^\bullet$ with $I^p \in \mathcal{I}$, with $M^p \to I^p$ injective for all p, and with $I^p = 0$ for all $p < a$.*

15.2 Diagram Chases

Lemma 15.12. (Five lemma) *Consider a commutative diagram of R-modules with exact rows*

$$
\begin{array}{ccccccccc}
M_1 & \longrightarrow & M_2 & \longrightarrow & M_3 & \longrightarrow & M_4 & \longrightarrow & M_5 \\
\downarrow u_1 & & \downarrow u_2 & & \downarrow u_3 & & \downarrow u_4 & & \downarrow u_5 \\
N_1 & \longrightarrow & N_2 & \longrightarrow & N_3 & \longrightarrow & N_4 & \longrightarrow & N_5.
\end{array}
$$

Assume that u_2 and u_4 are isomorphisms, u_1 is surjective and u_5 is injective. Then u_3 is an isomorphism.

Proof. This is a simple exercise of diagram chasing, which we leave to the reader. Alternatively, one can also work simply with the universal properties of kernel and cokernel. Such a proof then will also generalize to arbitrary abelian categories (see Definition 15.25). Let us illustrate this by proving that $\mathrm{Coker}(u_3) = 0$.

Denote by $v_i \colon M_i \to M_{i+1}$ and $w_i \colon N_i \to N_{i+1}$ the homomorphisms in the diagram. Let $n_3 \colon N_3 \to P$ be a homomorphism such that $n_3 \circ u_3 = 0$. We have to show that $n_3 = 0$. As $n_3 \circ w_2 \circ u_2 = n_3 \circ u_3 \circ v_2 = 0$ and as $\mathrm{Coker}(u_2) = 0$ we find $n_3 \circ w_2 = 0$. Hence $\mathrm{Ker}(w_3) = \mathrm{Im}(w_2) \subseteq \mathrm{Ker}(n_3)$ and there exists $n_4 \colon \mathrm{Im}(w_3) \to P$ with $n_4 \circ w_3 = n_3$. As $n_4 \circ u_4 \circ v_3 = 0$ we find

$$
\mathrm{Ker}(w_4 \circ u_4) = \mathrm{Ker}(u_5 \circ v_4) = \mathrm{Ker}(v_4) = \mathrm{Im}(v_3) \subseteq \mathrm{Ker}(n_4 \circ u_4)
$$

and hence $\mathrm{Im}(w_3) = \mathrm{Ker}(w_4) \subseteq \mathrm{Ker}(n_4)$ because u_4 is an isomorphism. Therefore $n_4 = 0$ and hence $n_3 = n_4 \circ w_3 = 0$. $\square$

Lemma 15.13. (Snake lemma) *A commutative diagram of R-modules with exact rows*

$$
\begin{array}{ccccccc}
M_1 & \longrightarrow & M_2 & \overset{p}{\longrightarrow} & M_3 & \longrightarrow & 0 \\
\downarrow u_1 & & \downarrow u_2 & & \downarrow u_3 & & \\
0 & \longrightarrow & N_1 & \overset{i}{\longrightarrow} & N_2 & \longrightarrow & N_3.
\end{array}
$$

induces an exact sequence

$$
\mathrm{Ker}(u_1) \to \mathrm{Ker}(u_2) \to \mathrm{Ker}(u_3) \overset{\partial}{\longrightarrow} \mathrm{Coker}(u_1) \to \mathrm{Coker}(u_2) \to \mathrm{Coker}(u_3),
$$

where "$\partial = i^{-1} \circ u_2 \circ p^{-1}$".

The informal definition of ∂ means that for $x \in \mathrm{Ker}(u_3)$ one first chooses $z \in M_2$ with $p(z) = x$ and then defines $\partial(x)$ as the image of an element $y \in N_1$ with $i(y) = u_2(z)$.

Note that if $M_1 \to M_2$ is injective, then $\mathrm{Ker}(u_1) \to \mathrm{Ker}(u_2)$ is injective. If $N_2 \to N_3$ is surjective, then $\mathrm{Coker}(u_2) \to \mathrm{Coker}(u_3)$ is surjective.

Proof. Diagram chase.   □

Lemma 15.14. *Let* $0 \to M^\bullet \xrightarrow{u} N^\bullet \xrightarrow{v} P^\bullet \to 0$ *be a short exact sequence of complexes of R-modules (i.e., for all* $i \in \mathbb{Z}$ *the sequence of R-modules* $0 \to M^i \to N^i \to P^i \to 0$ *is exact). Then there exist for all* $i \in \mathbb{Z}$ *homomorphisms* $\delta \colon H^i(P^\bullet) \to H^{i+1}(M^\bullet)$ *such that the sequence*

$$
\begin{aligned}
\ldots &\xrightarrow{\delta} H^i(M^\bullet) \xrightarrow{H^i(u)} H^i(N^\bullet) \xrightarrow{H^i(v)} H^i(P^\bullet) \\
&\xrightarrow{\delta} H^{i+1}(M^\bullet) \xrightarrow{H^{i+1}(u)} H^{i+1}(N^\bullet) \xrightarrow{H^{i+1}(v)} H^{i+1}(P^\bullet) \qquad (15.3) \\
&\xrightarrow{\delta} \ldots
\end{aligned}
$$

is exact and such that this long exact sequence is functorial for morphisms of exact sequences.

Proof. Applying the the snake lemma to

$$
\begin{array}{ccccccccc}
0 & \longrightarrow & M^{i-1} & \longrightarrow & N^{i-1} & \longrightarrow & P^{i-1} & \longrightarrow & 0 \\
 & & \downarrow{\scriptstyle d^{i-1}} & & \downarrow{\scriptstyle d^{i-1}} & & \downarrow{\scriptstyle d^{i-1}} & & \\
0 & \longrightarrow & M^i & \longrightarrow & N^i & \longrightarrow & P^i & \longrightarrow & 0
\end{array}
$$

shows that the rows of the following commutative diagram are exact:

$$
\begin{array}{ccccccc}
\mathrm{Coker}(d_M^{i-1}) & \longrightarrow & \mathrm{Coker}(d_N^{i-1}) & \longrightarrow & \mathrm{Coker}(d_P^{i-1}) & \longrightarrow & 0 \\
\downarrow{\scriptstyle d_M^i} & & \downarrow{\scriptstyle d_N^i} & & \downarrow{\scriptstyle d_P^i} & & \\
0 \longrightarrow \mathrm{Ker}(d_M^{i+1}) & \longrightarrow & \mathrm{Ker}(d_N^{i+1}) & \longrightarrow & \mathrm{Ker}(d_P^{i+1}). & &
\end{array} \qquad (*)
$$

Note that

$$
\begin{aligned}
\mathrm{Ker}(\mathrm{Coker}(d_M^{i-1}) \xrightarrow{d_M^i} \mathrm{Ker}(d_M^{i+1})) &= H^{i-1}(M^\bullet), \\
\mathrm{Coker}(\mathrm{Coker}(d_M^{i-1}) \xrightarrow{d_M^i} \mathrm{Ker}(d_M^{i+1})) &= H^i(M^\bullet).
\end{aligned}
$$

Hence applying the snake lemma to (*) yields an exact sequence

$$
H^i(M^\bullet) \to H^i(N^\bullet) \to H^i(P^\bullet) \xrightarrow{\delta} H^{i+1}(M^\bullet) \to H^{i+1}(N^\bullet) \to H^{i+1}(P^\bullet)
$$

and the long exact sequence (15.3) is obtained by pasting these sequences together. We omit the proof of the functoriality of δ in morphisms of exact sequences.   □

15.3 Injective Modules and K-injective Complexes

We will define the cohomology of sheaves – or more general derived functors – using injective objects and K-injective complexes. These are introduced here.

Definition 15.15 (Injective module). An R-module I is called *injective* if for every diagram of R-modules

$$M' \xrightarrow{\;i\;} M$$
$$\big\downarrow{\scriptstyle u}$$
$$I$$

with i injective, there exists a homomorphism of R-modules $\tilde{u} \colon M \to I$ such that $\tilde{u} \circ i = u$.

The notion of an injective module is dual to the notion of a projective module (Problem 14.4).

Proposition 15.16. *Let I be an R-module. Then the following assertions are equivalent:*

(i) *I is an injective R-module.*
(ii) *Every short exact sequence of R-modules of the form $0 \to I \to M \to M'' \to 0$ splits.*
(iii) *For every complex $M^\bullet$ of R-modules every quasi-isomorphism $s \colon I \to M^\bullet$ has a left inverse in $K(R)$ (i.e., there exists a morphism of complexes $r \colon M^\bullet \to I$, where I is considered as complex concentrated in degree 0, such that $r \circ s \simeq \mathrm{id}_{M^\bullet}$).*

Proof. "(i) $\Rightarrow$ (ii)". Let $0 \to I \xrightarrow{i} M \to M'' \to 0$ be an exact sequence. As I is injective, there exists a homomorphism $\tilde{u} \colon M \to I$ with $\tilde{u} \circ i = \mathrm{id}_I$. Hence the sequence splits.

"(ii) $\Rightarrow$ (iii)". Let $s \colon I \to M^\bullet$ be a quasi-isomorphism. In particular $H^a(M^\bullet) = 0$ for $a \neq 0$. Consider the complex $\tau_{\geq 0} M^\bullet$ given by

$$\cdots \longrightarrow 0 \longrightarrow \mathrm{Coker}(M^{-1} \to M^0) \longrightarrow M^1 \longrightarrow M^2 \longrightarrow \cdots$$

where the cokernel term is in degree 0. Then $M^\bullet \to \tau_{\geq 0} M^\bullet$ is a quasi-isomorphism because $H^a(M^\bullet) = 0$ for $a < 0$. Hence we may assume that $M^a = 0$ for $a < 0$. Then $s^0 \colon I \to M^0$ is injective and hence has a left inverse by applying (ii) to the exact sequence $0 \to I \xrightarrow{s^0} M^0 \longrightarrow \mathrm{Coker}(s^0) \to 0$. This yields a left inverse of s.

"(iii) $\Rightarrow$ (i)". Let $t^0 \colon M \to N^0$ be injective and $u \colon M \to I$ be arbitrary. Let $N^\bullet$ be the complex that is the canonical homomorphism $N^0 \to \mathrm{Coker}(i)$ in degree 0 and 1 and that vanishes in all other degrees. Then t^0 yields a quasi-isomorphism $t\colon M \to N^\bullet$. By Lemma 15.9 there exists in $K(R)$ a commutative diagram

$$
\begin{array}{ccc}
M & \xrightarrow{\ t\ } & N^\bullet \\
{\scriptstyle u}\downarrow & & \downarrow{\scriptstyle u'} \\
I & \xrightarrow{\ s\ } & P^\bullet,
\end{array}
$$

where s is a quasi-isomorphism. By (iii) there exists a left inverse r for s. Let $\tilde{u}\colon N^0 \to I$ be the map given by $r \circ u'$ in degree 0. We have $r \circ u' \circ t \simeq r \circ s \circ u \simeq u$ and hence $(r \circ u') \circ t = u$ because homotopic homomorphisms between complexes concentrated in degree 0 are equal. In particular we find $\tilde{u} \circ t^0 = u$. $\qquad\square$

Example 15.17.
1. If $R = k$ is a field, then every k-vector space is injective because every short exact sequence of k-vector spaces splits (by Example 14.6).
2. Let R be a principal ideal domain (e.g., $R = \mathbb{Z}$). Then an R-module M is injective if and only if $a \in R$, $a \neq 0$, the scalar multiplication $M \to M$, $m \mapsto am$, is surjective ([BouA3] §1.7, see also Problem 15.14).
   For instance, let $K = \operatorname{Frac} R$ be the field of fractions of R. Then K and K/R are injective R-modules.

Definition 15.18. A complex $I^\bullet$ of R-modules is called *K-injective* if for every complex $M^\bullet$ and for every quasi-isomorphism $u\colon I^\bullet \to M^\bullet$ there exists a left inverse in the category $K(R)$ (in other words, there exists a morphism of complexes $r\colon M^\bullet \to I^\bullet$ such that $r \circ u \simeq \operatorname{id}_{M^\bullet}$).

Proposition 15.19. *Let $I^\bullet$ be a complex that is bounded below and such that I^n is an injective R-module for all $n \in \mathbb{Z}$. Then the complex $I^\bullet$ is K-injective.*

Proof. If $I^\bullet$ is concentrated in degree 0, this is the implication "(i) $\Rightarrow$ (iii)" of Proposition 15.16. For the general case we refer to [Stacks] Tag 070J. $\qquad\square$

Lemma 15.20. *Consider a diagram in $K(R)$*

$$
\begin{array}{ccc}
M^\bullet & \xrightarrow[qis]{\ s\ } & I^\bullet \\
{\scriptstyle u}\downarrow & & \\
N^\bullet & \xrightarrow[qis]{\ } & J^\bullet,
\end{array}
$$

where the horizontal morphisms are quasi-isomorphisms and assume that $J^\bullet$ is K-injective. Then there exists a unique morphism $w\colon I^\bullet \to J^\bullet$ in $K(R)$ making the diagram commutative.

Proof. Existence. By Lemma 15.9 we find a diagram

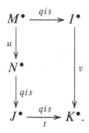

As $J^\bullet$ is K-injective, there exists a left inverse $r: K^\bullet \to J^\bullet$ of t in $K(R)$ and we may set $w := r \circ v$.

Uniqueness. For the uniqueness it suffices to show that if $u \simeq 0$ then one has necessarily $w \simeq 0$. But $u \simeq 0$ implies $w \circ s \simeq 0$ and hence there exists a quasi-isomorphism $t: J^\bullet \to P^\bullet$ such that $t \circ w \simeq 0$ by Lemma 15.10. As $J^\bullet$ is K-injective, there exists a left-inverse r of t. Hence $w \simeq (r \circ t) \circ w \simeq 0$. $\square$

15.4 Abelian Categories

All of the above notions for R-modules can be generalized to so-called abelian categories. These are categories that capture those properties of the category of R-modules that are essential to define most notions in homological algebra: One has an abelian group structure on the set of all morphisms between two objects, there exist finite direct sums and finite direct products, and for every morphism there exist kernels, cokernels (and hence images and coimages) and a morphism always induces an isomorphism from its coimage to its image ("fundamental homomorphism theorem").

Definition 15.21 (Preadditive category). A *preadditive category* is a category $\mathcal{A}$ together with the structure of an abelian group on $\operatorname{Hom}_{\mathcal{A}}(X, Y)$ for all objects X and Y such that the composition $\operatorname{Hom}_{\mathcal{A}}(X, Y) \times \operatorname{Hom}_{\mathcal{A}}(Y, Z) \to \operatorname{Hom}_{\mathcal{A}}(X, Z)$ is $\mathbb{Z}$-bilinear for all objects X, Y, and Z of $\mathcal{A}$.

A functor $F: \mathcal{A} \to \mathcal{B}$ between preadditive categories is called *additive* if the map $\operatorname{Hom}_{\mathcal{A}}(X, Y) \to \operatorname{Hom}_{\mathcal{B}}(F(X), F(Y))$ given by F is a homomorphism of abelian groups for all objects X and Y of $\mathcal{A}$.

The category of R-modules (R a fixed ring) together with the structure of abelian groups on the sets of homomorphisms defined by addition of R-linear maps is a preadditive category.

Definition and Remark 15.22 (Zero object). Let $\mathcal{A}$ be a preadditive category. An object Z of $\mathcal{A}$ is called a *zero object* if it satisfies the following equivalent properties:

(i) Z is an initial object in $\mathcal{A}$.

(ii) Z is a final object in $\mathcal{A}$.

(iii) $\mathrm{id}_Z = 0 \in \mathrm{Hom}_{\mathcal{A}}(Z, Z)$.

Indeed, clearly (i) and (ii) both imply that $\mathrm{Hom}_{\mathcal{A}}(Z, Z)$ consists of a single element, hence $\mathrm{id}_Z = 0$. Conversely, if (iii) holds, then for every morphism $u\colon X \to Z$ (respectively $v\colon Z \to Y$) one has by bilinearity of composition that $u = \mathrm{id}_Z \circ u = 0$ (respectively $v = v \circ \mathrm{id}_Z = 0$). Hence (ii) and (i) hold.

A zero object is unique up to unique isomorphism (if it exists) and is denoted by 0.

Condition (iii) shows that if $F\colon \mathcal{A} \to \mathcal{B}$ is an additive functor between preadditive categories, then F sends a zero object of $\mathcal{A}$ to a zero object of $\mathcal{B}$.

Next we define kernels, cokernels, images, and coimages in preadditive categories.

Remark 15.23 (Kernel, cokernel, image, and coimage). Let $\mathcal{A}$ be a preadditive category in which all finite limits and finite colimits exist, and let $u\colon X \to Y$ be a morphism in $\mathcal{A}$. Consider the following diagram in $\mathcal{A}$:

$$X \underset{0}{\overset{u}{\rightrightarrows}} Y.$$

Then its limit (respectively its colimit) is called the *kernel of u* (respectively *cokernel of u*), denoted by $\mathrm{Ker}(u) \to X$ (respectively $Y \to \mathrm{Coker}(u)$).

Kernels have the following universal property: A morphism $i\colon K \to X$ is a kernel of u if and only if $u \circ i = 0$ and if for any morphism $i'\colon K' \to X$ with $u \circ i' = 0$ there exists a unique morphism $v\colon K' \to K$ such that $i' = i \circ v$.

Dually, a morphism $p\colon Y \to X$ is a cokernel of u if and only if $p \circ u = 0$ and if for every morphism $p'\colon Y \to C'$ with $p' \circ u = 0$ there exists a unique morphism $v\colon C \to C'$ with $v \circ p = p'$.

The *image of u* is defined as $\mathrm{Im}(u) := \mathrm{Ker}(Y \to \mathrm{Coker}(u))$ and the *coimage of u* as $\mathrm{Coim}(u) := \mathrm{Coker}(\mathrm{Ker}(u) \to X)$.

If $\mathcal{A}$ is the category of R-modules for some ring R, then these notions of kernel and cokernel coincide with the usual notions by Example 14.2. Therefore the same holds for images and coimages.

Remark 15.24. Let $u\colon X \to Y$ be a morphism in a preadditive category $\mathcal{A}$ in which all finite limits and finite colimits exist. Then there is a unique factorization of u

$$X \longrightarrow \mathrm{Coim}(u) \overset{\bar{u}}{\longrightarrow} \mathrm{Im}(u) \longrightarrow Y. \tag{15.4}$$

Indeed, as $\mathrm{Ker}(u) \to X \to Y$ is zero, there exists a unique factorization $X \to \mathrm{Coim}(u) \to Y$ of u. The composition $\mathrm{Coim}(u) \to Y \to \mathrm{Coker}(u)$ is the unique morphism giving rise to $X \to X \to \mathrm{Coker}(u)$, which is zero. Hence $\mathrm{Coim}(u) \to \mathrm{Coker}(u)$ is zero and there exists a unique factorization $\mathrm{Coim}(u) \to \mathrm{Im}(u) \to Y$ of $\mathrm{Coim}(u) \to Y$.

Definition 15.25 (Abelian category). A preadditive category is called *abelian* if all finite limits and finite colimits exist and if for every morphism u the induced morphism $\mathrm{Coim}(u) \to \mathrm{Im}(u)$ is an isomorphism.

The hypothesis that all finite limits and colimits exist is highly redundant (see Problem 15.10).

We call a morphism $u: X \to Y$ in an abelian category *injective* or X a *subobject of* Y if $\mathrm{Ker}(u) = 0$ and write $X \subseteq Y$. The morphism u is called *surjective* or Y a *quotient of* X if $\mathrm{Coker}(u) = 0$. If $X \subseteq Y$ is a subobject, then we write $Y/X := \mathrm{Coker}(X \to Y)$.

Remark 15.26. Many notions like "(split) exact sequences" and results for R-modules work verbatim in the same way in an arbitrary abelian category. This holds in particular for all of the notions and results in Sects. 15.1–15.3. In particular we have the category $\mathrm{Com}(\mathcal{A})$ and the homotopy category $K(\mathcal{A})$ of complexes in $\mathcal{A}$ and the notions of injective objects in $\mathcal{A}$ and of K-injective complexes in $\mathcal{A}$. Note that all references given above for results that we did not prove formulate their results in arbitrary abelian categories.

Recall that we defined in Definition 13.48 a functor to be left exact (respectively right exact) if it commutes with finite limits (respectively finite colimits).

Remark 15.27 (Left exact functors). Let $F: \mathcal{A} \to \mathcal{B}$ be an additive functor between abelian categories. Then F is left exact (respectively right exact) if and only if for every exact sequence $0 \to X' \to X \to X''$ (respectively $X' \to X \to X'' \to 0$) in $\mathcal{A}$ the sequence $0 \to F(X') \to F(X) \to F(X'')$ (respectively $F(X') \to F(X) \to F(X'') \to 0$) is exact.

The easy proof is left to the reader (see also [KS2] Proposition 8.3.18).

One can show that every left or right exact functor is automatically additive (Problem 15.9).

Definition 15.28 (Existence of injective and K-injective resolutions). Let $\mathcal{A}$ be an abelian category. We say that $\mathcal{A}$ *has injective and K-injective resolutions* if for every complex $M^{\bullet}$ in $\mathcal{A}$ one has the following quasi-isomorphisms:

(a) There exists a quasi-isomorphism $s: M^{\bullet} \xrightarrow{qis} I^{\bullet}$, where $I^{\bullet}$ is a K-injective complex. This is called a *K-injective resolution of* $M^{\bullet}$.

(b) If there exists $a \in \mathbb{Z}$ such that $M^n = 0$ for all $n < a$, then one can in addition assume that I^n is an injective object in $\mathcal{A}$ for all $n \in \mathbb{Z}$, that $I^n = 0$ for all $n < a$, and that $s^n: M^n \to I^n$ is injective for all n. This is called an *injective resolution of* $M^{\bullet}$.

Every complex $I^\bullet$ as in (b) is K-injective by Proposition 15.19.

If $\mathcal{A}$ has injective and K-injective resolutions, then there exists in particular for every object X of $\mathcal{A}$ an exact sequence

$$0 \longrightarrow X \longrightarrow I^0 \longrightarrow I^1 \longrightarrow \ldots,$$

where I^p is an injective object, because such an exact sequence is the same as a quasi-isomorphism $X \xrightarrow{qis} I^\bullet$ with $I^p = 0$ for $p < 0$.

15.5   Problems

Problem 15.1. Let $\mathcal{A}$ be an abelian category and let $M^\bullet$ be a complex of objects in $\mathcal{A}$ with boundary maps $d^p_{M^\bullet}: M^p \to M^{p+1}$. For $k \in \mathbb{Z}$ define the *k-shifted complex* $M[k]^\bullet$ by $M[k]^p := M^{k+p}$ and $d^p_{M[k]^\bullet} := (-1)^k d^{k+p}_{M^\bullet}$. If $u: M^\bullet \to N^\bullet$ is a morphism of complexes, $u[k]: M[k]^\bullet \to N[k]^\bullet$ denotes the morphism of complexes with $u[k]^p = u^{k+p}$.

1. Show that $[k]$ defines a functor $\mathrm{Com}(\mathcal{A}) \to \mathrm{Com}(\mathcal{A})$ that induces a functor $[k]: K(\mathcal{A}) \to K(\mathcal{A})$.
2. Let $M^\bullet$ and $N^\bullet$ be complexes of objects in $\mathcal{A}$ and let $u: M^\bullet \to N^\bullet$ be a morphism of complexes. Show that a homotopy $u \simeq u$ is the same as a morphism of complexes $M^\bullet \to N[-1]^\bullet$.

Problem 15.2. Prove Lemma 15.11.

Problem 15.3. Let X_1 and X_2 be objects in a preadditive category $\mathcal{A}$. Show that the product $(p_i: X_1 \times X_2 \to X_i)_i$ exists if and only if the coproduct $(s_j: X_j \to X_1 \amalg X_2)_j$ exists. Show that in this case there is a unique morphism $r: X_1 \amalg X_2 \xrightarrow{\sim} X_1 \times X_2$ with $p_i \circ r \circ s_j = 0$ for $i \neq j$, and $p_i \circ r \circ s_j = \mathrm{id}_{X_i}$ for $i = j$. Show that r is an isomorphism.

Write $X_1 \oplus X_2$ for $X_1 \amalg X_2 = X_1 \times X_2$. A preadditive category in which all finite products exist is called an *additive category*.

Problem 15.4. Let $\mathcal{A}$ be a preadditive category, X and Y objects in $\mathcal{A}$, $u_1, u_2 \in \mathrm{Hom}_{\mathcal{A}}(X, Y)$ and assume that $X \oplus X$ and $Y \oplus Y$ exist (Problem 15.3). Show that $u_1 + u_2$ coincides with the composition

$$X \xrightarrow{\Delta_X} X \oplus X \xrightarrow{u_1 \oplus u_2} Y \oplus Y \xrightarrow{\Sigma_Y} Y,$$

where Δ_X is the diagonal and where Σ_Y is the unique morphism whose composition with the two component maps $Y \to Y \oplus Y$ is the identity.

Problem 15.5. Consider a complex $M^\bullet$ of R-modules as a $\mathbb{Z}$-diagram $n \mapsto M^n$ in the category of R-modules. What is $\lim_n M^n$ and $\operatorname{colim}_n M^n$?

Problem 15.6. Prove Lemma 15.12 and Lemma 15.13 for diagrams in arbitrary abelian categories.

Problem 15.7. Let $\mathcal{A}$ be an additive category (Problem 15.3) and let $F: \mathcal{A} \to \text{(Sets)}$ be a functor commuting with finite products. Show that there is a functor $\tilde{F}: \mathcal{A} \to \text{(Ab)}$ such that F is isomorphic to the composition of $\tilde{F}$ with the forgetful functor $\text{(Ab)} \to \text{(Sets)}$. Show that $\tilde{F}$ is unique up to unique isomorphism.
Hint: Problem 15.4.

Problem 15.8. Prove Remark 15.27.

Problem 15.9. Let $\mathcal{A}$ and $\mathcal{B}$ be additive categories (Problem 15.3) and let $F: \mathcal{A} \to \mathcal{B}$ be a functor. Show that F is additive if and only if F commutes with finite products.
Hint: To see that the condition is sufficient apply Problem 15.7 for all objects X to the functors $Y \mapsto \operatorname{Hom}_{\mathcal{A}}(X, Y)$ and $Y \mapsto \operatorname{Hom}_{\mathcal{B}}(F(X), F(Y))$.

Problem 15.10. Let $\mathcal{A}$ be an additive category (Problem 15.3) such that every morphism u has a kernel and a cokernel and such that the induced morphism $\operatorname{Coim}(u) \to \operatorname{Im}(u)$ is an isomorphism. Show that $\mathcal{A}$ is an abelian category.
Hint: Problem 13.10.

Problem 15.11. Let $\mathcal{A}$ be an abelian category. Show that every bounded below complex of objects in $\mathcal{A}$ has an injective resolution if and only if for every object X of $\mathcal{A}$ there exists an injective morphism $X \to I$, where I is an injective object of $\mathcal{A}$.

Problem 15.12. Let $\mathcal{A}$ be an abelian category and let $(I_\alpha)_\alpha$ be a family of objects of $\mathcal{A}$ such that the product $\prod_\alpha I_\alpha$ exists. Show that $\prod_\alpha I_\alpha$ is injective if and only if I_α is injective for all α.

Problem 15.13. Let $0 \to M' \xrightarrow{u} M \xrightarrow{v} M''$ be an exact sequence in an abelian category $\mathcal{A}$, and let $a': M' \to C'$ and $a'': M'' \to C''$ be injective morphisms in $\mathcal{A}$. Suppose that C' is an injective object in $\mathcal{A}$. Show that there exists an injective morphism $a: M \to C' \oplus C''$ such that the diagram

$$
\begin{array}{ccccc}
M' & \xrightarrow{u} & M & \xrightarrow{v} & M' \\
{\scriptstyle a'}\downarrow & & {\scriptstyle a}\downarrow & & \downarrow{\scriptstyle a''} \\
C' & \xrightarrow{i} & C' \oplus C'' & \xrightarrow{p} & C''
\end{array}
$$

commutes, where $i(c') = (c', 0)$ and $p(c', c'') = c''$.

Problem 15.14. Let R be a ring.

1. Show that an R-left module I is injective if and only if for every left ideal $\mathfrak{a}$ of R and for every R-linear map $u: \mathfrak{a} \to I$ there exists $x \in I$ such that $u(a) = ax$ for all $a \in \mathfrak{a}$. Deduce that an abelian group A is an injective $\mathbb{Z}$-module if and only if A is *divisible* (i.e., $A \to A$, $a \mapsto na$ is surjective for all $0 \neq n \in \mathbb{Z}$).

2. Show that the R-left module $I_R := \mathrm{Hom}_{\mathbb{Z}}(R, \mathbb{Q}/\mathbb{Z})$ is injective and a *cogenerator of the category of R-left modules* (i.e., for every R-left module M and for every $0 \neq m \in M$ there exists an R-linear map $u: M \to I_R$ with $u(m) \neq 0$).

3. For every R-module M set $I(M) := I_R^{\mathrm{Hom}_R(M, I_R)}$ and let $e: M \to I(M)$, $m \mapsto (u(m))_{u \in \mathrm{Hom}_R(M, I_R)}$. Show that $I(M)$ is an injective R-left module and that e is an injective R-linear map.

4. Deduce from Problem 15.11 that in the category of R-left modules every bounded below complex of objects has an injective resolution.

Appendix E: Local Analysis

16

In this appendix we recall some notions on differentiable and analytic functions of open subsets of finite-dimensional $\mathbb{K}$-vector spaces, where $\mathbb{K}$ either denotes the field of real numbers $\mathbb{R}$ or the field of complex numbers $\mathbb{C}$. These will be the local building blocks in the theory of manifolds. Most of the results are topics in any standard calculus class and we will give no proofs.

16.1 Differentiable Functions

Recall that all norms on a finite-dimensional $\mathbb{K}$-vector space V are equivalent and hence that there is a unique topology on V that is induced by some norm. If not stated otherwise, we will always endow all subsets of V with the induced topology. For $\mathbb{K}$-vector spaces V and W we denote by $\mathrm{Hom}_{\mathbb{K}}(V, W)$ the $\mathbb{K}$-vector space of $\mathbb{K}$-linear maps $V \to W$.

We recall some basic notions and definitions about differentiable functions. Our main reference is [La]. In the sequel let V and W be finite-dimensional $\mathbb{K}$-vector spaces and let $U \subseteq V$ be an open subset.

Definition 16.1 (Differentialbility). Let $f \colon U \to W$ be a map.

1. The map f is called *differentiable* or $\mathbb{K}$-*differentiable* if for all $u \in U$ there exists a (necessarily unique) $\mathbb{K}$-linear map $Df(u) \colon V \to W$ such that

$$\lim_{V \ni h \to 0} \frac{\|f(u + h) - \big(f(u) - Df(u)(h)\big)\|}{\|h\|} = 0,$$

where we choose some norms on V and W (the notion of differentiability is independent of the choice because all norms are equivalent). The map

$$Df \colon U \to \mathrm{Hom}_{\mathbb{K}}(V, W)$$

© Springer Fachmedien Wiesbaden 2016
T. Wedhorn, *Manifolds, Sheaves, and Cohomology*, Springer Studium Mathematik – Master,
DOI 10.1007/978-3-658-10633-1_16

is also independent of the choice of norms and is called the *derivative of f*. If we want to stress whether we are in the case $\mathbb{K} = \mathbb{R}$ or $\mathbb{K} = \mathbb{C}$, we use the notions "*real differentiable*" and "*complex differentiable*", respectively.

2. Let $f: U \to W$ be differentiable. For $v \in V$ we call

$$D_v f(u) := (Df(u))(v) = \lim_{\mathbb{K} \ni t \to 0} \frac{f(u + tv) - f(u)}{t}$$

the *partial derivative at u in direction v*. The map

$$D_v f : U \to W$$

is then called the *partial derivative in direction v*.

3. For $\alpha \in \mathbb{N}_0$ the notion of a C^α-map and its α-fold derivative $D^\alpha f$ is defined inductively. The map f is called a C^0-*map* or $\mathbb{K}$-C^0-*map* if f is continuous. We set $D^0 f := f$. For all $\alpha \in \mathbb{N}$ the map f is called a C^α-*map* or $\mathbb{K}$-C^α-*map* if f is $\mathbb{K}$-differentiable and $Df: U \to \mathrm{Hom}_{\mathbb{K}}(V, W)$ is a $\mathbb{K}$-$C^{\alpha-1}$-map. We set $D^\alpha f := D(D^{\alpha-1} f)$.

   If f is a C^α-map for all $\alpha \in \mathbb{N}_0$, then f is called a C^∞-*map* or a *smooth map*.

   If we want to stress whether we are in the case $\mathbb{K} = \mathbb{R}$ or $\mathbb{K} = \mathbb{C}$, we speak of *real* C^α-*maps* and *complex* C^α-*maps*, respectively.

If $(e_1, \ldots, e_n) \in V$ is a basis and $(x^1, \ldots, x^n) \in V^\vee$ is the dual basis, we also set

$$\partial_i f := \partial_{x^i} f := \frac{\partial f}{\partial x^i} := D_{e_i} f : U \to W. \tag{16.1}$$

Note that this is a misuse of notation because $\partial_i f = \frac{\partial f}{\partial x^i}$ depends for a fixed i only on the choice of e_i but not only on x^i (as e^i is not determined by x^i alone but by the whole basis $(x^1, \ldots, x^n)$). We usually use this notion only if $V = \mathbb{K}^n$, where we always choose the standard basis when we write $\partial_i f$.

Remark 16.2. Let $\alpha \in \mathbb{N}_0 \cup \{\infty\}$.

1. If $f: U \to W$ is differentiable, then f is continuous. If f is C^α, then f is also C^β for all $\beta \leq \alpha$.
2. The set of all C^α-functions $U \to W$ is denoted by $C^\alpha(U, W)$. It is a $\mathbb{K}$-subspace of the vector space of all maps $U \to W$ and for $f, g \in C^\alpha(U, W)$ and $a \in \mathbb{K}$ one has

$$D(af + g) = aDf + Dg.$$

Remark 16.3. Formally the real case $\mathbb{K} = \mathbb{R}$ and the complex case $\mathbb{K} = \mathbb{C}$ are very similar. Let V and W be finite-dimensional $\mathbb{C}$-vector spaces, $U \subseteq V$ open and let $f: U \to W$ be a map. Of course, we may consider V and W also as finite-dimensional $\mathbb{R}$-vector spaces. Then by definition f is complex differentiable if and only if f is real differentiable and $Df(u)$ is $\mathbb{C}$-linear for all $u \in U$.

On the other hand, the behavior of real and complex differentiable functions is very different – a topic that we will touch upon again and again. For instance, in the real case there exist for all $\alpha \in \mathbb{N}_0$ functions $f_\alpha: \mathbb{R} \to \mathbb{R}$ that are C^α but not $C^{\alpha+1}$, for instance

$$f_\alpha(x) := \begin{cases} -x^{\alpha+1}, & x < 0; \\ x^{\alpha+1}, & x \geq 0. \end{cases}$$

But from elementary complex analysis we know that for $\mathbb{K} = \mathbb{C}$ every differentiable map $f: U \to \mathbb{C}$, $U \subseteq \mathbb{C}$ open, is automatically a C^∞-map. In fact a much stronger assertion holds, namely f is automatically analytic (i.e., locally given by a convergent power series). We will make this more precise below.

Calculus (for instance by induction on [La] XIII, Theorem 7.1) tells us that real C^α can also be defined via partial derivations:

Proposition 16.4. *Let $\mathbb{K} = \mathbb{R}$, let $\alpha \in \mathbb{N}$, and let X be a generating system of the $\mathbb{R}$-vector space V (e.g., a basis or $X = V$). Then a map $f: U \to W$ is a (real) C^α-map if and only if for every tuple $(v_1, \ldots, v_\alpha) \in X^\alpha$ the iterated partial derivatives $D_{v_1} D_{v_2} \cdots D_{v_\alpha} f: U \to W$ exist and are continuous.*

In the complex case an analogous result is made superfluous by the fact that every complex differentiable map is automatically analytic (see Theorem 16.15).

The next two propositions are well known from elementary calculus – at least for $\mathbb{K} = \mathbb{R}$. We deduce the complex case from the real case by considering complex differentiable maps as real differentiable maps whose derivation has complex linear values (Remark 16.3).

Proposition 16.5 (Chain rule). *Let V_1, V_2, and V_3 be finite-dimensional $\mathbb{K}$-vector spaces, let $U_1 \subseteq V_1$ and $U_2 \subseteq V_2$ be open subsets and let $f: U_1 \to V_2$ and $g: U_2 \to V_3$ be maps with $f_1(U_1) \subseteq U_2$. If f and g are differentiable, then $g \circ f$ is differentiable and for $u_1 \in U_1$ and $u_2 := f(u_1)$ one has*

$$D(g \circ f)(u_1) = Dg(u_2) \circ Df(u_1).$$

If f and g are C^α-maps, then $g \circ f$ is a C^α-map.

Proposition 16.6. *Let $F: U \to W$ be a map. Then F is differentiable with $DF(u) = 0$ for all $u \in U$ if and only if F is locally constant.*

Example 16.7. Let $V, V_1, \ldots, V_r, W$ be finite-dimensional $\mathbb{K}$-vector spaces.

1. It follows immediately from the definition of the derivative that every $\mathbb{K}$-linear map $f: V \to W$ is differentiable and we have $Df(u) = f$ for all $u \in V$, i.e., $Df: V \to \mathrm{Hom}_{\mathbb{K}}(V, W)$ is the constant map with value f. In particular f is a C^∞-map with $D^r f = 0$ for all $r \geq 2$ by Proposition 16.6.
2. More generally, every r-multilinear map $\alpha: V := V_1 \times \cdots \times V_r \to W$ is a C^∞-map. For $u = (u_1, \ldots, u_r) \in V$ we have

$$(D\alpha(u))(v) = \sum_{i=1}^{r} \alpha(u_1, \ldots, u_{i-1}, v_i, u_{i+1}, \ldots, u_r)$$

   for all $v = (v_1, \ldots, v_r) \in V$.
3. A *homogeneous polynomial map* $p: V \to W$ *of degree* $r \in \mathbb{N}_0$ is by definition the composition of the linear map $V \to V \times \cdots \times V \to V, v \mapsto (v, \ldots, v)$ followed by an r-multilinear map $V \times \cdots \times V \to W$. Hence any *polynomial map* $V \to W$ (defined to be a sum of homogeneous polynomial maps, not necessarily of the same degree) is a C^∞-map.

Example 16.8. Let $F: U \to W$ be a C^α-map and let $L: W \to W'$ be a linear map between finite-dimensional $\mathbb{K}$-vector spaces. Then an easy induction and the chain rule shows that $D^r(L \circ F)(u) = L \circ D^r(u)$ for all $u \in U$ and $r \leq \alpha$.

16.2   Analytic Functions

We recall some facts on analytic functions, i.e., of maps that are locally given by convergent power series. For proofs we refer to [Ser] Part II, Chap. II.

Definition 16.9 (Analytic function). Let $U \subseteq \mathbb{K}^n$ be open, $u = (u_1, \ldots, u_n) \in U$. A function $f: U \to \mathbb{K}$ is called *analytic at* u if there exists an open neighborhood $\tilde{U}$ of u in U such that $f_{|\tilde{U}}$ is given by an absolutely convergent power series

$$f(x) = \sum_{(i_1, \ldots, i_n) \in \mathbb{N}_0^n} a_{i_1, \ldots, i_n} (x_1 - u_1)^{i_1} \ldots (x_n - u_n)^{i_n}, \qquad x \in \tilde{U}, \tag{16.2}$$

with $a_{i_1, \ldots, i_n} \in \mathbb{K}$ for all $x \in \tilde{U}$. Such a power series is called *power series expansion of* f *at* u.

The function $f: U \to \mathbb{K}$ is called *analytic* if it is analytic at every point u of U. A map $F: U \to \mathbb{K}^m$ is called *analytic* if each component is analytic.

Example 16.10. Every polynomial map is analytic, in particular every linear form $\mathbb{K}^n \to \mathbb{K}$ is analytic. Therefore every linear map $\mathbb{K}^n \to \mathbb{K}^m$ is analytic.

For $r \in (\mathbb{R}^{>0})^n$ and $u \in \mathbb{K}^n$ we call

$$P_r(u) := \{\, x \in \mathbb{K}^n \,;\, |x_i - u_i| < r_i \text{ for all } i = 1, \dots, n \}$$

the *polydisc of radius r around u*. Its closure $P_{\le r}(u)$ consists of those $x \in \mathbb{K}^n$ with $|x_i - u_i| \le r_i$ for all i.

Proposition 16.11. *Let $U \subseteq \mathbb{K}^n$ be open, $u \in U$. Let f be a power series as in (16.2). Let $r \in (\mathbb{R}^{>0})^n$ such that $\sum_{(i_1,\dots,i_n) \in \mathbb{N}_0^n} |a_{i_1,\dots,i_n}| r_1^{i_1} \dots r_n^{i_n}$ converges.*

1. *The power series f converges uniformly on $P_{\le r'}(u)$ for all $r' \in (\mathbb{R}^{>0})^n$ with $r_i' < r_i$ for all i and it converges absolutely in all points $x \in P_r(u)$. In particular it defines a continuous map $\tilde{f} \colon P_r(u) \to \mathbb{K}$.*
2. *The map $\tilde{f}$ is analytic and there is a unique power series expansion of $\tilde{f}$ at u, namely the one given by f.*
3. *The map $\tilde{f}$ is differentiable and for all $j = 1, \dots, n$ the partial derivative $\partial_j \tilde{f}$ is given by the absolutely convergent power series*

$$\partial_j \tilde{f}(x) = \sum_{(i_1,\dots,i_n) \in \mathbb{N}_0^n} a_{i_1,\dots,i_n} i_j (x_1 - u_1)^{i_1} \dots (x_{i_j} - u_{i_j})^{i_j - 1} \dots (x_n - u_n)^{i_n}. \quad (16.3)$$

In particular $\partial_j \tilde{f}$ is again analytic.

Corollary 16.12. *Let $U \subseteq \mathbb{K}^n$ be open. Every analytic map $F \colon U \to \mathbb{K}^m$ is a C^∞-function.*

Proposition 16.13. *Let $V \subseteq \mathbb{K}^m$ be open and let $f \colon U \to \mathbb{K}^m$ and $g \colon V \to \mathbb{K}^p$ be analytic maps such that $f(U) \subseteq V$. Then $g \circ f \colon U \to \mathbb{K}^p$ is analytic.*

As linear maps are analytic, we can define the notion of an analytic function also in a coordinate-free setting.

Remark and Definition 16.14. Let V and W be finite-dimensional $\mathbb{K}$-vector spaces, $U \subseteq V$ open. Then a map $f \colon U \to W$ is called *analytic* if for one choice (equivalently, for all choices) of $\mathbb{K}$-linear isomorphisms $x \colon V \xrightarrow{\sim} \mathbb{K}^n$ and $y \colon W \xrightarrow{\sim} \mathbb{K}^m$ the composition $y \circ f \circ x^{-1}|_{x(U)} \colon x(U) \to \mathbb{K}^m$ is analytic.

To see the mentioned equivalence suppose that $y \circ f \circ x^{-1}|_{x(U)}$ is analytic and let $\tilde{x} \colon V \xrightarrow{\sim} \mathbb{K}^n$ be another $\mathbb{K}$-linear isomorphism. As we have just remarked that compositions of analytic with linear maps are again analytic, we see that

$$y \circ f \circ \tilde{x}^{-1}|_{\tilde{x}(U)} = y \circ f \circ x^{-1}|_{x(U)} \circ x_{|U} \circ \tilde{x}^{-1}|_{\tilde{x}(U)}$$

is also analytic. The same argument works if one replaces y by another linear isomorphism.

Every analytic map $f: U \to W$ is a C^∞-map. The composition of analytic maps is again analytic.

Sometimes we will also call analytic maps C^ω-*maps*. Every C^ω-map is also a C^α-map for all $\alpha \in \mathbb{N}_0 \cup \{\infty\}$. To ease the notation we define

$$\widehat{\mathbb{N}}_0 := \mathbb{N}_0 \cup \{\infty, \omega\}, \qquad \widehat{\mathbb{N}} := \mathbb{N} \cup \{\infty, \omega\}.$$

We extend the usual total order on $\mathbb{N}_0$ to $\widehat{\mathbb{N}}_0$ by requesting $n < \infty < \omega$ for all $n \in \mathbb{N}_0$. We also define $\infty \pm \alpha := \infty$ and $\omega \pm \alpha := \omega$ for all $\alpha \in \mathbb{N}_0$.

Then every C^α-map is also a C^β-map for all $\alpha, \beta \in \widehat{\mathbb{N}}_0$ with $\beta \leq \alpha$. For $\mathbb{K} = \mathbb{R}$ there always exist functions that are C^β but not C^α: For $\alpha < \omega$ an example was given in Remark 16.3. A standard example of a real C^∞-function that is not analytic is given in Problem 16.1.

In the complex case, a map is C^1 if and only if it is C^ω. More precisely:

Definition and Theorem 16.15. *Let* $\mathbb{K} = \mathbb{C}$. *Then a map* $f: U \to W$ *is called* holomorphic *if the following equivalent assertions hold:*

 (i) f *is complex differentiable.*
 (ii) f *is analytic.*
 (iii) *There exists a basis* $(e_1, \ldots, e_n)$ *of* V *such that the partial complex derivatives* $D_{e_i} f$ *exist for all* $i = 1, \ldots, n$.
 (iv) *For all* $\alpha \in \mathbb{N}$ *and for all tuples* $(v_1, \ldots, v_\alpha)$ *of* $v_i \in V$ *the iterated partial derivatives* $D_{v_1} D_{v_2} \cdots D_{v_\alpha} f: U \to W$ *exist.*

For the proof we essentially refer to the literature.

Proof. The implications "(i) $\Rightarrow$ (iii)" and "(iv) $\Rightarrow$ (iii)" are trivial and the implications "(ii) $\Rightarrow$ (i)" and "(ii) $\Rightarrow$ (iv)" hold (for arbitrary $\mathbb{K}$) because analytic maps are C^∞ (Corollary 16.12). For the proof of the (difficult) implication "(iii) $\Rightarrow$ (ii)" we refer to [Hoe] Theorem 2.2.8. $\square$

Theorem 16.16 (Inverse function theorem). *Let* $\alpha \in \widehat{\mathbb{N}}$ *and let* $f: U \to W$ *be a* C^α-*map. Let* $\tilde{u} \in U$ *and assume that* $Df(\tilde{u}): V \to W$ *is bijective. Then there exist open neighborhoods* U_0 *of* $\tilde{u}$ *in* U, *and* W_0 *of* $\tilde{w} := f(\tilde{u})$ *in* W, *such that the restriction* $f_{|U_0}: U_0 \to W_0$ *is bijective and* $f_{|U_0}^{-1}: W_0 \to U_0$ *is a* C^α-*map with* $D(f_{|U_0}^{-1})(\tilde{w}) = Df(\tilde{u})^{-1}$.

Proof. For $\mathbb{K} = \mathbb{R}$ and $\alpha < \omega$ this is a standard result from calculus ([La] XIV Theorem 1.2). For $\mathbb{K} = \mathbb{C}$ we only have to prove the result for $\alpha = 1$ by Theorem 16.15. But then it follows from the real case, that $f_{|U_0}^{-1}: W_0 \to U_0$ exists and is a real C^1-map whose differential is complex linear at every point. Hence it is a complex C^1-map (Remark 16.3). Finally, for the case $\mathbb{K} = \mathbb{R}$ and $\alpha = \omega$ we refer to [KrPa] Theorem 2.5.1. $\square$

16.3   Higher Derivatives

Let us take a closer look at higher derivatives. If $f: U \to W$ is a C^α-map with $\alpha \geq 2$, then $D(Df)$ is a map $U \to \mathrm{Hom}_{\mathbb{K}}(V, \mathrm{Hom}_{\mathbb{K}}(V, W))$. More generally its r-th derivative (for $1 \leq r \leq \alpha$) is a map

$$U \to \mathrm{Hom}_{\mathbb{K}}(V, \mathrm{Hom}_{\mathbb{K}}(V, \mathrm{Hom}_{\mathbb{K}}(V, \ldots, \mathrm{Hom}_{\mathbb{K}}(V, W)\ldots)))$$

forming "$\mathrm{Hom}_{\mathbb{K}}(V, \cdot)$" r times. To ease the bookkeeping we first identify the right-hand side with $\mathrm{Hom}_{\mathbb{K}}(V^{\otimes r}, W)$ as follows.

For $\mathbb{K}$-vector spaces V_1, V_2, and W recall that by definition of the tensor product there is an isomorphism of $\mathbb{K}$-vector spaces, functorial in V_1, V_2, and W

$$\mathrm{Hom}_{\mathbb{K}}(V_1, \mathrm{Hom}_{\mathbb{K}}(V_2, W)) \xrightarrow{\sim} \{\, \beta: V_1 \times V_2 \to W \; ; \; \beta \text{ is } \mathbb{K}\text{-bilinear} \}$$
$$= \mathrm{Hom}_{\mathbb{K}}(V_1 \otimes_{\mathbb{K}} V_2, W) \tag{16.4}$$

given by attaching to the linear map $\varphi: V_1 \to \mathrm{Hom}_{\mathbb{K}}(V_2, W)$ the linear map $\beta_\varphi: v_1 \otimes v_2 \mapsto (\varphi(v_1))(v_2)$. More generally, there is a functorial isomorphism

$$\mathrm{Hom}_{\mathbb{K}}(V_1, \mathrm{Hom}_{\mathbb{K}}(V_2, \mathrm{Hom}_{\mathbb{K}}(V_3, \ldots, \mathrm{Hom}_{\mathbb{K}}(V_r, W)\ldots)))$$
$$\xrightarrow{\sim} \mathrm{Hom}_{\mathbb{K}}(V_1 \otimes_{\mathbb{K}} \cdots \otimes_{\mathbb{K}} V_r, W) \tag{16.5}$$

for $\mathbb{K}$-vector spaces $V_1, \ldots, V_r$ and W by identifying an element φ of the left-hand side with the $\mathbb{K}$-linear map

$$v_1 \otimes \cdots \otimes v_r \mapsto \bigl(\ldots \bigl((\varphi(v_1))(v_2)\bigr)\ldots(v_r)\bigr).$$

Finally recall that we write $V^{\otimes r} := V \otimes_{\mathbb{K}} \cdots \otimes_{\mathbb{K}} V$, r factors of V.

Hence if $f: U \to W$ is a C^α-map, we can (and will) consider for all $r \leq \alpha$ and $u \in U$ the r-th derivative $D^r f(u) \in \mathrm{Hom}_{\mathbb{K}}(V, \mathrm{Hom}_{\mathbb{K}}(V, \ldots, \mathrm{Hom}_{\mathbb{K}}(V, W)\ldots))$ as a $\mathbb{K}$-linear map

$$D^r f(u): V^{\otimes r} \to W. \tag{16.6}$$

Remark 16.17. Let $W = \prod_{i=1}^m W_i$ for (finite-dimensional) $\mathbb{K}$-vector spaces W_i. For a map $f: U \to W$ let $f_i: U \to W_i$ be the i-th component. Then f is a C^α-map if and only if f_i is a C^α-map for all i. In this case for $r \leq \alpha$ one has

$$D^r f(u) = (D^r f_1(u), \dots, D^r f_m(u)): V^{\otimes r} \longrightarrow W = \prod_{i=1}^m W_i.$$

In both the real and the complex case we can describe higher derivatives via partial derivatives as follows.

Remark 16.18. Let $f: U \to W$ be a C^α-map. Let $(e_1, \dots, e_n)$ be a $\mathbb{K}$-basis of V defining the partial derivative $\partial_j f$ in direction e_j for $j = 1, \dots, n$ (16.1). For $r \leq \alpha$ and $u \in U$ we consider $D^r f(u)$ as a map $V^{\otimes r} \to W$ via (16.5).

Let us first consider the case $V = \mathbb{K}^n$, $W = \mathbb{K}^m$ and $r = 1$. Then $Df(u): \mathbb{K}^n \to \mathbb{K}^m$ is given by the *Jacobian matrix*

$$J_f(u) := (\partial_j f_i(u))_{1 \leq i \leq m, 1 \leq j \leq n} \in M_{m \times n}(\mathbb{K}).$$

Now let $1 \leq r \leq \alpha$ be arbitrary. Then the $e_{j_1} \otimes \cdots \otimes e_{j_r}$ for $1 \leq j_1, \dots, j_r \leq n$ form a basis of $V^{\otimes r}$ and one has

$$D^r f(u)(e_{j_1} \otimes \cdots \otimes e_{j_r}) = (\partial_{j_r} \dots \partial_{j_1} f)(u) \in W. \tag{16.7}$$

Indeed, for $\mathbb{K} = \mathbb{R}$ this is a well-known result from calculus. It can be proved by induction on r using the description of the first derivative by the Jacobian matrix. The same proof works for $\mathbb{K} = \mathbb{C}$.

More generally, let $v_1, \dots, v_r \in V$ arbitrary vectors and $v_d = \sum_{i=0}^r a_{i,d} e_i$ with $a_{i,d} \in \mathbb{K}$. By linearity we deduce from (16.7)

$$D^r f(u)(v_1 \otimes \cdots \otimes v_r) = \sum_{(i_1, \dots, i_r)} a_{i_1, 1} \cdots a_{i_r, r} (\partial_{j_r} \dots \partial_{j_1} f)(u). \tag{16.8}$$

If $W = \mathbb{K}^m$ and $f_1, \dots, f_m: U \to \mathbb{K}$ are the components of f, then $(\partial_{j_r} \dots \partial_{j_1} f_i)(u)$, $i = 1, \dots, m$, are the components of $D^r f(u)(e_{j_1} \otimes \cdots \otimes e_{j_r})$.

Proposition 16.19. *Let V and W be finite-dimensional $\mathbb{K}$-vector spaces, let $U \subseteq V$ be open and let $f: U \to W$ be a C^α-map. For $r \leq \alpha$ consider the r-th derivative as a map $D^r f: U \to \operatorname{Hom}_{\mathbb{K}}(V^{\otimes r}, W)$. Then for every permutation $\sigma \in S_r$ one has*

$$Df(u)(v_{\sigma(1)} \otimes \cdots \otimes v_{\sigma(r)}) = Df(u)(v_1 \otimes \cdots \otimes v_r)$$

for all $u \in U$, $v_1, \dots, v_r \in V$. In other words $D^r f(u)$ induces a map

$$D^r f(u): \operatorname{Sym}^r(V) \to W.$$

Proof. In the case $\mathbb{K} = \mathbb{R}$ this is a standard fact from calculus (e.g., [La] XIII Theorem 6.2). In the complex case this follows from the fact that F is automatically analytic and from the concrete description of the derivative of a power series (16.3).   □

Every C^α-function can be approximated via its Taylor expansion by polynomials. For analytic functions (and hence for all differentiable functions in the case $\mathbb{K} = \mathbb{C}$) this is clear by definition. Hence we focus on $\mathbb{K} = \mathbb{R}$.

Proposition 16.20. *Suppose that* $\mathbb{K} = \mathbb{R}$. *Let* $f : U \to W$ *be a* C^α-map, $u \in U$ *and* $\rho \in \mathbb{R}^{>0}$ *such that* $B_\rho(u) \subseteq U$ *for some choice of a norm* $\| \ \|$ *on* V. *Let* $v \in V$ *such that* $\|v\| < \rho$. *Write* $v^{\otimes r}$ *for* $v \otimes \cdots \otimes v \in V^{\otimes r}$. *Then one has for all* $r \le \alpha$ *the Taylor expansion*

$$f(u + v) = \sum_{m=0}^{r} \frac{D^m f(u)(v^{\otimes m})}{m!} + R_{r,u}(v),$$

with

$$\lim_{y \to 0} \frac{R_{r,u}(y)}{\|y\|^r} = 0$$

locally uniformly in u.

Proof. This is again a standard fact from calculus (see [La] XIII Theorem 6.3, in particular the "Estimate of the Remainder" there).   □

Remark 16.21. Let $f : U \to W$ be a (real or complex) C^α-map, $u \in U$. Choose a basis $(e_1, \ldots, e_n)$ of V defining partial derivatives $\partial_i = \frac{\partial}{\partial x^i}$. The summands in the Taylor expansion have the following coordinate version (valid both for $\mathbb{K} = \mathbb{R}$ and $\mathbb{K} = \mathbb{C}$). We write $v = \sum_{i=1}^{n} v_i e_i$ with $v_i \in \mathbb{K}$. Then

$$\frac{D^r f(u)(v^{\otimes r})}{r!} = \sum_{i_1 + \cdots + i_n = r} \frac{1}{i_1! \cdots i_n!} v_1^{i_1} \cdots v_n^{i_n} \frac{\partial^r f}{\partial x_1^{i_1} \cdots \partial x_n^{i_n}}(u), \qquad (16.9)$$

where the sum is taken over all $(i_1, \ldots, i_n) \in \mathbb{N}_0^n$ whose sum is r. This follows from (16.8) and because

$$\frac{r!}{i_1! \ldots i_n!}$$

is the number of decompositions of a set E with r elements into a tuple $(E_1, \ldots, E_n)$ of disjoint subsets with $\#E_j = i_j$.

Now suppose that $f : U \to W$ is a real C^∞-map, let $\rho \in \mathbb{R}^{>0}$ such that $B_\rho(u) \subseteq U$ for some norm $\| \ \|$ on V and let $v = \sum_{i=1}^{n} v_i e_i \in V$ with $\|v\| < \rho$. Then

$$\sum_{m=0}^{\infty} \frac{D^m f(u)(v^{\otimes m})}{m!} = \sum_{(i_1, \ldots, i_n) \in \mathbb{N}_0^n} \frac{\partial_1^{i_1} \cdots \partial_n^{i_n} f(u)}{i_1! \cdots i_n!} v_1^{i_1} \cdots v_n^{i_n} \qquad (16.10)$$

is called the *Taylor series of* f *at* u.

Remark 16.22. For an analytic map $f : U \to W$ its power series expansion in some point $u \in U$ is its Taylor series in u ([Ser] Part II, Chap. II p.73).

16.4   Problems

Problem 16.1. Define $f : \mathbb{R} \to \mathbb{R}$ by $f(x) := 0$ for $x \leq 0$ and by $f(x) := \exp(-1/x)$ for $x > 0$. Show that all derivatives in 0 exist and are equal to zero. Deduce that f is C^∞ but not analytic.

Problem 16.2. Let V and W be finite-dimensional $\mathbb{K}$-vector spaces and let $r \in \mathbb{N}$. Show that the maps

$$T^r : \operatorname{Hom}_{\mathbb{K}}(V, W) \to \operatorname{Hom}_{\mathbb{K}}(T^r(V), T^r(W)), \qquad u \mapsto T^r(u)$$

$$\Lambda^r : \operatorname{Hom}_{\mathbb{K}}(V, W) \to \operatorname{Hom}_{\mathbb{K}}(\Lambda^r(V), \Lambda^r(W)), \qquad u \mapsto \Lambda^r(u)$$

$$\operatorname{Sym}^r : \operatorname{Hom}_{\mathbb{K}}(V, W) \to \operatorname{Hom}_{\mathbb{K}}(\operatorname{Sym}^r(V), \operatorname{Sym}^r(W)), \qquad u \mapsto \operatorname{Sym}^r(u)$$

are homogeneous polynomial maps of degree r and deduce that these maps are analytic.

Problem 16.3. Let $U \subseteq \mathbb{K}^n$ be open and let $f : U \to \mathbb{K}$ be analytic. Show that $\{x \in U \ ; \ D^m f(x) = 0 \text{ for all } m \geq 0\}$ is open and closed in U. Deduce that the following assertions are equivalent if U is connected:

(i)   There exists a point $x \in U$ such that $\partial_1^{i_1} \cdots \partial_n^{i_n} f(x) = 0$ for all $(i_1, \dots, i_n) \in \mathbb{N}_0^n$.
(ii)   There exists a non-empty open subset W of U such that $f_{|W} = 0$.
(iii) $f = 0$.

Problem 16.4. Let V be a finite-dimensional $\mathbb{C}$-vector space, let $U \subseteq V$ be open and connected, and let $f : U \to \mathbb{C}$ be a non-constant holomorphic map. Show that f is open. *Hint*: Deduce this from the classical result for $V = \mathbb{C}$.

References

[AJS] Alonso Tarrío L., López A.J., Souto Salorio M.J., *Localization in Categories of Complexes and Unbounded Resolutions*, Canad. J. of Math. **52** (2000), 225–247.

[AmEs3] Amann H., Escher J.: Analysis III. Birkhäuser (2009)

[BouA1] Bourbaki N.: Algebra, Chaps. 1–3. Springer (1989)

[BouA2] Bourbaki N., Algebra, Chaps. 4–7. Springer (2003)

[BouA3] Bourbaki N.: Algèbre, Chap. 10. Springer (2006)

[BouGT1] Bourbaki N.: General topology, Chaps. 1–4, 2nd printing. Springer (1989)

[BouGT2] Bourbaki N.: General topology, Chaps. 5–10, 2nd printing. Springer (1989)

[BouLie1] Bourbaki N.: Lie Groups and Lie Algebras, Chaps. 1–3. Springer (1989)

[BouTS] Bourbake N.: Theory of Sets. Springer (2008)

[Br1] Bredon G.E.: Topology and Geometry. Springer (1993)

[Br2] Bredon G.E.: Sheaf Theory, 2nd edition. Springer (1997)

[Car] Cartan H.: Variétés analytiques réelles et variétés analytiques complexes. Bull. de la SMF **85**, 77–99 (1957)

[GeMa] Gelfand S.I., Manin Yu.I.: Methods of Homological Algebra. Springer (1996)

[God] Godement R.: Topologie algébrique et théorie des faisceaux. Hermann (1958)

[Gra] Grauert H.: On Levi's problem and the imbedding of real analytic manifolds. Ann. of Math. **68**, 460–472 (1958)

[GuRo] Gunning R., Rossi H.: Analytic functions of several complex variables. Prentice-Hall (1965)

[HiNe] Hilgert J., Neeb K.-H., Structure and Geometry of Lie Groups. Springer (2012)

[Hoe] Hörmander L.: An Introduction to Complex Analysis in Several Variables. North-Holland (1990)

[Ive] Iversen B.: Cohomology of Sheaves. Springer (1968)

[Ker] Kervaire M.A.: A manifold which does not admit any differentiable structure. Comm. Math. Helvet. **34**, 257–270 (1960)

[KoPu] Koch W., Puppe D.: Differenzierbare Strukturen auf Mannigfaltigkeiten ohne abzählbare Basis. Arch. Math. (Basel) **19**, 95–102 (1968)

[KrPa] Krantz S.G., Parks H.R.: A Primer of Real Analytic Functions, 2nd edition. Birkhäuser (2002)

© Springer Fachmedien Wiesbaden 2016

T. Wedhorn, *Manifolds, Sheaves, and Cohomology*, Springer Studium Mathematik – Master,
DOI 10.1007/978-3-658-10633-1

[KS1] Kashiwara M., Schapira P.: Sheaves on Manifolds. Springer (1990)

[KS2] Kashiwara M., Schapira P.: Categories and Sheaves. Springer (2006)

[La] Lang S.: Real and Functional Analysis, 3rd edition. Springer (1993)

[LeeJe] Jeffrey Lee M.: Manifolds and Differential Geometry. Graduate Stud. in Math. **107**, AMS

[LeeJo] John Lee M.: Introduction to Smooth Manifolds. Graduate Texts in Math. **218**, Springer (2006)

[Mil] Milnor J.: On manifolds homeomorphic to the 7-sphere. Ann. of Math. **64**, 399–405 (1956)

[Sch] Schubert H.: Kategorien I. Akademie-Verlag Berlin (1970)

[Ser] Serre J.-P.: Lie Algebras and Lie Groups. Lect. Notes in Math. **1500**, Springer (1992)

[SGA4] Grothendieck A., Artin M., et. al.: Séminaire de Géometrie Algébrique du Bois-Marie, Théorie des Topos et cohomologie étale des schémas (1963–64). Lect. Notes in Math. **269**, **270**, **305**, Springer (1972/73)

[Spa] Spaltenstein N.: Resolutions of unbounded complexes, Comp. Math. **65**, 121–154 (1988)

[Stacks] de Jong J., et al.: The Stacks Project. http://stacks.math.columbia.edu

[Ste] Sternberg S.: Lectures on Differential Geometry. Prentice-Hall Mathematics Series (1964)

[tDi] tom Dieck T.: Algebraic topology. EMS Textbook in Mathematics (2008)

[Wel] Wells R.O.: Differential Analysis on Complex manifolds, 3rd edition. Springer

[Whi1] Whitney H.: Geometric Integration Theory. Princeton University Press (1957)

[Whi2] Whitney H.: Differentiable manifolds. Ann. of Math. **37**, 645–680 (1936)

Index

343

Printed in the United States
By Bookmasters